Spanish Relativity Meeting 2009

(ERE 2009)

AA002754

Journal of Physics: Conference Series Volume 229

Bilbao, Spain
7-11 September 2009

Editors:

Ruth Lazkoz **Raül Vera**

ISBN: 978-1-61738-623-7
ISSN: 1742-6588

Printed from e-media with permission by:

Curran Associates, Inc.
57 Morehouse Lane
Red Hook, NY 12571

Some format issues inherent in the e-media version may also appear in this print version.

Copyright© (2009) by the Institute of Physics
All rights reserved.

Printed by Curran Associates, Inc. (2010)

For permission requests, please contact the Institute of Physics
at the address below.

Institute of Physics
Dirac House, Temple Back
Bristol BS1 6BE UK

Phone: 44 1 17 929 7481
Fax: 44 1 17 920 0979

techtracking@iop.org

Additional copies of this publication are available from:

Curran Associates, Inc.
57 Morehouse Lane
Red Hook, NY 12571 USA
Phone: 845-758-0400
Fax: 845-758-2634
Email: curran@proceedings.com
Web: www.proceedings.com

TABLE OF CONTENTS

INVITED LECTURES

012001: Blackfolds 1
Roberto Emparan

012002: Dark Matters 7
Joseph Silk

CONTRIBUTIONS – PLENARY SESSIONS

012003: Bulk Flows in Inflation and in Lemaître-Tolman-Bondi Models 13
F. Atrio-Barandela, A. Kashlinsky, H. Ebeling, D. Kocevski

012004: Where Are the Trapped Surfaces? 19
Jan E. Åman, Ingemar Bengtsson, José M. M. Senovilla

012005: On Massive Gravitons in 2+1 Dimensions 25
Eric Bergshoeff, Olaf Hohm, Paul Townsend

012006: Nonlinear Electrodynamics and Cosmology 31
Nora Bretón

012007: On the Runaway Instability of Self-Gravitating Torus Around Black Holes 37
José A. Font, Pedro J. Montero, Masaru Shibata

012008: Gravitating Semilocal Strings 43
Betti Hartmann, Jon Urrestilla

012009: Static Vacuum Spacetimes with Prescribed Multipole Moments 49
Magnus Herberthson

012010: Physical Properties of Black Di-Ring 55
Hideo Iguchi, Takashi Mishima

012011: Towards Gravitational-Wave Astronomy of Core-Collapse Supernova Explosion 61
Kei Kotake

012012: H(z) Diagnostics to Discriminate Dark Energy Models 67
Sergio Del Campo, Ramón Herrera, Diego Pavón

012013: Penrose's Weyl Curvature Hypothesis and Conformally-Cyclic Cosmology 73
Paul Tod

CONTRIBUTIONS – PARALLEL SESSIONS

012014: Null Data Characterization of Asymptotically Flat Stationary Spacetimes 78
Andrés E. Aceña

012015: On the Linear Stability of a Scalar-Feld with Exponential Potential in a Flat Robertson-Walker Background 82
Artur Alho

012016: Relating the Newman-Penrose Constants to the Geroch–Hansen Multipole Moments 86
Thomas Bäckdahl

012017: Spacetime Brout–Englert–Higgs Effect in General Relativity Interacting with p-Brane Matter 90
Igor A. Bandos

012018: Semiclassical Instability of Warp Drives 94
C. Barceló, S. Finazzi, S. Liberati

012019: Dark Energy in Vector-Tensor Theories of Gravity 98
Jose Beltrán Jiménez, A. L. Maroto

012020: Racing a Quantum Computer Through Minkowski Spacetime 102
Jacob D. Biamonte

012021: Complete Classification of Second-Order Symmetric Spacetimes 106
Oihane F. Blanco, Miguel Sánchez, José M. M. Senovilla

012022: Gowdy Waves As a Test-Bed for Constraint-Preserving Boundary Conditions 110
Carles Bona, Carles Bona-Casas

012023: Black Hole Gas in TeV-Gravity Models 114
Mónica Borunda, Manuel Masip
012024: f(R) Brane Cosmology 118
Mariam Bouhmadi-López
012025: The Quadrupole Moment of Slowly Rotating Stars 122
M. Bradley, G. Fodor
012026: Spherically Symmetric Elasticity in Relativity 126
J. Carot, I. Brito, E. G. L. R. Vaz
012027: On a Recent Proposal of the Penrose Inequality 130
Alberto Carrasco, Marc Mars
012028: k-Essence and Extended Tachyons in Brane-Worlds 134
L. P. Chimento, Martín G. Richarte
012029: On the Local Existence of Maximal Slicings in Spherically Symmetric Spacetimes 138
Isabel Cordero-Carrión, José María Ibáñez, Juan Antonio Morales-Lladosa
012030: Maximal Analytic Extensions of the Emparan-Reall Black Ring 142
P. T. Chrusciel, J. Cortier
012031: The Gravitational Force on a Gyroscope and the Electromagnetic Force on a Magnetic Dipole as Analogous Tidal Effects 146
L. Filipe O. Costa, Carlos Herdeiro
012032: Getting Wahlquist's Metric from a Perturbation Theory Solution 150
J. E. Cuchí, J. Martín, A. Molina, E. Ruiz
012033: Black Holes in Modified Gravity Theories 154
A. De La Cruz-Dombriz, A. Dobado, A. L. Maroto
012034: Unleashing the Aether 158
Alberto Diez-Tejedor
012035: Petrov D Vacuum Spaces Revisited: Complete Tables and Invariant Classification by GHP Analysis 162
S. B. Edgar, Alfonso García-Parrado Gómez-Lobo, José M. Martín-García
012036: Using Black Hole Physics to Compute the Energy Loss of a Quark in Strongly Coupled QCD 166
Daniel Fernández
012037: Equation of State and Singularities in FLRW Cosmological Models 170
Leonardo Fernández-Jambrina, R. Lazkoz
012038: Spinor Calculus on 5-Dimensional Spacetimes 174
Alfonso García-Parrado Gómez-Lobo, José M. Martín-García
012039: Some Classical and Quantum Aspects of Cosmological Braneworlds with a Time-Dependent Extra Dimension 178
Suman Ghosh
012040: Conformastat Electrovacuum Spacetimes 182
Guillermo A. González, Raül Vera
012041: Dark Matter Halos: Consitency Between Rotational Curves and Gravitational Lenses 186
Alma X. González-Morales, Darío Núñez
012042: The Causal Boundary and Its Relations with the Conformal Boundary 190
J. Herrera
012043: Flat Synchronizations in Spherically Symmetric Space-Times 193
Alicia Herrero, Juan Antonio Morales-Lladosa
012044: Thermodynamic Structure of Gravitational Field Equations from Near-Horizon Symmetries 197
Dawood Kothawala
012045: Flat Deformation of a Spacetime with Two Killing Fields 201
Josep Llosa, J. Carot
012046: Harmonic Maps and Black Holes 205
João Lopes Costa
012047: Generating Black Strings 209
L. A. López
012048: On Minkowski-Like and De Sitter-Like Space-Times 213
Christian Lübbe
012049: Invariant Classification of Metrics Using Invariant Formalism 217
M. P. Machado Ramos, S. B. Edgar, M. Bradley
012050: On Interrelations Between Sibgatullin's and Alekseev's Approaches to the Construction of Exact Solutions of the Einstein-Maxwell Equations 221
F. J. Ernst, V. S. Manko, E. Ruiz
012051: Unifying Exact Scalar Field Cosmologies 225
José P. Mimoso, Tiago Charters

012052: From Causality to Time and Back 229
Ettore Minguzzi
012053: Creatable Universes: A New Approach 233
Ramon Lapiedra, Juan Antonio Morales-Lladosa
012054: Secondarily-Induced Gravitational Wave Background in Modified Gravity 237
Masaaki Morita, Hirotaka Takahashi
012055: On the Use of Curvature Invariants in Numerical Relativity 241
A. Nerozzi
012056: Stability of Black Holes from Hydrodynamics 245
Masato Nozawa
012057: Polarized Electrogowdy Spacetimes Censored 249
Ernesto Nungesser
012058: Inflation, Renormalization, and CMB Anisotropies 253
I. Agulló, J. Navarro-Salas, G. J. Olmo, L. Parker
012059: Numerical Treatment of the Light Propagation Problem in the Post-Newtonian Formalism 257
A. San Miguel, F. Vicente, J.-F. Pascual-Sánchez
012060: Critical Phenomena and Information Geometry in Black Hole Physics 261
Jan E. Åman, Narit Pidokrajt
012061: Study on Internal Structure of Maxwell-Gauss-Bonnet Black Hole 265
K. A. Rannu, S. O. Alexeyev, A. Barrau
012062: Physical Properties of the Double-Kerr Solution 269
Miguel S. Costa, Carlos Herdeiro, Carmen Rebelo
012063: Power Spectrum of Large-Scale Structure Cosmological Models in the Framework of Scalar-Tensor Theories 273
M. A. Rodríguez-Meza
012064: Late-Time Tails of Self-Gravitating Massless Fields 277
Andrzej Rostworowski
012065: Holographic Interacting Tachyon Model 281
Alberto Rozas-Fernández, David Brizuela, Norman Cruz
012066: Reconstructing Cosmological Solutions in F(R) Gravity. Towards a Unified Model of the Universe Evolution 285
Diego Sáez-Gómez
012067: Two Static Charged Masses in the External Field 289
J. Sánchez-Mondragón, V. S. Manko, E. Ruiz
012068: DBI Models for the Unification of Dark Matter and Dark Energy 293
L. P. Chimento, R. Lazkoz, I. Sendra
012069: Stellar Oscillations in TeVeS 297
Hajime Sotani
012070: Features of Gravitational Waves in Higher Dimensions 301
Otakar Svítek
012071: Integration Without Integration: New Killing Spinor Spacetimes 305
Norbert Van Den Bergh
012072: Black Holes in a Box 309
Helvi Witek, Vitor Cardoso, Leonardo Gualtieri, Carlos Herdeiro, A. Nerozzi, Ulrich Sperhake, Miguel Zilhão
012073: Maximally Inhomogeneous Gödel-Farnsworth-Kerr Generalizations 313
Lode Wylleman
012074: Numerical Relativity in Higher Dimensions 317
Miguel Zilhão, Helvi Witek, Ulrich Sperhake, Vitor Cardoso, Leonardo Gualtieri, Carlos Herdeiro, A. Nerozzi

POSTERS

012075: The Relativistic Brownian Motion: Interdisciplinary Applications 321
A. Aragonés-Muñoz, A. Sandoval-Villalbazo
012076: Geodesic Deviation Equation in Bianchi Cosmologies 324
D. L. Cáceres, L. Castañeda, J. M. Tejeiro
012077: Remarks on the Extreme Limit of the Double–Kerr Solution in the Equilibrium Case 327
I. Cabrera-Munguia, V. S. Manko, E. Ruiz
012078: An Anti-Schwarzshild Solution: Wormholes and Scalar-Tensor Solutions 330
José P. Mimoso, Francisco S. N. Lobo
012079: Polar Oscillations in Magnetars 333
Hajime Sotani, Kostas D. Kokkotas

012080: Black Hole Entropy for the General Area Spectrum 336
Tomo Tanaka, Takashi Tamaki
Author Index

PREFACE

Scientists working in Relativity, Gravitation and Cosmology in Spanish institutions have been organising the Spanish Relativity Meetings (ERE) for more than 30 years now, and 2009 was the turn of the group at the University of the Basque Country (UPV/EHU) to continue this well established tradition. There is no doubt these meetings have become a benchmark in the field not only because of the high scientific level reached in each edition, but also because of the almost legendary relaxed atmosphere.

The support of our sponsors --Spanish Ministry of Science and Innovation, through the funding actions FIS2008-03716-E and FIS2008-04494-E, Basque Government, UPV/EHU, and Spanish Relativity and Gravitation Society (SEGRE)-- and specially the scientific and social contributions of all the participants turned the ERE2009 a memorable edition for us, and we hope this is also the impression of everyone else involved in this meeting.

The organising committee was very pleased to present an excellent panel of invited speakers to dissert on four topics that are gathering enormous interest and activity: Quantum and thermodynamical effects in Gravity, Modern Cosmology, Numerical Relativity and Black holes. But the meeting also boasted a high scientific level thanks to the more than a hundred other participants from many different countries whose enthusiasm was an excellent proof of the good health of Relativity, Gravitation and Cosmology as a research area.

The scientific programme started on the 7th, with an opening by a representative of the Principal of the University of the Basque Country, along with the Deputy Mayor of Bilbao, and the Chair of the conference, José M.M. Senovilla. Lectures were delivered on the 7th, 8th, 9th and 11th of September 2009 following a scheme of plenary sessions in the morning (the first lecture always being an invited one) and two parallel sessions in the afternoon. There were of course coffee-breaks to refresh bodies and spirits and encourage the interaction between the participants and promote collaborative links, as scientific interaction is one of the main purposes of the meeting. The 10th was devoted to an emotional homage to Prof. Jesús Martín, with special talks to honour his scientific and academic careers. And last but not least, there was a quite plenty programme of social events, as has become customary in the ERE meetings; the programme started on the 6th with a friendly reception at the "Ein Prosit" bar, on the 7th we had a reception at the "Salón Árabe" of the Bilbao Town Hall hosted by the Deputy Mayor, the SEGRE public lecture in the Main Library of Bilbao on the 8th, a guided visit to the "Ría de Bilbao" on the Txinbito boat, and a closing dinner at the Aspaldiko restaurant in Loiu.

As editors of these proceedings and members of the Organizing Committee we want to take advantage of this opportunity to thank warmly everyone who made this conference possible and make the wish that in future editions to be held in Bilbao we will have again such a splendid support from the institutions and the scientific community.

Ruth Lazkoz and Raül Vera

Invited Speakers

- Roberto Emparan *(Universitat de Barcelona)*
- Frans Pretorius *(Princeton University)*
- Joseph Silk *(University of Oxford)*
- Robert M. Wald *(University of Chicago)*

Scientific Committee

- M. Alcubierre *(ICN, Universidad Nacional Autónoma de México)*
- R. Beig *(Institute of Theoretical Physics, University of Viena)*
- C. Cutler *(Jet Propulsion Laboratory, California Inst. of Technology)*
- T. Damour *(Institut des Hautes Études Scientifiques)*
- R. Maartens *(Institute of Cosmology and Gravitation, University of Portsmouth)*
- F. Quevedo *(DAMTP, University of Cambridge)*

Local Organising Committee

- J. Ibáñez
- R. Lazkoz
- J. M. M. Senovilla (Chair)
- R. Vera (Secretary and webmaster)

GOBIERNO DE ESPAÑA
MINISTERIO DE CIENCIA E INNOVACIÓN

EUSKO JAURLARITZA
GOBIERNO VASCO

eman ta zabal zazu

Universidad del País Vasco
Euskal Herriko Unibertsitatea

SEGRE

Bilbao
UDALA
AYUNTAMIENTO
KULTURA ETA HEZKUNTZA SAILA
ÁREA DE CULTURA Y EDUCACIÓN

Fisika Teorikoaren eta Zientziaren Historiaren Saila
Departamento de Física Teórica e Historia de la Ciencia

International Society on General Relativity and Gravitation

Journal of Physics: Conference Series **229** (2010) 012001 doi:10.1088/1742-6596/229/1/012001

Blackfolds

Roberto Emparan

Institució Catalana de Recerca i Estudis Avançats (ICREA), and Departament de Física Fonamental, Universitat de Barcelona, Marti i Franquès 1, E-08028 Barcelona, Spain

E-mail: `emparan@ub.edu`

Abstract. We review the blackfold approach, a recently developed effective worldvolume description of higher-dimensional black holes, that captures the long wavelength dynamics of black holes with horizons exhibiting two characteristic lengths of very different size. In this regime the black hole is regarded as a black brane curved into a submanifold of a background spacetime, and can be formulated in terms of an effective fluid that lives on a dynamical worldvolume. We discuss the resulting blackfold equations, which separate into a set of intrinsic and extrinsic equations.

1. Motivation

In recent years it has become increasingly apparent that General Relativity should not be regarded merely as a theory of gravity whose domain of application is confined to astrophysics and cosmology. Rather, it is a theoretical tool, much like quantum field theory, that can be used to describe the physics of systems that in some cases are far removed from the stars and the large-scale cosmos. A striking instance of this is provided by the AdS/CFT correspondence and its recent derivatives, AdS/QCD, AdS/QGP, AdS/cond-mat etc. Moreover, its importance to scenarios of physics beyond the Standard Model has only been underlined by the development of theories with TeV-scale gravity and by many of the most important advances in String Theory in the last decade.

This motivates us to develop the fundamentals of this theory, without specifically focusing on any particular application, with the idea that a better understanding of the basis of the theory will, in the long run, pay off in sometimes unexpected ways.

In this spirit, it makes sense to focus on the simplest, universal sector of the theory, that describes vacuum gravity,

$$R_{\mu\nu} = 0\,. \tag{1}$$

When trying to understand better a theory, it is usually a good idea to have some parameter that can be varied in order to analyze its behavior in different regimes. For the vacuum Einstein equations, the only parameter that seems to be available is D, the number of spacetime dimensions in which the theory is formulated. Thus, in this talk, we will be considering the theory in an arbitrary number of dimensions.

It is also a good idea to focus on the simplest objects that the theory describes that also capture its non-trivial physics in way that is both tractable and illustrative — the 'hydrogen atoms' of the theory. In the case of General Relativity, these are obviously its black hole solutions.

General Relativity and its black holes in four or less dimensions have been extensively studied in the past and it is fair to say that they are very well understood. In contrast, despite the intense interest in string theory over the last 25 years, surprisingly little was known about the properties of vacuum black holes in $D > 4$ until about a decade ago.

The current interest in the subject was sparked by two developments: first, a reassessment of the possible evolution of the Gregory-Laflamme instability of black branes [1] and the subsequent discovery of inhomogeneous black brane phases [2, 3]; and second, the discovery of an exact black ring solution in five-dimensional gravity, which exemplified the possibility of non-spherical horizon topologies and absence of uniqueness [4]. Both discoveries pointed to the possibility that the physics of higher-dimensional black holes is richer than the strikingly simple picture that uniqueness theorems imply for four-dimensional black holes. We have been discovering that in fact there is a remarkably wide variety of new behaviors for black holes in $D > 4$. But it also turns out that the two discoveries mentioned above contain in its simplest form the essence of all the novel physics of higher-dimensional black holes that has been discovered so far — and we conjecture that this may in fact be the key to all of it.

Underlying this talk will be the theme that the main novel feature of higher-dimensional black holes is the existence of regimes with two different characteristic length scales along the horizon. This provides a framework for organizing the landscape of new solutions according to the hierarchical relation between these two scales. This organization will also indicate what kind of physics is expected in each regime, and the tools that will be appropriate to investigate them.

This talk is based on [5, 6], and we refer the reader to these papers for more details and references on this subject.

2. Organization in scales

We restrict ourselves here, for simplicity, to the case of neutral, vacuum black holes, where the two length scales are associated with the mass and angular momentum

$$\ell_M \sim (GM)^{\frac{1}{D-3}}\,, \qquad \ell_J \sim \frac{J}{M}\,. \tag{2}$$

In the four-dimensional case these length scales are of the same order, due to the Kerr-bound $J \leq GM^2$. In higher-dimensional spacetimes, however, it is known that black hole solutions exist for which the angular momentum for given mass can be arbitrarily high, so that $\ell_J \gg \ell_M$. In particular, this is observed in five dimensions for ultraspinning black rings and in six and higher dimensions for ultraspinning Myers-Perry black holes. The ability to have these two separated lengths is the origin of essentially all novel features of these higher-dimensional black holes, and suggests therefore an organization according to a hierarchy of scales:

(i) $\ell_J \lesssim \ell_M$: black holes behave qualitatively similarly to the four-dimensional Kerr black hole.
(ii) $\ell_J \approx \ell_M$: threshold of new black hole dynamics.
(iii) $\ell_J \gg \ell_M$: the separation of scales enables an effective description of long-wavelength dynamics in terms of blackfolds.

The focus in this lecture is on the third regime, where the existence of a small parameter ℓ_M/ℓ_J allows the introduction of efficient approximate analytical methods, which we call the "blackfold approach". The effective description of black holes in this regime is based on the idea that in the limit $\ell_M/\ell_J \to 0$, keeping the horizon size finite, the black hole geometry approaches that of an infinitely extended black brane, possibly boosted along some of its worldvolume directions. This has been observed in all presently known exact solutions that admit an ultraspinning limit. In particular, in five dimensions the black ring becomes thin in the ultraspinning limit and the geometry approaches that of a boosted black string. In particular, the thickness r_0 of the ring is much smaller than its radius R in this limit, i.e., $\ell_M/\ell_J \sim r_0/R \to 0$. In six and higher dimensions Myers-Perry black holes pancake along the plane of rotation in the ultraspinning limit and the geometry approaches that a black brane. In this case this means that the thickness r_0 of the pancake is much smaller than its radius R.

The connection to (boosted) black branes observed in these examples gives the crucial hint to identify the variables for a general effective description of higher-dimensional black holes in this regime. Namely, the black hole is regarded as *black* brane whose worldvolume spans a curved submani*fold* of a background spacetime, referred to as a *blackfold*.

The blackfold approach draws heavily from the beautiful theory of classical brane dynamics developed by Carter in [7], which generalizes the dynamics of a particle traversing a worldline in a background spacetime to that of branes tracing out embedded worldvolumes. Another central ingredient of the approach, regarded as a long-wavelength effective theory, is that part of its dynamics take the form of the dynamics of a fluid that lives on a dynamical worldvolume. In this context, black branes correspond to specific types of fluid, which are to leading order of the perfect fluid form. The blackfold equations then reduce to a set of intrinsic equations, the Euler equations of the fluid, and a set of extrinsic equations for the worldvolume embedding, which is a generalized geodesic equation for the motion of branes. The effective theory of black holes when $\ell_M/\ell_J \ll 1$ is thus a theory that describes how to bend the worldvolume of a black brane in a background spacetime.

3. Effective worldvolume theory

3.1. Collective coordinates for a black brane

We start by schematically splitting the degrees of freedom of General Relativity into long and short wavelength components,

$$g_{\mu\nu} = \{g_{\mu\nu}^{(\mathrm{long})}, g_{\mu\nu}^{(\mathrm{short})}\}\,, \tag{3}$$

where $\mu, \nu = 0, \ldots, D-1$ are spacetime indices. The Einstein-Hilbert action is then approximated as

$$I_{EH} = \frac{1}{16\pi G}\int d^D x\sqrt{-g}R \approx \frac{1}{16\pi G}\int d^D x\sqrt{-g^{(\mathrm{long})}}R^{(\mathrm{long})} + I_{\mathrm{eff}}[g_{\mu\nu}^{(\mathrm{long})}, \phi]\,, \tag{4}$$

where $I_{\mathrm{eff}}[g_{\mu\nu}^{(\mathrm{long})}, \phi]$ is an effective action obtained after integrating-out the short-wavelength gravitational degrees of freedom. What is meant by this will be explained below, but we first wish to identify the 'collective coordinates' ϕ that enter this effective action, which describes the resulting coupling to the long-wavelength component of the gravitational field.

Since the limit $\ell_M/\ell_J \to 0$ of known black holes results in flat black branes, the effective theory is taken to describe the collective dynamics of a boosted black p-brane.

$$ds^2_{p-\mathrm{brane}} = \left(\eta_{ab} + \frac{r_0^n}{r^n}u_a u_b\right)d\sigma^a d\sigma^b + \frac{dr^2}{1-\frac{r_0^n}{r^n}} + r^2 d\Omega^2_{n+1}\,, \tag{5}$$

with r_0 the horizon radius and u^a the velocity field, satisfying $u^a u^b \eta_{ab} = -1$. Here, and in the following we set the spacetime dimension to be $D = 3 + p + n$ and denote the worldvolume coordinates by $\sigma^a = (t, z^i)$, $a, b = 0, \ldots, p$.

The D collective coordinates of the black brane are then

$$\phi(\sigma^a) = \{X^\perp(\sigma^a), r_0(\sigma^a), u^i(\sigma^a)\}\,, \tag{6}$$

where r_0 is the 'horizon thickness', u^i the p independent components of the velocity field and $X^\perp$ the $D-p-1$ position of the brane in directions transverse to the worldvolume. In the

Journal of Physics: Conference Series **229** (2010) 012001 doi:10.1088/1742-6596/229/1/012001

long-wavelength effective theory these are allowed to vary slowly along the worldvolume $\mathcal{W}_{p+1}$, over a length scale R much longer than the size-scale of the black brane,

$$R \gg r_0 \,. \tag{7}$$

Here, the scale R is typically set by the smallest intrinsic or extrinsic curvature radius of the worldvolume.

In order to preserve manifest diffeomorphism invariance it is convenient to introduce some gauge redundancy and enlarge the set of embedding coordinates of the worldvolume of the black brane to include all the spacetime coordinates $X^\mu(\sigma^a)$. From this embedding one can compute an induced metric

$$\gamma_{ab} = g^{(\mathrm{long})}_{\mu\nu} \partial_a X^\mu \partial_b X^\nu \,, \tag{8}$$

which is the geometry induced on the worldvolume of the brane by the far-zone ($r \gg r_0$) background metric $g^{(\mathrm{long})}_{\mu\nu}$. The near-zone ($r \ll R$) metric is the solution (5) in the strict $R \to \infty$ limit. However, for R large but finite, the collective coordinates depend on σ, so that the near-zone metric for the black brane is of the form

$$ds^2_{(\mathrm{short})} = \left(\gamma_{ab}(\sigma) + \frac{r_0^n(\sigma)}{r^n} u_a(\sigma) u_b(\sigma) \right) d\sigma^a d\sigma^b + \frac{dr^2}{1 - \frac{r_0^n(\sigma)}{r^n}} + r^2 d\Omega^2_{n+1} + \dots \,. \tag{9}$$

which is an approximate solution to Einstein equations in the limit $r_0/R \ll 1$. Furthermore, the long and short degrees of freedom interact together in the 'overlap' or 'matching-zone' $r_0 \ll r \ll R$, where the metrics $g^{(\mathrm{long})}_{\mu\nu}$ and $g^{(\mathrm{short})}_{\mu\nu}$ must match.

3.2. *Effective stress tensor*

If we consider the effective equations of motion resulting from (4) by varying with respect to the long-wavelength metric $g^{(\mathrm{long})}_{\mu\nu}$, one finds Einstein equations for $g^{(\mathrm{long})}_{\mu\nu}$, sourced by an effective worldvolume stress tensor

$$T^{\mathrm{eff}}_{\mu\nu} = - \frac{2}{\sqrt{-g_{(\mathrm{long})}}} \frac{\delta I_{\mathrm{eff}}}{\delta g^{\mu\nu}_{(\mathrm{long})}} \bigg|_{\mathcal{W}_{p+1}} \,. \tag{10}$$

The appropriate notion of this effective stress-tensor can be argued to be the quasilocal stress-energy tensor introduced by Brown and York. For ease of notion, we drop from now on the superscript ^{eff} on the stress-tensor $T_{\mu\nu}$ and $^{(\mathrm{long})}$ from the background metric $g_{\mu\nu}$.

The effective stress tensor is computed in the zone $r_0 \ll r \ll R$, where the gravitational field is weak and the quasilocal stress tensor $T_{ab} = \partial_a X^\mu \partial_b X^\nu T_{\mu\nu}$ is, to leading order in r_0/R, the same as the ADM stress tensor. It is not difficult to compute this for the boosted black p-brane (5), and introducing a slow variation of the collective coordinates the stress tensor becomes

$$T^{ab}(\sigma) = \frac{\Omega_{(n+1)}}{16\pi G} r_0^n(\sigma) \left(n u^a(\sigma) u^b(\sigma) - \gamma^{ab}(\sigma) \right) + \dots \tag{11}$$

Here the dots stand for terms with gradients of $\ln r_0$, u^a, and γ_{ab}, which are taken to be small and are neglected to the order we are working in.

3.3. *Fluid perspective*

We note that the effective stress-tensor (11) is of the perfect fluid form

$$T^{ab} = (\varepsilon + P) u^a u^b + P \gamma^{ab} \,, \tag{12}$$

with energy density ε and pressure P

$$\varepsilon = \frac{\Omega_{(n+1)}}{16\pi G}(n+1)r_0^n\,, \qquad P = -\frac{1}{n+1}\,\varepsilon\,, \tag{13}$$

and velocity u^a satisfying $u^a u^b \gamma_{ab} = -1$. This is in fact expected on general grounds, since the the long-wavelength effective theory for any kind of brane will take the form of a derivative expansion for an effective stress-energy tensor that satisfies the conservation equations $D_a T^{ab} = 0$ (D_a is the worldvolume covariant derivative). To lowest derivative order the stress tensor will then be that of an isotropic perfect fluid. Going beyond the perfect fluid approximation (12), the stress tensor will acquire dissipative terms proportional to gradients of $\ln r_0$, u^a, γ^{ab}. These are not discussed here, and are at any rate absent for stationary configurations.

4. Blackfold equations

From the induced metric (8) one introduces the first fundamental form of the submanifold and perpendicular projector

$$h^{\mu\nu} = \partial_a X^\mu \partial_b X^\nu \gamma^{ab}\,, \qquad \perp_{\mu\nu} = g_{\mu\nu} - h_{\mu\nu}\,. \tag{14}$$

The tensor $h^\mu{}_\nu$ projects onto the worldvolume $\mathcal{W}_{p+1}$, while $\perp^\mu{}_\nu$ projects along directions orthogonal. From this one computes the extrinsic curvature tensor

$$K_{\mu\nu}{}^\rho = h_\mu{}^\sigma \overline{\nabla}_\nu h_\sigma{}^\rho\,, \qquad \overline{\nabla}_\mu = h_\mu{}^\nu \nabla_\nu\,, \tag{15}$$

which is tangent to $\mathcal{W}_{p+1}$ along its (symmetric) lower indices μ, ν, and orthogonal to $\mathcal{W}_{p+1}$ along ρ. For use below, we also note that its trace is the mean curvature vector

$$K^\rho = h^{\mu\nu} K_{\mu\nu}{}^\rho = \overline{\nabla}_\mu h^{\mu\rho}\,. \tag{16}$$

Given a brane stress-energy tensor with support on the $p+1$-dimensional worldvolume $\mathcal{W}_{p+1}$ one then requires it to obey the conservation equation

$$\overline{\nabla}_\mu T^{\mu\rho} = 0\,. \tag{17}$$

These equations separate into $D-p-1$ equations in directions orthogonal to $\mathcal{W}_{p+1}$ and $p+1$ equations parallel to $\mathcal{W}_{p+1}$,

$$T^{\mu\nu} K_{\mu\nu}{}^\rho = 0 \qquad (\textit{extrinsic equations})\,, \tag{18}$$

$$D_a T^{ab} = 0 \qquad (\textit{intrinsic equations})\,. \tag{19}$$

One may now apply these equations to the specific stress tensor (13) of a neutral black brane, in which case the equations (18), (19) become after some algebra

$$K^\rho = n \perp^\rho{}_\mu \dot{u}^\mu\,, \tag{20}$$

$$\dot{u}_a + \frac{1}{n+1} u_a D_b u^b = \partial_a \ln r_0\,. \tag{21}$$

This set of *blackfold equations* describes the general collective dynamics of a neutral black brane. Physically, these equations show that the force that accelerates an element of the fluid is given along worldvolume directions by pressure gradients (Euler equation) and in directions transverse to the worldvolume by the extrinsic curvature.

Acknowledgments

I am grateful to the organizers of the ERE2009 for the invitation to this stimulating conference. The work presented here was done in collaboration with Troels Harmark, Vasilis Niarchos and Niels Obers. The author was supported by DURSI 2005 SGR 00082 and 2009 SGR 168, MEC FPA 2007-66665-C02 and CPAN CSD2007-00042 Consolider-Ingenio 2010.

References

[1] Horowitz G T and Maeda K 2001 Fate of the black string instability *Phys. Rev. Lett.* **87** 131301
[2] Gubser S S 2002 On non-uniform black branes *Class. Quantum Grav.* **19** 4825
[3] Wiseman T 2003 Static axisymmetric vacuum solutions and non-uniform black strings *Class. Quantum Grav.* **20** 1137
[4] Emparan R and Reall H S 2002 A rotating black ring in five dimensions *Phys. Rev. Lett.* **88** 101101
[5] Emparan R, Harmark T, Niarchos V and Obers N A 2009 World-volume effective theory for higher-dimensional black holes. (Blackfolds) *Phys. Rev. Lett.* **102** 191301
[6] Emparan R, Harmark T, Niarchos V and Obers N A 2009 Essentials of Blackfold Dynamics *Preprint* arXiv:0910.1601
[7] Carter B 2001 Essentials of classical brane dynamics *Int. J. Theor. Phys.* **40** 2099–2130

Spanish Relativity Meeting (ERE 2009) IOP Publishing
Journal of Physics: Conference Series **229** (2010) 012002 doi:10.1088/1742-6596/229/1/012002

Dark Matters

Joseph Silk

Department of Physics, University of Oxford, Oxford OX1 3RH, UK

E-mail: silk@astro.ox.ac.uk

Abstract. One of the greatest mysteries in the cosmos is that it is mostly dark. That is, not only is the night sky dark, but also most of the matter and the energy in the universe is dark. For every atom visible in planets, stars and galaxies today there exists at least five or six times as much "Dark Matter" in the universe. Astronomers and particle physicists today are seeking to unravel the nature of this mysterious but pervasive dark matter, which has profoundly influenced the formation of structure in the universe. Dark energy remains even more elusive, as we lack candidate fields that emerge from well established physics. I will describe various attempts to measure dark matter by direct and indirect means, and discuss the prospects for progress in unravelling dark energy.

1. Introduction

The nature of dark matter remains one of the greatest enigmas in astrophysics. We know much about its distribution but have yet to find a resolution. The problem may be summarised as follows: the matter budget is $\Omega_m = 0.21; \Omega_b = 0.04; \Omega_* = 0.005$, referring to dark matter, baryons and stars respectively [1].

The baryonic component is confirmed via three independent techniques: light element nucleosynthesis, cosmic microwave background acoustic peaks and intergalactic hydrogen measured via the Lyman alpha absorption line forest. The ionized intergalactic medium is also detected via absorption of highly ionized species (Fe, O etc) both in clusters and in the diffuse component. The unambiguous conclusion via measurements at three epochs: three minutes, 3×10^5 yr and 3×10^9 yr, is that 90% of the dark matter is nonbaryonic.

We can say more about the dark matter. The cosmic microwave background fluctuations confirm not only that dark matter is the dominant component in the universe but that it is weakly interacting with the baryonic matter. Otherwise the fluctuations would be strongly damped. Simulations of large-scale structure have had great success in reproducing the three-dimensional galaxy distribution. These models are predicated on growth by gravitational instability of the small fluctuations in the radiation, seen on the last scattering surface of the CMB, that are weakly coupled to the baryons. One more property is tested: the dark matter must not only be non-baryonic, weakly interacting matter, but it must also be thermally and dynamically cold. This occurs naturally if thermal and kinetic decoupling of the dark matter occurred in the very early universe. One can be more precise on this latter point: dark matter decoupling must have occurred before the epoch of primordial nucleosynthesis at $T \sim 0.1$ MeV.

This is the good news. There is some bad news too. While cold dark matter provides a near perfect fit to large-scale structure as measured by statistical indicators (such as correlation functions and gaussianity probes) and on the reasonable assumption that light traces mass, the fit is far from perfect on galactic and sub-galactic scales. Many galaxies have dark matter cores rather than the predicted cusps [2]. While this conclusion is not easily traceable to the initial dark matter distribution in massive disk galaxies because of the uncertain role of baryonic contraction and non-axisymmetric gravitational instabilities in modifying the dark matter profile, the situation is far simpler for low surface brightness dwarfs that are dominated by dark matter. Here there is little doubt that cores are prevalent [3]. An equally serious problem is the large number of predicted dwarfs that are massive enough, more than $\sim 10^6 M_\odot$, to have undergone baryonic dissipation and formed stars. These are problems directly attributable to the dark matter. There also are problems associated with luminous galaxies that are a consequence of the hierarchical formation of dark matter halos and the associated galaxies. These are more astrophysical in nature, and centre around such issues as the epoch, duration, ordering and efficiency of galaxy formation. Astrophysical feedback from some combination of supernovae, reionization, dynamical relaxation, supermassive black holes and gravitational accretion may provide a resolution to some or all of these matters, e.g. [4], [5], [6], [7], [8]. I will subsequently take cold dark matter as given, and review prospects for its detection if it consists of massive weakly interacting particles.

The remarkable success of distant supernova surveys over the past decade has provided strong evidence for the acceleration of the distant universe, and dark energy, more specifically, the cosmological constant, provides by far our best empirical model for understanding the data. Even before the acceleration was observationally confirmed, there was circumstantial evidence from the shortfall of dark matter relative to the critical density of the universe favoured by some, although not all, astronomical measurements for the presence of a dominant dark energy component. The case for dark energy is further reinforced by additional evidence for acceleration from the effect of the varying gravitational potential of the universe on photon propagation from the last scattering surface of the cosmic microwave background, measured as the integrated Sachs-Wolfe effect. However there is no theory for the dark energy. We have no explanation of the small value of the vacuum energy nor of its relatively recent dominance: were it dominant at early epochs, galaxies would not have formed. Theorists have retreated towards anthropic selection as providing a resolution of the dark energy dilemma. I will briefly describe the unsatisfactory situation at present with regard to possible alternatives.

2. Dark matter: neutrinos

The most desirable non-baryonic matter candidate has to be the neutrino. At least two of the neutrino species are guaranteed to have mass from the mass differences measured in solar and atmospheric neutrino oscillation experiments. Of course these masses are far from summing up to the $\sim$10 eV in total neutrino mass required for the dark matter. In fact while direct mass measurements only constrain the electron neutrino mass to be below this range, such an option is ruled out by cosmology. Measurements of the abundance of x-ray clusters and the primordial density fluctuation power spectrum jointly constrain the sum of the neutrino masses to not exceed $\sim$1 eV [9]. Future cosmological experiments will improve these limits by an order of magnitude, and will be complemented by direct measurements of the electron neutrino mass from tritium end-point decay (KATRIN) as well as neutrinoless double beta decay and accelerator-based oscillation experiments [10].

Sterile neutrinos present another option. These right-handed neutrinos decouple much earlier than left-handed neutrinos, and while the density is suppressed, much higher masses are possible. One candidate has a mass of a few keV, and has attracted interest for two reasons: decay into x-ray photons occurs over $\sim 10^{26}$ sec, leading to a potential signal, and the associated free streaming occurs at a comoving scale of order a kpc, resulting in a preferred scale for the smallest dark halos of some millions of solar masses. Whether observations of the recently discovered dark-matter-dominated dim dwarfs in the MWG require such an interpretation in terms of a preferred mass scale is under intense current debate[11]. An alternative possibility is that reionization may provide a threshold capable of imprinting a preferred epoch, and hence density scale [12]. .

3. Dark matter: neutralinos

By far the favoured non-baryonic candidate is motivated by the lightest SUSY relic. Dubbed a Weakly Interacting Massive Particle (or WIMP), the relic abundance is obtained via freeze-out for a cross-section typical of weak interactions: $\langle\sigma v\rangle \approx 3\times 10^{-26}\mathrm{cm}^3/\mathrm{s}\ 1/\Omega_\chi$ yields $\Omega_\chi \sim 0.2$ for the lightest stable relic, a massive neutralino. Of course there are some 120 free parameters in SUSY, and one has to perform Monte Carlo Markov Chain searches over this parameter space to obtain specific predictions of the relevant cross-sections needed for experiments (elastic for direct detection, annihilation for indirect detection). Usually, minimal SUSY is adopted, and in this case the mass range is expected to lie in the range of 100-1000 proton masses. The lower bound is determined by the SUSY scale, the upper bound by unitarity scaling ($\sigma_{ann} \propto m_\chi^{-2}$) leading to overproduction of relic particles.

The ideal solution would be direct detection of the flux of some $10^7 m_{\chi,GeV}^{-1}\mathrm{cm}^{-2}\mathrm{s}^{-1}$ passing through any terrestrial detector. Many experiments are currently underway using techniques that search for nuclear recoils by some combination of ionization, fluorescence or phonon production in detectors that typically are deep underground to provide shielding from cosmic rays. To date, no definitive detections have been found, the most promising so far being two events in cryogenic germanium reported by the CDMS2 collaboration [13]. However the background level is comparable to the signal.

Indirect detection is being vigorously pursued of annihilation products from the rare annihilations that occur in the halo and even in the sun. High energy neutrinos are generated by WIMPs annihilating in the sun and are observable with downward-looking neutrino telescopes such as ANTARES and ICECUBE. To date, only limits have been set. Overall, in our galaxy, there is expected to be $\sim 10^{39}$GeV/s in total annihilation power detectable via several channels, including energetic $\gamma, e^+e^-, \bar{p}, \nu$. Experiments currently taking data include the PAMELA and FERMI satellites. Following earlier detections by the HEAT balloon-borne detectors, the Italian-Russian PAMELA satellite has measured a rise in the cosmic ray positron fraction up to 100 GeV [14]. This was predicted to be a consequence of a neutralino WIMP with mass $\sim$ 100 GeV. FERMI and the HESS telescope in Namibia have measured the high energy electron plus positron flux up to 300 GeV and 1 TeV respectively, and found an excess relative to pre-existing models of secondary production combined with propagation calculations. These measurements are local, to the solar neighbourhood.

Gamma rays provide a non-local constraint, both on the dark matter concentration in the inner galaxy and on dark matter in the outer halo as well as on isotropic extragalactic backgrounds. The FERMI angular resolution is 5 arc min at 300 GeV, reducing to 5 degrees at 0.02 GeV. The dark matter annihilation gamma rays include the direct annihilation products and inverse Compton generation by the high energy leptons. FERMI detects a quasi-spherical haze in the central kpc of the galaxy that has been attributed to inverse Compton scattering by high energy electrons on starlight [18]. Previously, analysis of WMAP foregrounds found a central haze at 50-90 GHz, due to synchrotron emission by an electron population of similar energy and spatial distribution. The coincidence with the dark matter distribution is suggestive of a common annihilation source, but far from conclusive.

Astrophysics explanations of the PAMELA/HESS/FERMI/WMAP data include the following; nearby electron-positron pair sources, most notably pulsar winds, and cosmic ray acceleration of leptons as well as hadrons in supernova remnants. Particle physics solutions include both annihilating and decaying dark matter. For the case of annihilating WIMPs, the cross-section associated with the massive neutralino is far too small since the predicted positron flux is proportional to ρ/m_χ^2. The necessary boost of order ~ 1000 for TeV dark matter particle masses may be provided by the Sommerfeld effect, especially if the effects are included of cold dark matter substructure in the halo [15], [16]. In the case of decaying dark matter, the massive neutralino requires a decay time of order $\sim 10^{26}$ sec [17]. All predictions are compounded by the propagation uncertainties

Here are some predictions that may help distinguish between these alternative explanations of the data. Low surface brightness dwarf galaxies should be detectable in the case of annihilating dark matter because of the cold substructure boost, as seen in the latest CDM simulations with $\sim 1000 M_\odot$ resolution and weighted by ρ^2/v^2 as predicted for the Sommerfeld effect boost [19]. If neutralino masses extend to a TeV, as inferred from HESS, the Planck satellite should detect a corresponding high frequency synchrotron haze above 100 GHz. The Fermi gamma rays should include prompt emission from the annihilations, characterised by a hard spectrum. Fermi spectral resolution is $\Delta E/E \sim 0.1$ and possible annihilation line features may be detectable. These would constitute gamma ray smoking guns of annihilation. Decay of dark matter necessarily involves a spatial distribution that is less concentrated than that predicted for annihilating dark matter, and refined propagation models may be expected to yield distinct predictions for FERMI and PLANCK. On the astrophysical front, supernova acceleration of cosmic rays involves some spallation which may have observable consequences in terms of cosmic ray abundance ratios [20]. The pulsar wind interpretation requires that nearby sources dominate, and anisotropy in the positron flux may be detectable in the direction of nearby pulsars [21].

4. Dark energy

Dark energy may be said to have originated with Einstein's introduction of the cosmological constant to account for the stability of a static universe. Rapidly abandoned by 1930 when Hubble's law of expansion began to be widely accepted, the cosmological constant was reincarnated into dark energy half a century later by M. Turner into its new parametrization $p = w\rho$ and $w = -1$. Strong evidence for dark energy is provided by observations of Type Ia supernovae that exploded when the Universe was 2/3 of its present size at $z \sim 1$. The data scatter is greatly reduced by applying an empirical correlation between maximum brightness and decay time. The supernovae are found to be $\sim 25\%$ fainter than expected from local light curve observations. From the supernovae, the acceleration yields the difference between the densities of dark matter and dark energy. The dark matter density is inferred from large-scale structure observations, including gravitational lensing, baryonic acoustic oscillations and galaxy cluster counts. Finally, the CMB peaks yield the curvature of the universe, effectively the sum of dark matter and dark energy, once degeneracies, e.g. with the age of the universe, are controlled. From the WMAP5 data, combined with the other data sets, the current cosmological parameters take on the values $\Lambda = 0.75$, $\Omega_m = 0.21$, and $\Omega_b = 0.04$. Also $w \approx -1 \pm 0.01$. The cosmological constant is favoured by the current data. However the errors are still dominated by systematics.

Confirmation of the cosmological constant has led to one of the greatest mysteries in physics. The predicted value, that expected for the vacuum energy at the Planck scale, exceeds the measured value by some 120 orders of magnitude. This leads to the questions: why so small? And why is it emerging so recently as a cosmological force (or field)? One class of explanations seeks to explain why our universe is so special by postulating the existence of the multiverse. We live in the unique corner of multiverse space where the repulsive force associated with dark energy is small enough to allow galaxies to form. This is anthropic reasoning and has been strongly advocated by many (but not all) eminent string theorists.

The multiverse idea has great explanatory power: it is motivated by string theory and by eternal inflation, and does provide an empirically based rationalization for fine tuning as a development from known physical principles. However its weakness is that successful scientific theories make predictions. The multiverse theory is hitherto not verifiable either by observation or by theory. The latter would require the multiverse to be an implication of well established physics, and one can hardly say that landscape theory or eternal inflation are well established. The multiverse makes no predictions because it can be tuned to explain whatever we have discovered or whatever we may discover, indeed it can explain anything at all. Even the value of the (hopefully to be imminently discovered) Higgs boson can be explained by landscape pressure in the multiverse scenario.

There are two alternative approaches to anthropic reasoning. One is to hope for the emergence of a Theory of Everything. We are very far from this, although a glimmer of such a future is given by an example: reduction from 10^{500} Calabi-Yau manifolds to 3 may be achieved by requiring topological simplicity: the surviving manifolds with the smallest Hodge numbers in a 3 generation example are just three, coinciding (in number) to the three families of the standard model of particle physics [22].

Another, more astrophysical, approach is to postulate that we live near the centre of a giant void, of order a Gpc in radius, Provided the universe is asymptotically weakly positively curved, the need for dark energy is removed. The transition gives distant acceleration, and the global properties reproduce those of the standard cosmological model without any need for dark energy [23], [24].

5. Summary

Dark matters are challenged by cosmology. Explanations are sparse. Changing fundamental physics provides one possible way forward. Attempts to modify the nature of dark matter include introducing self-interacting dark matter and shadow sector dark matter. Attempts to modify Einstein-Newton gravity are more motivated, but have not had much success because of the Bayesian penalty involved in introducing free parameters, such as the scalar, vector and tensor fields in Bekenstein's TeVeS theory of gravity. Concerning dark energy, the principal alternative at hand to anthropic selection in the landscape involves a giant void-dominated universe whose fine-tuning, while large, may still be less than any of the alternatives on offer.

Evidently, we need experimental progress. We can hope that detection in multiple windows will provide the "smoking gun" that will elucidate the nature of dark matter. Experiments are being proposed at vast cost to search for hints of new physics by reducing the error bars on Λ and w. Some view this as an act of desperation which should at least be complemented by supporting experiments directed towards more tangible goals, such as identification of the first stars in the universe, as well as theory efforts directed towards elucidating a new theory of dark energy. Considering the enormous progress made in cosmology in the last century one can reasonably expect that new ideas will emerge in the future that render our current theories resemble those of a blind man exploring his limited world.

References

[1] Dunkley J *et al* 2009 *Astrophys. J. S.* **180** 306
[2] Navarro J F *et al* 2009 *MNRAS* 1918
[3] de Blok W J G 2010 *Advances in Astronomy* **2010** 5
[4] Mashchenko S, Wadsley J and Couchman H M P 2008 *Science* **319** 174
[5] Jardel J R and Sellwood J A 2009 *Astrophys. J.* **691** 1300
[6] Silk J and Norman C 2009 *Astrophys. J.* **700** 262
[7] Khochfar S and Ostriker J P 2008 *Astrophys. J.* **680** 54
[8] Macciò A V, Kang X, Fontanot F, Somerville R S, Koposov S and Monaco P 2009 *MNRAS* 1907
[9] Vikhlinin A *et al* 2009 *Astrophys. J.* **692** 1060
[10] Fogli G *et al* 2009 *Nucl. Phys. Proc. Suppl.* **188** 27-30
[11] Adén D, Wilkinson M I, Read J I, Feltzing S, Koch A, Gilmore G F, Grebel E K and Lundström I 2009 *Astrophys. J. L.* **706** L150
[12] Koposov S E, Yoo J, Rix H -W, Weinberg D H, Macciò A V and Escudé J M 2009 *Astrophys. J.* **696** 2179
[13] The CDMS Collaboration and Ahmed Z 2009 *Preprint* arXiv:0912.3592
[14] Boezio M *et al* 2009 *New Journal of Physics* **11** 105023

[15] Lattanzi M and Silk J 2009 *Phys. Rev.* D **79** 083523
[16] Cholis I, Dobler G, Finkbeiner D P, Goodenough L and Weiner N 2009 *Phys. Rev.* D **80** 123518
[17] Ibarra A, Tran D and Weniger C 2010 *J. Cosmol. Astropart. P.* JCAP01(2010)009
[18] Dobler G, Finkbeiner D P, Cholis I, Slatyer T R and Weiner N 2009 *Preprint* arXiv:0910.4583
[19] Kuhlen M, Madau P and Silk J 2009 *Science* **325** 970
[20] Mertsch P and Sarkar S 2009 *Phys. Rev. Lett.* **103** 081104
[21] Grasso D *et al* 2009 *Astroparticle Physics* **32** 140
[22] Braun V, Candelas P and Davies R 2009 *Preprint* arXiv:0910.5464
[23] Garcia-Bellido J and Haugbølle T 2008 *J. Cosmol. Astropart. P.* JCAP04(2008)003
[24] Clifton T, Ferreira P G and Land K 2008 *Phys. Rev. Lett.* **101** 131302

Bulk flows in inflation and in Lemaître-Tolman-Bondi models

F Atrio-Barandela[1]*, A Kashlinsky[2], H Ebeling[3] and D Kocevski[4]

[1]Universidad de Salamanca, Spain.
[2]Goddard Space Flight Center, Maryland.
[3]Institute of Astronomy, University of Hawaii.
[4]Dpt of Physics, UC Davis.

E-mail: atrio@usal.es, Alexander.Kashlinsky@nasa.gov, ebeling@ifa.hawaii.edu, kocevski@physics.ucdavis.edu

Abstract. We describe the recent measurements of bulk flows of unprecedented scale and amplitude. These results are not easily accommodate in the standard ΛCDM cosmology. We study two possible solutions: preinflationary superhorizon sized perturbation and void models of the Lemaître-Tolman-Bondi type and discuss their viability.

1. Introduction

In standard gravitational instability picture, peculiar velocities are departures from the Hubble flow due to local inhomogeneities. Gravitational instability states that the present large scale structure of the Universe originates from the growth of small density perturbations on an initially weakly inhomogeneous Friedmann-Robertson-Walker model. Then, they trace directly the matter density field. Peculiar velocities of individual galaxies are very difficult to measure and only moments of the velocity field have been derived from the data. One such moment is the bulk flow, the centre of mass motion of the matter contained on a given volume. For all scales larger than Matter-Radiation equality [$\sim$ 100 Mpc/h], bulk flows probe the primordial power spectrum of matter density perturbations. The expected amplitude is

$$P(k) \propto k \qquad \Rightarrow \qquad V_{rms} \simeq \left(\frac{r}{100\mathrm{Mpc/h}}\right)^{-1} 250\mathrm{km/s} \tag{1}$$

Measuring peculiar velocities on scales larger than 100Mpc/h offers a test of inflation and of the gravitational instability paradigm.

Determination of peculiar velocity using galaxies requires to subtract the velocity due to the Hubble expansion. Since uncertainties on distance estimators increase with distance, measurement of peculiar velocities has been limited to scales $\leq$ 100Mpc/h. In [1] we proposed to use clusters as tracers of the velocity field using the temperature anisotropies induced on the Cosmic Microwave Background (CMB). The proposed method, based on the Sunyaev-Zel'dovich (SZ) effect [2], has systematics that are very different from those of galaxy distance estimators. Physically, the hot intracluster gas distorts the black body spectrum of the CMB photons due to the *thermal* motion of electrons in the potential wells of clusters and to the *kinematic* motion of the cluster as a whole with respect to the matter rest frame [2]. The CMB temperature anisotropies generated by the thermal SZ effect are proportional to the electron pressure integrated along the line of sight $\hat{n}$. The anisotropies produced by kinematic SZ effect depend on the electron density n_e and the projection of the peculiar velocity $\vec{v}_{cl}$ of the cluster along the line of sight. If clusters are isothermal, then

$$\left(\frac{\Delta T}{T}\right)_{tSZ}(\hat{n}) = G(\nu)\tau\frac{k_B T_x}{m_e c^2}, \qquad \left(\frac{\Delta T}{T}\right)_{kSZ}(\hat{n}) = \tau\frac{\hat{n}\cdot\vec{v}_{cl}}{c}. \tag{2}$$

In eq. (2) $k_B T_x$ is the cluster X-ray temperature, $m_e c^2$ the electron annihilation temperature and $\tau = \int n_e(\vec{r})dl$ is the optical thickness of the cluster along the line of sight, integrated over the cluster volume. $G(\nu)$ gives the frequency dependence of tSZ effect. The SZ effect is very useful to identify clusters at high redshifts since, being a distortion of the CMB black body spectrum, it is independent of distance. In the Rayleigh-Jeans part of the spectrum, $G(\nu) \simeq -2$ and clusters appear as temperature decrements in CMB maps and can be distinguished from other foreground contributions.

Since clusters can be detected with the SZ effect and its effect is redshift independent, the kSZ effect is a very promising tool to trace the velocity field to very high redshifts. But, similarly to the case of galaxies, measurements of the peculiar velocity of a single cluster are dominated by the cosmological CMB anisotropy. The peculiar velocity of a sample of clusters can be measured with much better accuracy. In [1] we proposed that by measuring the dipole components a_{1m} on CMB maps at *the position of clusters* one would determine the bulk flow of the cluster sample. The signal is:

$$a_{1m} = 1\mu K\frac{v_{bulk}}{300km/s} \pm 3\mu K\left[\frac{N_{cl}}{1000}\right]^{1/2} \pm 0.6\mu K\left[\frac{N_{pixels}}{10000}\right]^{1/2} \pm 0.2\mu K\left[\frac{N_{cl}}{1000}\right]^{1/2} \tag{3}$$

The error bar on the measurement contains contributions that do not have a dipole pattern: noise, that integrates with the number of pixels and frequency bands, and intrinsic CMB and thermal SZ, that integrate as the number of clusters. The error bar is dominated by the sampling variance of the CMB cosmological signal, resulting from evaluating a dipole on the fraction of the sky occupied by clusters. To reduce this contribution, in [1] we also proposed to filter the intrinsic cosmological anisotropy. We showed how the velocity field could be probed using the Wilkinson Microwave Anisotropy Probe (WMAP) and PLANCK satellites data.

2. Uncovering the 'Dark Flow'.

The technique proposed by us in [1] was used in [3] to uncover a large scale flow of clusters of galaxies on the unprecedented scale $\sim$ 300Mpc/h. To carry out the measurement we assembled a catalogue of $\sim$ 1000 clusters with well measured redshifts, X-ray luminosities and electron densities. To remove the intrinsic CMB signal, a filter was constructed that minimized the difference $\langle(\delta T - \delta T_{\Lambda\mathrm{CDM}})^2\rangle$. The filter effectively removed the CMB signal down to the limit imposed by cosmic variance [4]. By evaluating the dipole contribution at cluster locations, we measured a flow of amplitude $600 - 1000$km/s in the direction $(l, b) = (283^o; 11^o) \pm 30^o$ -in galactic coordinates-, coincident with the direction of the Local Group with respect to Isotropic CMB frame: $(276^o \pm 3^o; 30^o \pm 3^o)$ [5].

In Figure 1 we plot the amplitude of the bulk flow for different scales. We selected cluster subsamples of our main catalogue, containing all clusters with redshift $z \leq [0.04, 0.08, 0.12, 0.16, 0.2, 0.3]$. The data correspond to the analysis of WMAP 3 year data. The error has two main contributions: noise and intrinsic CMB residuals. In [4] we showed that the error bars of the analysis of WMAP 3-year data were dominated by CMB residuals, correlated

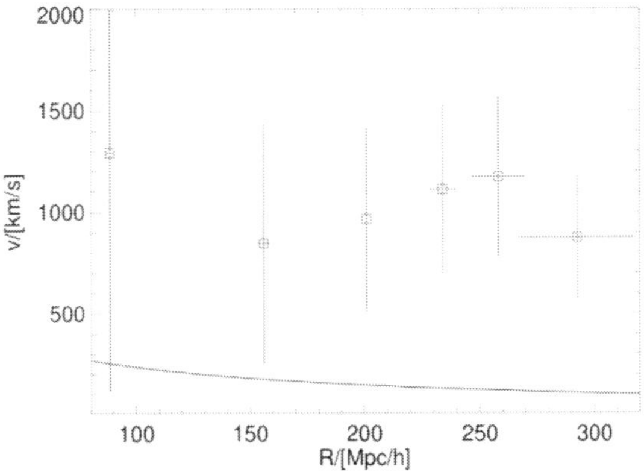

Figure 1. Amplitude of the bulk flow on the given scale. Vertical lines give the error bars on the measured amplitude and horizontal error bars join the mean and median redshift of the cluster sample. The solid line represents the expected amplitude on the standard ΛCDM model.

between the 8 WMAP Differencing Assemblies, and not by noise. The errors in Figure 1 have been corrected accordingly.

At present, our formalism does not allow us to determine the scale length of the flow. As an order of magnitude estimate in Figure 1 we took the middle point between the mean and the median redshift of the cluster subsample. The horizontal error bar, stretching from the mean to the median redshift, provides an estimate of the uncertainty on the scale of the flow. The solid line represents the prediction of the concordance ΛCDM model. Even though the measurement has large errors, it is clear that the measured amplitude, at the largest scale probed by our cluster sample, is much larger than the concordance model prediction.

Let us summarize the different pieces of evidence that support the idea of a large scale flow:

(a) The motion is found at cluster positions. The error bars in Figure 1 are computed by placing clusters at random positions on the filtered data excising the pixels occupied by real clusters and by the mask. A more extensive discussion can be found in [4].

(b) The signal is persistent when we include all clusters with redshifts $z \leq 0.12$ to $z \leq 0.3$, where the number of clusters varies from ~ 150 clusters to ~ 700. Any undiagnosed systematic effect present in the data would dilute the signal when more clusters were added.

(c) The dipole kSZ signal is measured when the thermal SZ monopole is effectively zero. The thermal and kinematic components are both generated by the X-ray gas. We demonstrated that the dipole contribution was not generated by the thermal motion of the intra cluster gas: If clusters were isothermal, the kinematic and thermal SZ effects would be proportional to each other (see eq. 2). It was believed that the kinematic SZ effect could only be measured in experiments with extensive frequency coverage [6]. But in [7] we showed that clusters are not isothermal. Rather, they follow a Navarro, Frenk and White profile, with temperature and electron density falling with increasing distance from the centre. Then, on the cluster outskirts the kinematic component dominates over the thermal and the former signal is present when the latter is effectively zero.

(d) Since the motion of clusters induces secondary anisotropies on the CMB, if all clusters in a sample move at a similar speed then the more massive clusters would generate larger anisotropies, larger dipoles. This prediction was confirmed in [8] where we showed that the thermal SZ monopole of a cluster sample correlates with the dipole. Larger tSZ signal, i.e. more

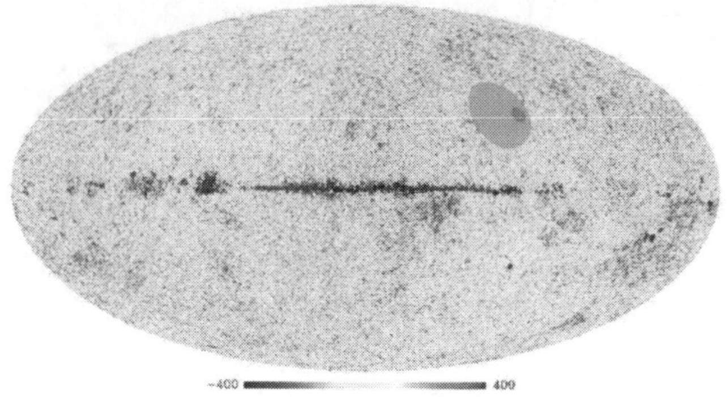

Figure 2. Map of the WMAP 3-year data Q1 Differencing Assembly. The larger circle give the measured bulk flow direction and its 1σ uncertainty. The smaller circle gives the same magnitudes of the motion of the Local Group [5].

massive clusters, have systematically larger dipoles indicating that all the clusters in the sample share the same motion.

Briefly, (a-d) are strong indications that the flow is real and affects all the clusters of our sample. The main limitation of the method is the conversion from the measured signal in units of temperature to a velocity. This conversion factor can be uncertain by up to a 50% and could lead to an overestimate of the amplitude of the flow but would not affect the direction. In Figure 2, we superposed two circular shaded region on WMAP Q1 Differencing Assembly 3-year data. The outer circle gives the direction of the measured bulk flow and its 1σ uncertainty. The inner circle gives the direction of the CMB dipole derived from the Dipole anisotropy of COBE Differential Microwave Radiometers [5].

3. Origin of the 'Dark Flow'

The amplitude and coherence of the flow has strong implications for the present gravitational instability picture. Several explanations have been put forward. The flow could be an indication of a tilt created by the pre-inflationary inhomogeneous structure of space-time [9] that might provide an indirect probe of the landscape produced in certain variants of string cosmology [10, 11] or be an indication of a higher dimensional structure of gravity [12].

Alternatively, it could have a 'local' origin, due to a void or large scale overdensity [13]. Large scale voids of $\sim 1Gpc$ with an underdensity of $\delta \sim -0.3$ have been advocated to explain the cold spot on the cosmic microwave background on scales of $\sim 5^o$. An order of magnitude of the effect can be easily calculated. At a distance R from the centre of a Lemaître-Tolman-Bondi void:

$$v_p = \frac{1}{6}\Omega^{0.6} H_o R\delta = 600km/s\left(\frac{R}{1Gpc}\right)\left(\frac{\delta}{0.1}\right) \qquad (4)$$

Such voids will generate a peculiar velocity field with a measurable shear. At present, our results have not enough statistical power to determine the shear of the flow. However, our latest results [8] indicate the flow extends out to $\sim$ 750Mpc, without showing signs of convergence, which cast doubts on this type of explanation.

The scale and coherence of the flow supports a different and more radical solution requiring 'global' motion: the dipole anisotropy of the CMB is intrinsic. The observers at rest with the matter distribution in the Universe do not coincide with the observers where the CMB is isotropic. If the CMB dipole anisotropy is intrinsic, then viewed from the isotropic CMB frame the Universe appears to be tilted, with galaxies moving from one side of our Hubble volume to the other with speeds of the order of 600 km s^{-1}.

An intrinsic anisotropy on the last scattering surface requires a superhorizon isocurvature perturbations of wavelength L much larger than the Hubble radius cH_o^{-1}. First, the perturbation can not be adiabatic since it would produce the same dipole on all energy densities. Second, this perturbation must be a preinflationary remnant since its amplitude is not given by the Harrison-Zel'dovich prescription. The amplitude δ_L of this primordial perturbation could be of order unity. It will generate a horizon flow D, also reflected in the quadrupole Q and higher order multipoles [10, 14]:

$$D \sim \frac{v}{c} \sim \left(\frac{cH_o^{-1}}{L}\right)\delta_L \sim 2\times 10^{-3} \qquad Q \sim \left(\frac{cH_o^{-1}}{L}\right)^2 \delta_L \sim 4\times 10^{-6} \tag{5}$$

A pre-inflationary remnant of $\delta_L \sim 1$ would produce a dipole of the required amplitude if $L \sim 500cH_o^{-1}$ [3]. However, it will also generate a quadrupole of similar magnitude to the one measured [10]. Higher order multipoles are strongly suppressed and are subdominant with respect to the anisotropies generated during inflation.

The quadrupole estimate of eq. (5) would add linearly to the quadrupole generated by perturbations on horizon scales and this could be a testable prediction. As is well known, the quadrupole in COBE and WMAP data is smaller than the ΛCDM predicted value. Depending on the exact structure of the superhorizon isocurvature perturbation, the intrinsic and inflation generated quadrupoles could add or subtract, preferentially along the direction of the motion, aligning the quadrupole with the dipole and removing power, bringing the measured quadrupole in better agreement with the ΛCDM model.

4. Conclusions.

We have developed a new method to measure the bulk flow velocity of clusters of galaxies. The method is based on measuring the dipole anisotropy at cluster locations generated by the kinematic SZ effect. These measurements have very different systematics than peculiar velocity surveys. We have determined a coherent bulk flow on the direction of the CMB dipole on a scale of $\sim$ 300Mpc/h, with amplitude of 600 – 1000km/s. The amplitude of the flow is almost one order of magnitude larger than the prediction of the standard cosmological model. This flow could be generated by $\sim 1Gpc$ void or by density perturbations on superhorizon scales. Measuring the shear and other higher order moments could discriminate between a 'local' or a 'global' origin of the flow.

Other observational results are consistent with the measurements presented here, although at a much smaller scale. The dipole anisotropy of the all sky X-ray selected clusters of galaxies indicates that an important contribution to the velocity of the Local Group is due to overdensities $>$ 150Mpc/h [15]. Measurements using galaxy peculiar velocities or reconstructions of the density field suggest similar flows on scales of 50 – 100Mpc/h [16]. A more recent analysis suggest the source of the flow to be an effective distance larger than 200Mpc/h [17], providing evidence in favour of a superhorizon perturbation.

Acknowledgments

This work was supported by NASA ADP grants NNG04G089G and 09-ADP09-0050 FAB acknowledges grants FIS2006-05319/GR-234 from Spanish Ministerio de Educación y Ciencia and Junta de Castilla y León.

References

[1] Kashlinsky A and Atrio-Barandela F 2000 *Astrophys. J.* **536** L67

Journal of Physics: Conference Series **229** (2010) 012003 doi:10.1088/1742-6596/229/1/012003

[2] Sunyaev R and Zel'dovich I 1972 *Comments on Astrophys. Space Phys.* **4** 173
Sunyaev R and Zel'dovich I 1980 *Mon. Not. R.A.S.* **190** 413
[3] Kashlinsky A, Atrio-Barandela F, Kocevski D and Ebeling H 2008 *Astrophys. J.* **686** L49
Kashlinsky A, Atrio-Barandela F, Kocevski D and Ebeling H 2009 *Astrophys. J.* **691** 1479
[4] Atrio-Barandela F, Kashlinsky A, Ebeling H, Kocevski D and Edge A 2009 The error budget of the Dark Flow measurement *Preprint* ArXiv:1001.1261
[5] Kogut A et al 1993 *Astrophys. J.* **419** L1
[6] Carlstrom J, Holder G and Reese E 2002 *Annu. Rev. Astron. Astrophys.* **40** 643
[7] Atrio-Barandela F, Kashlinsky A, Kocevski D and Ebeling H 2008 *Astrophys. J.* **675** L57
[8] Kashlinsky A, Atrio-Barandela F, Ebeling H, Edge A and Kocevski D 2009 A new measurement of the bulk flow of X-ray luminous clusters of galaxies *Preprint* ArXiv:0910.4958
[9] Turner M 1991 *Phys. Rev.* D **44** 3737
[10] Kashlinsky A, Tkachev I and Frieman J 1994 *Phys. Rev. Lett.* **73** 1582
[11] Mersini-Houghton L and Holman R 2009
[12] Afshordi N, Geshnizjani G and Khoury J 2009
Beltrán-Jimenez J and Maroto A 2009 *J. Cosmol. Astropart. P.* JCAP03(2009)015
Khoury J and Wyman M 2009 *Phys. Rev.* D **80** 064023
[13] García-Bellido J and Haugboelle T 2008 *J. Cosmol. Astropart. P.* JCAP09(2008)016
[14] Grishchuk L and Zel'dovich Y 1978 *Sov. Astron.* **22** 125
[15] Kocevski D and Ebeling H 2006 *Astropys. J.* **645** 1043
[16] Watkins R, Feldman H and Hudson M 2009 *MNRAS* **392** 743
Lavaux G et al 2010 *Astrophys. J.* **709** 483
[17] Feldman H, Watkins R and Hudson M 2009. *Preprint* ArXiv:0911.5516

Where are the trapped surfaces?

Jan E Åman[1], Ingemar Bengtsson[1]* and José M M Senovilla[2]

[1] Fysikum, Stockholms Universitet, S-106 91 Stockholm, Sweden
[2] Física Teórica, Universidad del País Vasco, Apartado 644, 48080 Bilbao, Spain

E-mail: ingemar@physto.se

Abstract. We discuss the boundary of the spacetime region through each point of which a trapped surface passes, first in some simple soluble examples, and then in the self-similar Vaidya solution. For the latter the boundary must lie strictly inside the event horizon. We present a class of closed trapped surfaces extending strictly outside the apparent horizon.

1. Introduction

Roger Penrose tells the story of how the importance of closed trapped surfaces occurred to him while crossing a street in London. He then promptly forgot about the matter, but—fortunately—recalled it later in the day [1]. One can make a case for trapped surfaces being the single most important ingredient in gravitational collapse. Their popular acclaim has been clouded by the fact that they remain hidden behind the event horizon, but numerical relativists do encounter them, while the very existence of the event horizon may be in some doubt from quantum gravity.

Can one *define* a black hole using its trapped surfaces [2]? An important notion here is the dynamical horizon, a spacelike hypersurface foliated by marginally trapped surfaces [3]. But not enough is known about the uniqueness of such objects [4]. Indeed little is actually known about where trapped surfaces occur even in the simple Vaidya solution, which describes the collapse of a spherically symmetric cloud of incoherent radiation (or "null dust") [5]. This is the situation that we have tried to remedy, with partial success.

Any dynamical horizon lies inside the trapping boundary, defined as the boundary of an inextendible spacetime region for which each point lies on some trapped surface [2]. In very simple cases this boundary is easily found. In the Vaidya spacetime it must lie somewhere between the event horizon and a spherically symmetric dynamical horizon, but coincides with neither [6, 7]. Here we will present a family of closed trapped surfaces that extend into the no-man's land between them. The surfaces we consider are genuinely trapped, in that both of their null expansions are non-positive, whereas much of the recent literature concerns outer trapped surfaces for which no condition is imposed on the inner expansion. There are good reasons for this [8], but there are reasons for our choice too [2, 3]. The reader must not confuse the two. Eardley conjectured that the boundary of the region through which closed outer trapped surfaces pass is the event horizon [9], and for the Vaidya case this was proven by Ben-Dov [10].

2. Simple examples and how to address them

A spacetime where the trapping boundary can be identified almost by inspection is the recollapsing $k = 1$ Friedmann model. It is foliated by round 3-spheres and has a moment of time symmetry for which the trace K of the second fundamental form vanishes. Any equatorial sphere in such a 3-sphere is minimal within the 3-sphere, and trapped if $K < 0$. The null expansions vanish at the moment of time symmetry, so we expect this to be the trapping boundary. But how do we prove that a closed trapped surface cannot extend partly below it?

Let e_i^a project the cotangent space onto the cotangent space of some spacelike surface of codimension two. Let $\bar{\xi}^i$ be the projection of an arbitrary spacetime vector field ξ^a, and let $K_{ij}{}^a$ be the shape tensor of the surface. Then

$$e_i^a e_j^b \nabla_a \xi_b = \bar{\nabla}_i \bar{\xi}_j + K_{ij}{}^a \xi_a \ . \tag{1}$$

The derivative on the right hand side uses the Levi-Civita connection of the first fundamental form. If we symmetrise and then contract with the first fundamental form we obtain

$$\frac{1}{2}\gamma^{ij} e_i^a e_j^b \mathcal{L}_\xi g_{ab} = \bar{\nabla}_i \bar{\xi}^i + H^a \xi_a \ , \tag{2}$$

where $H^a = \gamma^{ij} K_{ij}{}^a$ is the mean curvature vector. This is future directed and timelike for a trapped surface. Suppose that the vector field ξ^a is future directed and timelike too, so that $H^a \xi_a < 0$, and suppose that ξ^a is such that the left hand side of this equation is non-negative. If we integrate over a closed surface the divergence goes away and we have a contradiction. Either the surface extends into a region where ξ^a is spacelike, or else it is not trapped [11].

If the vector field ξ^a is hypersurface forming we can say more. Assume that there exist functions $F > 0$ and τ such that

$$\xi_a = -F \nabla_a \tau \ . \tag{3}$$

On a closed surface the function τ will assume a minimum. At that minimum

$$\bar{\nabla}_i \xi^i = -\bar{\nabla}_i (F \bar{\nabla}^i \tau) = -F \bar{\nabla}_i \bar{\nabla}^i \tau - \bar{\nabla}_i F \bar{\nabla}^i \tau = -F \bar{\nabla}_i \bar{\nabla}^i \tau < 0 \ . \tag{4}$$

If ξ^a is such that the left hand side of eq. (2) is positive semidefinite we conclude that

$$H^a \xi_a > 0 \tag{5}$$

at the minimum of τ. Since ξ^a is timelike, the mean curvature vector H^a cannot be, and the surface cannot be trapped at the minimum of τ.

To return to the Friedmann model, it admits a future directed conformal Killing vector field for which the left hand side of eq. (2) is positive prior to the moment of time symmetry, and this is what we need in order to prove that the moment of time symmetry is indeed the trapping boundary. Through every point of this trapping boundary there passes a marginally trapped and indeed minimal closed surface. Examples where this is not so include open recollapsing Friedmann models where the trapping boundary is filled by minimal surfaces that are not closed. An intermediate case is provided by the BTZ wormhole, where minimal closed surfaces are dense in the trapping boundary, but do not pass through every point there [12].

The minimum argument can be used to give a stronger conclusion about the trapping boundary. At a minimum in τ

$$\gamma^{ij} e_i^a e_j^b \nabla_a \xi_b = -F \bar{\nabla}_i \bar{\nabla}^i \tau + H^a \xi_a \ . \tag{6}$$

The left hand side is a projection of the second fundamental form of a hypersurface orthogonal to ξ^a. The Laplacian is positive at the minimum, and the scalar product is negative if the surface is trapped, so the right hand side is negative. Therefore the second fundamental form must have at least one negative eigenvalue if a trapped surface otherwise to its future touches a hypersurface in a point. In 2+1 dimensional Minkowski space this is easy to see: A trapped "surface" is then a spacelike curve bending downwards in inertial time, and a surface with positive definite second fundamental form is like a hyperboloid bending upwards in time. Obviously a trapped curve otherwise in its future cannot touch it in a point.

3. The self-similar Vaidya solution

We now turn to Vaidya's spacetime, with the metric

$$ds^2 = -\left(1 - \frac{2m(v)}{r}\right) dv^2 + 2dvdr + r^2 d\theta^2 + r^2 \sin^2\theta d\phi^2 \ . \tag{7}$$

Einstein's equations read

$$G_{ab} = 8\pi T_{ab} = \frac{2\dot{m}}{r^2}\nabla_a v \nabla_b v \ . \tag{8}$$

The energy conditions are obeyed provided that $\dot{m} \geq 0$, where the dot denotes differentiation with respect to v. Otherwise the rate at which radiation comes in is at our disposal. We choose

$$m = \begin{cases} 0 & , \quad v \leq 0 \\ \mu v & , \quad 0 \leq v \leq M/\mu \\ M & , \quad v \geq M/\mu \end{cases} \ . \tag{9}$$

This describes a spherically symmetric shell of incoherent radiation entering flat spacetime from past null infinity, ending in a Schwarzschild black hole when the inflow stops—provided that $\mu > 1/16$, otherwise the result is a naked singularity [13, 14]. Other choices of mass function have been studied [15]. Our choice is special because the Vaidya region of the solution has a homothetic Killing vector

$$\eta = v\partial_v + r\partial_r \ , \tag{10}$$

whose flow lines are confined to hypersurfaces with constant x, where $x = v/r$. This extra symmetry will be quite helpful to us. We expect that the behaviour for a somewhat different mass function would be qualitatively similar but harder to get. The round spheres at $r = 2m$ are marginally trapped and foliate a regular dynamical horizon. We refer to it as the apparent horizon, because it is the apparent horizon in a spherically symmetric slicing of spacetime, and we do not wish to prejudge the issue whether there are other, non-spherically symmetric dynamical horizons present.

In a spherically symmetric spacetime the coordinate r has a meaning related to the area of the preferred round spheres. There is also a very special vector field known as the Kodama vector field ξ [16]. In our case it is $\xi = \partial_v$. It is not a Killing vector field in general, but it does define a direction in which the area of the preferred round spheres is constant. It is hypersurface forming, and future directed and timelike outside the apparent horizon. Through eq. (3) it defines a "Kodama time" τ in this region, and this can be used in the minimum argument presented in section 2. A trapped surface sticking out of the apparent horizon at some value of τ can only reach higher values in its exterior. Therefore trapped surfaces at values of τ smaller than the smallest value τ_Σ assumed on the apparent horizon itself are excluded. Setting $\tau = \tau_\Sigma$ defines a spacelike hypersurface Σ, touching the apparent horizon just where the latter joins the event horizon. It is spherically symmetric and defined by a function $v = v(r)$ obeying

$$\frac{dv}{dr} = \frac{1}{1 - \frac{2m}{r}} \ . \tag{11}$$

This is easily solved for in terms of elementary functions in the self-similar case [14]. In the flat region constant Kodama time means constant inertial time, and the hypersurface Σ will enter the flat region if and only if $\mu > 1/8$.

Is it possible that Σ is the trapping boundary we look for? In fact it cannot be, since the eigenvalues of its second fundamental form are $(k_1, k_2, k_2) = (k_1, 0, 0)$ where $k_1 > 0$. But we have already argued that such a hypersurface cannot be touched by marginally trapped surfaces. There must be a region to the future of Σ which is free of trapped surfaces, but at least we have constrained the trapping boundary from below. We have also proved that the latter must be spacelike close to the event horizon, since it is squeezed from below by the spacelike Σ.

4. Tongues sticking out of the apparent horizon

Is it possible that the trapping boundary coincides with the apparent horizon? In fact no [6]. An argument due to Galloway and Wald shows that marginally trapped round spheres in a spherically symmetric dynamical horizon can always be perturbed in such a way that they extend partly outside it, and such that the perturbation causes them to become trapped [17]. But their argument does not show how far below the apparent horizon these trapped surfaces extend. Here we will present a—non-optimal—construction which allows us to find closed trapped surfaces extending a finite distance away from the apparent horizon.

Our surfaces are defined as cross sections of a cone, by means of the two equations

$$v = \frac{k}{2\mu}r - v_0 \, , \qquad v = \frac{1}{2\mu}r + a(\theta) \, . \tag{12}$$

When $a = 0$ this surface sits on the apparent horizon. When $|a|$ is large the surface may extend into the Schwarzschild or Minkowski regions of the solution, but we stick to the above definition—it defines a smoothly embedded surface in $\mathbf{R}^4$, although its first fundamental form will not be smooth and its null expansions will jump at the boundaries between different regions.

The condition for outer trapping is the somewhat lengthy inequality

$$\begin{aligned} NH(k_+) = -2\alpha r^3 + 2r^3\left(1 - \frac{2m}{r}\right) + \frac{\alpha k - 2\mu}{k-1}r^2(a'' + a'\cot\theta) + \\ + \frac{a'^2}{(k-1)^2}\left(4\mu k m + 8\mu^2 r + \alpha k^2 m - 8\mu\alpha k r - k^2\dot{m}r\right) + \\ + \frac{a'^2}{(k-1)^2}\left(1 - \frac{2m}{r}\right)\left(4\mu k r + \alpha k^2 r - k^2 m - r\left(1 - \frac{2m}{r}\right)k^2\right) + \\ + \frac{ka'^2}{(k-1)^2}\left(4\mu - k\left(1 - \frac{2m}{r}\right)\right)\frac{\alpha k - 2\mu}{k-1}a'\cot\theta < 0 \, , \end{aligned} \tag{13}$$

where $H(k_+)$ is the outer null expansion, N is a positive normalisation factor, and

$$\alpha = \frac{2\mu}{k} + \frac{(k-1)^2 r^2}{k^2 a'^2}\left(1 - \sqrt{1 + \frac{4\mu k a'^2}{(k-1)^2 r^2} - \left(1 - \frac{2m}{r}\right)\frac{k^2 a'^2}{(k-1)^2 r^2}}\right) \, . \tag{14}$$

To simplify the calculation we set

$$k = 1 + l(l+1) \, , \qquad a(\theta) = a_0 + a_l P_l(\cos\theta) \, , \tag{15}$$

where P_l is a Legendre polynomial. Together with eqs. (12) this defines a set of "tongues" that—as we will see—stick partly out of the apparent horizon. To first order in a the trapping condition becomes simply $a_0 > 0$, and the tongue extends partly outside the apparent horizon if there is a θ such that $a(\theta) < 0$. This is the perturbation considered by Galloway and Wald [17].

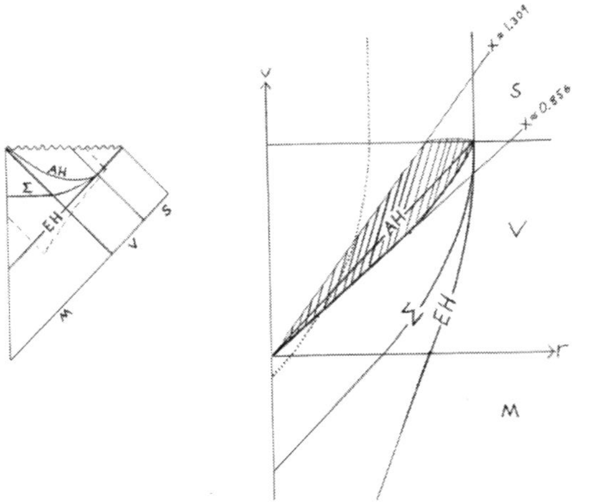

Figure 1. To the left is a Penrose diagram of a Vaidya spacetime. The region inside the dashed curve is shown as a $v-r$-diagram to the right, and includes a part of the event horizon (EH), the spacelike part of the apparent horizon (AH), and the hypersurface Σ below which no trapped surfaces can extend. The tongues considered in the text are confined between two lines of constant x, and can extend only a little bit into the Schwarzschild region S. The trapped surfaces discussed earlier [6] are to the left of the dotted curve. The figure is for $\mu = 1/2$ only.

Our aim is now to adjust the parameters a_0 and a_l for every given v_0 in such a way that the extent to which the tongue sticks out of the horizon is maximised. To deal with the inequality (13) we had to fall back on Mathematica. General considerations about when the cone becomes tangential to hypersurfaces of constant Kodama time show that the possible extent of the tongue is larger the smaller l is, so we confined ourselves to $l = 1$ and $l = 2$. Then we investigated the two cases $\mu = 1/2$ and $\mu = 2$ in detail and found—for the case when the tongues remain in the Vaidya region of spacetime—that the most stringent condition from the inequality comes at the tips of the tongue, where $a' = 0$. But once this is so the inequality (13) is easily dealt with analytically. For $m = \mu v$, $l = 1$, the trapping conditions at the tips hold as long as

$$\mu = \frac{1}{2} \ : \qquad 0.856 < x = \frac{v}{r} < 1.309 \tag{16}$$

$$\mu = 2 \ : \qquad -0.417 < x = \frac{v}{r} < 0.75 \ . \tag{17}$$

This is then how far a maximally extended closed trapped tongue extends. The inner trapping condition holds throughout the allowed region. Thus the region occupied by trapped tongues is bounded by hypersurfaces of constant x. Given that the trapping condition becomes critical at the tips of the tongues this could have been predicted, because it is known that if the outer null expansion vanishes at a point it remains zero if the surface is moved by a homothety [18]. In the self-similar Vaidya solution this is generated by the vector field η, see eq. (10).

For the two values of μ that we investigated in detail we also found that the intrinsic curvature R of the tongue obeys $R > 0$ throughout the region where it is trapped, thus disproving a conjecture by Hayward [19].

So far we have avoided the complication that the tongues may extend into the Minkowski or Schwarzschild regions, where the mass function is different. This must now be dealt with. It is no longer true that the trapping condition is always at its most stringent at the tips of the tongue. After a lengthy Mathematica calculation we found, however, that the upper Schwarzschild tip is usually the place where trapping fails first. The result is that the region into which the tongues can extend shrinks—as indeed it must, since its boundary must reach the point where the dynamical horizon meets the event horizon. See Fig. 1 for a summary of the calculation in the $\mu = 1/2$ case. For $\mu = 2$ the tongues can extend into the flat region.

It is not the case though that the Schwarzschild region makes life more difficult for all kinds of closed trapped surfaces. Indeed the flat and Vaidya region has locally trapped surfaces in the equatorial plane $\theta = \pi/2$. Topologically they are open disks meeting the Schwarzschild boundary in a circle. If they are carefully adjusted, and do not extend to large values of r, it is possible to close them in the Schwarzschild region while keeping them trapped, so the Schwarzschild region in a sense creates closed trapped surfaces that could not exist without it [6].

5. So where is the trapping boundary?

A large part of the no-man's land between the dynamical horizon and the event horizon has now been occupied by the trapped region, but we have not been able to pin down its boundary precisely. Some things are clear: in a spherically symmetric spacetime the trapping boundary must itself be spherically symmetric, it cannot contain any marginally trapped surfaces, it cannot have a positive definite second fundamental form, and it must be spacelike close to the event horizon because it is squeezed from below by the spacelike hypersurface Σ. From our picture for $\mu = 1/2$ it is tempting to conjecture that a part of the boundary sits at $x =$ constant. This part would be determined entirely by the behaviour of trapped surfaces inside the self-similar Vaidya region. But for $\mu = 2$ the trapping boundary is determined largely by what happens inside the Schwarzschild region—not by the local physics in the Vaidya region.

Our calculations have not addressed the question to what extent the dynamical horizon is unique. This also remains as an interesting question.

Acknowledgements

IB thanks Greg Galloway and Catherine Williams for discussions, and the Swedish Research Council VR for support. JMMS is supported by grants FIS2004-01626 (MICINN) and GIU06/37 (UPV/EHU). JÅ and IB found the Spanish Relativity Meeting as stimulating as ever.

References

[1] Penrose R 1989 *The Emperor's New Mind* (Oxford: Oxford University Press)
[2] Hayward S A1994 *Phys. Rev.* D **49** 6467
[3] Ashtekar A and Krishnan B 2004 *Living Rev. Relativity* **7** 10
[4] Ashtekar A and Galloway G 2005 *Adv. Theor. Math. Phys.* **9** 1
[5] Vaidya P C 1951 *Proc. Indian Acad. Sci.* A **33** 264
[6] Bengtsson I and Senovilla J M M 2009 *Phys. Rev.* D **79** 024027
[7] Senovilla J M M 2009 *Physics and Mathematics of Gravitation, Encuentros Relativistas Españoles* ed Kunze K *et al* (New York: American Institute of Physics) p 72
[8] Andersson L 2009 *Physics and Mathematics of Gravitation, Encuentros Relativistas Españoles* ed Kunze K *et al* (New York: American Institute of Physics) p 3
[9] Eardley D M 1998 *Phys. Rev.* D **57** 2299
[10] Ben-Dov I 2007 *Phys. Rev.* D **75** 064007
[11] Mars M and Senovilla J M M 2003 *Class. Quant. Grav.* **20** L93
[12] Åminneborg S, Bengtsson I, Brill D, Holst S, and Peldán P 1998 *Class. Quant. Grav.* **15** 627
[13] Hiscock W A, Williams L G, and Eardley D M 1982 *Phys. Rev.* D **26** 751
[14] Papapetrou A 1985 *A Random Walk in Relativity and Cosmology: Essays in Honor of P. C. Vaidya and A. K. Raychaudhuri* ed Dadhich N *et al* (New Delhi: Wiley Eastern) pp 184-191
[15] Schnetter E and Krishnan B 2006 *Phys. Rev.* D **73** 021502(R)
[16] Kodama H 1980 *Prog. Theor. Phys.* **63** 1217
[17] Galloway G and Wald R M private communication.
[18] Carrasco A and Mars M 2009 *Class. Quant. Grav.* **26** 175002
[19] Hayward S A 2008 *Preprint* arXiv:0810.0923

Spanish Relativity Meeting (ERE 2009) IOP Publishing
Journal of Physics: Conference Series **229** (2010) 012005 doi:10.1088/1742-6596/229/1/012005

On massive gravitons in 2+1 dimensions

Eric Bergshoeff[1]⋆**, Olaf Hohm**[2] **and Paul Townsend**[3]

[1] Centre for Theoretical Physics, University of Groningen, Nijenborgh 4, 9747 AG Groningen, The Netherlands
[2] Center for Theoretical Physics, Massachusetts Institute of Technology, Cambridge, MA 02139, USA
[3] Department of Applied Mathematical Physics and Theoretical Physics, Centre for Mathematical Sciences, University of Cambridge, Wilberforce Road, Cambridge, CB3 0WA, U.K.

E-mail: e.a.bergshoeff@rug.nl; ohohm@mit.edu; P.K.Townsend@damtp.cam.ac.uk

Abstract. The Fierz-Pauli (FP) free field theory for massive spin-2 particles can be extended, in a spacetime of (1+2) dimensions (3D), to a generally covariant parity-preserving interacting field theory, in at least two ways. One is "new massive gravity" (NMG), with an action that involves curvature-squared terms. Another is 3D "bigravity", which involves non-linear couplings of the FP tensor field to 3D Einstein-Hilbert gravity. We review the proof of the linearized equivalence of both "massive 3D gravity" theories to FP theory, and we comment on their similarities and differences.

1. Introduction: Fierz-Pauli and NMG

There has been increased interest in recent years in the possibility of massive gravitons. This is motivated in part by the discovery of cosmic acceleration, which might be explainable in terms of an infra-red modification of general relativity that gives the graviton a small mass (e.g. [1]), and in part by the idea that some theory involving massive gravitons could be the low energy limit of a non-critical string-theory underlying QCD (e.g. [2]). As for so many other gravitational physics issues, it may be useful to consider the possibilities for massive gravitons in the simpler context of a three-dimensional (3D) spacetime.

The standard free-field description of massive gravitons, i.e. massive particles of spin-2, is that due to Fierz and Pauli (FP) in terms of a symmetric tensor field, which we will call $f_{\mu\nu}$. The FP action is

$$S_{FP}[f] = S^{(2)}_{EH}[f] - \frac{1}{4}m^2 \int d^3x \, \left(f^{\mu\nu} f_{\mu\nu} - f^2\right) , \tag{1}$$

where $f = \eta^{\mu\nu} f_{\mu\nu}$ for Minkowski metric η, which we assume to have "mostly plus" signature, and

$$S^{(2)}_{EH}[f] = -\frac{1}{2} \int d^3x \, f^{\mu\nu} G^{(lin)}_{\mu\nu}(f) , \tag{2}$$

where $G^{(lin)}_{\mu\nu}(f)$ is the linearized Einstein tensor expressed as a self-adjoint differential operator acting on the tensor field f; the FP action therefore reduces to the quadratic approximation to the Einstein-Hilbert (EH) action in the massless limit. The field equations are equivalent to

$$\left(\Box - m^2\right) f_{\mu\nu} = 0 , \qquad \partial^\mu f_{\mu\nu} = 0 , \qquad f = 0 . \tag{3}$$

The differential subsidiary condition is needed to remove ghost modes and the tracefree condition is needed to remove lower spin modes. In 4D the FP equations propagate the 5 helicity states of a massive spin 2 particle. In 3D they propagate the two ± 2 helicity states of a massive spin 2 particle.

The main issue of interest is whether one can generalize the FP free field theory to a consistent interacting theory. One could try to couple $f_{\mu\nu}$ to some linear combination of the stress tensor and its trace but this leads to difficulties (see [3] for a recent review). In 3D there is another option. To see this, we first solve the differential subsidiary condition by writing

$$f_{\mu\nu} = -\frac{1}{2}\epsilon_\mu{}^{\eta\rho}\epsilon_\nu{}^{\tau\sigma}\partial_\eta\partial_\tau h_{\rho\sigma} \equiv G^{(\text{lin})}_{\mu\nu}(h) \tag{4}$$

for some symmetric tensor potential h. The remaining FP equations become

$$\left(\Box - m^2\right) G^{(\text{lin})}_{\mu\nu}(h) = 0\,, \qquad R^{(\text{lin})}(h) = 0\,, \tag{5}$$

where $R^{(\text{lin})}(h)$ is the trace of the linearized Ricci tensor $R^{(\text{lin})}_{\mu\nu}(h)$. These equations are derivable from the quadratic action

$$S^{(2)}_{NMG}[h] = \int d^3x \left\{\frac{1}{2}h^{\mu\nu}G^{(\text{lin})}_{\mu\nu}(h) + \frac{1}{m^2}G^{\mu\nu}_{(\text{lin})}(h)S^{(\text{lin})}_{\mu\nu}(h)\right\}, \tag{6}$$

where

$$S^{(\text{lin})}_{\mu\nu}(h) = R^{(\text{lin})}_{\mu\nu}(h) - \frac{1}{4}\eta_{\mu\nu}R^{(\text{lin})}\,. \tag{7}$$

This new quadratic action for symmetric tensor potential h has an obvious extension to a non-linear generally covariant action for metric $g_{\mu\nu} = \eta_{\mu\nu} + h_{\mu\nu}$; this yields "new massive gravity" (NMG) [4].

It is convenient to give the NMG action in a slightly more general form involving a dimensionless parameter σ; in appropriate units, this action is [4]

$$S_{NMG}[g] = \int d^3x\,\sqrt{-\det g}\,\left[\sigma R + m^{-2}K\right]\,, \tag{8}$$

where

$$K := R^{\mu\nu}R_{\mu\nu} - \frac{3}{8}R^2 \;=\; G^{\mu\nu}S_{\mu\nu}\,, \qquad S_{\mu\nu} := R_{\mu\nu} - \frac{1}{4}g_{\mu\nu}R\,. \tag{9}$$

The tensor $S_{\mu\nu}$ is the 3D Schouten tensor, which plays an important role in conformal gravity as the gauge potential for conformal boosts[1]. To recover the quadratic action (6) we should set $\sigma = -1$, which means that the EH term has the 'wrong' sign (relative to the EH action of general relativity).

We have arrived at the action for NMG by a method [5] that guarantees its on-shell equivalence to the FP theory in the linearized limit. This method is a very general one [6] but it does not guarantee *off-shell* equivalence. However, this can be proved very simply for NMG by means of the following alternative action involving an 'auxiliary' tensor field f [4]:

$$S_{NMG}[g,f] = \sigma S_{EH}[g] + \int d^3x\sqrt{-\det g}\left[f^{\mu\nu}G_{\mu\nu} - \frac{1}{2}m^2 g^{\mu[\rho}g^{\nu]\sigma}f_{\mu\nu}f_{\rho\sigma}\right]. \tag{10}$$

This action, which is only 2nd order in derivatives, is quadratic in the tensor $f_{\mu\nu}$. The f field equation is $f_{\mu\nu} = (2/m^2)S_{\mu\nu}$; using this equation to eliminate $f_{\mu\nu}$ (or integrating it out by

[1] This is true in any dimension but the expression for the Schouten tensor in terms of the Ricci tensor is dimension dependent. In D spacetime dimensions it is defined as $S_{\mu\nu} = (D-2)^{-1}R_{\mu\nu} - [2(D-1)(D-2)]^{-1}R\,g_{\mu\nu}$.

Gaussian functional integration in the path integral) one recovers the action (8) in terms of the metric alone. This is true for any value of σ, so for $\sigma = 0$ the action (10) constitutes a 2nd order version of the "pure-K" model studied in [7]. Focusing on NMG proper, we set $\sigma = -1$. To linearize about the Minkowski vacuum, we now write

$$g_{\mu\nu} = \eta_{\mu\nu} + h_{\mu\nu} - f_{\mu\nu} \tag{11}$$

and expand the action in powers of the Minkowski space tensors (h, f). To quadratic order we find that

$$S^{(2)}_{NMG}[h, f] = -S^{(2)}_{EH}[h] + S_{FP}[f]\,. \tag{12}$$

Since the quadratic EH action for h propagates no modes, the modes propagated by the linearized NMG theory are the same as those of the FP theory[2].

Higher-derivative extensions of 4D general relativity have been investigated since Weyl's 1918 attempt at a unified theory of gravity and electromagnetism. A special feature of curvature-squared invariants is that they contribute to the quadratic approximation in an expansion about a Minkowski vacuum, so they contribute to the linearized field equations, which become fourth order rather than second order. Equations of higher than second order generally propagate ghosts (modes of negative energy) in addition to physical modes, and this is what happens here, in 4D: in addition to the massless graviton, there is also a massive scalar mode and a massive spin-2 ghost [10]. The 3D case is different because the EH action by itself does not propagate any modes, so the massless graviton is now absent. This means that we are free to change the overall sign of the action to arrange for the massive spin-2 modes to be physical, i.e. not ghosts; we can then ensure that they are not tachyons by a choice of relative sign for the EH and curvature-squared terms. However, there is still the massive scalar mode, which is now a ghost. Remarkably, the mass of this scalar ghost mode goes to infinity as the relative coefficient of the two independent curvature-squared terms is tuned to the value that yields the 'K-combination' of (9), so we end up with the unitary massive pure spin-2 model that is NMG. All of this can be read off from the results of Nishino and Rajpoot [11] for the general 3D model with EH and curvature-squared invariants but these authors considered only the 'right-sign' EH term and (presumably as a consequence) did not draw attention to the special features of the K-combination.

One motivation for considering the EH action with the 'wrong' sign is that this is a feature of "topologically massive gravity" (TMG) [12], which is also "higher-derivative" because it involves the third order, and parity violating, Lorentz Cherns-Simons (LCS) term. TMG propagates only a single spin 2 mode, which is possible in 3D if parity is violated. This possibility can also be realized by a first-order "self-dual" spin-2 theory that can be thought of as the "square-root" of the 3D FP theory [13]. If the differential subsidiary condition of this model is solved, the remaining equations become precisely those of linearized TMG, so TMG can be deduced by the same procedure that we used above to arrive at NMG. Moreover, in this case the on-shell equivalence extends immediately to an off-shell equivalence because with only one propagating mode one can always choose the overall sign of the action so as to ensure that it is physical rather than a ghost. Unitarity for NMG is a more delicate matter because there are two propagating modes. Unitarity follows from the proof of off-shell equivalence to the FP theory reviewed above, but it can also be checked explicitly via a canonical analysis [7]. A nice feature of this method is that one sees explicitly the cancellation of the higher time derivatives; in other words, the action

[2] In [4] we set $g_{\mu\nu} = \eta_{\mu\nu} + h_{\mu\nu}$ and used the h equation to eliminate h from the action. This step was criticized in [7] because it involves a back-substitution that is not generally permissible. However, the legitimacy of this back-substitution is manifest in the new basis implicit in (11) because it just eliminates the quadratic EH term, which propagates no modes. This change of basis was first made in [8]; it had occurred independently to us and we used it in our general analysis of unitarity for the 'cosmological' extension of NMG [9].

is actually second order in time derivatives despite being fourth order in space derivatives. This is reminiscent of Hořava gravity [14] but without any violation of Lorentz invariance!

There is a natural generalization of NMG that is suggested by TMG: one simply adds a LCS term to the action. This yields what we have called "general massive gravity" (GMG) [4]. There are now two mass parameters, which can be traded for the two masses $m_\pm$ of the two spin-2 modes of helicities ± 2, which we may assume to be such that $m_- \geq m_+ > 0$. As parity flips the sign of helicities, a model with $m_+ \neq m_-$ breaks parity. When $m_+ = m_-$ the parity-violating LCS term is absent and we recover NMG. The limit $m_- \to \infty$ for fixed m_+ yields TMG, in which the curvature-squared term is absent, so both NMG and TMG are special cases of GMG. Conversely, it is possible to recover GMG by a "soldering" of two TMG models [15], one propagating a helicity $+2$ mode with mass m_+ and the other propagating a helicity -2 mode with mass m_-.

2. 3D bigravity

In the case of 4D massless gravitons, it is known that any theory of interacting gravitons must reduce to general relativity at low energy [16, 17, 18]. Even if these low energy theorems continue to apply in 3D, they cannot restrict the interactions of massive gravitons because these will decouple at sufficiently low energy. There is therefore no obvious reason why NMG should be the unique interacting extension of the 3D FP theory. Indeed, an alternative parity-preserving generally covariant massive gravity model was recently proposed by Bañados and Theisen [19]; their "bigravity" model for massive 3D gravitons is a 3D version of the 4D "f-g gravity" of Isham *et al.* [20].

The Bañados-Theisen (BT) model involves a dimensionless parameter, which we shall call α, and two tensor fields; we shall use a different basis for these two fields and we call them $(\tilde{g}, f)$. Also, we omit here the cosmological terms included in [19]; we will later discuss briefly the cosmological extension. For $\alpha(\alpha+1) > 0$, the BT action is

$$S_{BT}[\tilde{g}, f] = \alpha S_{EH}[\tilde{g} + \beta f] + S_{EH}[\tilde{g} - \alpha\beta f] - \int d^3x\, \mathcal{U}(\tilde{g}, f) \tag{13}$$

where

$$\beta = 1/\sqrt{\alpha(\alpha+1)}\,. \tag{14}$$

The last term in the action is the potential term, defined by some choice of scalar density $\mathcal{U}$ constructed from the metric tensor and the f-tensor, without derivatives. The choice made in [19] has the form

$$U(\tilde{g}, f) = \frac{1}{2}m^2\sqrt{-\det\tilde{g}}\; \tilde{g}^{\mu[\rho}\tilde{g}^{\nu]\sigma} f_{\mu\nu} f_{\rho\sigma} + \mathcal{O}\left(f^3\right)\,. \tag{15}$$

A feature of the BT action is that it assumes invertibility of both $\tilde{g}$ and $\tilde{g}+\beta f$. We can interpret the invertibility constraint on, say, $\tilde{g}$ as the geometrical condition that $\tilde{g}$ be a metric tensor but invertibility of $\tilde{g}+\beta f$ then puts a constraint on the tensor field f that is not so obviously geometrical.

The proof that linearized bigravity is equivalent to the FP theory is simple. We write

$$\tilde{g}_{\mu\nu} = \eta_{\mu\nu} + h_{\mu\nu} \tag{16}$$

and expand the action (13) in powers of the Minkowski space tensors (h, f). The result to quadratic order is

$$S^{(2)}_{BT}[h, f] = (\alpha+1)\, S^{(2)}_{EH}[h] + S_{FP}[f]\,. \tag{17}$$

This is equivalent to the FP theory for any α, because the quadratic EH action for h propagates no modes, although we should recall the constraint $\alpha(\alpha+1) > 0$.

If β is initially taken as a free parameter then one finds that unitarity is violated if $\alpha(\alpha+1) < 0$ irrespective of the choice of β, and that (14) can be imposed without loss of generality when $\alpha(\alpha+1) > 0$. When $\alpha = 0$ the action (13) is equivalent to the EH action, so we can ignore this case. When $\alpha = -1$, the action (13) is trivial but it is also not equivalent to the action considered in [19]; in this case we should instead consider the action

$$S_{BT}[\tilde{g}, f]|_{\alpha=-1} = -S_{EH}[g] + S_{EH}[g-f] - \int d^3x\, \mathcal{U}(\tilde{g}, f)\,. \tag{18}$$

Linearization now yields the quadratic action

$$S^{(2)}_{BT}[h, f]\Big|_{\alpha=-1} = \int d^3x \left\{ h^{\mu\nu} G^{(\mathrm{lin})}_{\mu\nu}(f) - \frac{1}{4} m^2 \left(f^{\mu\nu} f_{\mu\nu} - f^2 \right) \right\}, \tag{19}$$

but this is precisely what one gets from linearization of (10) when $\sigma = 0$. As observed in [9], the h tensor is now a Lagrange multiplier imposing a constraint that has the solution $f_{\mu\nu} = 2\partial_{(\mu} A_{\nu)}$. Using this in the f equation, we find that

$$G^{(\mathrm{lin})}_{\mu\nu}(h) = m^2 \left[\partial_{(\mu} A_{\nu)} - \eta_{\mu\nu} (\partial \cdot A) \right]. \tag{20}$$

This equation determines the metric fluctuation tensor h in terms of the source, which is itself constrained by the linearized Bianchi identity; this constraint is just the 3D Maxwell equation $\partial^\mu F_{\mu\nu} = 0$, where $F_{\mu\nu} = 2\partial_{[\mu} A_{\nu]}$. It is also true that the action (19) reduces to the 3D Maxwell action upon making the substitution $f_{\mu\nu} = 2\partial_{(\mu} A_{\nu)}$. Either way, we see that a single massless mode is propagated, of undefined spin because spin is not defined for massless particles in 3D. This is consistent with the canonical analysis of the "pure-K" model [7].

3. Cosmological NMG and 3D bigravity

We conclude with a few brief observations on the cosmological extension of NMG and 3D bigravity. For NMG one may add the cosmological term

$$S_{cos}[g] = -2\lambda m^2 \int d^3x \sqrt{-\det g} \tag{21}$$

where λ is a new dimensionless parameter. It is convenient in this context to also allow the parameter σ to be arbitrary, and to allow for $m^2 < 0$. Anti-de Sitter (adS) vacua are possible for $\lambda \neq 0$, and in such cases one may study the nature of the associated 2D conformal field theory (CFT) [21]. It turns out that this has a negative central charge whenever the quadratic approximation to the bulk theory (expanded about the adS vacuum) is unitary, and vice versa [9]. There is one exception, for $\lambda = 3$, in which the massive gravitons are replaced by massive spin 1 modes and the central charge of the CFT is zero; in this case the CFT is of logarithmic type [22].

In the bigravity case we may add the cosmological term

$$S_{cos}[\tilde{g}] = \frac{2(\alpha+1)}{\ell^2} \int d^3x \sqrt{-\det \tilde{g}}\,. \tag{22}$$

This allows an adS vacuum of radius ℓ for the metric $\tilde{g}$, with $f = 0$. Linearization about adS was considered in [19], and also in [23]. 'Bulk' unitarity requires either that $\alpha \leq -1$ or that $\alpha \geq 0$. From our understanding of the results of [19], the central charge of the associated 2D CFT is (in our notation) a positive factor times $(\alpha+1)$, which would mean that it is negative for $\alpha < -1$ but positive for $\alpha > 0$. It would appear from this that 3D cosmological bigravity is analogous to cosmological NMG for $\alpha < -1$ but not for $\alpha > 0$. This can be seen in vestigial form from the Minkowski space actions: when $\alpha < -1$ the quadratic actions (17) are related by a rescaling (which is trivial at $\alpha = -2$) but not when $\alpha > 0$ because of the different sign of the quadratic EH term; although this does not affect the conclusions concerning propagating modes, and may continue to be irrelevant at the non-linear level in a Minkowski vacuum, it makes a significant difference in an adS vacuum.

In conclusion, both NMG and 3D bigravity constitute generally covariant interacting extensions of the free-field 3D Fierz-Pauli model for massive particles of spin 2. Each will likely have its advantages and disadvantages. Whether either will provide insight into the problem of massive 4D gravity remains to be seen.

Acknowledgments

EB thanks the organizers of the 2009 Spanish Relativity Meeting ERE2009 for a stimulating environment. PKT is supported by an EPSRC Senior Fellowship. The work of OH is supported by the DFG – The German Science Foundation and in part by funds provided by the U.S. Department of Energy (DoE) under the cooperative research agreement DE-FG02-05ER41360.

References

[1] Eckhardt D H, Pestana J L G and Fischbach E 2010 Cosmological implications of massive gravitons *New Astron.* **15** 175
[2] 't Hooft G 2007 Unitarity in the Brout-Englert-Higgs Mechanism for Gravity *Preprint* arXiv:0708.3184
[3] Goldhaber A S and Nieto M M 2008 Photon and Graviton Mass Limits *Preprint* arXiv:0809.1003
[4] Bergshoeff E A, Hohm O and Townsend P K 2009 Massive Gravity in Three Dimensions *Phys. Rev. Lett.* **102** 201301
[5] Andringa R, Bergshoeff E A, de Roo M, Hohm, Sezgin E and Townsend P K 2010 Massive 3D Supergravity *Class. Quantum Grav.* **27** 025010
[6] Bergshoeff E A, Hohm O and Townsend P K 2009 On Higher Derivatives in 3D Gravity and Higher Spin Gauge Theories *Preprint* arXiv:0911.3061
[7] Deser S 2009 Ghost-free, finite, fourth order D=3 (alas) gravity *Phys. Rev. Lett.* **103** 101302
[8] Nakasone M and Oda I 2009 On Unitarity of Massive Gravity in Three Dimensions *Prog. Theor. Phys.* **121** 1389
[9] Bergshoeff E A, Hohm O and Townsend P K 2009 More on Massive 3D Gravity *Phys. Rev.* D **79** 124042
[10] Stelle K S 1978 Classical Gravity With Higher Derivatives *Gen. Rel. Grav.* **9** 353
[11] Nishino H and Rajpoot S 2006 (Curvature)^2-terms for supergravity in three dimension *Phys. Lett.* B **639** 110
[12] Deser S, Jackiw R and Templeton S 1982 Topologically massive gauge theories *Annals Phys.* **140** 372
[13] Aragone C and Khoudeir A 1986 Selfdual massive gravity *Phys. Lett.* B **173** 141
[14] Horava P 2009 Quantum Gravity at a Lifshitz Point *Phys. Rev.* D **79** 084008
[15] Dalmazi D and Mendonça E L 2009 Generalized soldering of ± 2 helicity states in $D = 2 + 1$ *Phys. Rev.* D **80** 025017
[16] Weinberg S 1965 Photons and gravitons in perturbation theory: Derivation of Maxwell's and Einstein's equations *Phys. Rev.* **138** B988
[17] Deser S 1970 Self-interaction and gauge invariance *Gen. Rel. Grav.* **1** 9
[18] Boulware D G and Deser S 1975 Classical General Relativity Derived From Quantum Gravity *Annals Phys.* **89** 193
[19] Banados M and Theisen S 2009 Three-dimensional massive gravity and the bigravity black hole *J. High Energy Phys.* JHEP11(2009)033
[20] Isham C J, Salam A and Strathdee J A 1971 F-dominance of gravity *Phys. Rev.* D **3** 867
[21] Brown J D and Henneaux M 1986 Central Charges in the Canonical Realization of Asymptotic Symmetries: An Example from Three-Dimensional Gravity *Commun. Math. Phys.* **104** 207
[22] Grumiller D and Hohm O 2009 $AdS_3/LCFT_2$ - Correlators in New Massive Gravity *Preprint* arXiv:0911.4274
[23] Afshar H R, Alishahiha M and Naseh A 2009 On three dimensional bigravity *Preprint* arXiv:0910.4350

Nonlinear electrodynamics and cosmology

Nora Bretón

Physics Dep., Centro de Investigación y de Estudios Avanzados del I.P.N.

E-mail: nora@fis.cinvestav.mx

Abstract. Nonlinear electrodynamics (NLED) generalizes Maxwell's theory for strong fields. When coupled to general relativity NLED presents interesting features like the non-vanishing of the trace of the energy-momentum tensor that leads to the possibility of violation of some energy conditions and of acting as a repulsive contribution in the Raychaudhuri equation. This theory is worth to study in cosmological and astrophysical situations characterized by strong electromagnetic and gravitational fields.

1. Introduction

Seventy-five years ago Max Born and Leopold Infeld proposed a new theory, a fully relativistic and gauge invariant nonlinear electrodynamics (1934) [1]. The origins of the nonlinear electrodynamics of Born and Infeld (BI) can be traced to the work of Gustav Mie (1909), who made the first attempt to construct a purely electromagnetic theory of charged particles. But the question why a charged particle does not explode under the repulsion of the Coulomb forces acting between its constituents had been asked even before.

The aim of BI was to cure the divergence of the field and energy of the point charge at the charge position. BI proposed a Lagrangian that depended in nonlinear way of the two electromagnetic invariants. It is the simplest possible Lagrangian: the square root of the determinant of a second rank covariant tensor. Besides some reasonable physical requirements and so constructed to preserve the finiteness of fields, the proposal was arbitrary.

The development of quantum electrodynamics (QED) soon after (1936) showed that nonlinear effects arise in the interaction of electromagnetic fields with vacuum. In fact, the effective Lagrangian that takes into account the vacuum polarization by a constant electric field, to first order in the radiative contributions, has the same form as the BI Lagrangian [2], [3].

A renewed interest in BI theory arose in connection with string theory and p-branes. It turns out that Lagrangians with determinants, equivalent to the BI Lagrangian, frequently appear in these theories [4].

When BI theory is coupled to general relativity (GR), interesting features arise that make worth to apply it to systems with both strong electromagnetic and gravitational fields. The Weyl anomaly is one of this features that is due to the non-vanishing of the trace of the energy-momentum tensor. It leads to the violation of some energy conditions as well as to avoid the focussing of neighbouring geodesics [5].

Another effect is that photon trajectories are modified due to the presence of the nonlinear electromagnetic field. The consequence is that light trajectories are not null geodesics of the background metric but null geodesics of an effective metric that depends on the nonlinear electromagnetic field [6].

2. Born Infeld's theory and quantum electrodynamics

The form of the Lagrangian that BI proposed, as the theory that generalized Maxwell's, was inspired in a finiteness principle for the electromagnetic field magnitude, analogous to the special relativity theory that assumed an upper limit to the velocity of light. For a maximal field strength b, the proposed Lagrangian that depend non-linearly on the electromagnetic invariants, $F = \frac{1}{2}(B^2 - E^2)$ and $G = E{\cdot}B$, has the form

$$L_{BI} = b^2\sqrt{1 - \frac{(B^2 - E^2)}{b^2} + \frac{(E{\cdot}B)^2}{b^4}} - b^2. \tag{1}$$

2.1. The static spherically symmetric solution in BI theory

One of the achievements of BI theory is that it solves the point charge singularity. Solving for the static spherically symmetric case, without magnetic field, the BI equations are

$$\begin{aligned}
&\vec{B} = \vec{H} = 0, \quad \nabla \times \vec{E} = 0, \quad \nabla \cdot \vec{D} = 0, \\
&D_r = \frac{e}{r^2}, \quad E_r = -\phi_{,r}(r), \quad \phi(r) = \frac{e}{r_0^2} f(\frac{r}{r_0}), \quad r_0^2 = \frac{e}{b}, \\
&f(x) = \int_x^\infty \frac{dy}{\sqrt{1+y^4}}.
\end{aligned} \tag{2}$$

The electrostatic potential $\phi(r)$ is finite everywhere since it is given in terms of an elliptic integral that is finite at the origin. This makes the difference from the Coulombian potential that diverges as one approaches the origin. Moreover, using the charge and classical radius of the electron (r_0) BI obtained an estimation for the maximum field strength: $b = \frac{e}{r_0^2} = 1.8 \times 10^{18}$ V/cm.

For fields smaller than b, $E < b$, $B < b$, expanding the square root in Eq. (1) and keeping terms up to order $O(b^{-2})$, we obtain

$$L_{BI} \approx \frac{(E^2 - B^2)}{2} + \frac{(E{\cdot}B)^2}{2b^2} - \frac{(E^2 - B^2)^2}{8b^2} + \mathrm{O}(b^{-4}). \tag{3}$$

Taking into account that the maximal field strength in BI theory is $b = 10^{18}$ Volt/cm, such fields are not necessarily small. In the linear limit, obtained when $b \to \infty$, one recovers the Maxwell Lagrangian, $L = -F$.

2.2. QED Lagrangian

Exploring quantum mechanical corrections for electromagnetic processes Weisskopf (1936) [2], and later, in a covariant form, Schwinger (1951) [3] deduced that nonlinear effects take place due to the probability of electron-positron pair creation (vacuum polarization). The approach of QED is based on a gauge invariant effective action which takes into account modifications of Maxwell electrodynamics induced by several processes: quantum vacuum responds as if it were a classical medium. Similar effects appear inside a dielectric medium that responds non-linearly to an external stimulus. In fact such nonlinear electromagnetic effects as photon-photon scattering or vacuum polarization have been observed in laboratory at fields of the critical order of $E_{crit} \approx 10^{16}$ Volt/m [7].

The one-loop effective Lagrangian for vacuum polarization due to a constant external electromagnetic field depends on the invariants of the electromagnetic field in the same way as BI Lagrangian Eq. (3). It is given by

$$\begin{aligned} L_{eff} &= L_M + L^{(1)}, \\ L^{(1)} &= a(E^2 - B^2)^2 + a^\star(E \cdot B)^2 \cdots, \\ a &= \frac{2\alpha^2}{45 m_e^4}, \quad a^\star = \frac{7}{4}a, \end{aligned}$$

where α is the fine structure constant, m_e is the rest mass of the electron, $L_M = -F$, is the Maxwell Lagrangian and $L^{(1)}$ is the part of the radiative corrections to first order. The previous Lagrangian was derived for slowly varying fields, $\frac{\hbar}{m_e c}|\nabla F| << |F|$ and $\frac{\hbar}{m_e c^2}|\frac{\partial F}{\partial t}| << |F|$. Moreover, it is assumed that the production of real electron-positron pairs is negligible, i.e. the threshold for energies is $|E| < \frac{m_e^2}{|e|} = \frac{m_e^2 c^3}{|e|\hbar}$.

This result gives support to the explicit form of the Lagrangian of Born-Infeld among the wide variety of nonlinear electromagnetic Lagrangians that one can propose.

3. Coupling gravity with nonlinear electromagnetic fields

For situations where strong electromagnetic and gravitational fields occur, as early universes or close to stars with strong magnetic fields, it makes sense to couple gravitation to NLED. The field equations are derived from the action,

$$S = \int d^4x\sqrt{-g}\{\frac{R}{16\pi} - L_{\rm NLED}\}, \tag{4}$$

where R denotes the scalar curvature, $g := \det|g_{\mu\nu}|$ and $L_{\rm NLED}$ is the electromagnetic part, that depends in nonlinear way on the invariants of the electromagnetic field. Being the Born-Infeld Lagrangian a particular case, in what follows we consider the general case, $L = L(F, G)$. The corresponding stress-energy tensor and the scalar curvature are given by

$$\begin{aligned} 4\pi T_{\mu\nu} &= -L_{,F}F_{\mu}^{\alpha}F_{\alpha\nu} + (GL_{,G} - L)g_{\mu\nu}, \\ R &= 8\pi(L - FL_{,F} - GL_{,G}) = -8\pi T. \end{aligned} \tag{5}$$

where $L_{,x} = dL/dx$.

3.1. Breakdown of conformal invariance

From Eq. (5) it is evident that conformal invariance of Maxwell equations is broken by the non-vanishing of the trace of the energy-momentum tensor,

$$T_{\alpha}^{\alpha} = L_{,F}F + GL_{,G} - L, \tag{6}$$

The Weyl (trace or conformal) anomaly arises also when a scalar field (inflaton, dilaton, pseudo-Goldstone bosons, axions) is included in the action. Also when adding extra dimensions or couplings to curvature terms. The interaction of photons with other fields breaks conformal invariance as well. Remind that only equations describing massless particles are conformally invariant.

However, conformal anomaly can be advantageous in certain scenarios, for instance in inflationary epochs. If conformal invariance is broken during inflation, then inflation provides a mechanism to amplify perturbations, like primordial magnetic seed fields [8], [9], [10]. If we want that large scale correlated fields be produced during inflation, conformal invariance should be broken, otherwise they are not conveniently enlarged during inflation.

3.2. NLED energy conditions

Let us explore what kind of energy conditions arise from a stress-energy tensor derived from a nonlinear electromagnetic theory.

Consider a timelike vector, V^{α}, $V_{\alpha}V^{\alpha} < 1$. Assuming that the local energy density should be non-negative (weak energy condition, WEC),

$$T_{\mu\nu}V^{\mu}V^{\nu} \geq 0, \tag{7}$$

and the requirement that the local energy flow vector be non-spacelike (DEC) amounts to,

$$T_{\alpha\beta}T^{\alpha}_{\gamma}V^{\beta}V^{\gamma} \leq 0. \tag{8}$$

In terms of the Lagrangian $L(F, G)$ both conditions hold provided

$$L_{,F} > 0, \quad (L - FL_{,F} - GL_{,G}) \geq 0. \tag{9}$$

While for the strong energy condition (SEC),

$$R_{\mu\nu}V^{\mu}V^{\nu} \geq 0, \tag{10}$$

using the Einstein equations it can be settled as,

$$R_{\mu\nu}V^{\mu}V^{\nu} = 8\pi(T_{\mu\nu}V^{\mu}V^{\nu} + \frac{T}{2}) \geq 0. \tag{11}$$

From Eq. (11) note that NLED matter can violate SEC if the trace T is negatively enough. Moreover, this fact has consequences in the focussing of neighbouring geodesics as the analysis of the Raychaudhuri equation shows.

The rate of change of expansion of timelike congruences $\Theta(\lambda)$, is ruled out by the Raychaudhuri equation,

$$\frac{d\Theta(\lambda)}{d\lambda} = -R_{\mu\nu}V^{\mu}V^{\nu} + 2\omega^2 - 2\sigma^2 - \frac{\Theta^2}{3} + \dot{V}^{\alpha}_{;\alpha}. \tag{12}$$

Let us consider a geodesic congruence, $V^{\alpha} = 0$, without vorticity $\omega = 0$, then the expansion $\Theta(\lambda)$ will monotonically decrease along a geodesic, since $-R_{\mu\nu}V^{\mu}V^{\nu} < 0$ induces contraction of the geodesic lines. Then focussing of neighbouring geodesics is unavoidable if SEC is fulfilled, since the other terms on the r. h. s. of Eq. (12) are also negative. The situation is different if SEC is violated, $R_{\mu\nu}V^{\mu}V^{\nu} < 0$, provided

$$-R_{\mu\nu}V^{\mu}V^{\nu} > 2\sigma^2 + \frac{\Theta^2}{3}, \tag{13}$$

the possibility exists of avoiding singularities, i.e. stopping the focusing of geodesics (non existence of trapped surfaces) [11], [12]. This effect can also be obtained from the inclusion of exotic matter, phantom matter, a cosmological constant or fluids with negative pressure, all of them generate a repulsive effect. In here there is no exotic matter, the question is if the produced repulsive effect is enough to some purposes. For wormholes this aspect has been explored [13], as well as in cosmologies [14]. In Fig. 1 it is shown the behaviour of BI radiation in a Bianchi space [15].

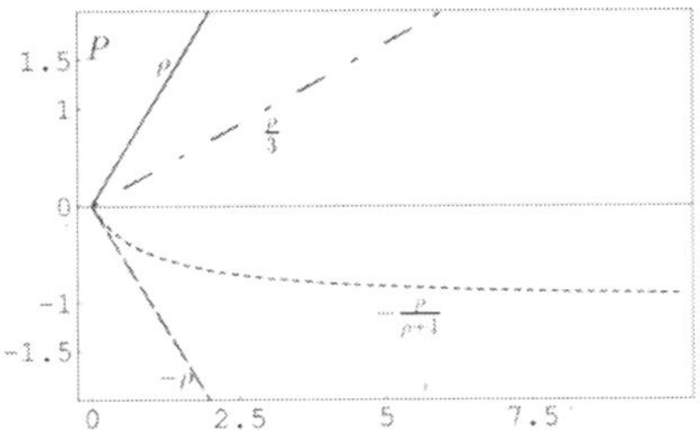

Figure 1. Behaviour of radiation in a Bianchi space [15]. It is shown the pressure P as function of the energy density ρ. Nonlinear electromagnetic radiation corresponds to the dashed curve $P = -\rho/(\rho+1)$ compared with vacuum-like state, $P = -\rho$ (broken curve) and normal radiation, $P = \rho/3$.

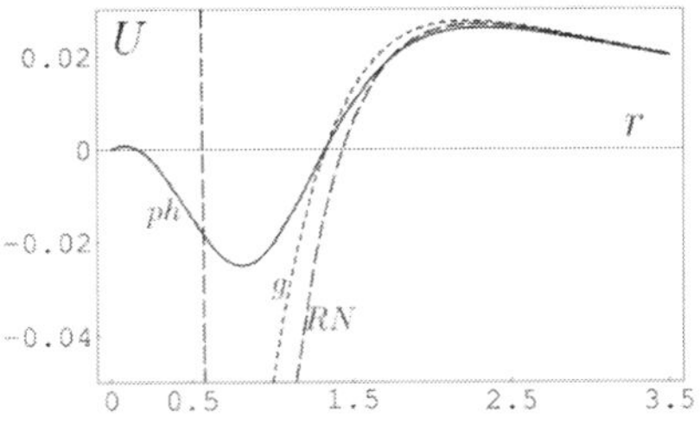

Figure 2. The potential $U(r)$ corresponding to null geodesics in Reissner-Nordstrom (RN) and Born-Infeld black holes [18] is shown. For the Born-Infeld black hole the potentials for null geodesics (g, dashed curve) and the light trajectories (ph, full curve) are not the same.

3.3. Photons obey an effective geometry

Another effect produced in spacetime by a nonlinear electromagnetic field is the modification of the geometry as *seen* by travelling photons; indeed, the velocity of light is different respect to the propagation in vacuum [16], [17]. From BI equations it is evident since they resemble Maxwell's equations in a material medium.

To determine the effective metric for light, we need to analyse the characteristic surfaces, these are surfaces that propagate the field discontinuities, i.e. they are like the wavefronts. For a curved spacetime the equation for the characteristic surfaces S is

$$g^{\mu\nu}S_{,\mu}S_{,\nu} = 0. \tag{14}$$

Locally these surfaces are normal to the light rays. In NLED the Eq. (14) is modified, in such a way that photons propagate in an effective metric, $\gamma_{\mu\nu}$. For the particular case of BI electrodynamics characteristic surfaces are given by,

$$(g^{\mu\nu} + \frac{4\pi}{b^2}T^{\mu\nu})S_{,\mu}S_{,\nu} = \gamma^{\mu\nu}S_{,\mu}S_{,\nu} = 0. \tag{15}$$

Clearly the responsible of that deviation is the stress-energy tensor, $T^{\mu\nu}$. Hence, as far as b is finite, the propagation of gravitational and electromagnetic discontinuities do not occur at the same velocity. For a black hole with a Born-Infeld electromagnetic field, actually both trajectories converge at the horizon [18]; in Fig. 2 are shown the potentials for massless particles in such a BI black hole, compared with the potential for the Reissner-Nordstrom (RN) back hole, its linear counterpart.

The corresponding equation for the light propagation vectors k^{μ} in a spacetime with an arbitrary NLED, $L(F)$, is given by

$$(L_F g^{\mu\nu} - 4L_{FF}F^{\mu\alpha}F^{\nu}_{\alpha})\, k_{,\mu}k_{,\nu} = \gamma^{\mu\nu}k_{,\mu}k_{,\nu} = 0. \tag{16}$$

In ordinary optics ($L = -F, g_{\mu\nu} = \eta_{\mu\nu}$) Eq. (16), this is known as the *eikonal* equation. In the case that both invariants $F \neq 0, G \neq 0$, $L(F,G)$, the photon trajectories follow the effective metric $\gamma^{\mu\nu}_{\pm}$ with

$$\gamma_{\pm}^{\mu\nu} = L_{,F} g^{\mu\nu} - 4[(L_{,FF} + \Omega_{\pm} L_{,FG}) F^{\mu}_{\lambda} F^{\lambda\nu} + (L_{,FG} + \Omega_{\pm} L_{,GG}) F^{\mu}_{\lambda} \hat{F}^{\lambda\nu}], \tag{17}$$

the function $\Omega_{\pm}$ satisfies a quadratic equation [6] thus two possibilities are allowed, γ_{+} and γ_{-} then there are two possible trajectories for light and birefringence arises. The only nonlinear electrodynamics in which the speed of light does not depend on the polarization is the Born-Infeld theory [16].

4. Conclusions

Nowadays the motivations to study nonlinear generalizations of Maxwell electrodynamics is different from the one that inspired Born and Infeld. The interest is now related to cosmological and astrophysical observations, to understand the intense magnetic fields in neutron stars, the origin of the observed large-scale correlated magnetic fields, to favour inflation or as a mechanism to avoid singularities, these effects have been roughly explained in this contribution.

References

[1] Born M and Infeld L 1934 *Proc. Roy. Soc. Lond.* A **144** 425
Born M and Infeld L 1934 *Proc. Roy. Soc. Lond.* A **143** 410
Born M and Infeld L 1934 *Proc. Roy. Soc. Lond.* A **147** 522
[2] Weisskopf W S 1939 *Phys. Rev.* **56** 72
[3] Schwinger J 1951 *Phys. Rev.* **82** 664
[4] Fradkin E S and Tseytlin A A 1983 *Nucl. Phys.* B **161** 123
[5] García-Salcedo R and Bretón N 2005 *Class. Quantum Grav.* **22** 4783 (*Preprint* gr-qc/0410142)
[6] Plebański J F 1970 *Lecture Notes on Nonlinear Electrodynamics* (Copenhagen: NORDITA)
Gutiérrez S A, Dudley A L and Plebański J F 1981 *J. Math. Phys.* **22** 2835
[7] Blaschke D B, Prozorkevich A V, Roberts C D, Schmidt S M and Smolyansky S A 2006 *Phys. Rev. Lett.* **96** 140402
Burke D L *et al* 1997 *Phys. Rev. Lett.* **79** 1626
[8] Kunze K E 2008 *Phys. Rev.* D **77** 023530
[9] Campanelli L, Cea P, Fogli G L 2008 *Phys. Rev.* D **77** 043001
[10] Mosquera-Cuesta H J and Lambiase G 2009 *Phys. Rev.* D **80** 023013
[11] Novello M, Perez Bergliaffa S E and Salim J M 2000 *Class. Quantum Grav.* **17** 3821
[12] García-Salcedo R and Bretón N 2005 *Class. Quantum Grav.* **22** 4783 (*Preprint* gr-qc/0410142)
[13] Arellano A V B and Lobo F S N 2006 *Class. Quantum Grav.* **23** 5811
[14] Novello M, Bergliaffa S E P and Salim J 2004 *Phys. Rev.* D **69** 127301 (*Preprint* astro-ph/0312093)
Moniz P V 2002 Quintessence and Born-Infeld cosmology *Phys. Rev.* D **66** 103501
García-Salcedo R and Bretón N 2000 Born-Infeld Cosmologies *Int. J. Mod. Phys.* A **15** 4341 (*Preprint*: gr-qc/0004017)
Dyadichev V V, Gal'tsov D V, Zorin A G and Zotov M Y 2002 Non-Abelian Born–Infeld cosmology *Phys. Rev.* D **65** 084007 (*Preprint* hep-th/0111099)
Vollick D N 2003 Anisotropic Born-Infeld Cosmologies *Gen. Rel. Grav.* **35**, 1511 (*Preprint* hep-th/0102187)
[15] García-Salcedo R and Bretón N 2003 *Class. Quantum Grav.* **20** 5425 (*Preprint* gr-qc/0210142)
[16] Boillat G 1972 *Lett. Nuovo Cim.* **4** 274
[17] Cho Y M, Pak D G and Walker M L 2006 *Phys. Rev.* D **73** 065014
[18] Bretón N 2002 *Class. Quantum Grav.* **19** 601 (*Preprint* gr-qc/0210142)

On the runaway instability of self-gravitating torus around black holes

José A Font[1]*, Pedro J Montero[2] and Masaru Shibata[3]

[1] Departamento de Astronomía y Astrofísica, Universidad de Valencia, Dr. Moliner 50, 46100 Burjassot (Valencia), Spain

2 Max-Planck-Institute für Astrophysik, Karl-Schwarzschild-Str. 1, 81748, Garching bei München, Germany

3 Yukawa Institute of Theoretical Physics, Kyoto University, Kyoto 606-8502, Japan

E-mail: `j.antonio.font@uv.es`

Abstract. Black holes surrounded by self-gravitating tori are astrophysical systems which may naturally form following the core collapse of a massive star or the merger of two neutron stars. We present here results from fully general relativistic numerical simulations of such systems in order to assess the influence of the torus self-gravity on the onset of the so-called runaway instability. This instability, which might drive the rapid accretion of the disk on shorter timescales than those required to power a relativistic fireball, potentially challenges current models of gamma-ray bursts. Our simulations indicate that the self-gravity of the torus does not actually favour the onset of the instability.

1. Introduction

Relativistic self-gravitating tori orbiting black holes (BHs) are expected to form in a number of scenarios, such as in the merger of a binary system formed by either a BH and a neutron star (NS) or by two NSs (see e.g. [1] and references therein), and in the aftermath of the gravitational collapse of the rotating core of massive stars [2, 3]. Such tori have large average rest-mass densities, highly super-Eddington mass fluxes, and the angular momentum obeys sub-Keplerian distributions. State-of-the-art numerical simulations have recently started to provide quantitative estimates of the viability of such systems to form (see e.g. [4, 1]). As such systems are regarded as the central engine for gamma-ray bursts [5], understanding its formation, stability and dynamics is a stimulating task.

In this respect, the so-called runaway instability [6] may play a significant role as it could destroy the torus on dynamical timescales. In a marginally stable torus, the radial pressure gradient may drive the transfer of mass towards the BH through the cusp-like inner edge of the torus. Accretion increases the BH mass and spin and changes the gravitational field of the system. If the cusp moved deeper into the torus, mass accretion would increase, turning the accretion process runaway unstable. The runaway instability has so far been investigated using various approximations (see e.g. [6, 7, 8, 9]). Studies based on stationary models show that the self-gravity of the disk favours the instability, while there are additional parameters which have a stabilizing effect, namely the rotation of the BH and the radial distribution of specific angular momentum. Time-dependent, general relativistic hydrodynamical (GRHD) axisymmetric simulations of the runaway instability of tori around BHs have been performed by [10, 11], treating the dynamics of the gravitational field in an approximative way and neglecting the self-gravity of the torus. Overall, [10, 11] found that tori with constant distribution of specific angular momentum were unstable while non-constant (power law) angular momentum disks were stable.

In this work we present results which extend previous studies by incorporating an important missing ingredient in the modelling - the self-gravity of the torus. These results have also been discussed in [12], upon which the present article is based. In our work we use the initial data for a BH and a self-gravitating torus in the puncture framework derived by [13] and carry out fully relativistic simulations to investigate the onset of the runaway instability.

2. Basic equations

We follow the so-called BSSN formulation of the Einstein equations [14, 15]. Spacetime is hence foliated into a set of non-intersecting spacelike hypersurfaces, leading to the following line element $ds^2 = -(\alpha^2 - \beta_i\beta^i)dt^2 + 2\beta_i dx^i dt + \gamma_{ij}dx^i dx^j$, where α, β^i and γ_{ij} are the lapse function, the shift vector, and the 3-metric, respectively. BSSN makes use of a conformal decomposition of the 3-metric, $\tilde{\gamma}_{ij} = e^{-4\phi}\gamma_{ij}$ and the conformal-traceless extrinsic curvature $\tilde{A}_{ij} = e^{-4\phi}(K_{ij} - \gamma_{ij}K/3)$, with the conformal factor ϕ satisfying $e^{4\phi} = \gamma^{1/3} \equiv \det(\gamma_{ij})^{1/3}$. In addition to evolution equations for $\tilde{\gamma}_{ij}$ and $\tilde{A}_{ij}$, there are evolution equations for the conformal factor ϕ (or $\chi \equiv e^{-4\phi}$), the trace of the extrinsic curvature K, and the "conformal connection functions" $\tilde{\Gamma}^i$ [15]. Moreover, there are two kinematical additional variables left undetermined, α and β^i. These are computed using the so-called "moving puncture gauge", a combination of the "1+log" condition [16] for the lapse, and the "Gamma-freezing" condition for the shift vector [17].

The GRHD equations are the local conservation laws of momentum and energy, encoded in the stress-energy tensor $T^{\mu\nu}$, and of the matter density, J^μ (the continuity equation), i.e., $\nabla_\mu T^{\mu\nu} = 0$ and $\nabla_\mu J^\mu = 0$, where ∇_μ stands for the 4-dimensional covariant derivative. The density current is given by $J^\mu = \rho u^\mu$, where u^μ is the 4-velocity of the fluid and ρ its rest-mass density. Following [18] the GRHD equations are written in flux-conservative form in cylindrical coordinates. Since the Einstein equations are solved only in the $y = 0$ plane with Cartesian coordinates, the hydrodynamic equations are rewritten in Cartesian coordinates for $y = 0$. By defining the vector of unknowns, $\mathbf{U}$, fluxes $\mathbf{F}^x$ and $\mathbf{F}^z$ along the x and z directions, and the vector of sources $\mathbf{S}$ the set of hydrodynamic equations can be cast in conservative form as

$$\partial_t \mathbf{U} + \partial_x \mathbf{F}^x + \partial_z \mathbf{F}^z = \mathbf{S}, \tag{1}$$

We refer to [18] for further details on these equations, in particular for the expressions of the various vectors appearing in Eq. (1). To close the system of equations, we choose two possible equations of state, the so-called Γ-law equation of state (ideal fluid) given by $P = (\Gamma - 1)\rho\epsilon$, where ϵ is the specific internal energy, and a polytropic equation of state $P = \kappa\rho^\Gamma$. Here κ is the polytropic constant, $\Gamma = 1 + 1/N$ and N is the polytropic index.

3. Numerical approach and initial data

The numerical simulations are performed with the `nada` code [19] which solves the (BSSN) Einstein equations in conjunction with the "cartoon" method and the moving puncture approach. Correspondingly, the GRHD equations are solved with high-resolution shock-capturing schemes employing the HLLE approximate Riemann solver. The time update of the evolution equations is done using a 4th-order Runge-Kutta method of lines. The code has recently been validated against a number of stringent tests in vacuum and non-vacuum spacetimes [19]. For the simulations reported here we use an equally spaced (x, z) grid with a grid spacing $\Delta x = \Delta z = 0.05$ and $N_x \times N_z = 600 \times 600$ points to cover a computational domain, $0 \leq x \leq L$ and $0 \leq z \leq L$, with $L = 30$.

Table 1. Main properties of the equilibrium models in units of $c = G = M_\odot = 1$ (unless shown otherwise). From left to right the columns report the name of the model, the type of specific angular momentum distribution, the torus-to-BH mass ratio, the position of the maximum density point $r_{\rm max}$, the position of the inner and outer radii of the torus $r_{\rm in}$ and $r_{\rm out}$, the maximum rest-mass density and the orbital period at the center of the torus $t_{\rm orb}$.

Model	j − law	$M_t/M_{\rm BH}$	$r_{\rm max}$	$r_{\rm in}$	$r_{\rm out}$	$\rho_{\rm max}$ (g/cm^3)	$t_{\rm orb}$
E1	const	0.1	7.10	9.81	14.17	1.233×10^{14}	223.81
M1	const	0.1	7.17	4.92	10.17	3.189×10^{14}	147.81
M2	const	1.0	8.87	4.02	19.97	2.202×10^{14}	199.54
M3	non-const	0.1	10.47	4.92	19.97	3.902×10^{13}	245.37
M4	non-const	0.5	10.02	4.07	19.97	1.538×10^{14}	229.91

The initial data for the numerical simulations of the system formed by a BH and a self-gravitating torus in axisymmetric equilibrium have been computed by [13], to which the reader is addressed for further details. The equilibrium configurations for the matter fields are obtained by assuming a perfect fluid stress-energy tensor, and adopting a polytropic equation of state. Initial configurations can be constructed with either constant or non-constant specific angular momentum distributions, defined as $j \equiv hu_\varphi$, where h is the enthalpy. For the simulations reported in this work we have considered tori around a Schwarzschild BH (of mass $M_{\rm BH} = 1$). The adiabatic index is $\Gamma = 4/3$ to mimic a degenerate relativistic electron gas, and the polytropic constant κ is chosen such that the torus-to-BH mass ratio, $M_t/M_{\rm BH}$, is 0.1, 0.5 or 1, depending on the model. A list of the models along with their main features is given in Table 1.

4. Results

We first test the ability of the code to keep the stationarity of a fluid configuration initially in equilibrium, model E1. This test is performed both in a fixed spacetime, which nevertheless has the contribution from the self-gravity of the torus, and in a fully dynamical spacetime. The accuracy check of the evolution is performed by comparing the stationarity and conservation of different fluid quantities over a timescale which is several times the dynamical one. Results for the fixed spacetime simulation show that at the end of the simulation ($t = 1000M$), the difference between the final total rest-mass with respect to the initial value is less than 10^{-3} while the central rest-mass density only differs by 1% with the initial one. Correspondingly, for the dynamical spacetime evolution the difference between the final total rest mass with respect to the initial value is less than 0.1%. Although it is due to accumulated numerical error during the evolution, this violation of the rest mass does not affect in any significant way the stability of the torus for the duration of our simulation. Similar results are obtained for the time evolution of the central rest-mass density and the total angular momentum. These two tests provide reassuring evidence of the ability of the code to keep the torus in equilibrium for evolutions longer than the characteristic dynamical timescales of these objects.

Next, we introduce a perturbation on the (marginally stable) equilibrium models models M1 to M4 to evaluate the appearance of the runaway instability. This is done by perturbing the v^x component of the 3-velocity of the torus, which drives a small mass transfer through the inner edge of the tori (as was also done in [20]). For each model, the numerical simulations are stopped at $t \sim 2000$, which corresponds to ~ 10 ms (between 8 to 10 orbital periods depending on the model), since after this value the growth of the Hamiltonian constraint violation would lead to a loss of accuracy for the spacetime evolution. Nevertheless, the considered timescale

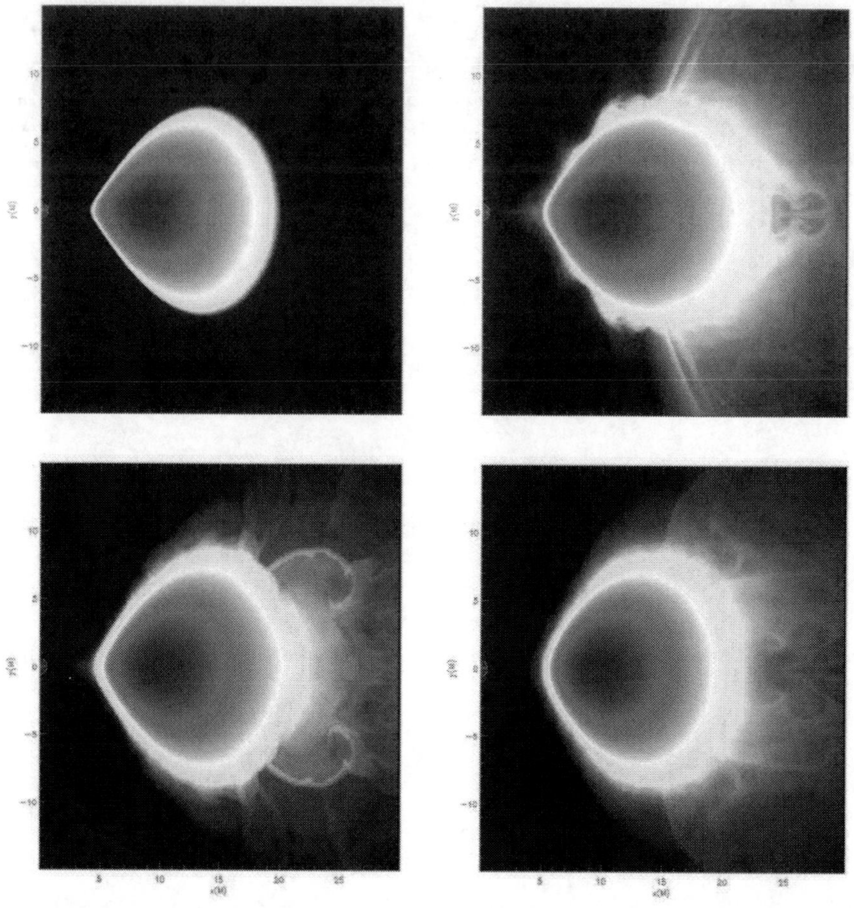

Figure 1. Colour coded isodensity contours of the perturbed model M2 at four diferent stages of its evolution, from initial (top-left panel) to final time (bottom-right panel), $t \sim 10t_{\rm orb}$. Despite the Kelvin-Helmholtz-like dynamics in the disk-external medium boundary the location of the torus remains firmly fixed in space.

would safely allow to identify the runaway instability, if present, since, according to [10, 11], it could even take place within one orbital period for the more massive models M2 and M4.

Figure 1 shows four different snapshots of the evolution of model M2, up to a a final time of $t \sim 10t_{\rm orb}$ (bottom-right panel). The evolution of this model typifies the characteristic evolution of all models studied. The initial perturbation triggers the accretion of mass and angular momentum through the cusp and on to the BH, leading to a global readjustment of the effective (gravitational and centrifugal) field. This is sufficient to momentarily reduce the amplitude of the mass flux and stop the disk from further in-falling towards the BH. As a new equilibrium is reached the disk departs from the BH leading to a long-lasting cycle of oscillations. Our simulations show that such a quasi-periodic oscillatory behaviour, which had already been found in the test-fluid simulations of non-self-gravitating disks performed by [20], is also present when further physics is incorporated in the numerical modelling, namely self-gravity and fully dynamical spacetime and hydrodynamical evolutions. The colour coded isodensity contours

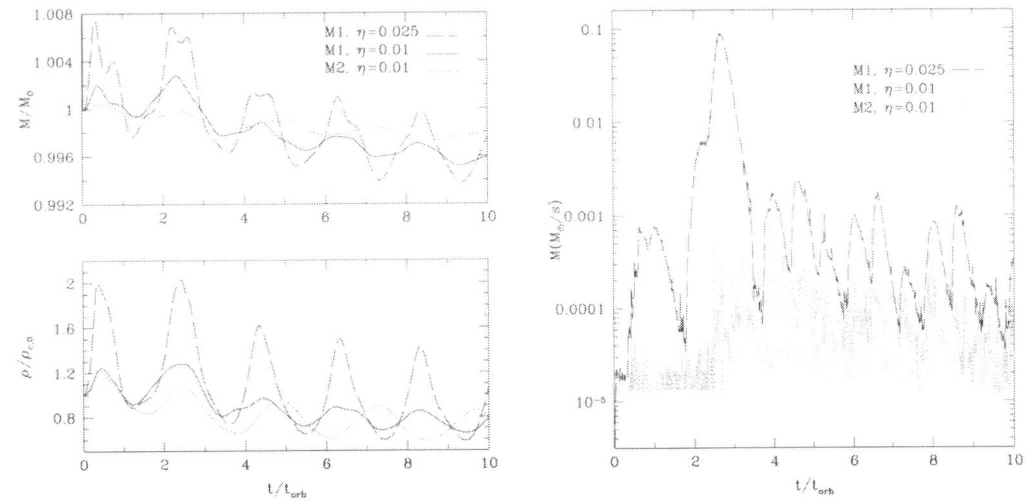

Figure 2. Left panel: Time evolution of the total rest-mass (top) and central rest-mass density, each of them normalized to its initial value, for the evolution of models M1 and M2. Right panel: Mass accretion rate evolution for models M1 and M2.

displayed in Fig. 1 also show the interesting dynamics at the boundary region separating the disk from the external medium. In particular, Kelvin-Helmhotz-driven eddies are seen being shed downwind from the edge of the disk during each oscillation (this is specially noticeable when these data are visualized in an animation).

To further analyze the system dynamics we plot in Figure 2 time evolutions of global quantities. The left panel of this figure displays the evolution of the normalized total rest-mass and central density for models M1 (solid and dashed lines) and M2 (dotted line). M1 is a j-constant model with an initial rest-mass of $M_t = 0.1 M_{\rm BH}$, thus representing a model with torus-to-BH mass ratio in broad agreement with the simulations of NS/NS or BH/NS mergers [1, 4, 21]. For this initial model, we consider two different perturbation amplitudes, $\eta = 0.01$ (solid line) and $\eta = 0.025$ (dashed line). As mentioned before, the perturbation triggers a phase of axisymmetric oscillations of the torus around its equilibrium which are present throughout the simulation. Such oscillations induce a small outflow of matter through the cusp towards the BH but the total rest-mass of the torus is not reduced in a significant way, as inferred from the upper panel of Figure 2. At the end of the simulation ($t \sim 10 t_{\rm orb}$) the rest-mass of the torus M1 is conserved up to about 1%. Therefore, the BH mass and spin do not increase notably, the torus showing no sign of the runaway instability. Further information is obtained from the right panel of Figure 2 which shows the time evolution of the mass accretion rate. This rate is larger the larger the initial perturbation is. For $\eta = 0.025$, the initial stage of very small mass transfer through the cusp of the torus is followed by a stage in which the oscillatory behavior of the mass flux, a signature of the induced oscillations, is obvious throughout the numerical evolution. Interestingly, during the oscillation phase, the mass flux does not increase in amplitude with time, as one would expect prior to the onset of the runaway instability (see [10, 11]). Instead, it reaches a maximum of about $\dot{M} \sim 0.1\ {\rm M}_\odot/{\rm s}$ after the second orbital period to later decrease and oscillate around a lower value, never showing any sign of exponential growth. The effects of the self-gravity are of course more important for model M2 for which $M_t = M_{\rm BH}$. Despite being a more massive disk the overall dynamics of M2 is very similar to that of the less massive torus, M1 (see dotted lines in Fig. 2). Again, the amplitude in the evolution of the mass flux does not increase with time and does not lead to the runaway instability. Finally, to check the influence of the rotation law on the runaway instability [10] we also carry out two additional numerical evolutions for two j-non-constant models, M3 and M4, which have torus-to-BH mass ratios of 0.1 and 0.5 respectively. Despite the difference in the rotation law with respect to models M1 and M2, the dynamics is very similar to that reported in of Figure 2: disks are stable and show no exponential growth of the mass flux for timescales larger than the dynamical timescale.

5. Summary

We have presented results from fully general relativistic numerical simulations of a system formed by a BH surrounded by a self-gravitating torus in equilibrium (see also [19, 12]). Test simulations using fixed spacetime evolutions show that the torus remains in equilibrium around its initial configuration for more than 5 dynamical timescales, when the simulation was stopped. Correspondingly, by simulating the same model in a dynamical spacetime we found that our numerical code `nada` is able to keep the torus in equilibrium for the several hundred M that lasted the simulation. In addition, results from numerical simulations of marginally-stable self-gravitating tori, aimed at evaluating the influence of the torus self-gravity on the onset of the runaway instability, have also been presented. These simulations show that all models exhibit a persistent phase of axisymmetric oscillations around their equilibria for several dynamical timescales without the appearance of the runaway instability. As a result, the self-gravity of the torus does not play a critical role favoring the appearance of the runaway instability.

Acknowledgments

Work supported by the Collaborative Research Center on *Gravitational Wave Astronomy* of the Deutsche Forschungsgesellschaft (DFG SFB/Transregio 7), the Spanish *Ministerio de Educación y Ciencia* (AYA 2007-67626-C03-01) and the Grant-in-Aid for Scientific Research (21340051) and the Grant-in-Aid for Scientific Research on Innovative Area (20105004) of the Japanese Monbukagakusho.

References

[1] Baiotti L, Giacomazzo B and Rezzolla L 2008 *Phys. Rev.* D **78** 084033
[2] Woosley S E 1993 *Astrophys. J.* **405** 273
[3] Paczynski B 1998 *Astrophys. J.* Lett. **494** L45
[4] Shibata M and Taniguchi K 2006 *Phys. Rev.* D **73** 064027
[5] Meszaros P 2006 *Rep. Prog. Phys.* **69** 2259
[6] Abramowicz M A, Calvani M and Nobili L 1983 *Nature* **302** 597
[7] Khanna R and Chakrabarti S K 1992 *MNRAS* **259** 1
Masuda N, Nishida S and Eriguchi Y 1998 *MNRAS* **297** 1139
[8] Nishida S, Lanza A, Eriguchi Y and Abramowicz M A 1996 *MNRAS* **278** L41
[9] Wilson D B 1984 *Nature* **312** 620
[10] Font J A and Daigne F 2002 *MNRAS* **334** 383
[11] Daigne F and Font J A 2004 *MNRAS* **349** 841
[12] Montero P J, Font J A and Shibata M 2009 submitted
[13] Shibata M 2007 *Phys. Rev.* D **76** 064035
[14] Shibata M and Nakamura T 1995 *Phys. Rev.* D **52** 5428
[15] Baumgarte T W and Shapiro S L 1999 *Phys. Rev.* D **59** 024007
[16] Bona C, Massó J, Seidel E and Stela J 1997 *Phys. Rev.* D **56** 3405
[17] Alcubierre M, Brügmann B, Diener P, Koppitz M, Pollney D, Seidel E and Takahashi R 2003 *Phys. Rev.* D **67** 084023
[18] Shibata M 2003 *Phys. Rev.* D **67** 024033
[19] Montero P J, Font J A and Shibata M 2008 *Phys. Rev.* D **78** 064037
[20] Zanotti O, Rezzolla L and Font J A 2003 *MNRAS* **341** 832
[21] Shibata M, Kyutoku K, Yamamoto T and Taniguchi K 2009 *Phys. Rev.* D **79** 044030

Journal of Physics: Conference Series **229** (2010) 012008 doi:10.1088/1742-6596/229/1/012008

Gravitating semilocal strings

Betti Hartmann[1*] and Jon Urrestilla[2,3]

[1]School of Engineering and Science, Jacobs University Bremen, 28759 Bremen, Germany
[2]Department of Theoretical Physics, University of the Basque Country, UPV-EHU, 48080 Bilbao, Spain
[3]Department of Physics and Astronomy, University of Sussex, Brighton BN1 9QH, United Kingdom

E-mail: `b.hartmann@jacobs-university.de`, `jon.urrestilla@ehu.es`

Abstract. We discuss the properties of semilocal strings minimally coupled to gravity. Semilocal strings are solutions of the bosonic sector of the Standard Model in the limit $\sin^2\theta_W = 1$ (where θ_W is the Weinberg angle) and correspond to embedded Abelian–Higgs strings for a particular choice of the scalar doublet. We focus on the limit where the gauge boson mass is equal to the Higgs boson mass such that the solutions fulfil the Bogomolnyi-Prasad-Sommerfield (BPS) bound.

1. Introduction

Semilocal strings [1, 2] are solutions to the bosonic sector of the Standard Model in the limit $\sin^2\theta_W = 1$, where θ_W is the Weinberg angle. This model has $SU(2) \times U(1)$ symmetry, where the $SU(2)$ is global, while the $U(1)$ symmetry is a local, i.e. gauge symmetry. This model can be seen as an extension of the $U(1)$ Abelian–Higgs model in which the complex scalar field of the Abelian–Higgs model is replaced by a complex scalar doublet, i.e. the number of scalar degrees of freedom is doubled. Abelian–Higgs strings, however, have rather different properties than semilocal strings. As for the strings obtained in [3], the stability of semilocal strings does not follow from the topology of the vacuum manifold (as it does for Abelian–Higgs strings), but from dynamical arguments. The ratio between the gauge and Higgs boson mass governs the stability of the semilocal strings: for Higgs boson mass larger (smaller) than the gauge boson mass semilocal strings are unstable (stable) and in the case of equality between the masses (the BPS limit), a degenerate one-parameter family of stable solutions exists. The parameter corresponds roughly to the width of the strings. In other words, in the BPS limit semilocal strings of arbitrary width have the same energy. Therefore, there is a zero mode associated with the width of the semilocal strings. It was shown in [4] that when this zero mode gets excited, it always leads to the growth of the string core.

It is interesting to note that semilocal strings share some properties with some new BPS strings. In [3] the conjecture that (stringy) D-strings are related to D-term BPS strings in the four dimensional field theory description was studied. New BPS cosmic strings of finite energy coupled to an axionic field were presented. The model has three different types of strings: tachyonic, axionic and hybrid strings. These objects are different from usual cosmic strings: among other things, a zero mode is present in the model, too. This zero mode can also be excited [5] leading to decompactification effects.

Semilocal strings have been studied in cosmological settings both in the context of their formation [6], network properties [7] and their CMB implications [8]. Due to an expected lower density of semilocal strings (as opposed to Abelian–Higgs strings) and due to the aforementioned tendency to become "fatter", these strings are interesting as possible solutions to relax constraints in inflationary models that predict too many cosmic strings [9].

In this paper we couple semilocal strings minimally to gravity. Besides obtaining the gravitating solutions, we will carefully study the properties of different members of the one-parameter families. As already mentioned, all solutions have the same (global) energy, but they have different energy densities. Therefore we expect to obtain differences in the local curvature of space–time. For example, in the supermassive case (when the symmetry breaking scale of the theory is larger than the Planck mass) Abelian–Higgs strings only exist up to some maximal value of the radial coordinate [10]. For the one-parameter families of semilocal strings, this maximal value of the radial coordinate will change when varying the width of the string (and the gravitational coupling); and in the limit there will be a solution for all space.

2. The model

The action reads:

$$S = \int d^4x \sqrt{-g}\left(\frac{R}{16\pi G} + \mathcal{L}_M\right) , \tag{1}$$

where R denotes the Ricci scalar, g is the determinant of the metric tensor, G is Newton's constant and $\mathcal{L}_M$ denotes the Lagrangian density of the semilocal model that has $SU(2)\times U(1)$ symmetry:

$$\mathcal{L}_m = -\frac{1}{4}F_{\mu\nu}F^{\mu\nu} + (D_\mu\Phi)^\dagger D^\mu\Phi - \frac{\lambda}{2}\left(\Phi^\dagger\Phi - \eta^2\right)^2 . \tag{2}$$

Here $F_{\mu\nu} = \partial_\mu A_\mu - \partial_\nu A_\mu$ is the field strength tensor of the $U(1)$ gauge field and $D_\mu\Phi = (\partial_\mu - ieA_\mu)\Phi$ is the covariant derivative of the complex scalar field doublet $\Phi = (\phi_1, \phi_2)^T$. e denotes the gauge coupling, λ the self-coupling, η the vacuum expectation value and G Newton's constant. There are three mass scales in the theory: the Planck mass $M_{Pl} = G^{-1/2}$, the Higgs boson mass $M_H = \sqrt{2\lambda}\eta$ and the gauge boson mass $M_W = \sqrt{2}e\eta$. In the following, we will only be interested in the case with $M_H^2 = M_W^2$, i.e. $e^2 = \lambda$. This is the Bogomolnyi–Prasad–Sommerfield (BPS) limit, that has been studied extensively for $G = 0$ in [1, 2]. In the following, we will keep λ and e^2 explicitly, but keep in mind that $e^2 = \lambda$.

2.1. Ansatz

In this paper, we study solutions that are cylindrically symmetric, i.e. solutions that have a rotational symmetry in the x-y-plane and do not explicitly depend on z.

The Ansatz for the metric in cylindrical coordinates (t, ρ, φ, z) is given by

$$ds^2 = N^2(\rho)dt^2 - d\rho^2 - L^2(\rho)d\varphi^2 - N^2(\rho)dz^2 , \tag{3}$$

where we have chosen $g_{tt} = -g_{zz}$ due to the boost symmetry of the solution. Note that for BPS solutions, the Einstein equations tell us that $N(\rho) \equiv 1$ which is essentially related to the fact that the sum of the energy density $\epsilon = T_0^0$ and the pressure components p_ρ, p_φ, p_z is zero in this case. We will thus set $N(\rho) \equiv 1$ in the following - unless otherwise stated.

The Ansatz for the matter fields is [1, 2]:

$$A_\varphi = \frac{n}{e}a(\rho) \ , \ \phi_1 = \eta f_1(\rho)e^{in\varphi} \ , \ \phi_2 = \eta f_2(\rho) \ . \tag{4}$$

By using the rescalings $x = e\eta\rho$ and $L(x) = e\eta L(\rho)$ the energy density $\epsilon = T_0^0$ reads:

$$\frac{T_0^0}{e^2\eta^4} = \frac{1}{2}\frac{1}{L^2}n^2(a')^2 + (f_1')^2 + (f_2')^2 + \frac{1}{L^2}n^2 f_1^2(1-a)^2 + \frac{1}{L^2}n^2a^2f_2^2 + \frac{\lambda}{2e^2}(f_1^2 + f_2^2 - 1)^2 , \tag{5}$$

where now and in the following the prime denotes the derivative with respect to x. The solutions fulfil an energy bound, such that the energy per unit length μ is given by :

$$\mu = \int d^2x\sqrt{-g}T_0^0 = 2\pi \int dx LT_0^0 = 2\pi n\eta^2 \ . \tag{6}$$

The deficit angle is related to μ by $\delta = 8\pi G\mu = 8\pi G(2\pi n\eta^2)$ and is given by:

$$\delta = 2\pi(1 - L'|_{\rho=\infty}) \ . \tag{7}$$

2.2. The equations

The BPS equations read

$$f_1' + \frac{n(a-1)}{L}f_1 = 0 \ , \ \ f_2' + \frac{na}{L}f_2 = 0 \ , \ \ \frac{na'}{L} + (f_1^2 + f_2^2 - 1) = 0 \tag{8}$$

and the Einstein equation is

$$\frac{L''}{L} = -\alpha^2\left((f_1^2 + f_2^2 - 1)^2 + \frac{2n^2(1-a)^2}{L^2}f_1^2 + \frac{2n^2a^2}{L^2}f_2^2\right) , \tag{9}$$

where $\alpha^2 = (4\pi M_W^2)/(e^2M_{Pl}^2) = 8\pi G\eta^2$. Note that with this abbreviation, the deficit angle is given by $\delta/(2\pi) = \alpha^2 n$. For $\alpha = 0$, the equations allow for a direct relation between the two scalar field functions f_1, f_2 [1, 2]. The relation for $\alpha \neq 0$ reads:

$$f_1 = c \cdot \exp\left(\int \frac{n}{L}dx\right) \cdot f_2 \ . \tag{10}$$

To solve the differential equations numerically, we have to impose appropriate boundary conditions. The requirement of finiteness of the energy and regularity at the origin leads to the following conditions:

$$f_2'(0) = 0 \ , \ \ a(0) = 0 \ , \ \ L(0) = 0 \ , \ \ L'(0) = 1 \ , \ \ a(\infty) = 1 \ . \tag{11}$$

Note that we could also choose another set of boundary conditions for the matter fields by e.g. imposing boundary conditions on f_1. These conditions are however equivalent to each other due to the relation between the functions f_1 and f_2 (see below). The set of boundary conditions given in (11) has proven convenient in numerical simulations.

3. Embedded Abelian–Higgs strings

In the limit $f_2(\rho) \equiv 0$ the equations of motion (8), (9) reduce to the equations of the $U(1)$ Abelian–Higgs model minimally coupled to gravity. Hence, the semilocal solutions correspond to embedded Abelian–Higgs strings. In contrast to the solutions of the original $U(1)$ Abelian–Higgs model [11], the embedded Abelian–Higgs strings are however unstable in flat space–time for $M_H^2 > M_W^2$ [2]. In any case, the embedded strings share many properties with the usual $U(1)$ strings, it is therefore interesting to summarize the properties of gravitating $U(1)$ Abelian–Higgs strings. In [12] it was shown that the solutions in curved space–time fulfil a BPS bound for $M_W = M_H$ that is essentially the same bound as that in flat space–time [13]. The explicit solutions have been constructed numerically in [14, 15]. It has been observed that there are different types of solutions depending on the choice of the gravitational coupling α.

For $\alpha^2 n \leq 1$, the deficit angle is smaller than 2π and globally regular solutions are possible. For $\alpha^2 n > 1$, the deficit angle is larger than 2π and the metric function L vanishes at a finite

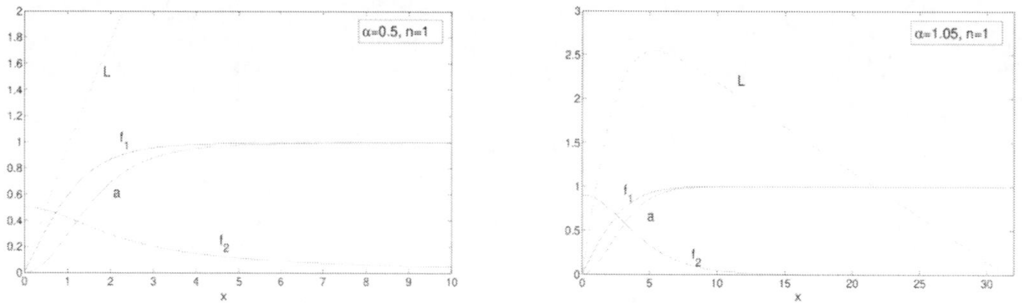

Figure 1. The profiles of a typical gravitating semilocal string with $\alpha = 0.5$ and $n = 1$ (left) and of a typical supermassive semilocal string with $\alpha = 1.05$ and $n = 1$ (right).

value of the radial coordinate. Since the deficit angle can only be larger than 2π if the symmetry breaking scale η is larger than the Planck scale, these solutions were called "supermassive" strings [10].

Away from the BPS limit, the metric function $N(x)$ is non-trivial and it was realized in [14, 15] that the string-like solutions have shadow solutions for the same choice of Higgs to gauge boson mass ratio and gravitational coupling. For globally regular solutions, there are Melvin solutions in addition to the string solutions which have a different asymptotic behaviour $N(x \to \infty) \propto x^{2/3}$, $L(x \to \infty) \propto x^{-1/3}$, while for the supermassive solutions there are shadow solutions in the form of Kasner solutions. For the latter, the metric function $N(x)$ vanishes at a finite value of the radial coordinate, while $L(x)$ diverges there. Note that Melvin and Kasner solutions do not exist in the BPS limit since $N(x) \equiv 1$.

4. Semilocal strings

In flat space-time, it was shown that solutions with $f_2(x) \neq 0$ exist only in the BPS limit $M_W = M_H$ [1, 2]. In fact, there exists a family of solutions that can be parametrized by the value of $f_2(x)$ at the origin, $f_2(0)$, which can be interpreted as a condensate on the string. For any choice of $f_2(0)$, the solutions fulfil the energy bound. They are hence degenerate in energy and this degeneracy is directly linked to the existence of a zero mode. Recently, semilocal strings coupled to so-called dark strings [16] have been studied [17] and it was shown that semilocal solutions with $f_2(x) \neq 0$ exist also away from the BPS limit and that in particular, the solutions with $f_2(x) \neq 0$ are lower in energy than the corresponding embedded Abelian–Higgs strings.

4.1. Globally regular semilocal strings

We have solved the equations (8), (9) subject to the boundary conditions (11) numerically using the ODE solver COLSYS [18].

The total energy and deficit angle don't change when varying the value of $f_2(0)$ for fixed α and n. This is related to the zero mode present in the model that persists to exist when studying the solutions in curved space-time. A typical gravitating semilocal string solution is shown in Fig. 1 (left) for $\alpha = 0.5$ and $n = 1$.

We have studied the dependence of the constant c appearing in (10) on the value of the condensate at the origin, $f_2(0)$. For a fixed value of $f_2(0)$, we observe that c decreases (slightly) for increasing α and that for $\alpha \to 1$, c tends to a finite value. Note that for $\alpha > 1$, the solutions become supermassive with $L(x_0) = 0$ at a finite value of the radial coordinate such that the integral in (10) diverges. For a fixed value of α, the value of c depends strongly on the choice of

$f_2(0)$. For $f_2(0) \to 0$, the value of c tends to infinity, while for $f_2(0) \to 1$ it tends to zero. This can be understood by looking at the non-gravitating ($n = 1$) counterpart [1], where :

$$f_2(x) = \frac{1}{c}\frac{f_1(x)}{x} \tag{12}$$

For $f_2(0) \to 0$ we have $f_2(x) \equiv 0$ and recover the embedded Abelian–Higgs solutions. The relation (12) can only be fulfilled if $1/c \to 0$. For $f_2(0) \to 1$ we have $f_2(x) \equiv 1$, hence for the requirement of finiteness of the energy $f_1(x) \equiv 0$. From (12) we find that $c \to 0$ in this limit.

4.2. Supermassive semilocal strings

For $\alpha^2 n > 1$, the metric function L possesses a zero at a finite distance x_0 from the string core such that the solution exists only on the interval $x \in [0 : x_0]$. A typical solution is shown in Fig. 1 (right) for $\alpha = 1.05$, $n = 1$ and the value of the condensate $f_2(0) = 0.9$. In this case $x_0 \approx 31.4$.

We have studied the dependence of x_0 on α and the choice of the value of the condensate on the string, $f_2(0)$. Our results are shown in Fig.2 for $n = 1$. For comparison, we also give the values for the embedded Abelian case with $f_2(0) = 0$, i.e. $f_2(x) \equiv 0$. For all choices of $f_2(0)$, x_0 tends to infinity for $\alpha \to 1$, indicating that for $\alpha^2 < 1$, the solutions are globally regular. For increasing α, x_0 decreases. We observe that for a fixed value of α the value of x_0 increases with increasing $f_2(0)$, i.e. the larger the condensate the larger the range of x on which the supermassive solutions exist.

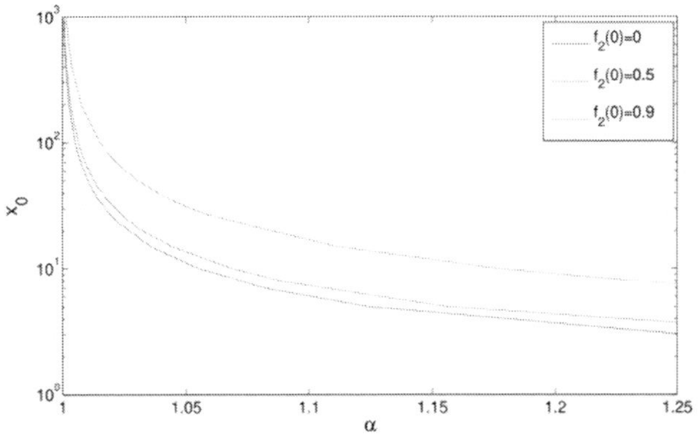

Figure 2. The value of the radial coordinate at which the metric function L vanishes, $L(x_0) = 0$ in dependence on the gravitational coupling α for $n = 1$ and different choices of the value of the condensate $f_2(0)$.

In the limit $f_2(0) = 1$, the solutions are such that $f_1(x < \infty) \equiv 0$, $f_2(x < \infty) \equiv 1$, $a(x < \infty) \equiv 0$, while $f_1(x = \infty) = 1$, $f_2(x = \infty) = 0$, $a(x = \infty) = 1$. This means that we have an infinite contribution to the energy density due to the infinite derivatives of the matter functions at infinity. Hence, we need to require $L(x = \infty) = 0$ in order to get finite energy solutions. The increase of the condensate to $f_2(0) = 1$ thus allows to make the solutions regular on the interval $x \in [0 : \infty[$ independent of the choice of the gravitational coupling α.

5. Conclusions

We have studied the gravitational properties of semilocal strings in the BPS limit. Like their non-gravitating counterparts, gravitating semilocal strings fulfil an energy bound and a direct relation between the two components of the scalar doublet exists, which now depends on the metric. The interval of the radial coordinate on which semilocal strings with energy per unit length (roughly) larger than the Planck mass $\mu > M_{Pl}^2/4$ exist increases with increasing value of the condensate and extends to infinity in the limit where the condensate tends to unity.

Acknowledgments

We thank the organisers of the ERE2009 for their hospitality. This work was partially supported by the Basque Government (IT-357-07), the Spanish Ministerio de Ciencia e Innovación (FPA2009-10612) and the Spanish Consolider-Ingenio 2010 Programme CPAN (CSD2007-00042) (J.U.).

References

[1] Vachaspati T and Achúcarro A 1991 *Phys. Rev.* D **44**, 3067
Vachaspati T and Achúcarro A 2000 *Phys. Rept.* **327** 347
[2] Hindmarsh M 1992 *Phys. Rev. Lett.* **68** 1263
Hindmarsh M 1993 *Nucl. Phys.* B **392** 461
[3] Blanco-Pillado J, Dvali G and Redi M 2005 *Phys. Rev.* D **72** 105002
[4] Leese R A 1992 *Phys. Rev.* D **46** 4677
[5] Achúcarro A and Sousa K 2006 *Phys. Rev.* D **74** 081701
[6] Achúcarro A, Kuijken K, Perivolaropoulos L and Vachaspati T 1992 *Nucl. Phys.* B **388** 435
Achúcarro A, Borrill J and Liddle A R 1998 *Phys. Rev.* D **57** 3742
Achúcarro A, Borrill J and Liddle A R 1999 *Phys. Rev. Lett.* **82** 3742
Urrestilla J, Achúcarro A, Borrill J and Liddle A R 2002 *J. High Energy Phys.* JHEP08(2002)033
[7] Achúcarro A, Salmi P and Urrestilla J 2007 *Phys. Rev.* D **75** 121703
[8] Urrestilla J, Bevis N, Hindmarsh M, Kunz M and Liddle A R 2008 *J. Cosmol. Astropart. P.* JCAP07(2008)010
[9] Urrestilla J, Achúcarro A and Davis A C 2004 *Phys. Rev. Lett.* **92** 251302
[10] Garfinkle D and Laguna P 1989 *Phys. Rev.* D **39** 1552
Ortiz M 1991 *Phys. Rev.* D **43** 2521
[11] Nielsen H and Olesen P 1973 *Nucl. Phys. B* **61** 45; Reprinted in Rebbi C and Soliani G 1985 *Solitons and Particles* (Singapore: World Scientific)
[12] Linet B 1987 *Phys. Lett.* A **124** 240.
[13] Bogomolny E 1976 *Sov. J. Nucl. Phys.* **24** 449; Reprinted in Rebbi C and Soliani G 1985 *Solitons and Particles* (Singapore: World Scientific).
[14] Christensen M, Larsen A and Verbin Y 1999 *Phys. Rev.* D **60** 125012
[15] Brihaye Y and Lubo M 2000 *Phys. Rev.* D **62** 085004
[16] Vachaspati T 2009 *Phys. Rev.* D **80** 063502
[17] Brihaye Y and Hartmann B 2009 The effect of dark strings on semilocal strings *Preprint* arXiv:0907.3233
[18] Ascher U, Christiansen J and Russell R 1979 *Math. of Comp.* **33** 659
Ascher U, Christiansen J and Russell R 1981 *ACM Trans.* **7** 209

Spanish Relativity Meeting (ERE 2009) IOP Publishing
Journal of Physics: Conference Series **229** (2010) 012009 doi:10.1088/1742-6596/229/1/012009

Static vacuum spacetimes with prescribed multipole moments

Magnus Herberthson
Department of Mathematics, Linköping University, SE-581 83 Linköping, Sweden.
E-mail: maher@mai.liu.se

Abstract. This paper gives sufficient conditions on a sequence of multipole moments for a static spacetime to exist with precisely these moments. We outline the proof, which is constructive in the sense that a metric having prescribed multipole moments up to a given order can be calculated. These sufficient conditions agree with already known necessary conditions, and hence this completes the proof of a long standing conjecture due to Geroch.

1. Introduction

There are various definitions of multipole moments, [6]. For static asymptotically flat spacetimes, the standard definition was given by Geroch [1], and this definition was later extended to the stationary case by Hansen [3]. In [1], Geroch made two conjectures, where the second reads

Conjecture 2: Given any set of multipole moments, subject to the appropriate convergence condition, there exists a static spacetime of Einstein's equations having precisely those moments.

This conjecture, see theorem 1 below, was proved in [5]; here we only give the most important lemmas used and a short outline of the arguments used in the proof. For earlier and related results, we also refer to [5].

Basically, the result stated in theorem 1 is that the necessary conditions in [4] are also sufficient, i.e., if they are satisfied, there exists a static asymptotically flat vacuum spacetime with these moments. In short, this was possible to prove by casting the metric in a special form, which ensured both correct moments and also uniqueness of relevant variables. This determines the metric formally, and by referring to a result by Friedrich, [2], theorem 1 follows.

2. Multipole moments of stationary spacetimes

Since we refer to a result concerning stationary spacetimes from [4], we will quote the definition of multipole moments given by Hansen in [3], which is an extension to stationary spacetimes of the definition for static spacetimes by Geroch [1]. We thus consider a stationary spacetime (M, g_{ab}) with signature $(-,+,+,+)$ and with a timelike Killing vector field ξ^a. We let $\lambda = -\xi^a\xi_a$ be the norm, and define the twist ω through $\nabla_a\omega = \epsilon_{abcd}\xi^b\nabla^c\xi^d$. If V is the 3-manifold of trajectories, the metric g_{ab} induces the positive definite metric

$$h_{ab} = \lambda g_{ab} + \xi_a\xi_b$$

on V. It is required that V is asymptotically flat, i.e., there exists a 3-manifold $\hat{V}$ and a conformal factor Ω satisfying

(i) $\hat{V} = V \cup i^0$, where i^0 is a single point

(ii) $\hat{h}_{ab} = \Omega^2 h_{ab}$ is a smooth metric on $\hat{V}$

(iii) At i^0, $\Omega = 0, \hat{D}_a\Omega = 0, \hat{D}_a\hat{D}_b\Omega = 2\hat{h}_{ab}$,

where $\hat{D}_a$ is the derivative operator associated with $\hat{h}_{ab}$. i^0 is referred to as spacelike infinity[1]. On M, and/or V, one defines the scalar potential

$$\phi = \phi_M + i\phi_J, \quad \phi_M = \frac{\lambda^2 + \omega^2 - 1}{4\lambda}, \phi_J = \frac{\omega}{2\lambda}. \tag{1}$$

The multipole moments of M are then defined on $\hat{V}$ as certain derivatives of the scalar potential $\hat{\phi} = \phi/\sqrt{\Omega}$ at i^0. More explicitly, following [3], let $\hat{R}_{ab}$ denote the Ricci tensor of $\hat{V}$, and let $P = \hat{\phi}$. Define the sequence $P, P_{a_1}, P_{a_1a_2}, \ldots$ of tensors recursively:

$$P_{a_1\ldots a_n} = C[\hat{D}_{a_1}P_{a_2\ldots a_n} - \frac{(n-1)(2n-3)}{2}\hat{R}_{a_1a_2}P_{a_3\ldots a_n}], \tag{2}$$

where $C[\,\cdot\,]$ stands for taking the totally symmetric and trace-free part. The multipole moments of M are then defined as the tensors $P_{a_1\ldots a_n}$ at i^0.

In [1], a slightly different setup and a different potential is used, but it is known [7] that the potential used there and (1) with $\omega = 0$ produce the same moments.

3. Static spacetimes with prescribed multipole moments

Consider $\mathbf{R}^3$ with Cartesian coordinates $\mathbf{r} = (x, y, z) = (x^1, x^2, x^3)$, and let $\alpha = (\alpha_1, \alpha_2, \alpha_3)$ be a multi-index. With the convention that, in terms of components, $P^0_\alpha = P^0_{\underbrace{11\cdots1}_{\alpha_1}\underbrace{22\cdots2}_{\alpha_2}\underbrace{33\cdots3}_{\alpha_3}}$, we have the following theorem.

Theorem 1. *Let $P^0, P^0_{a_1}, P^0_{a_1a_2}, \ldots$ be a sequence of real valued totally symmetric and trace free tensors on $\mathbf{R}^3$, and let $P^0, P^0_{i_1}, P^0_{i_1i_2}, \ldots$ be the corresponding components with respect to the Cartesian coordinates $\mathbf{r} = (x, y, z)$. If $u(\mathbf{r}) = \sum_{|\alpha|\geq 0} \frac{\mathbf{r}^\alpha}{\alpha!}P^0_\alpha$ converges in a neighbourhood of the origin in $\mathbf{R}^3$, there exists a static asymptotically flat vacuum spacetime having the moments $P^0, P^0_{a_1}, P^0_{a_1a_2}, \ldots$.*

Note that we do not require the monopole $P^0 = \hat{\phi}(0)$ to be non-zero[2]. The proof, however, will first be carried out under the assumptions $P^0 \neq 0$, and this condition will then be relaxed in Section 3.7.

3.1. Outline of the proof

As mentioned in Section 1, it is possible to reduce the coordinate freedom in the metric components, and simultaneously establish a direct link between the potential $\hat{\phi}$ and the desired moments. This will be done in Section 3.4, where the link is expressed in theorem 6.

In Section 3.5, we will address the conformal field equations from Section 3.3, and further restrict the form of the metric. This restriction ensures that the resulting conformal field equations, (4), (after splitting into regular and non-regular parts) have smooth extensions to i^0.

[1] i^0 is also used in a four-dimensional context.

[2] Due to the definition of ϕ, $P^0 = -m$, where m is the mass of the spacetime.

These tensor equations will then be written as a set of scalar equations, and if a certain subset of these equations are satisfied, the full metric will be formally determined (see lemma 7).

In Section 3.6 we will address the full set of equations, as well as the issue of convergence of the power series derived. By referring to a result by Friedrich, [2], convergence of the power series will be concluded. It will also be seen that the full set of equations are satisfied.

Finally, in Section 3.7, the condition that the monopole is nonzero will be relaxed.

3.2. Notation

Small Latin letters $a, b, \ldots$ refer to abstract indices, as in Section 2. Small Latin letters $i, j, k, \ldots$ are numerical indices and refer to components with respect to the normal coordinates $(x, y, z) = (x^1, x^2, x^3)$ introduced below. Since these components refer to this particular coordinate system only, the equations will not be tensor equations. With this said, we will still use $=$ instead of $\doteq$.

Almost all variables[3] are assumed to be formally analytic, i.e., they admit a formal power series expansion (again in terms of the chosen coordinates), so that, for a tensor, A_{ijk} say, we can write

$$A_{ijk} = \sum_{n=0}^{\infty} A_{ijk}^{[n]}$$

where $A_{ijk}^{[n]}$ denotes polynomials in (x, y, z) which are homogeneous of order n (and where the summation may be formal). Both η_{ij} and η^{ij} denote the identity matrix, and by $[A_{ij}]$ we denote the trace of A_{ij}, i.e., $\eta^{ij} A_{ij}$. We also use $\partial_i = \frac{\partial}{\partial x^i}$ and the operator $D = x^j \frac{\partial}{\partial x^j}$.

With $r = \sqrt{x^2 + y^2 + z^2}$, $f_1 \equiv f_2 \pmod{r^2}$ means that $f_1(x, y, z) - f_2(x, y, z) = r^2\, g(x, y, z)$ for some (formally analytic) function g. When $f \equiv 0 \pmod{r^2}$, so that $f = r^2\, g$ for some formally analytic function g, we also use the shorter notation $r^2 | f$.

In the proof of lemma 8 we will refer to [2] and hence use some of the notation there.

3.3. The field equations

Apart from having the correct multipole moments, we must ensure that the metric describes a static vacuum spacetime. We will formulate our equations on the 3-manifold $(\hat{V}, \hat{h}_{ab})$ defined in Section 2, starting with the field equations from [1]. However, the 3-manifolds in [1] and [3] are defined slightly differently, and the relations imply the following.

Starting with a static spacetime (M, g_{ab}) with timelike killing vector ξ^a, we put $0 < \lambda = -\xi^a \xi_a$ and $\Psi = 1 - \sqrt{\lambda}$. From [1] we consider[4] a 3-surface V_G orthogonal to ξ^a. In terms of the metric on V_G: $(h_G)_{ab} = g_{ab} + \xi_a \xi_b / \lambda$, the field equations are, [1],

$$\left\{ \begin{array}{c} (D_G)^a (D_G)_a \Psi = 0 \\ (R_G)_{ab} = \frac{1}{\Psi - 1} (D_G)_a (D_G)_b \Psi \end{array} \right. \Leftrightarrow \left\{ \begin{array}{c} (R_G)^a{}_a = 0 \\ (R_G)_{ab} = \frac{1}{\sqrt{\lambda}} (D_G)_a (D_G)_b \sqrt{\lambda} \end{array} \right. , \tag{3}$$

where $(D_G)_a$ is the derivative operator and $(R_G)_{ab}$ is the Ricci tensor associated with $(V_G, (h_G)_{ab})$.

On the 3-manifold V on the other hand, the metric is $h_{ab} = \lambda g_{ab} + \xi_a \xi_b$, i.e., $h_{ab} = \lambda (h_G)_{ab}$, and with $\hat{h}_{ab} = \Omega^2 h_{ab}$, this implies that $\hat{h}_{ab} = (\sqrt{\lambda}\Omega)^2 (h_G)_{ab}$. We can now express equations (3) on $(\hat{V}, \hat{h}_{ab})$ using as conformal factor $\hat{\Omega} = \sqrt{\lambda}\Omega$. Using the properties of conformal transformations,

[3] The exception is $r = \sqrt{x^2 + y^2 + z^2}$

[4] In [1], the potential $\psi = -\Psi$ is used.

[9], we find that (3) becomes

$$\begin{array}{rcl} \hat{R}+4\hat{h}^{ab}\hat{D}_a\hat{D}_b\ln\hat{\Omega}-2\hat{h}^{ab}\hat{D}_a\ln(\hat{\Omega})\hat{D}_b\ln\hat{\Omega} & = & 0 \\ \hat{R}_{ab}+\hat{D}_a\hat{D}_b\ln\hat{\Omega}+\hat{h}_{ab}\hat{h}^{cd}\hat{D}_c\hat{D}_d\ln\hat{\Omega}+ & & \\ \hat{D}_a\ln(\hat{\Omega})\hat{D}_b\ln\hat{\Omega}-\hat{h}_{ab}\hat{h}^{cd}\hat{D}_c\ln(\hat{\Omega})\hat{D}_c\ln\hat{\Omega} & = & \\ \frac{1}{\sqrt{\lambda}}[\hat{D}_a\hat{D}_b\sqrt{\lambda}+\hat{D}_a\ln(\hat{\Omega})\hat{D}_b\sqrt{\lambda}+ & & \\ \hat{D}_b\ln(\hat{\Omega})\hat{D}_a\sqrt{\lambda}-\hat{h}_{ab}\hat{h}^{cd}\hat{D}_d\ln(\hat{\Omega})\hat{D}_c\sqrt{\lambda}]. & & \end{array} \quad (4)$$

Thus the task is to find a conformal factor $\hat{\Omega}$ and a metric $\hat{h}_{ij}$, defined in a neighbourhood of i^0, which both produces the desired multipole moments and also satisfies (4).

3.4. Prescribed moments and canonical form of the metric

In this section we will, in a sense, reverse the arguments from [4]. The key point is to work in normal coordinates, in which the Ricci tensor $\hat{R}_{ab}$ takes a special form. This will follow from the following sequence of lemmas, where lemma 3 is taken from [8] (and formulated for the case when $\hat{h}_{ij}(i^0)=\eta_{ij}$).

Lemma 2. *Let (x,y,z) be Cartesian coordinates on $\mathbf{R}^3$, let $\mathbf{r}^a=x(\frac{\partial}{\partial x})^a+y(\frac{\partial}{\partial y})^a+z(\frac{\partial}{\partial z})^a$, and put $r=\sqrt{x^2+y^2+z^2}$. Then, for any tensor $T_{a_1\ldots a_n}$, $\mathbf{r}^{a_1}\ldots\mathbf{r}^{a_n}C[T_{a_1\ldots a_n}]\equiv\mathbf{r}^{a_1}\ldots\mathbf{r}^{a_n}T_{a_1\ldots a_n}$* $(\mathrm{mod}\ r^2)$

Lemma 3. $(x,y,z)=(x^1,x^2,x^3)$ *are normal coordinates on $(\hat{V},\hat{h}_{ab})$, if and only if $x^i\hat{h}_{ij}(x^k)=x^i\hat{h}_{ij}(i^0)$, i.e., $x^i\hat{h}_{ij}=x^i\eta_{ij}$.*

Lemma 4. *Suppose that $\hat{h}_{ij}$ is given in normal coordinates, so that $A_{ij}=\hat{h}_{ij}-\eta_{ij}$ satisfies $A_{ij}x^i=0$. Suppose further that A_{ij} has the property that $r^2|\eta^{ij}A_{ij}$.*

Then the corresponding property holds for the inverse $\hat{h}^{ij}$, i.e., with $B^{ij}=\hat{h}^{ij}-\eta^{ij}$, we have $r^2|\eta_{ij}B^{ij}$ and $B^{ik}\hat{h}_{jk}x^j=B^{ik}\eta_{jk}x^j=0$.

Lemma 5. *Suppose that (x^1,x^2,x^3) are normal coordinates on $(\hat{V},\hat{h}_{ab})$. Then $r^2|\eta^{ij}(\hat{h}_{ij}-\eta_{ij})\Rightarrow r^2|\mathbf{r}^a\mathbf{r}^b\hat{R}_{ab}$.*

We will now return to the original recursion (2). By combining the previous lemmas, and using Taylor's theorem, we have the following theorem, which allows for the direct connection between the moments $P_{a_1\ldots a_n}$ and the potential $P=\hat{\phi}=\sum_{|\alpha|\geq 0}\frac{\mathbf{r}^\alpha}{\alpha!}P^0_\alpha$.

Theorem 6. *Let (x^1,x^2,x^3) be normal coordinates on $\hat{V}$, $\hat{h}_{ij}-\eta_{ij}$ satisfy $r^2|\eta^{ij}(\hat{h}_{ij}-\eta_{ij})$, and let $P_{a_1\ldots a_n}$ be defined by the recursion* (2). Then

$$x^{i_1}\ldots x^{i_n}P_{i_1\ldots i_n}\equiv x^{i_1}\ldots x^{i_n}\partial_{i_1}\cdots\partial_{i_n}P \quad (\mathrm{mod}\ r^2)$$

In short, if the coordinates (x,y,z) are such that $\hat{h}_{ij}$ satisfies the conditions of lemma 4, and if the potential $P=\hat{\phi}$ is expressed in these coordinates, it is possible to specify the multipole moments.

3.5. The conformal field equations and uniqueness of the metric

In this section, we will address the conformal field equations (4), using a metric satisfying that conditions of lemma 4. Since (4) contains λ which not formally smooth, it can be split into a regular and non-regular part (again, see [5]). To ensure smooth extensions of these to i^0, the form of $\hat{h}_{ij}$ will be restricted further, namely into the form

$$\hat{h}_{ij}=\eta_{ij}+f(x,y,z)(x_ix_j-r^2\eta_{ij})+r^2\gamma_{ij},\quad \gamma_{ij}x^i=0. \quad (5)$$

Under this metric, a set of scalar equations equivalent to (4) can be derived. The scalar components formed from (4), which are defined in [5], are denoted by $S, t_{00}, t_{11}, t_{22}, t_{33}, t_{01}, t_{02}, t_{03}$, and the vanishing of these is equivalent to (4) being satisfied.

In terms of these scalar quantities, one can prove the following lemma.

Lemma 7. *Suppose that the metric components $\hat{h}_{ij}$ takes the form (5). Then the equations*

$$S = 0, \quad t_{11} = 0, \quad t_{22} = 0, \quad t_{33} = 0$$

determines the metric $\hat{h}_{ij}$ as formal power series.

It is noteworthy that these equations give the metric components of the appropriate order explicitly as an polynomial expression in the desired/prescribed multipole moments and metric components of lower order.

3.6. Convergence of the metric

In this section we will assure that the series expansion for $\hat{h}_{ij}$, which was found in the previous section, converges and also solves the full conformal field equations. This is the content of the following lemma.

Lemma 8. *Suppose that $\hat{h}_{ij}$ is a formal power series for the metric of the form (5), producing the moments of theorem 1. If*

$$S = 0, \quad t_{11} = 0, \quad t_{22} = 0, \quad t_{33} = 0,$$

then

$$t_{00} = 0, \quad t_{0k} = 0, k = 1, 2, 3$$

and the power series is convergent in a neighbourhood of i^0.

This will conclude the proof of theorem 1 when the monopole is nonzero.

To prove of this lemma, relies on a result by Friedrich, namely theorem 1.1 in [2], where a related property is investigated. In [2] it is proven that under rather similar settings, there exists a metric which instead of prescribed multipole moments, produces a set of prescribed *null data*, where also growth conditions for the null data in order for the series expansion of the metric to converge are given. These null data are proven to be in a one-to-one correspondence with the family of multipole moments with non-zero monopole, but since this correspondence is rather implicit, the actual conditions on the multipole moments (as required by the conjecture by Geroch) are not clear.

Using the notation in [2], the condition is roughly that there exist M, r such that

$$|C[\nabla_{\mathbf{a_q}} \cdots \nabla_{\mathbf{a_3}} R_{\mathbf{a_2 a_1}}](i^0)| \leq \frac{Mp!}{r^p}, \quad \mathbf{a_p}, \dots \mathbf{a_1}, \mathbf{b}, \mathbf{c} = 1, 2, 3, \quad p = 0, 1, 2, \dots \tag{6}$$

where $R_{\mathbf{a_2 a_1}}$ is the Ricci tensor of [2], and the bold indices indicate that we are evaluating tensor expressions with respect to a certain frame. It is possible to prove that the conformal factor relating $\hat{h}_{ab}$ and the metric used in [2] is smooth, and this also, starting with $\hat{\phi}$, allows us to specify a potential with desired moments in [2]. Finally, using induction, it is possible to show that the estimates which are known to hold for $\hat{\phi}$, namely that for some $\tilde{M}, \tilde{r}$, $|\partial_\alpha \hat{\phi}(0)| \leq \frac{p! \tilde{M}}{\tilde{r}^p}$, $p = |\alpha| = 0, 1, 2, \dots$, implies the estimates in (6), for some constants M, r. From this lemma 8 and hence theorem 1 follow when $P^0 \neq 0$.

3.7. The case $P^0 = 0$

To complete the proof, the condition that $P^0 \neq 0$, i.e., $m \neq 0$ must be relaxed. This will be possible due to the fact that the conformal factor $\hat{\Omega}$ is well behaved for $m = 0$. Thus, suppose that we have a sequence of totally symmetric and trace-free tensors $P^0, P^0_{i_1}, P^0_{i_1 i_2}, \ldots$ as in theorem 1, and where $P^0 = 0$. In particular, it is assumed that $u(\mathbf{r}) = \sum_{|\alpha|\geq 0} \frac{\mathbf{r}^\alpha}{\alpha!} P^0_\alpha$ converges in some polydisc $U : \{(x, y, z), |x| < d, |y| < d, |z| < d\}$. By replacing only the monopole, i.e., putting $P^0 = m_0 > 0$, the corresponding sequence $m_0, P^0_{i_1}, P^0_{i_1 i_2}, \ldots$ corresponds to the function $\tilde{u}(\mathbf{r}) = m_0 + \sum_{|\alpha|>0} \frac{\mathbf{r}^\alpha}{\alpha!} P^0_\alpha$, which also converges in U. Thus, by the arguments given so far there exists convergent, in a polydisc V say, power series

$$\hat{h}_{ij} = \sum_{|\alpha|\geq 0} (c_{ij})_\alpha \mathbf{r}^\alpha$$

for the metric components; furthermore we also know that there is a static spacetime having the multipole moments $m_0, P^0_{i_1}, P^0_{i_1 i_2}, \ldots$. Now, from the recursion producing the metric, it is seen that each coefficient $(c_{ij})_\alpha$ is a polynomial in m_0, and hence the metric components h_{ij} can be regarded as a power series in the four variables (m, x, y, z):

$$\hat{h}_{ij} = \sum_{|\beta|\geq 0} (d_{ij})_\beta m^{\beta_1} x^{\beta_2} y^{\beta_3} z^{\beta_4},$$

where the multi-index $\beta = (\beta_1, \beta_2, \beta_3, \beta_4)$. Since this series converges for $m = m_0$, $(x, y, z) \in V$, it also converges for $|m| < m_0, (x, y, z) \in V$. We can now choose $m = 0$ and still have convergence of $\hat{h}_{ij}$, κ and g_{ij}. In particular, with $m = 0$ the multipole moments will be the initially desired sequence $P^0, P^0_{i_1}, P^0_{i_1 i_2}, \ldots$ where $P^0 = 0$.

4. Discussion

In this paper, we have outlined the proof of a long standing conjecture by Geroch, [1]. The full arguments are found in [5]. The conditions given are simple and natural. In essence, each allowed set of multipole moments is connected to a harmonic function on $\mathbf{R}^3$, defined in a neighbourhood of $\mathbf{0}$. The proof is constructive in the sense that an explicit metric having prescribed moments up to a given order can be calculated.

References

[1] Geroch R 1970 Multipole Moments. II. Curved Space *J. Math. Phys.* **11** 2580
[2] Friedrich H 2007 Static Vacuum Solutions from Convergent Null Data Expansions at Space-Like Infinity *Ann. Henri Poincaré* **8** 817-884
[3] Hansen R O 1974 Multipole moments of stationary spacetimes *J. Math. Phys.* **15** 46
[4] Bäckdahl T and Herberthson M 2006 Calculation of, and bounds for, the multipole moments of stationary spacetimes *Class. Quantum Grav.* **23** 5997
[5] Herberthson M 2009 Static spacetimes with prescribed multipole moments: a proof of a conjecure by Geroch *Class. Quantum Grav.* **26** 215009
[6] Quevedo H 1990 Multipole Moments in General Relativity - Static and Stationary Vacuum Solutions *Fortschritte der physik* **38** 733
[7] Simon W and Beig R 1983 The multipole structure of stationary spacetimes *J. Math. Phys.* **24** 1163
[8] Thomas T Y 1961 *Concepts from Tensor Analysis and Differential Geometry* (Academic Press)
[9] Wald R M 1984 *General Relativity* (Chicago: The University of Chicago Press)

Spanish Relativity Meeting (ERE 2009) IOP Publishing
Journal of Physics: Conference Series **229** (2010) 012010 doi:10.1088/1742-6596/229/1/012010

Physical properties of black di-ring

Hideo Iguchi* **and Takashi Mishima**

Laboratory of Physics, College of Science and Technology, Nihon University, Narashinodai, Funabashi, Chiba 274-8501, Japan

E-mail: iguchi.h@phys.ge.cst.nihon-u.ac.jp, tmishima@phys.ge.cst.nihon-u.ac.jp

Abstract. We study the stationary asymptotically flat five-dimensional vacuum solution describing black di-ring: two concentric black rings. We give an outline of solution generating methods to construct the black di-ring and introduce the rod structures of the black di-ring and the seed solutions. We investigate the phase structure of black di-ring to show the infinite non-uniqueness of the solution. We analyze the distributions of black di-rings whose Komar angular momentum of inner ring vanishes and whose angular velocity of inner ring vanishes in the phase diagram. We also investigate temperatures of horizons for these two special families to consider the existence of the thermal equilibrium black di-rings.

1. Introduction

One of the most important recent findings of the higher-dimensional general relativity is a single-rotational black ring solution by Emparan and Reall [1]. (See also [2].) This solution is a vacuum, axially symmetric and asymptotically flat solution of the five-dimensional general relativity. The topology of the event horizon is $S^1 \times S^2$. The black ring rotates along the direction of S^1. The balanced black ring which has no conical singularity has a minimum of angular momentum for a fixed mass parameter. When the angular momentum is near this minimum, there are two different black rings with the same angular momentum. They are called fat and thin black rings according to their shapes. In addition we have a single-rotational spherical black hole [3] with the same asymptotic parameters. As a result, there can be three different configurations for the fixed mass and angular momentum parameter. This fact is often called the discrete non-uniqueness of the five dimensional vacuum solutions.

In this several years, we found important five-dimensional vacuum solutions by using solitonic solution generating techniques. The solution of black ring with S^2 rotation was constructed by the Bäcklund transformation [4]. The same solution was found by another method [5]. It was shown that the solution of S^1 rotating black ring can be generated by the solitonic technique [6, 7]. The black ring solution with two angular momentum components was also constructed by the inverse scattering method [8]. In five dimensions, in addition to the solutions with single horizon, there exist solutions with disconnected event horizons. Black Saturn which is a spherical black hole surrounded by a black ring was constructed by the inverse scattering method [9]. It was shown that the black rings can be superposed concentrically [10, 11] and orthogonally [12, 13].

The existence of multi-black hole configurations implies continuous non-uniqueness of five-dimensional black holes. The phase diagram of the black Saturn was investigated in [9, 14]. The plot of random sets of points in the phase diagram showed that the black Saturn covers the wide region of the phase diagram. The phases of black Saturn were investigated based on the thin and long ring approximation in which the black Saturn can be modelled as a simple superposition of an MP black hole and a black ring [9]. It was argued that the configurations that approach maximal entropy for fixed mass and angular momentum are black Saturns with a nearly static black hole and a very thin black ring.

The black di-ring also indicates infinite non-uniqueness [10]. It was confirmed that there are infinite number of black di-rings for same mass and same angular momentum. Distributions of black di-rings in the phase diagram have not been fully investigated. When we approximate the black di-ring as a simple superposition of two concentric black rings, we can roughly estimate the region covered by the black di-ring in the phase diagram. The maximum of the area would be smaller than the one of black Saturn for the same mass and angular momentum. Because of the strong non-linearity, however, we need precise analyses for the distributions of black di-ring in the phase diagram for the decisive conclusion.

In this article we analyze the phase structure of black di-ring by numerical plots in the phase diagram. In addition, we investigate the phases of two particular family of the black di-ring. One is the solution whose Komar angular momentum of inner ring is zero. The other is the solution whose angular velocity of inner ring is zero.

2. Black di-ring solution

The black di-ring solution was generated by using the Bäcklund transformation [10] and reconstructed by the inverse scattering method [15]. The solution has two disconnected $S^1 \times S^2$ event horizons with single angular momentum component. In this section we briefly review the rod structure analysis and shortly see the solution generating methods which were used to generate the black di-ring.

2.1. Rod structure analysis

The rod structure is a powerful tool to describe an axially symmetric vacuum solutions. In the recent investigations for the uniqueness theorem of higher dimensional black holes, the rod structure plays an important role. Here we give a brief explanation of the rod structure analysis elaborated by Harmark [16]. See [16] for complete explanations.

We denote the D-dimensional axially symmetric stationary metric as

$$ds^2 = G_{ij}dx^i dy^j + e^{\nu}(d\rho^2 + dz^2) \tag{1}$$

where G_{ij} and ν are functions only of ρ and z and $i, j = 0, 1, \ldots, D-3$. The $D-2$ by $D-2$ matrix field G satisfies the following constraint

$$\rho = \sqrt{|\det G|}. \tag{2}$$

The equations for the matrix field G can be derived from the Einstein equation $R_{ij} = 0$ as

$$G^{-1}\nabla G = (G^{-1}\nabla G)^2, \tag{3}$$

where the differential operator ∇ is the gradient in three-dimensional unphysical flat space with metric

$$d\rho^2 + \rho^2 d\omega^2 + dz^2. \tag{4}$$

Because of the constraint $\rho = \sqrt{|\det G|}$, at least one eigenvalue of $G(\rho, z)$ goes to zero for $\rho \to 0$. However it was shown that if more than one eigenvalue goes to zero as $\rho \to 0$, we have a curvature singularity there. Therefore we consider solutions which have only one eigenvalue goes to zero for $\rho \to 0$, except at isolated values of z. Denoting these isolated values of z as

Journal of Physics: Conference Series **229** (2010) 012010 doi:10.1088/1742-6596/229/1/012010

$a_1, a_2, \ldots, a_N$, we can divide the z-axis into the $N+1$ intervals $[-\infty, a_1]$,$[a_1, a_2]$,...,$[a_N, \infty]$, which is called as rods. These rods correspond to the source added to the equation (3) at $\rho = 0$ to prevent the break down of the equation there.

The eigenvector for the zero eigenvalue of $G(0, z)$

$$\mathbf{v} = v^i \frac{\partial}{\partial x^i}, \tag{5}$$

which satisfies

$$G_{ij}(0, z) v^i = 0, \tag{6}$$

determines the direction of the rod. If the value of $\frac{G_{ij} v^i v^j}{\rho^2}$ is negative (positive) for $\rho \to 0$ the rod is called timelike (spacelike). Each rod corresponds to the region of the translational or rotational invariance of its direction. The timelike rod corresponds to a horizon. The spacelike rod corresponds to a compact direction.

The rod structure of black di-ring is composed of two finite timelike rods, two finite spacelike rods and two semi-infinite spacelike rods. Both timelike rods have non-trivial direction vectors. For the single spin black di-ring, there are six parameters, of which four are the lengths of the finite rods and two are the non-trivial components of direction vectors of timelike rods. Figure 1 shows the rod structure of the single spin black di-ring.

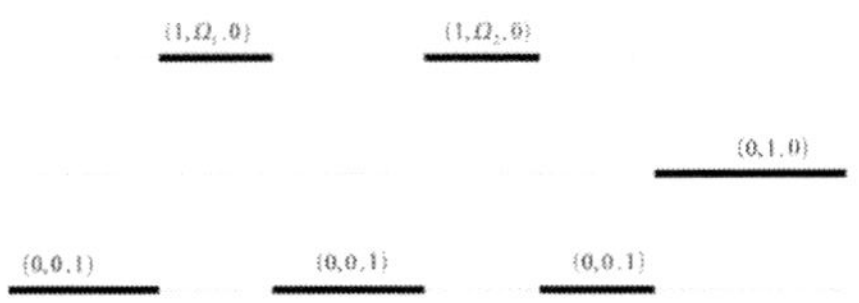

Figure 1. Rod structure of black di-ring with single angular momentum component.

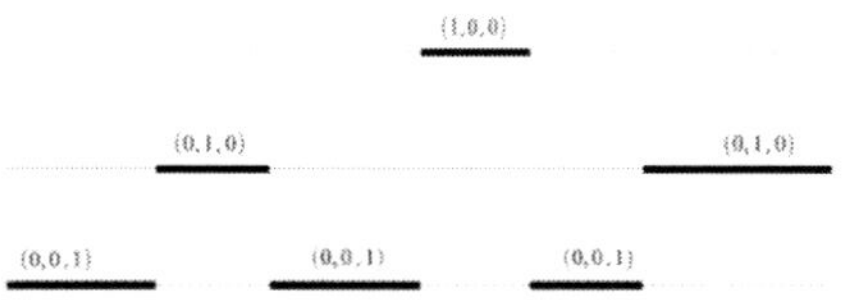

Figure 2. Rod structure of seed solution to generate a black di-ring by Bäcklund transformation.

2.2. Solution generating methods

To solve the five-dimensional Einstein equations for the axially symmetric vacuum spacetime with single spin, we use the fact that a part of the equations can be reduced to the Ernst equation in the four dimensions. A solution for the Ernst equation can be obtained from the Bäcklund transformation for an seed solution. To generate the black di-ring solution by the Bäcklund transformation [10], we start from the static seed solution whose rod structure is described in Fig.2. By the Bäcklund transformation, the finite spacelike rod with direction vector $(0, 1, 0)$ is transformed to the timelike rod with non-trivial direction vector $(1, \Omega_1, 0)$. In addition the trivial direction vector of finite timelike rod of seed solution is transformed to the non-trivial one. See [10] for the details of the methods and the exact expression of the solution.

The black di-ring solution was regenerated by the inverse scattering method [15]. In the analysis the seed solution denoted in Fig. 3 was used. To construct the black di-ring from this seed, two (anti-)solitons with trivial vectors are removed at first. Next two (anti-)solitons with non-trivial vectors are added at the same points. Another type of the seed solution can become a seed of the black di-ring. For example, we can construct the black di-ring from the seed solution whose rod structure is given by Fig.4. The expressions of these two black di-ring solutions are different each other. However it can be shown that these two solutions exactly coincide with each other. The expression by Evslin and Krishnan [15] is simpler than the others. We can easily make a correlation between the parameters of the solution constructed by the Bäcklund transformation and the one constructed by inverse scattering method from the seed of Fig 4.

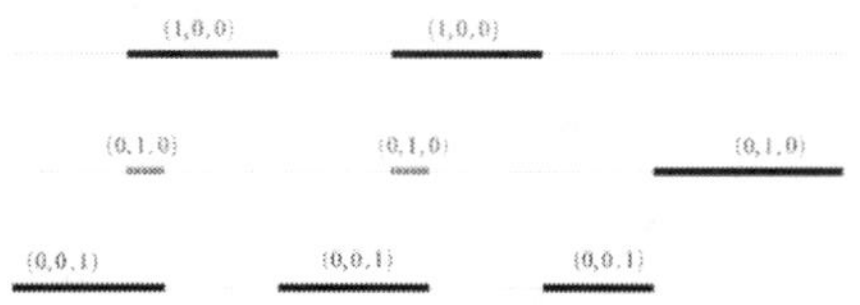

Figure 3. Rod structure of the seed metric to generate the black di-ring by the inverse scattering method in [15]. Two finite spacelike rods with direction vectors $(0,1,0)$ have negative line density.

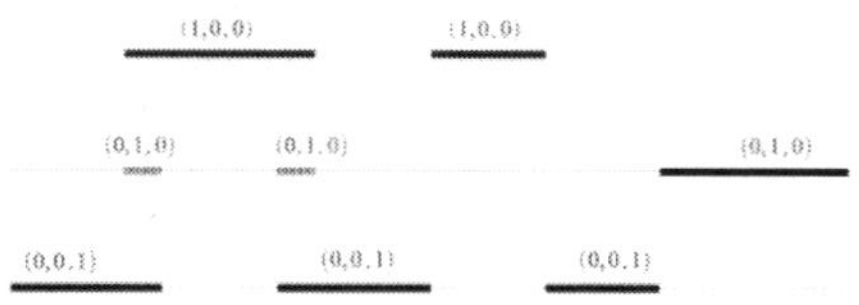

Figure 4. Rod structure of the seed metric to generate the black di-ring by the inverse scattering method. The position of the rod with negative line density is different from the seed of Evslin and Krishnan [15].

3. Phase structure of black di-ring

We can calculate physical variables of black di-ring from the exact expressions of the solution. To investigate the phase structure of the black di-ring, we define the following dimensionless variables,

$$j^2 = \frac{27\pi}{32G}\frac{J^2}{M^3}, \quad a_h = \frac{3}{16}\sqrt{\frac{3}{\pi}}\frac{A_h}{(GM)^{3/2}}, \quad \tau_i = \sqrt{\frac{32\pi}{3}}T_i(GM)^{1/2}. \tag{7}$$

Here M and J are the ADM mass and angular momentum of the black di-ring, respectively. A_h is a total area of the event horizons and T_i is a temperature of i's horizon.

To analyze the phase structure of black di-ring, we solve the two balance conditions to decide the appropriate parameters for the regular solution, e.g., six lengths of finite rods in Fig. 3 and plot the corresponding point in the phase diagram. Figure 5 shows a part of the plots for about 25000 sets of parameters. The black di-ring can become larger entropy than the black ring with the same mass and angular momentum. There is a black di-ring with zero angular momentum as similar as the black Saturn.

In Fig. 6, we plot the relation between the reduced angular momentum j and the reduced area a_h of the black di-ring whose inner ring has zero Komar angular momentum. In Fig. 7, we plot the relation between the reduced temperatures of outer ring τ_1 and inner ring τ_2 for the same black di-rings. We can see that there is a thermal equilibrium black di-ring whose temperature of both horizons are equivalent with each other. It can be shown that there is no solution whose angular velocities of event horizons are equivalence with each other in this plot.

In Fig. 8, we plot the relation between the reduced angular momentum j and the reduced area a_h of the black di-ring whose inner ring has zero angular velocity. In Fig. 9, we plot the relation between the reduced temperatures of outer ring τ_1 and inner ring τ_2 for the same black di-rings. We can see that there is a thermal equilibrium black di-ring whose temperature of both horizons are equivalent with each other. It can be shown that there is no solution whose angular velocities of event horizons are equivalent with each other in this plot.

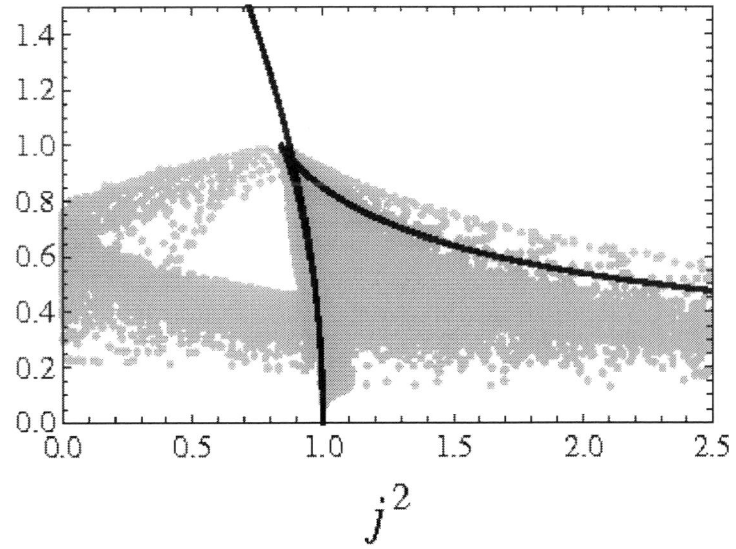

Figure 5. The distribution of black di-rings in the phase diagram. Black bold lines shows the behaviors of Myers-Perry black hole and black ring.

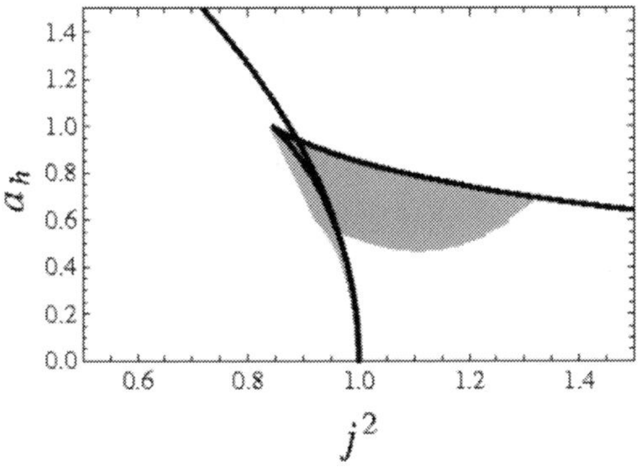

Figure 6. Distribution of black di-ring for zero inner angular momentum. Black bold lines shows the behaviors of Myers-Perry black hole and black ring.

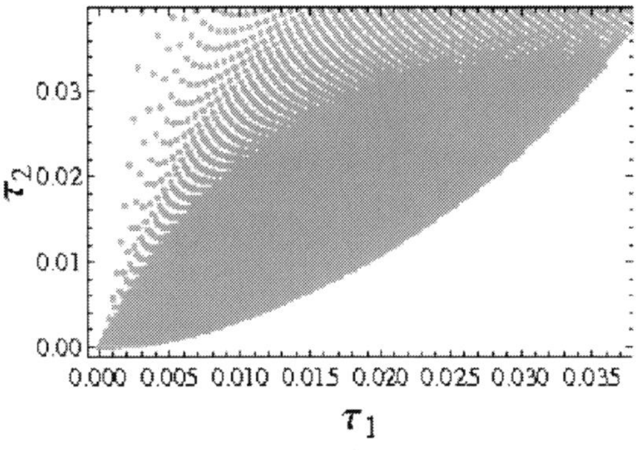

Figure 7. Relation of reduced temperatures of horizons for zero inner angular momentum.

4. Concluding remarks

We analyzed the physical properties of black di-ring. We numerically solved the balance conditions to decide the parameters set for the regular solution and plotted the corresponding point in the phase diagram. We reconfirmed the infinite non-uniqueness of the black di-ring. We confirmed that the black di-ring can satisfy the balance conditions even if the total angular momentum is zero. We also investigated the phase structures of black di-rings for two special cases. One is the black di-ring with zero inner angular momentum and the other is zero inner angular velocity. We showed that the temperatures of inner and outer ring can become equivalent for the both cases. The angular velocities of the horizons can not become equivalent with each other in these cases.

We have recently confirmed the existence of the thermodynamically equilibrium black di-ring

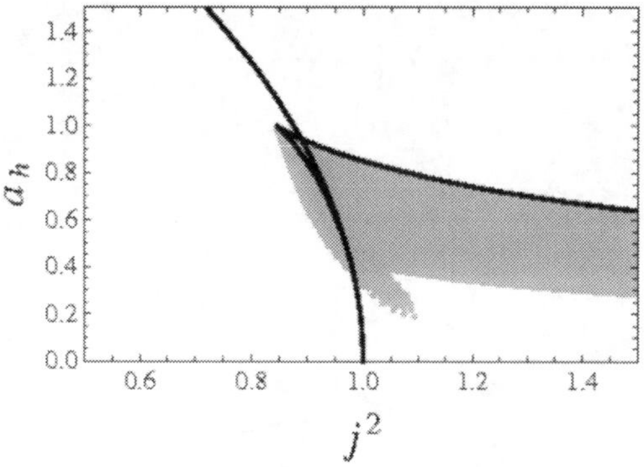

Figure 8. Distribution of black di-ring for zero inner angular velocity. Black bold lines shows the behaviors of Myers-Perry black hole and black ring.

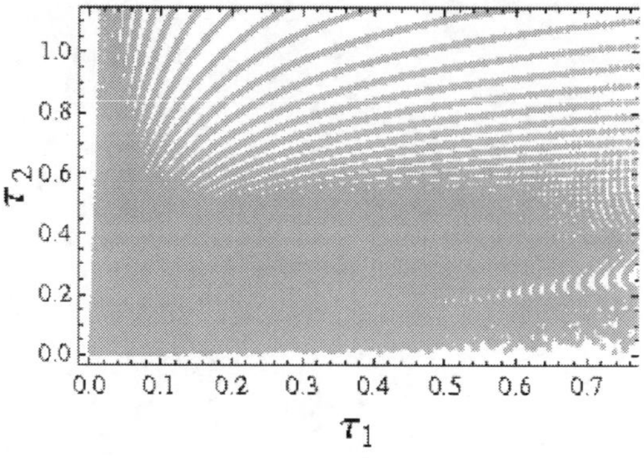

Figure 9. Relation of reduced temperatures of horizons for zero inner angular velocity.

whose temperatures and angular velocities of horizons are equivalent with each other. This configuration can not be expected from a simple superposition of black rings because fixing temperature and angular velocity uniquely determines a black ring. There seems little chance to impose equality of all temperature and all angular velocities on solutions of multi black rings without non-linearity of the solution. The existence of black di-ring in thermodynamical equilibrium may suggest the strong coupling of the rings in the configuration.

Acknowledgments

This work is partially supported by Grant-in-Aid for Young Scientists (B) (No. 20740143) from Japanese Ministry of Education, Science, Sports, and Culture.

References

[1] Emparan R and Reall H S 2002 *Phys. Rev. Lett.* **88** 101101 (*Preprint* hep-th/0110260)
[2] Emparan R and Reall H S 2006 *Class. Quantum Grav.* **23** R169 (*Preprint* hep-th/0608012)
[3] Myers R C and Perry M J 1986 *Ann. Phys.* **172** 304
[4] Mishima T and Iguchi H 2006 *Phys. Rev.* D **73** 044030 (*Preprint* hep-th/0504018)
[5] Figueras P 2005 *JHEP* **07** 039 (*Preprint* hep-th/0505244)
[6] Iguchi H and Mishima T 2006 *Phys. Rev.* D **73** 121501 (*Preprint* hep-th/0604050)
[7] Tomizawa S and Nozawa M 2006 *Phys. Rev.* D **73** 124034 (*Preprint* hep-th/0604067)
[8] Pomeransky A A and Sen'kov R A 2006 (*Preprint* hep-th/0612005)
[9] Elvang H and Figueras P 2007 *JHEP* **05** 050 (*Preprint* hep-th/0701035)
[10] Iguchi H and Mishima T 2007 *Phys. Rev.* D **75** 064018 (*Preprint* hep-th/0701043)
[11] Iguchi H and Mishima T 2007 *J. Phys. Conf. Ser.* **66** 012056
[12] Izumi K 2008 *Prog. Theor. Phys.* **119** 757–774 (*Preprint* 0712.0902)
[13] Elvang H and Rodriguez M J 2008 *JHEP* **04** 045 (*Preprint* 0712.2425)
[14] Elvang H, Emparan R and Figueras P 2007 *JHEP* **05** 056 (*Preprint* hep-th/0702111)
[15] Evslin J and Krishnan C 2009 *Class. Quantum Grav.* **26** 125018 (*Preprint* 0706.1231)
[16] Harmark T 2004 *Phys. Rev.* D **70** 124002 (*Preprint* hep-th/0408141)

Journal of Physics: Conference Series **229** (2010) 012011 doi:10.1088/1742-6596/229/1/012011

Towards gravitational-wave astronomy of core-collapse supernova explosion

Kei Kotake

Division of Theoretical Astronomy/Center for Computational Astrophysics, National Astronomical Observatory of Japan, 2-21-1, Osawa, Mitaka, Tokyo, 181-8588, Japan

E-mail: `kkotake@th.nao.ac.jp`

Abstract. We study properties of gravitational waves based on the three-dimensional (3D) simulations, which demonstrate the neutrino-driven explosions aided by the standing accretion shock instability (SASI). Pushed by evidence supporting slow rotation prior to core-collapse, we focus on the asphericities in neutrino emissions and matter motions outside the protoneutron star. By performing a ray-tracing calculation in 3D, we estimate accurately the gravitational waveforms from anisotropic neutrino emissions. In contrast to the previous work assuming axisymmetry, we find that the gravitational waveforms vary much more stochastically because the explosion anisotropies depend sensitively on the growth of the SASI which develops chaotically in all directions. Our results show that the gravitational-wave spectrum has its peak near $\sim$ 100 Hz, reflecting the SASI-induced matter overturns of $\sim O(10)$ ms. We point out that the detection of such signals, possibly visible to the LIGO-class detectors for a Galactic supernova, could be an important probe into the long-veiled explosion mechanism.

1. Introduction

The gravitational-wave astronomy is now becoming reality. In fact, significant progress has been made on the gravitational wave detectors, such as LIGO,VIGRO,GEO600,TAMA300, and AIGO with their international network of the observatories (e.g., [1] for a review). For the detectors, core-collapse supernovae especially in our Galaxy, have been supposed as one of the most plausible sources of gravitational waves (see, for example, [2, 3] for recent reviews).

Traditionally, most of the model calculations of gravitational waves (GWs) have focused on the bounce signals in the context of rotational and magnetorotational core-collapse (e.g., see [4, 5, 2, 3] and references therein). However recent stellar evolution calculations suggest that such rapid rotation assumed in most of the previous studies, albeit attracting much attention currently as a relevance to collapsar [6], is not canonical for progenitors of core-collapse supernovae with neutron star formations [7, 8]

In the case of the slowly rotating supernova cores, two other ingredients are expected to be important in the much later phases after core bounce, namely convective motions and anisotropic neutrino emissions. Thus far, various physical ingredients for producing asphericities and the resulting GWs in the post-bounce phase, have been studied such as the roles of precollapse density inhomogeneities [9, 10], moderate rotation of the iron core [11], g-mode oscillations of protoneutron stars (PNSs) [12], and the standing-accretion-shock-instability (SASI) [13, 14].

However, most of them have been based on two-dimensional (2D) simulations that assume axisymmetry. Then, the growth of SASI (e.g., [15, 16, 17, 18]) and the large-scale convection, both of which are now considered to generically develop in the post-bounce phase and to help the neutrino-driven explosions [20, 21], develop along the symmetry axis preferentially, thus suppressing the anisotropies in explosions. So far very few three-dimensional (3D) studies have been conducted [10, 19]. Moreover, neither the growth of the SASI nor its effects on the GWs have been studied yet because SASI is suppressed artificially owing to the limited computational domain [10] or to the early shock-revival [19], which has not been discovered by other supernova simulations.

In this contribution, we report the properties of the gravitational radiation based on the 3D simulations, which demonstrate the neutrino-driven explosions aided by SASI [22]. Supported by the evidence of the slow rotation prior to core-collapse [7, 8], we focus on the asphericities outside the protoneutron stars, which are produced by the growth of SASI. By performing a ray-tracing calculation in 3D [23], we estimate accurately the gravitational waveforms from anisotropic neutrino emissions. We show that the features of the gravitational waveforms are significantly different than the ones in the axisymmetric cases, which should tell us the necessity of the 3D supernova modelling.

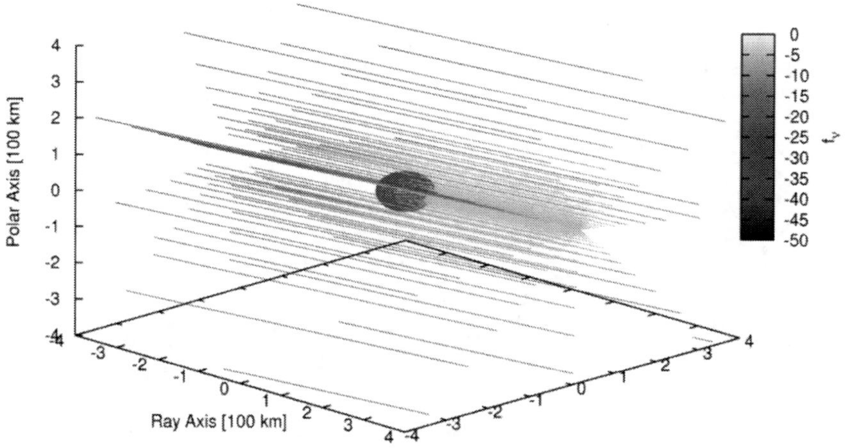

Figure 1. An example of ray-tracing of neutrinos for estimating $dl_\nu(\mathbf{\Omega})/d\Omega$ towards a given direction of $\mathbf{\Omega}$. The central region colored by red represents the surface of the protoneutron star (PNS) (located at 50 km in radius), which is the inner boundary of our computation. The color-scale on the rays indicates the values of f_ν, the neutrino occupation probability, which is calculated by the line integral of the neutrino transport equations. For this snapshot taken from our simulations, the higher values of f_ν are seen to come just in front of the PNS (yellow on the rays), while f_ν becomes smaller in the distant regions from the PNS.

2. Computation of Gravitational Waves from Anisotropic Neutrino Emission

Following the method in [24, 10], we derived the two modes of the GWs from anisotropic neutrino emissions as follows,

$$\begin{aligned} h_+ &= C\int_0^t \int_{4\pi} d\Omega'(1 + s(\theta')c(\phi')s(\xi) + c(\theta')c(\xi)) \times \\ &\quad \frac{(s(\theta')c(\phi')c(\xi) - c(\theta')s(\xi))^2 - s^2(\theta')s^2(\phi')}{[s(\theta')c(\phi')c(\xi) - c(\theta')s(\xi)]^2 + s^2(\theta')s^2(\phi')} \frac{dl_\nu(\Omega', t')}{d\Omega'}, \end{aligned}$$

and

$$\begin{aligned} h_\times &= 2C\int_0^t \int_{4\pi} d\Omega'(1 + s(\theta')c(\phi')s(\xi) + c(\theta')c(\xi)) \times \\ &\quad \frac{s(\theta')s(\phi')(s(\theta')c(\phi')c(\xi) - c(\theta')s(\xi))}{[s(\theta')c(\phi')c(\xi) - c(\theta')s(\xi)]^2 + s^2(\theta')s^2(\phi')} \frac{dl_\nu(\Omega', t')}{d\Omega'}, \end{aligned}$$

where $s(A) \equiv \sin(A), c(B) \equiv \cos B$, $C \equiv 2G/(c^4R)$ with G, c and R, being the gravitational constant, the speed of light, the distance of the source to the observer respectively, $dl_\nu/d\Omega$ represents the direction-dependent neutrino luminosity emitted per unit of solid angle into direction of $\mathbf{\Omega}$, and ξ is the viewing angle (e.g., [23]). For simplicity, we consider here two cases, in which the observer is assumed to be situated along 'polar' ($\xi = 0$) or 'equatorial'($\xi = \pi/2$) direction. To determine $dl_\nu/d\Omega$, we perform a ray-tracing calculation (see Figure 1). Since the regions outside the PNSs are basically thin to neutrinos, we solve the neutrino transport equations in a post-processing manner to estimate the neutrino anisotropy, in which neutrino absorptions and emissions by free nucleons, dominant outside the PNSs, are taken into account (see [23] for more details). To get the numerical convergence, we need to set 45000 rays to determine the luminosity for each given direction of $\mathbf{\Omega}$. The GW amplitudes from mass motions are extracted by using the standard quadrupole formula (e.g., [10]), and their spectra with the FFT techniques.

3. Results

Figure 2 depicts the 3D hydrodynamics features of SASI from the early phase of the non-linear regime of SASI (top left) until the shock break-out (bottom right) with the gravitational waveform from neutrinos inserted in each panel.

After about 100 ms, the deformation of the standing shock becomes remarkable marking the epoch when the SASI enters the non-linear regime (top left of Figure 2). At the same time, the gravitational amplitudes begin to deviate from zero. Comparing the top two panels in Figure 3, which shows the total amplitudes (top) and the neutrino contribution only (bottom), it can be seen that the gross structures of the waveforms are predominantly determined by the neutrino-originated GWs with the slower temporal variations ($\gtrsim 30 - 50$ ms), to which the GWs from matter motions with rapid temporal variations ($\lesssim 10$ ms) are superimposed. So, we first pay attention to the neutrino GWs in the following and discuss the importance of the matter GWs later in the spectrum analysis.

As seen from the top right through bottom left to right panels of Figure 2, the major axis of the growth of SASI is shown to be not aligned with the symmetric axis (:Z axis in the figure) and the flow inside the standing shock wave is not symmetric with respect to this major axis (see the first and third quadrant in Figure 2). This is a generic feature in the computed 3D models, which is in contrast to the axisymmetric case. The GW amplitudes from SASI in 2D showed an increasing trend with time due to the symmetry axis, along which SASI can develop preferentially [13, 23]. Free from such a restriction, a variety of the waveforms is shown to appear (see waveforms inserted in Figure 2). Furthermore, the 3D standing shock can also oscillate in

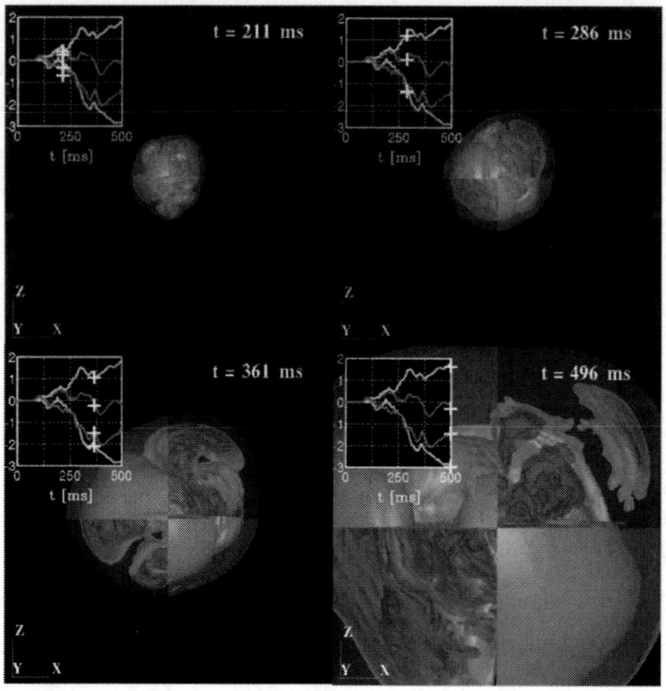

Figure 2. Four snapshots of the entropy distributions of a representative 3D supernova explosion model. The second and fourth quadrant of each panel shows the surface of the standing shock wave. In the first and third quadrant, the profiles of the high entropy bubbles (colored by red) inside the section cut by the ZX plane are shown. The side length of each plot is 1000km. The insets show the gravitational waveforms from anisotropic neutrino emissions, with '+' on each curves representing the time of the snapshot. Note that the colors of the curves are taken to be the same as the top panel of Figure 3.

all directions, which leads to the smaller explosion anisotropy than 2D. With these two factors, the maximum amplitudes seen either from the equator or the pole becomes smaller than 2D. On the other hand, their sum in terms of the total radiated energy are found to be almost comparable between 2D and 3D models, which is likely to imply the energy equipartition with respect to the spatial dimensions.

Figure 3 shows the gravitational waveforms for different luminosity models. To trigger explosions, we employed the so-called light-bulb approximation (see, e.g., [17]) and adjust the neutrino luminosities from the PNSs. The input luminosity for the two pair panels differs only 0.5%. Despite the slight difference in the luminosities, the waveforms of each polarization are shown to exhibit no systematic similarity when seen from the pole or equator. This is found to be due to the chaotic nature of SASI influenced by small differences.

The GW spectrum of model A, being a representative 3D explosion model, is presented in Figure 4. It can be seen that the neutrino GWs, albeit dominant over the matter GWs in the lower frequencies below $\sim$ 10 Hz, become very difficult to detect for ground-based detectors whose sensitivity is limited mainly by the seismic noises at such lower frequencies (e.g., [1] for a review about detectors). On the other hand, the gravitational-wave spectra from matter motions peak near $\sim$ 100 Hz, reflecting the growth of the $\ell = 2$ mode of SASI with timescales of $O(10)$ ms. Such signals from a Galactic supernova are probably within the detection limits of the LIGO-class detectors, and seem surely visible to the next-generation detectors such as

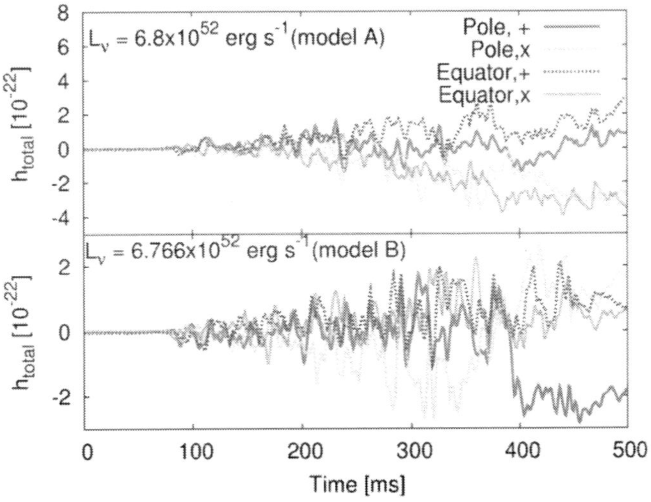

Figure 3. Gravitational waveforms from neutrinos (bottom) and from the sum of neutrinos and matter motions (top), seen from the polar axis and along the equator (indicated by 'Pole' and 'Equator') with polarization (+ or × modes) for models A and B. The distance to the SN is assumed to be 10 kpc.

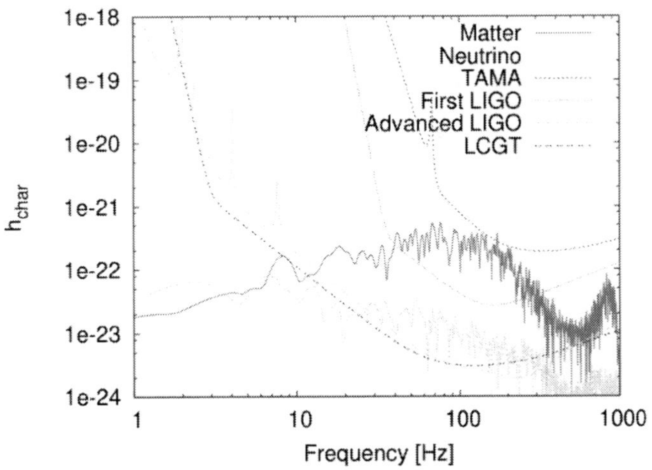

Figure 4. Characteristic gravitational-wave spectra from anisotropic neutrino emissions ('Neutrino') and matter motions ('Matter') with optimal detection limits of TAMA, the first LIGO, advanced LIGO, and LCGT for a supernova at a distance of 10 kpc.

the advanced LIGO and LCGT. It is noted that another peak in the GW spectra near $\sim$ 1kHz is from the rapidly varying matter motions of O(ms), induced by the local hydrodynamical instabilities. These gross properties in the GW spectra are found to be common to the other luminosity models. Thus the peak in the spectra near $\sim$ 100Hz is a characteristic feature for the 3D models studied in this paper.

4. Discussion

Here it should be noted that the simulations highlighted in this paper are only a very first step towards realistic 3D modelling of the supernova explosions. The approximations adopted in this paper, such as the replacement of the PNS by the fixed inner boundary and the light-bulb approach with the constant neutrino luminosity, should be improved. The exploration of these phenomena will require consistent simulations covering the entire stellar core and starting from gravitational collapse with neutrino transfer coupled to the 3D magnetohydrodynamics. Besides, the effects of rotation on the SASI are yet to be investigated [26], although the construction of the rotating steady accretion flows is still a major undertaking.

Bearing these caveats in mind, the stochastic nature of the GWs, which is illuminated here by the 3D models, is qualitatively new. It should be also mentioned that the GW spectra for the acoustic mechanism showed a very strong peak near $\sim$ kHz, reflecting the g-modes oscillations of the PNS [12], and are significantly different from the waveforms discussed here for the SASI-aided neutrino-driven models. This means that the successful detection of the GW signals will supply us a powerful tool to probe the explosion mechanism. Recently the next-generation detectors, by which the GWs studied here could be surely visible for a Galactic source, are planned to be on line soon (2014!) [27]. Towards the gravitational-wave astronomy to unveil the supernova mechanism, it is an urgent task for theorists to make precise predictions of the GW signals based on more sophisticated 3D supernova modelling.

It is gratefully acknowledged that the results presented in this article are the outcome of the collaborations with W. Iwakami, N. Ohnishi and S. Yamada. Numerical computations were carried on XT4 at CfCA of the NAOJ. This study was supported in part by the Grants-in-Aid for the Scientific Research from the Ministry of Education, Science and Culture of Japan (Nos. 19540309 and 20740150).

References

[1] Hough J and Rowan S 2005 *Journal of Optics A: Pure and Applied Optics* **7** 257
[2] Kotake K, Sato K and Takahashi K 2006 *Reports of Progress in Physics* **69** 971
[3] Ott C D 2009 *Class. Quantum Grav.* **26** 063001
[4] Kotake K, Yamada S and Sato K 2003 *Phys. Rev.* D **68** 044023
[5] Kotake K, Yamada S, Sato K, Sumiyoshi K, Ono H and Suzuki H 2004 *Phys. Rev.* D **69** 124004
[6] Woosley S E and Heger A 2006 *Astrophys. J.* **637** 914
[7] Heger A, Woosley S E and Spruit H C 2005 *Astrophys. J.* **626** 350
[8] Ott C D, Burrows A, Thompson T A, Livne E and Walder R 2006 *Astrophys. J. S.* **164** 130
[9] Burrows A and Hayes J 1996 *Phys. Rev. Lett.* **76** 352
[10] Müller E and Janka H-T 1997 *A.& A.* **317** 140
[11] Müller E, Rampp M, Buras R, Janka H-T and Shoemaker D H 2004 *Astrophys. J.* **603** 221
[12] Ott C D, Burrows A, Dessart L and Livne E 2006 *Phys. Rev. Lett.* **96** 201102
[13] Kotake K, Ohnishi N and Yamada S 2007 *Astrophys. J.* **655** 406
[14] Marek A, Janka H-T and Müller E 2009 *A. & A.* **496** 475
[15] Blondin J M, Mezzacappa A and DeMarino C 2003 *Astrophys. J.* **584** 971
[16] Scheck L, Plewa T, Janka H-T, Kifonidis K and Müller E 2004 *Phys. Rev. Lett.* **92** 011103
[17] Ohnishi N, Kotake K and Yamada S 2006 *Astrophys. J.* **641** 1018
[18] Foglizzo T, Galletti P, Scheck L and Janka H-T 2007 *Astrophys. J.* **654** 1006
[19] Fryer C L, Holz D E and Hughes S A 2004 *Astrophys. J.* **609** 288
[20] Bethe H A 1990 *Reviews of Modern Physics* **62** 801
[21] Marek A and Janka H-T 2009 *Astrophys. J.* **694** 664
[22] Iwakami W, Kotake K, Ohnishi N, Yamada S and Sawada K 2008 *Astrophys. J.* **678** 1207
[23] Kotake K, Iwakami W, Ohnishi N and Yamada S 2009 *Astrophys. J.* **704** 951
[24] Epstein R 1978 *Astrophys. J.* **223** 1037
[25] Burrows A, Livne E, Dessart L, Ott C D and Murphy J 2006 *Astrophys. J.* **640** 878
[26] Iwakami W, Kotake K, Ohnishi N, Yamada S and Sawada K 2008 *Preprint* arXiv:0811.0651
[27] Lantz B 2006 *APS Meeting Abstracts* http://www.ligo.caltech.edu/advLIGO/

Spanish Relativity Meeting (ERE 2009) IOP Publishing
Journal of Physics: Conference Series **229** (2010) 012012 doi:10.1088/1742-6596/229/1/012012

$H(z)$ diagnostics to discriminate dark energy models

Sergio del Campo[1], **Ramón Herrera**[1] **and Diego Pavón**[2★]

[1] Instituto de Física, Pontificia Uiversidad Católica de Valparaíso, Chile

[2] Departamento de Física, Universidad Autónoma de Barcelona, Spain

E-mail: diego.pavon@uab.es

Abstract. To extract the invaluable information on the nature of dark energy hidden in the history of the Hubble factor we must have at our disposal simple and practical criteria to be used on the data. Here, we propose some diagnostics based on $H(z)$ that may help discern whether the dark energy interacts with dark matter or not and whether it is phantom or quintessence in nature.

1. Introduction

The nature of the dark energy driving the present accelerated expansion of the Universe constitutes one of the biggest mysteries in cosmology. Within Einstein gravity dark energy may be simply a highly fine-tuned cosmological constant or something more complex, as an evolving scalar or a tachyon field with constant or variable equation of state, w. In either case, dark energy may interact, or not, with dark matter and if it does, the interaction may take a variety of expressions. Interaction with baryonic matter is rather unlikely in view of the severe constraints arising from local gravity experiments [1, 2].

Given the ample manifold of possibilities, diagnostics to tell apart between them should be welcome especially if they are simple and based on the less possible number of assumptions.

In a recent work Sahni *et al.* [3] introduced the redshift dependent function $Om(x) := (E^2(x)-1)/(x^3-1)$, with $E(x) := H(x)/H_0$ and $x := 1+z$, which for spatially flat Friedmann-Robertson-Walker cosmologies ($\Omega_{m0}+\Omega_{x0}=1$) reduces to $Om(x) = \Omega_{m0} + (1-\Omega_{m0})(x^{3(1+w)} - 1)(x^3-1)^{-1}$ with the interesting feature that for the ΛCDM model it yields $Om(x) = \Omega_{m0}$, whereas for quintessence fields ($-1 < w < -1/3$) and phantom fields ($w < -1$) it gives $Om(x) > \Omega_{m0}$ and $Om(x) < \Omega_{m0}$, respectively. Because of this diagnostic does not rely on the density parameters Ω_{m0} and Ω_{x0} it avoids the drawbacks inherent to our comparatively poor knowledge of them. In fact, it is expected to be useful when accurate data of the Hubble factor, $H(z)$, at different redshifts become available. This seems likely to come sooner than a substantial improvement in our knowledge of Ω_{m0} and Ω_{x0}. However, this criterion is to fail if dark energy happens to interact non-gravitationally with dark matter. This is interesting because while the concordance ΛCDM model fits most cosmological data, within statistical errors, rather well recent data from 185 supernovae Ia, at redshifts as low as $z < 0.08$ [4, 5], appear to cast serious doubts on the validity of the said model (even though w, when considered constant, results very close to -1) and may suggest some non-gravitational interaction between both dark components [6]. Further, the dynamics of nearly one hundred relaxed galaxy clusters lends substantial support to a continuous transfer from dark energy to dark matter [7], something at variance with the concordance model.

Here we propose several criteria to discriminate interacting from non-interacting dark energy models based solely on the history of the Hubble factor. The underlying assumption of these diagnostics is the validity Friedmann-Robertson-Walker metric, they do not presuppose any specific cosmological model. Obviously, accurate data about $H(z)$ are needed for the criteria to be useful. Regrettably, present data are too scarce and exhibit large bar errors. Nevertheless the situation is expected to improve in a significant manner comparatively soon (tough likely not up to the desired level) thanks to the Baryon Oscillation Spectroscopy Survey (BOSS) 5-years project [8] which aspires to measure the absolute cosmic distance scale and expansion rate with percent-level precision at redshifts $z < 0.7$ and $z \simeq 2.5$. Likewise, $H(z)$ will be obtained with unprecedent precision, in a considerable redshift range, (i) by measuring the time drift of the redshift of cosmological objects. In particular, Liske *et al* [9] suggested that thanks to Mc Vittie's formula, $\dot{z} \simeq H_0[1 + z - E(z)]$ [10], the next generation of Extremely Large Telescopes will be able to determine the redshift drift of high redshifts quasars in the interval $2 < z < 5$ using the absorption lines of the Lyman-α forest of these far away objects. (ii) By observationally determining the the time shift of the angle, θ, subtended by clusters of galaxies and using the formula $H(z) \simeq (1+z)(\mid \delta\theta/\theta \mid)^{\cdot}$ [11]. This may be achieved in the near future by means of micro-arcsecond astrometry [12].

We shall restrict ourselves to homogeneous and isotropic universes with flat space sections dominated by baryons, cold dark matter and dark energy (subscripts b, c, and x, respectively) satisfying the conservation equations

$$\dot{\rho}_b + 3H\rho_b = 0\,, \tag{1}$$

$$\dot{\rho}_c + 3H\rho_c = Q\,, \tag{2}$$

$$\dot{\rho}_x + 3H(1+w)\rho_x = -Q\,, \tag{3}$$

where Q stands for the interaction (coupling) term. In what follows, the relationship $\rho_b + \rho_c = \rho_m$ is to be understood. It is worthy of note that when $Q > 0$ -as we shall consider throughout- the coincidence problem is alleviated since ρ_c decreases more slowly with expansion and ρ_x more quickly, and it may even be solved in full [13, 14]. This is true no matter whether the dark energy corresponds to a scalar or tachyonic field or a Chaplygin gas [15]. Besides, if Q were negative, the second law of thermodynamics could be violated [16] and one of the energy densities could become negative either at high or low redshifts.

Lacking a fundamental theory the quantity Q cannot be derived from first principles. Nevertheless, we may surmise likely expressions for it by noting that Q must be small[1] (at least lower than $3H\rho_m$) and depend on the energy densities multiplied by a quantity with units of inverse of time -the natural choice being the Hubble factor. Therefore, we will consider that $Q = Q(H\rho_c,\, H\rho_x)$. By power law expanding this function and keeping the first term only, one follows $Q \simeq \epsilon_c\, H\rho_c + \epsilon_x\, H\rho_x$. Given the absence of information about the coupling, it appears advisable to work with just one parameter rather than two. Thence, three different options arise: $\epsilon_c = 0$, $\epsilon_x = 0$, and $\epsilon_c = \epsilon_x$. It should be noted that these expressions are obtainable from the scalar-tensor theory of gravity of Kaloper *et al* [17] -see e.g. [18, 19]. Below we shall consider successively three expressions for Q with $0 < \epsilon << 1$, while keeping $w =$ constant, and look for diagnostics -based in the history of $H(z)$- that may discriminate interacting from non-interacting models.

Interactions terms of the type $Q = \Gamma_c\, \rho_c + \Gamma_x\, \rho_x$, where the Γ_i $(i = c, x)$ quantities stand for constant rates, have also been proposed in the literature [20]. Unfortunately, either ρ_c or ρ_x becomes negative at early or late times and, in general, they cannot solve the coincidence

[1] If it were large and positive, dark matter would dominate the expansion today. On the other hand, if Q were large and negative, the Universe would have been dominated by dark energy practically from the outset.

problem. This is why we do not study them here (however, diagnostics for these cases can be found in [21]).

2. Diagnostics

In this section we set up criteria to tell apart interacting models with $Q = Q(H\rho_c, H\rho_x)$ from non-interacting models and from decaying cosmological constant (DCC) models [22]. In all the cases w is assumed constant.

2.1. Interaction term proportional to $H\,\rho_c$

We begin by considering that the interaction term takes the form

$$Q = 3\epsilon H \rho_c\,, \tag{4}$$

where the factor 3 was introduced for mathematical convenience.

By integrating (1)-(3) and using Friedmann's equation, $3H^2 = \kappa^2(\rho_b + \rho_c + \rho_x)$, one follows

$$E^2(x) = \Omega_{b0}\, x^3 \,+\, \Omega_{c0}\, x^{3(1-\epsilon)} \,+\, \Omega_{x0}\, x^{3(1+w)} \,+\, \left(\frac{\epsilon}{\epsilon + w}\right) \Omega_{c0} \left[x^{3(1+w)} - x^{3(1-\epsilon)}\right]. \tag{5}$$

Consequently,

$$\begin{aligned} \frac{dE^2(x)}{dx^3} &= \Omega_{b0} + \Omega_{c0}(1-\epsilon)x^{-3\epsilon} + \Omega_{x0}(1+w)x^{3w} \\ &+ \left(\frac{\epsilon}{\epsilon + w}\right) \Omega_{c0} \left[(1+w)x^{3w} \,-\, (1-\epsilon)x^{-3\epsilon}\right], \end{aligned} \tag{6}$$

and

$$\begin{aligned} \frac{d^2E^2(x)}{d(x^3)^2} &= -(1-\epsilon)\,\Omega_{c0}\, x^{-3(1+\epsilon)} \,+\, w(1+w)\,\Omega_{x0}\, x^{3(w-1)} \\ &+ \left(\frac{\epsilon}{\epsilon + w}\right) \Omega_{c0} \left[w(1+w)x^{3(w-1)} \,+\, \epsilon(1-\epsilon)x^{-3(1+\epsilon)}\right]. \end{aligned} \tag{7}$$

In view of the hypothesis $0 < \epsilon << 1$ and the fact that observation reveals that $|1+w| << 1$, the right hand side of (6) is dominated by its two first terms. Likewise, the first term dominates the right hand side of (7). Therefore, for not large redshifts (i.e., $z < 3$) the normalized Hubble function, $E^2(x)$, is growing and concave. Thus, its shape will not tell a decaying energy field (with $w \neq -1$) from non-interacting quintessence or phantom models with constant equation of state. However, discrimination will likely come at higher redshifts since the first derivative may tell interacting from non-interacting cosmologies depending on whether $dE^2(x)/dx^3$ tends to Ω_{b0} or Ω_{c0} at large redshifts. Since, observationally, Ω_{b0} and Ω_{m0} are separated by a non-small gap (the latter is about six or seven times larger than the former) this may serve to discriminate interacting from non-interacting models.

It should be noted that the $Om(x)$ diagnostic would mistake an interacting dark energy model for a noninteracting quintessence (phantom) model if applied at low (high) redshifts.

Recalling that Ω_{b0} is reasonably well determined by nucleosynthesis measurements[2], another criterion can be established by means of the ratio

$$R_\Delta := \frac{\left[\Delta(x_1, x_2) \,-\, \Omega_{b0}\,(x_2^3 - x_1^3)\right]_A}{\left[\Delta(x_1, x_2) \,-\, \Omega_{b0}\,(x_2^3 - x_1^3)\right]_B}\,, \tag{8}$$

[2] $0.036 \leq \Omega_{b0} \leq 0.047$, see, e.g., Ref. [23].

where $\Delta(x_1, x_2) := E^2(x_2) - E^2(x_1)$. In the present case (8) reduces to

$$R_\Delta \simeq \frac{x_2^{3(1-\epsilon_A)} - x_1^{3(1-\epsilon_A)}}{x_2^{3(1-\epsilon_B)} - x_1^{3(1-\epsilon_B)}} . \tag{9}$$

In DCC models an extra $(1 - \epsilon_B)/(1 - \epsilon_A)$ factor appears multiplying the right hand side [21]. Thanks to this, we might hope to discriminate DCC models from interacting dark energy models with w close to the cosmological constant value, -1, and also discriminate models featuring the interaction (4).

2.2. *Interaction term proportional to $H\,\rho_x$*

We now assume that the interaction term takes the form

$$Q = 3\epsilon H \rho_x \tag{10}$$

which may explain the non-vanishing temperature of the gas of sterile neutrinos [24, 25].

Then,

$$E^2(x) = x^3 - \Omega_{x0}\left(\frac{w}{w+\epsilon}\right)\left[x^3 - x^{3(1+w+\epsilon)}\right] . \tag{11}$$

For $w = -1$ (i.e., DCC with interaction given by Eq. (10)), it reduces to

$$E^2(x) = x^3 - \frac{\Omega_{\Lambda 0}}{1-\epsilon}\left[x^3 - x^{3\epsilon}\right] . \tag{12}$$

Again, $E^2(x)$ is growing and concave whence its graph will tell neither DCC models with interaction term given by (10) from DCC models with interaction term given by (4) nor these models from non-interacting quintessence models with $w =$ constant.

Expression (12) will help constrain ϵ without need of knowing $\Omega_{\Lambda 0}$ since the ratio

$$\frac{[E^2(x) - x^3]_{x=x_1}}{[E^2(x) - x^3]_{x=x_2}} = \frac{x_1^3 - x_1^{3\epsilon}}{x_2^3 - x_2^{3\epsilon}} \tag{13}$$

does not depend on that quantity.

Unlike the previous case, as inspection of the right hand side of Eq. (11) shows, the behavior at high redshifts of $dE^2(x)/dx^3$ will not discriminate interacting from non-interacting cosmological models. For this to be feasible ϵ should take unrealistic high values.

By deriving (11) twice we get $d^2E^2(x)/d(x^3)^2 = \Omega_{x0}\, w(1 + w + \epsilon)\, x^{3(w+\epsilon-1)}$. If eventually this quantity is accurately determined, it will enable us to set useful constraints on w and ϵ since its ratio for two interacting dark energy models (say A and B) does not depend on the Ω parameters,

$$\frac{d^2E^2/d(x^3)^2|_A}{d^2E^2/d(x^3)^2|_B} = \frac{w_A(1 + w_A + \epsilon_A)}{w_B(1 + w_B + \epsilon_B)} \frac{x^{3(w_A+\epsilon_A)}}{x^{3(w_B+\epsilon_B)}} . \tag{14}$$

Once again $E^2(x)$, given by (11), is a growing function of redshift, concave for quintessence fields and convex for phantom fields. Accordingly, the shape of $E^2(x)$ will tell coupled quintessence fields from coupled phantom fields but not from non-interacting dark energy fields.

2.3. Interaction term proportional to $H\,(\rho_c + \rho_x)$

Assuming the interaction term

$$Q = 3\epsilon H(\rho_c + \rho_x)\,, \tag{15}$$

widely considered in the literature -see, e.g. [7, 13, 14, 20, 26]-, we get

$$E^2(x) = \Omega_{b0}\,x^3 + \frac{1}{2\,\epsilon}\left\{\Omega_{C_1}[2\epsilon - (A+B)] - \Omega_{C_2}(A-B)\,x^{-3B}\right\}x^{\gamma_1} + \Omega_{C_2}\,x^{\gamma_2}\,, \tag{16}$$

where the constants appearing in this expression are all positive-definite and depend on ϵ, w, ρ_{c0}, and ρ_{x0} only [21].

Further, E^2 is a growing function of x^3, concave for quintessence and convex for phantom (Fig.1) which will help discriminate phantom from quintessence dark energy in interacting models with Q given by Eq. (15). However, as is apparent, for $x^3 \geq 5$ the graphs show a nearly straight line behavior whereby one must focus on redshifts below, say, 1.7.

Inspection of the right hand side of Eq. (16) shows that for interacting models ($\epsilon > 0$) the derivative $dE^2(x)/dx^3$ tends to Ω_{b0} at high redshifts. Thus if observation reveals that behavior, it would be suggestive of interaction.

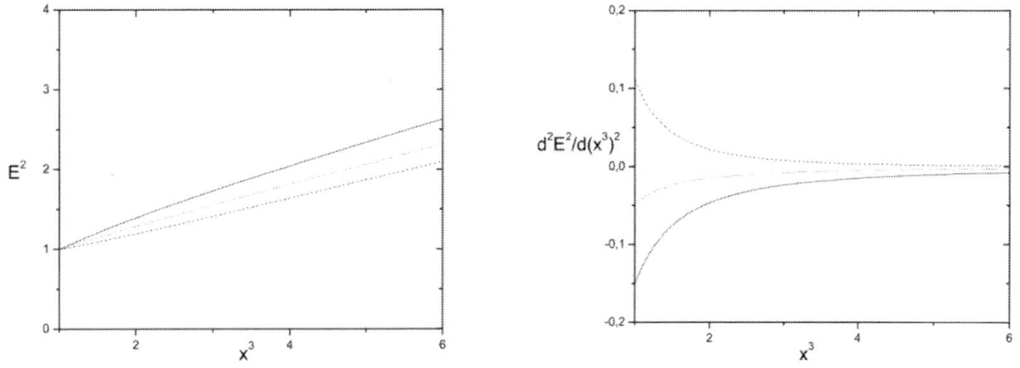

Figure 1. Left panel: Graphs of E^2 vs. x^3 as given by Eq. (16) for three values of w. Right panel: Graphs of the second derivative of E^2 with respect to x^3 for the same three values of w. Solid, dotted, and dashed lines are for $w = -0.8$, -1.0, and -1.2, respectively. Note that only phantom models present convex curves for reasonable values of the parameters. In drawing the graphs we took $\Omega_{b0} = 0.04$, $\Omega_{c0} = 0.26$, $\Omega_{x0} = 0.70$, and $\epsilon = 0.05$.

3. Discussion

To extract the invaluable information hidden in the history of the Hubble factor on the nature of dark energy one must first set up simple and practical criteria to be used on the data. We proposed several criteria, of notable mathematical simplicity, based on an accurate knowledge of the said history -something we might hope to have at our disposal comparatively soon.

Specifically if $d^2E^2/d(x^3)^2$ results positive (negative), then the dark energy is of phantom type (either quintessence or a decaying cosmological constant). Only if the dark energy is a conserved cosmological constant, the graph of E^2 vs. x^3 will result in a straight line.

To discern whether dark energy is interacting or not the behavior of $dE^2(x)/dx^3$ at high redshifts must be studied. If this derivative tends to the present value of the fractional baryon density, Ω_{b0}, then there will be grounds to believe that is it interacting with the coupling term Q given either by Eqs. (4) or (15). If it tends to Ω_{m0}, then we may strongly suspect that it is not interacting; and, in particular, if it coincides with Ω_{m0} independently of redshift, then it will most likely be a conserved cosmological constant -see [21] for details. Finally, if the redshift function $dE^2(x)/dx^3$ does not tend to any of these two quantities, we may conclude that either dark energy is interacting with Q given by Eq.(10) or some other law not considered here, or that in reality w is not a constant but a function of redshift. Obviously, this would complicate matters very much because it would call for the introduction of further parameters in the analysis.

We restricted ourselves to constant w and vanishing spatial curvature. While this automatically limits the scope of our work we feel it does not do it seriously, at least no more than otherwise since it would introduce additional unknown parameters that would compromise the said scope in other ways.

Acknowledgments

This work was supported from the "Comisión Nacional de Ciencias y Tecnología" (Chile) through the FONDECYT Grant No. 1070306 (SdC), No. 1090613 (RH and SdC) and the Spanish Ministry of Education and Science under Grant FIS2006-12296-C02-01. D P acknowledges "FONDECYT-Concurso incentivo a la cooperación internacional" No. 7090081.

References

[1] Peebles P J E and Ratra B 2003 *Rev. Mod. Phys.* **75** 559
[2] Hagiwara K *et al* 2002 *Phys. Rev.* D **66** 010001
[3] Sahni V, Shafieloo A and Starobinsky A A 2008 *Phys. Rev.* D **78** 103502
[4] Hicken M *et al* 2009 *Astrophys. J.* **700** 331
[5] Hicken M *et al* 2009 *Astrophys. J.* **700** 1097
[6] Shafieloo A, Shani V and Starobinsky A A 2009 *Phys. Rev.* D **80** 101301(R)
[7] Abdalla E, Abramo L R and de Souza J C C 2009 Signature of the interaction between dark energy and dark matter in observations *Preprint* arXiv:0910.5236 [gr-qc]
[8] Schlegel D, White M and Eisenstein D 2009 The Baryon Oscillation Spectroscopic Survey: Precision measurements of the absolute distance scale *Preprint* arXiv:0902.4680 [astro-ph.CO]
[9] Liske J *et al* 2008 *Mon. Not. R. Astron. Soc.* **386** 1192
[10] Mc Vittie G C 1962 *Astrophys. J.* **136** 334
[11] Zhang H and Zhu Z-H 2009 *J. High Energy Phys.* (in press)
[12] Brown AGA 2008 Getting ready for the micro-arcsecond era *Preprint* arXiv:0802.3568 [astro-ph]
[13] del Campo S, Herrera R and Pavón D 2009 *Phys. Rev.* D **78** 021302(R)
[14] del Campo S, Herrera R and Pavón D 2009 *J. Cosmol. Astropart. P.* JCAP01(2009)020
[15] Chimento L P, Jakubi A S and Pavón D 2003 *Phys. Rev.* D **67** 087302
[16] Pavón D and Wang B, 2009 *Gen. Rel. Grav.* **41** 1
[17] Kaloper N and Olive K A 1988 *Phys. Rev.* D **57** 811
[18] Curbelo R, González T and Quirós I 2006 *Class. Quantum Grav.* **23** 1585
[19] Zhang H and Zhu Z-H 2006 *Phys. Rev.* D **73** 043518
[20] Caldera-Cabral G, Maartens R and Ureña-López L A 2009 *Phys. Rev.* D **79** 063518
[21] del Campo S, Herrera R and Pavón D 2009 $H(z)$ diagnostics on the nature of dark energy *(work in progress)*
[22] Overduin J M and Cooperstock F I 1998 *Phys. Rev.* D **58** 043506
[23] Olive K A 2003 TASI lectures on dark matter *Preprint* arXiv:astro-ph/0301505.
[24] Hansen S H, Lesgourgues J, Haehnelt M G, Matarrese S and Riotto A 2005 *Phys. Rev.* D **71** 063534
[25] Zhou J, Wang B, Pavón D and Abdalla E 2009 *Mod. Phys. Lett.* A **24** 1689
[26] Olivares G, Atrio-Barandela F and Pavón D 2005 *Phys. Rev.* D **71** 063523

Penrose's Weyl curvature hypothesis and conformally-cyclic cosmology

Paul Tod
Mathematical Institute, St Giles', Oxford OX1 3LB, UK
E-mail: tod@maths.ox.ac.uk

Abstract. I discuss two ideas of Roger Penrose and the relation between them: his Weyl curvature hypothesis from thirty years ago and his recent idea for a conformally-cyclic universe, that is a cosmological model with positive cosmological constant in which the conformal metric is cyclic - it can be continued through the future null infinity of one aeon to the big bang of the following aeon.

1. Introduction

Penrose's Weyl curvature hypothesis was proposed by him in [1] thirty years ago. Since that time, it has provided the motivation for a series of works in mathematical general relativity, so that one has a good knowledge of cosmological models satisfying the hypothesis. More recently, he has described his idea of a conformally-cyclic cosmology in numerous lectures over the past five years or so, with the fullest account of this circle of ideas so far in his recent article [2]. In this talk, I shall sketch a review of both sets of ideas, emphasising how the second fills a gap in the first.

2. The Weyl Curvature Hypothesis

The starting point for the WCH is the observation that the Big Bang singularity was very special. Penrose argues that the universe must have been in a low entropy state initially in order for there now to be a second law of thermodynamics. Assuming, as is commonly done, that the matter content of the universe was in thermal equilibrium near to the bang, and therefore in a state of high entropy, one needs the contribution to the entropy from the rest of physics, which means the contribution form gravity or equivalently geometry, to be low. That is, the geometry must be highly ordered.

Penrose gives a quantitative form to this argument as follows: the entropy of the contents of the past light cone of an observer now can be estimated as 10^{88} (this estimate from the earlier article [1] is revised up to 10^{101} in [2] to take account of the large black holes now known to exist in galactic centres); however if all the matter in this past light cone was in black holes the entropy would rise to 10^{123}, which is probably the maximum possible; interpreting entropy as the logarithm of phase-space volume, this means that the actual universe lies in an unimaginably small proportion $10^{10^{101}}/10^{10^{123}}$ of the available phase-space volume, which is very special indeed.

There is no generally accepted definition of gravitational entropy but Penrose argues that low gravitational entropy must mean small Weyl tensor. Why the Weyl tensor? One answer is that this is the 'free gravitational field' since, in GR, the Ricci tensor is fixed point-wise by the matter. Another is that, according to the BKL picture, singularities of the vacuum Einstein equations, which are necessarily singularities of the Weyl tensor, can be extraordinarily complicated, probably space-time chaotic. To avoid this chaos the Big Bang, while being a cosmological or curvature singularity, needs to have a non-singular Weyl tensor.

In [1], as a result of these arguments, Penrose proposed that *the Weyl tensor vanishes at the Big Bang singularity*. A weaker hypothesis which he allowed is simply that the Weyl tensor is finite there. In either case one has the problem of identifying finiteness of some components of the Riemann tensor, namely those corresponding to the Weyl tensor, simultaneously with other components, belonging to the Ricci tensor, being singular.

One way to deal with this problem is to make a stronger hypothesis, based on the conformal rescaling properties of the curvature. If two metrics $\breve{g}_{ab}$ and g_{ab} are conformally related:

$$\breve{g}_{ab} = \breve{\Omega}^2 g_{ab}, \tag{1}$$

then their Weyl tensors are related by

$$\breve{C}_{abc}{}^{d} = C_{abc}{}^{d}. \tag{2}$$

Now suppose that $\breve{\Omega} = 0$ on a smooth space-like hypersurface Σ in the manifold M with metric g_{ab}, then the metric $\breve{g}_{ab}$ on the submanifold $\breve{M}$ of M on which $\breve{\Omega} > 0$ will have a curvature singularity at Σ, but one at which the Weyl tensor is finite, by virtue of (2). We shall call a cosmological singularity with this character a *conformal gauge singularity* [3] since the conformal metric is regular and the singularity is due to the choice of metric in the conformal class. Singularities like this have also been called *isotropic* [4] and *conformally compactifiable* [5].

To investigate the existence of cosmological models with conformal gauge singularities, one may seek a Cauchy problem for the unphysical metric g_{ab} and conformal factor $\breve{\Omega}$ with data at Σ, whose solution gives a solution $\breve{g}_{ab}$ of the Einstein equations with a suitable matter source in $\breve{M}$. This programme has been carried out with the following results: such a Cauchy problem is well-posed for

- perfect fluids with equation of state $p = (\gamma - 1)\mu$, $1 < \gamma \leq 2$, and for dust assuming also spatial homogeneity [5] (see the earlier work [6] for $\gamma = 4/3$); the data consist of just the unconstrained 3-metric of Σ; if the space-time Weyl tensor vanishes at Σ then it vanishes throughout the space-time, which is then FLRW;
- massless Einstein-Vlasov [7] (see [8] for the spatially-homogeneous case); the data consist of just the initial distribution function, a non-negative function of six variables subject to an integral 'vanishing dipole' condition; now the space-time Weyl tensor can vanish at Σ but be nonzero in the space-time;
- spatially-homogeneous metrics with cosmological constant and either perfect fluids with equation of state $p = (\gamma - 1)\mu$, $1 \leq \gamma \leq 2$, or massive Einstein-Vlasov; data as before [9].

There is also some evidence, without a complete existence proof, for well-posedness with other matter models including scalar fields and Einstein-Yang-Mills-Vlasov [10].

This work provides many examples of cosmological models which satisfy the Weyl curvature hypothesis, based on the stronger assumption of a cosmological singularity which is a conformal gauge singularity. One may therefore ask for conditions on a cosmological singularity for it to be a conformal gauge singularity. This is a difficult question but there exist partial answers [3], [11]. One needs more tools from conformal geometry, in particular the notions of conformal geodesic and a conformally-invariant propagation along conformal geodesics. The theory can be succinctly expressed in terms of the conformally-invariant tractor calculus, [12]. The result is as follows: given an incomplete conformal geodesic γ running into a space-time singularity which is not 'conformally-infinitely remote' (in a way which can be made precise), then there will exist a conformal rescaling of a neighbourhood U of a final segment of γ for which the rescaled metric extends, with any desired degree of smoothness, to a larger neighbourhood *provided* the tractor curvature and its tractor derivatives up to a certain order are bounded. The tractor curvature essentially consists of the Weyl curvature and its divergence, so that the theorem takes the form that a conformal extension of a neighbourhood of γ is possible given suitable boundedness conditions on the Weyl tensor and its derivatives.

This is a *local* extension theorem. One would like to have a *global* extension theorem, yielding an extension through all space-time singular points at once, given some boundedness conditions on the Weyl tensor and its derivatives. This has been done in the case of spatial homogeneity [11] but is outstanding in general.

3. Conformally-cyclic cosmology

The starting point for the second set of ideas under review is the emerging consensus that the expansion of the universe is accelerating in a way consistent with a positive cosmological constant, $\Lambda > 0$. Assuming there is no recollapse, the universe will be asymptotically-de Sitter in the remote future in the sense that the physical space-time metric, which we shall now write as $\hat{g}_{ab}$ with the physical space-time as $\hat{M}$, will have a conformal rescaling

$$\hat{g}_{ab} = \hat{\Omega}^2 g_{ab}$$

in terms of an unphysical metric g_{ab} on unphysical M, and a conformal factor $\hat{\Omega}$ which grows exponentially with physical proper-time t,

$$\hat{\Omega} \sim \exp Ht, \text{ where } \Lambda = 3H^2.$$

The unphysical metric is smooth at future null infinity $\mathcal{I}^+$, which is the space-like surface at $t = \infty$ or at $\hat{\Omega}^{-1} = 0$ in M. It necessarily follows that the Weyl tensor $C_{abc}{}^d$ of M is zero at $\mathcal{I}^+$, but the 'spin-two field' $K_{abc}{}^d = \hat{\Omega} C_{abc}{}^d$ has a nonzero limit there, representing gravitational radiation in the remote future.

In [2], Penrose presents a picture of the very remote future with positive Λ as a physical world in which proper-time plays no role. He remarks that all stars will have completed their evolution and either collapsed to form black holes or been swallowed up by the massive black holes at the centres of galaxies. Black holes themselves will eventually decay by the Hawking process and the content of the universe will very largely be just electromagnetic and gravitational radiation, both of them massless fields. To complete the picture of a world from which proper-time has vanished, Penrose hypothesises that all massive particles eventually either decay to radiation or lose their mass in some unspecified way. Of course according to some theories protons do decay, but electrons present a challenge to this view. However, this doesn't seem to me to be crucial for the next step, which is Penrose's 'outrageous' proposal (his adjective): the stronger version of the Weyl curvature hypothesis proposes that the space-time metric can be rescaled to extend the conformal structure through the Big Bang, which is then a surface, say Σ, at which the Weyl curvature vanishes; a positive Λ leads to a universe in which the space-time metric can be rescaled to extend the conformal metric through $\mathcal{I}^+$, which is then another surface, Σ' at which the Weyl curvature vanishes; these surfaces, with their conformal metrics, can then be identified. One arrives at a description of the universe in terms of a conformal manifold with a conformal metric which is cyclic – a conformally-cyclic cosmology. The physical or space-time metric is not cyclic in the sense of being smooth through Σ, as the conformal factor giving the physical metric from the unphysical cyclic metric runs from zero at the Big Bang to infinity at $\mathcal{I}^+$, returning to zero again for the next cycle, which Penrose calls an aeon.

Journal of Physics: Conference Series **229** (2010) 012013 doi:10.1088/1742-6596/229/1/012013

Note that the electromagnetic and gravitational radiation present in the remote future in one aeon propagate through Σ as primordial radiation in the next aeon. There will also be a scalar field after the Big Bang. This is mentioned in [2] but the following argument may not be exactly what Penrose had in mind: if the physical metrics before and after are $\hat{g}_{ab}$ and $\check{g}_{ab}$, while the unphysical cyclic metric is g_{ab} so that

$$\hat{g}_{ab} = \hat{\Omega}^2 g_{ab}, \quad \check{g}_{ab} = \check{\Omega}^2 g_{ab},$$

then the corresponding physical (before and after) Ricci tensors are related by

$$\Phi^2 \hat{R}_{ab} = \Phi^2 \check{R}_{ab} + 2\Phi \check{\nabla}_a \Phi_b - 4\Phi_a \Phi_b + \check{g}_{ab}(\Phi \check{\Box} \Phi + |\check{\nabla}\Phi|^2), \tag{3}$$

where $\Phi = \hat{\Omega}\check{\Omega}^{-1}$. The trace of (3) gives

$$\check{\Box}\Phi + \frac{1}{6}\check{R}\Phi = \frac{1}{6}\hat{R}\Phi^3, \tag{4}$$

so that, assuming constant $\check{R}$ and $\hat{R}$, Φ satisfies the conformally-invariant wave equation (and diverges at Σ).

Now suppose the matter content in $\hat{M}$ before Σ is radiation plus cosmological constant so that

$$\hat{R}_{ab} = -\hat{T}_{ab} + \frac{1}{4}\hat{R}\hat{g}_{ab},$$

where $\hat{R}$ is constant, and

$$\hat{T}_{ab} = \frac{1}{3}\hat{\rho}(4\hat{u}_a\hat{u}_b - \hat{g}_{ab}).$$

The rescaling which preserves the conservation equation is

$$\check{T}_{ab} = \Phi^2 \hat{T}_{ab},$$

which again has the radiation form with, in particular, $\check{\rho} = \Phi^4 \hat{\rho}$. Then (3) becomes

$$-\check{T}_{ab} = \Phi^2 \check{R}_{ab} + 2\Phi \check{\nabla}_a \Phi_b - 4\Phi_a \Phi_b + \check{g}_{ab}(|\check{\nabla}\Phi|^2 - \frac{1}{6}\check{R}\Phi^2 - \frac{1}{12}\hat{R}\Phi^4). \tag{5}$$

With our assumptions, this is the Einstein field equation in $\check{M}$, where Φ satisfies the conformally-invariant wave equation (4), and then the right-hand-side in (5) is recognisable as the 'new improved energy-momentum tensor' for such a Φ [13].

Whether or not this is exactly Penrose's route to Φ, part of his circle of ideas is that the dark matter in $\check{M}$ is described by a scalar field Φ arising after the Bang.

One observational consequence of conformally-cyclic cosmology (CCC) would be a calculation of the spectrum of density perturbations to be expected after the Bang. This hasn't been done, but it is likely to be scale-free as in the calculation from inflation since, in a sense, one can think of CCC in terms of inflation happening 'before the bang'. Penrose [2] has drawn attention to another consequence of CCC which gives the possibility of a crucial test: approaching $\mathcal{I}^+$, late on in conformal time in $\hat{M}$ (rather than in proper time, which has an infinite range so that there is no meaning to 'late on') the principal physical events will be black hole mergers; these will give a characteristic gravitational radiation pattern of a burst concentrated near a single light cone, which will in turn give characteristic spherical ripples on Σ; these ripples should appear as circles of perturbation ΔT in CMB maps, the detailed distribution of which would depend on the late-time history of the previous aeon.

References

[1] Penrose R 1979 'Singularities and time-asymmetry' in *General relativity: an Einstein centennary volume* eds S W Hawking and W Israel (Cambridge: Cambridge University Press)

[2] Penrose R 2008 Causality, quantum theory and cosmology in *On space and time* ed S Majid (Cambridge: Cambridge University Press)

[3] Luebbe C and Tod K P 2009 An extension theorem for conformal gauge singularities *J. Math. Phys.* **50** 112501, DOI: 10.1063/1.3239509

[4] Goode S J and Wainwright J 1985 Isotropic singularities in cosmological models *Class. Quantum Grav.* **2** 99-115

[5] Anguige K and Tod K P 1999 Isotropic cosmological singularities I: Polytropic perfect fluid spacetimes *Ann. Physics* **276** 257-293

[6] Newman R P A C 1993 On the structure of conformal singularities in classical general relativity *Proc. Roy. Soc. London* A **443** 473-492

[7] Anguige K 2000 Isotropic cosmological singularities. III. The Cauchy problem for the inhomogeneous conformal Einstein-Vlasov equations *Ann. Physics* **282** 395-419

[8] Anguige K and Tod K P 1999 Isotropic cosmological singularities II: The Einstein-Vlasov system *Ann. Physics* **276** 294-320

[9] Tod K P 2008 Isotropic cosmological singularities in spatially homogeneous models with a cosmological constant *Class. Quantum Grav.* **24** 2415-2432

[10] Tod K P 2003 Isotropic cosmological singularities: other matter models *Class. Quantum Grav.* **20** 521-534

[11] Luebbe C and Tod K P 2008 A global conformal extension theorem for perfect fluid Bianchi space-times *Ann.Phys.* **323** 2905-2912

[12] Bailey T N, Eastwood M G, and Gover A R 1994 Thomas's structure bundle for conformal, projective and related structures *Rocky Mountain J. Math.* **24** 1191-1217.

[13] Penrose R and Rindler W 1986 *Spinors and space-time 2: Spinor and twistor methods in space-time geometry* (Cambridge: Cambridge University Press)

Null data characterization of asymptotically flat stationary spacetimes

Andrés E Aceña

Max Planck Institute for Gravitational Physics, Am Mühlenberg 1, D-14476 Golm, Germany

E-mail: acena@aei.mpg.de

Abstract. Using two sequences of symmetric trace free tensors, called the 'null data', we give a complete characterization of the asymptotics of all asymptotically flat, stationary solutions with non-vanishing ADM mass to Einstein's vacuum field equations. We present necessary and sufficient growth estimates on the null data for the existence of such a solution.

1. Introduction

Since Penrose's [16] proposal to use conformal compactification as a way to describe isolated systems in General Relativity a lot of effort has been devoted to understand which are the spacetimes that belong to the group of asymptotically flat spacetimes, which are the properties of those spacetimes, and to know if this group includes the spacetimes that are physically relevant as isolated systems.

Considering in particular the time independent solutions of Einstein's vacuum field equations, Geroch [11] has given a definition of multipoles for static asymptotically flat spacetimes, which was later generalized by Hansen [13] to the stationary case. Beig and Simon [4], [5] and Kundu [15] add to this construction by showing that static or stationary solutions to Einstein's vacuum field equations that satisfy some decay rate at infinity are analytic even at infinity. These results show that the given definitions of the multipoles make sense. They also show that given two solutions, if they have the 'same' multipoles, then the solutions are isomorphic, thus representing the 'same' spacetime. Despite these important results the question about existence of a spacetime given the multipoles plus some convergence criteria on them remained open.

Taking a different approach, Reula [17] has shown existence and uniqueness of asymptotically flat stationary spacetimes in terms of a boundary value problem, when data are prescribed in the sphere that encompasses the asymptotic end. However, in order to be able to control the precise asymptotic behaviour of the spacetime, it would be convenient to have a complete description of the asymptotically flat stationary spacetimes in terms of asymptotic quantities, in the spirit of Hansen's multipoles.

Some progress towards such a result have been done by Bäckdahl and Herberthson [3], who found necessary bounds on the multipoles. Bäckdahl [2] has also found necessary and sufficient bounds on the multipoles in the axisymmetric case.

Concerned with the question of existence of static solutions with a given asymptotic behaviour, Friedrich [10] has used as data a sequence of trace-free symmetric tensors defined at infinity, referred to as *null data*, which are different but related to the multipoles. He has shown that imposing certain types of estimates on the null data is necessary and sufficient for the existence of asymptotically flat static spacetimes. This result has been used by Herberthson [14] to find necessary and sufficient bounds on Geroch's multipoles for the existence of a static solution.

In this article we present results obtained in [1], where generalizing Friedrich's null data [10] we use two sequences of trace-free symmetric tensors, referred again to as null data, to characterize all asymptotically flat stationary vacuum solutions of Einstein's field equations. We present necessary and sufficient conditions for the null data to determine apart from gauge conditions unique real analytic solutions and thus to provide a complete characterization of all possible asymptotically flat solution to the stationary vacuum field equations. For technical details and the proof of the result the reader is referred to [1].

2. Asymptotically flat stationary spacetimes

We consider a stationary vacuum spacetime $(\tilde{M}, \tilde{g}_{\mu\nu}, \xi^\mu)$. Here $\tilde{M}$ is a four-dimensional manifold, $\tilde{g}_{\mu\nu}$ is a Lorentzian metric with signature $(+---)$ that satisfy Einstein's vacuum equations $Ric[\tilde{g}] = 0$, and ξ^μ is a time-like Killing vector field with complete orbits. As shown by Geroch [12] the description of this spacetime can be done in terms of fields defined on the quotient space of $\tilde{M}$ with respect to the trajectories of ξ^μ, $\tilde{N}$. As we are interested in the asymptotics of the spacetime at spatial infinity, $\tilde{N}$ is assumed to be diffeomorphic to the complement of a closed ball $\bar{B}_R(0)$ in $\mathbb{R}^3$. This assumption is natural in the present context, it corresponds to the idea of an isolated system, where the material sources are confined to a bounded region outside of which is vacuum. Then, following Hansen [13], it is possible starting from $(\tilde{M}, \tilde{g}_{\mu\nu}, \xi^\mu)$ to define a negative definite metric $\tilde{h}_{ab}$ and two potentials $\tilde{\phi}_M$, $\tilde{\phi}_S$ on $\tilde{N}$. Einstein's vacuum field equations on $\tilde{M}$ imply then the following equations for the potentials and the metric on $\tilde{N}$,

$$\Delta_{\tilde{h}}\tilde{\phi}_A = 2R[\tilde{h}]\tilde{\phi}_A, \quad A = M, S, \tag{1}$$

$$R_{ab}[\tilde{h}] = 2[(\tilde{D}_a\tilde{\phi}_M)(\tilde{D}_b\tilde{\phi}_M) + (\tilde{D}_a\tilde{\phi}_S)(\tilde{D}_b\tilde{\phi}_S) - (\tilde{D}_a\tilde{\phi}_K)(\tilde{D}_b\tilde{\phi}_K)], \tag{2}$$

where $\tilde{\phi}_K = \left(\frac{1}{4} + \tilde{\phi}_M^2 + \tilde{\phi}_S^2\right)^{\frac{1}{2}}$. Reciprocally, if one has a solution $(\tilde{N}, \tilde{h}_{ab}, \tilde{\phi}_M, \tilde{\phi}_S)$ of (1), (2) then a stationary spacetime $(\tilde{M}, \tilde{g}_{\mu\nu}, \xi^\mu)$ can be constructed (cf. [9] for a detailed discussion). As it is equivalent but simpler to work in the quotient manifold $\tilde{N}$ than in the spacetime $\tilde{M}$, we are looking for solutions of (1) and (2).
We also consider $(\tilde{N}, \tilde{h}_{ab})$ to be asymptotically flat, i.e., we assume $(\tilde{N}, \tilde{h}_{ab})$ to admit a smooth conformal extension in the following way: there exist a smooth Riemannian manifold (N, h_{ab}) and a function $\Omega \in C^2(N) \cap C^\infty(\tilde{N})$ such that $N = \tilde{N} \cup \{i\}$, where i is a single point,

$$\Omega > 0 \text{ on } \tilde{N}, \qquad h_{ab} = \Omega^2 \tilde{h}_{ab} \text{ on } \tilde{N},$$
$$\Omega|_i = 0, \quad D_a\Omega|_i = 0, \quad D_aD_b\Omega|_i = -2h_{ab}|_i,$$

where D is the covariant derivative operator defined by h. This makes N diffeomorphic to an open ball in $\mathbb{R}^3$, with centre at the point i, which represents space-like infinity.

3. The null data characterization

As said, it would be convenient to have a complete description of the asymptotically flat solutions to (1), (2) in terms of asymptotic quantities. In our attempt for such a characterization it turns out convenient to use as asymptotic quantities a generalization of Friedrich's static null data [10] to the stationary case. Using as conformal factor

$$\Omega = \tfrac{1}{2}m^{-2}\left[\left(1 + 4\tilde{\phi}_M^2 + 4\tilde{\phi}_S^2\right)^{\frac{1}{2}} - 1\right], \tag{3}$$

where m is a non-zero constant and defining the conformal potentials as

$$\phi_M = \Omega^{-\frac{1}{2}}\tilde{\phi}_M, \qquad \phi_S = \Omega^{-\frac{1}{2}}\tilde{\phi}_S,$$

we define the following two sequences of trace-free symmetric tensors at infinity

$$\mathcal{D}_n^\phi = \{\mathcal{C}(D_{a_1}\phi)(i), \mathcal{C}(D_{a_2}D_{a_1}\phi)(i), \mathcal{C}(D_{a_3}D_{a_2}D_{a_1}\phi)(i), \ldots\},$$
$$\mathcal{D}_n^S = \{S_{a_2a_1}(i), \mathcal{C}(D_{a_3}S_{a_2a_1})(i), \mathcal{C}(D_{a_4}D_{a_3}S_{a_2a_1})(i), \ldots\},$$

where $\phi = \phi_S$ and S_{ab} is the trace free part of the Ricci tensor of h. These two sequences are referred to as the *null data.*
The natural question now is to what extent do the null data determine the metric h and the potentials ϕ_M, ϕ_S, or even the definition of the null data makes sense. For this to be the case the metric and the potentials should be real analytic even at i in suitable coordinates and conformal rescaling. Beig and Simon [5] (using as conformal factor (3)) and Kundu [15] have shown that the metric and the potentials do extend in a suitable gauge as real analytic fields to i if it is assumed that

$$(\phi_M(i))^2 + (\phi_S(i))^2 \neq 0.$$

As explained in [18] (cf. also [6]), in order for a solution of (1), (2) to lead to an asymptotically flat spacetime M it is necessary that $\phi_S(i) = 0$. So, we assume from now on that

$$\phi_M(i) \neq 0, \qquad \phi_S(i) = 0.$$

We express now the tensors in $\mathcal{D}_n^\phi$, $\mathcal{D}_n^S$ in terms of an h-orthonormal frame $c_{\mathbf{a}}$, $\mathbf{a} = 1, 2, 3$, at i. Denoting by $D_{\mathbf{a}}$ the covariant derivative in the direction of $c_{\mathbf{a}}$,

$$\mathcal{D}_n^{\phi*} = \{\mathcal{C}(D_{\mathbf{a}_1}\phi)(i), \mathcal{C}(D_{\mathbf{a}_2}D_{\mathbf{a}_1}\phi)(i), \mathcal{C}(D_{\mathbf{a}_3}D_{\mathbf{a}_2}D_{\mathbf{a}_1}\phi)(i), ...\},$$
$$\mathcal{D}_n^{S*} = \{S_{\mathbf{a}_2\mathbf{a}_1}(i), \mathcal{C}(D_{\mathbf{a}_3}S_{\mathbf{a}_2\mathbf{a}_1})(i), \mathcal{C}(D_{\mathbf{a}_4}D_{\mathbf{a}_3}S_{\mathbf{a}_2\mathbf{a}_1})(i), ...\}.$$

These tensors are defined uniquely up to rigid rotations in $\mathbb{R}^3$.
From Cauchy estimates on analytic functions it is possible to show that if the metric h and the potential ϕ are real analytic near i, then there exist constants $M, r > 0$ such that the components of these tensors satisfy the estimates

$$|\mathcal{C}(D_{\mathbf{a}_p}...D_{\mathbf{a}_1}\phi)(i)| \leq \frac{Mp!}{r^p}, \quad \mathbf{a}_p, ..., \mathbf{a}_1 = 1, 2, 3, \quad p \geq 0,$$
$$|\mathcal{C}(D_{\mathbf{a}_p}...D_{\mathbf{a}_1}S_{\mathbf{bc}})(i)| \leq \frac{Mp!}{r^p}, \quad \mathbf{a}_p, ..., \mathbf{a}_1, \mathbf{b}, \mathbf{c} = 1, 2, 3, \quad p \geq 0.$$

The main result to be presented in the following theorem is the statement that these estimates are not only necessary but also sufficient to have an analytic solution of the stationary field equations.

Theorem. *Suppose* $m \neq 0$ *and*

$$\hat{\mathcal{D}}_n^\phi = \{\psi_{\mathbf{a}_1}, \psi_{\mathbf{a}_2\mathbf{a}_1}, \psi_{\mathbf{a}_3\mathbf{a}_2\mathbf{a}_1}, ...\},$$
$$\hat{\mathcal{D}}_n^S = \{\Psi_{\mathbf{a}_2\mathbf{a}_1}, \Psi_{\mathbf{a}_3\mathbf{a}_2\mathbf{a}_1}, \Psi_{\mathbf{a}_4\mathbf{a}_3\mathbf{a}_2\mathbf{a}_1}, ...\},$$

are two infinite sequences of symmetric, trace free tensors given in an orthonormal frame at the origin of a 3-dimensional Euclidean space. If there exist constants $M, r > 0$ *such that the components of these tensors satisfy the estimates*

$$|\psi_{\mathbf{a}_p...\mathbf{a}_1}| \leq \frac{Mp!}{r^p}, \quad \mathbf{a}_p, ..., \mathbf{a}_1 = 1, 2, 3, \quad p \geq 1,$$
$$|\Psi_{\mathbf{a}_p...\mathbf{a}_1\mathbf{bc}}| \leq \frac{Mp!}{r^p}, \quad \mathbf{a}_p, ..., \mathbf{a}_1, \mathbf{b}, \mathbf{c} = 1, 2, 3, \quad p \geq 0,$$

then there exists an analytic, asymptotically flat, stationary vacuum solution $(\tilde{h}, \tilde{\phi}_M, \tilde{\phi}_S)$ *with mass monopole* m *and zero angular momentum monopole, unique up to isometries, so that the null data implied by* $h = \frac{1}{4}m^{-4}[(1+4\tilde{\phi}_M^2+4\tilde{\phi}_S^2)^{\frac{1}{2}}-1]^2\tilde{h}$ *and* $\phi_S = 2^{\frac{1}{2}}m[(1+4\tilde{\phi}_M^2+4\tilde{\phi}_S^2)^{\frac{1}{2}}-1]^{-\frac{1}{2}}\tilde{\phi}_S$ *in a suitable frame* $c_{\mathbf{a}}$ *as described above satisfy*

$$\mathcal{C}(D_{\mathbf{a}_q}...D_{\mathbf{a}_1}\phi_S)(i) = \psi_{\mathbf{a}_q...\mathbf{a}_1}, \quad \mathbf{a}_q, ..., \mathbf{a}_1 = 1, 2, 3, \quad q \geq 1,$$
$$\mathcal{C}(D_{\mathbf{a}_q}...D_{\mathbf{a}_3}S_{\mathbf{a}_2\mathbf{a}_1})(i) = \Psi_{\mathbf{a}_q...\mathbf{a}_1}, \quad \mathbf{a}_q, ..., \mathbf{a}_1 = 1, 2, 3, \quad q \geq 2.$$

The type of estimates imposed here on the abstract null data does not depend on the orthonormal frame in which they are given. Since these estimates are necessary as well as sufficient, all possible asymptotically flat solutions of the stationary vacuum field equations are characterized by the null data.

4. Conclusions

The result presented closes in some sense the quest to find a characterization of asymptotically flat stationary vacuum spacetimes.
This work contains the static case as a particular case, and is a generalization of Friedrich's work [10] from the static to the stationary case.
Corvino and Schoen [8] and Chruściel and Delay [7] have proven that it is possible to deform given general asymptotically flat vacuum data in an annulus in order to glue that data to stationary vacuum data in the asymptotic region. In relation with those works, as they need a family of asymptotically flat stationary solutions to perform the gluing procedure, our result gives a complete survey of the possible stationary asymptotics that can be attained beyond the known exact solutions.
With respect to Hansen's multipoles, as both, the multipoles and the null data, determine the metric and the potentials, there is a bijective map between them. Therefore the multipoles and the null data are from a mathematical point of view equivalent, although the equivalent result as the one here presented has so far not been found in terms of the multipoles.

References

[1] Aceña A 2009 Convergent null data expansions at space-like infinity of stationary vacuum solutions *Annales Henri Poincaré* **10** 275–337
[2] Bäckdahl T 2007 Axisymmetric stationary solutions with arbitrary multipole moments *Class. Quantum Grav.* **24** 2205–15
[3] Bäckdahl T and Herberthson M 2006 Calculation of, and bounds for, the multipole moments of stationary spacetimes *Class. Quantum Grav.* **23** 5997–6006
[4] Beig R and Simon W 1980 Proof of a multipole conjecture due to Geroch *Commun. Math. Phys.* **78** 75–82
[5] Beig R and Simon W 1981 On the multipole expansion for stationary space-times *Proc. R. Soc. Lond. A* **376** 333–41
[6] Beig R and Simon W 1983 The multipole structure of stationary space-times *J. Math. Phys.* **24** 1163–71
[7] Chruściel P and Delay E 2003 On mapping properties of the general relativistic constraints operator in weighted function spaces, with applications *Mémoires de la SMF* **94** vi–103
[8] Corvino J and Schoen R 2006 On the asymptotics for the vacuum Einstein constraint equations *J. Differ. Geom.* **73** 185–217
[9] Dain S 2001 Initial data for stationary spacetimes near spacelike infinity *Class. Quantum Grav.* **18** 4329–38
[10] Friedrich H 2007 Static vacuum solutions from convergent null data expansions at space-like infinity *Ann. Henri Poincaré* **8** 817–84
[11] Geroch R 1970 Multipole moments. II. Curved space *J. Math. Phys.* **11** 2580–8
[12] Geroch R 1971 A method for generating solutions of Einstein's equations *J. Math. Phys.* **12** 918–24
[13] Hansen R 1974 Multipole moments of stationary space-times *J. Math. Phys.* **15** 46–52
[14] Herberthson M 2009 Static spacetimes with prescribed multipole moments: a proof of a conjecture by Geroch Class. Quantum Grav. **26** 215009
[15] Kundu P 1981 On the analyticity of stationary gravitational fields at spatial infinity *J. Math. Phys.* **22** 2006–11
[16] Penrose R 1963 Asymptotic properties of fields and space-times *Phys. Rev. Lett.* **10** 66–8
[17] Reula O 1989 On existence and behaviour of asymptotically flat solutions to the stationary Einstein equations *Commun. Math. Phys.* **122** 615–24
[18] Simon W 1980 *Die Multipolstruktur stationärer Räume in der allgemeinen Relativitätstheorie* PhD thesis, Formal- und Naturwissenschaftlichen Fakultät der Universität Wien

Journal of Physics: Conference Series **229** (2010) 012015 doi:10.1088/1742-6596/229/1/012015

On the linear stability of a scalar-field with exponential potential in a flat Robertson-Walker background

Artur Alho
Centro de Matemática, Universidade do Minho, 4710-057 Braga, Portugal
E-mail: artur.alho@gmail.com

Abstract. We study covariant and gauge-invariant linear scalar perturbations of a scalar-field with positive exponential potential in a flat Robertson-Walker background. By applying a dynamical systems approach we investigate how the phase of the perturbations evolves, finding bounds on the wave-number in terms of the slope parameter, for which the perturbations decays when approaching the inflationary solution.

1. Introduction

When modelling structure formation in the universe, an important role is often played by the study of scalar perturbations of spatially homogeneous and isotropic spacetimes [1]. In this context, the linear perturbation framework put forward by Ellis and Bruni [2] has been widely used. The latter consists on considering the *threading* of the spacetime from a preferred vector-field associated with the matter content, in order to obtain a closed set of evolution and constraint equations, which can be linearized about a Robertson-Walker (RW) background, c.f [3]. The advantage of this approach relies on the perturbation variables being gauge-invariant by construction. Formally, the evolution of the scalar variables is given by a second order partial differential equation (PDE), or a system of PDEs in a multicomponent universe model. Often an expansion into spatial harmonics is performed in order to get a second-order linear homogeneous ordinary differential equation (ODE) for the normal modes

$$\ddot{\Delta}_{(n)} + \mathcal{A}(t)\dot{\Delta}_{(n)} + \mathcal{B}(t)\Delta_{(n)} = 0, \tag{1}$$

where $\Delta_{(n)}$ is the scalar harmonic and $\mathcal{A}$ and $\mathcal{B}$ depend on the background quantities only. When it is not possible to solve exactly such equation, a qualitative analysis for the late-time behaviour of the perturbations can be used (see e.g. [4, 5]). This consists on using expansion normalized variables for the background together with the logarithmic time variable τ defined by $\frac{d\tau}{dt} = H$, where H is the Hubble parameter, and then by defining the *phase of perturbation* variable

$$\mathcal{U}_{(n)} := \frac{\Delta'_{(n)}}{\Delta_{(n)}} \tag{2}$$

which satisfies the first-order non-linear ODE

$$\mathcal{U}'_{(n)} = -\mathcal{U}^2_{(n)} - \mathcal{A}(\tau)\mathcal{U}_{(n)} - \mathcal{B}(\tau) \tag{3}$$

and is coupled to an evolution equation for these normalized background variables. The equilibrium points of $\mathcal{U}$ are given by $\mathcal{U} = \frac{1}{2}\left(-\mathcal{A} \pm \sqrt{\mathcal{A}^2 - 4\mathcal{B}}\right)$, and they are real if and only if $\mathcal{A}^2 - 4\mathcal{B} \geq 0$. From the definition (2), the interpretation of $\mathcal{U}$ during the evolution is easily given by the following conditions: (i) if $\mathcal{U} < 0$, the perturbation decays; (ii) if $\mathcal{U} > 0$, perturbation grows; (iii) if $\mathcal{U} = 0$, the perturbation has a constant value.

2. Background dynamics

We will assume as the background a RW spacetime with a scalar-field and its self-interaction potential of exponential type

$$\mathcal{V}(\phi) = \Lambda \exp \lambda\phi, \quad \lambda \in \Re \tag{4}$$

where Λ is a positive cosmological constant. We will use expansion-normalized variables defined by $\Psi := \frac{\dot{\phi}}{\sqrt{6}H}$, $\Phi := \frac{\sqrt{\mathcal{V}(\phi)}}{\sqrt{3}H}$ and $K := -\frac{{}^3R}{6H^2}$. Using these variables the system of non-linear ODEs governing the background dynamics with exponential potential becomes [6]

$$\Psi' = (q-2)\,\Psi - \frac{\sqrt{6}}{2}\lambda\Phi^2 \tag{5}$$

$$\Phi' = \left(q + 1 + \frac{\sqrt{6}}{2}\lambda\Psi\right)\Phi \tag{6}$$

where the deceleration parameter is given by $q = 2\Psi^2 - \Phi^2$ and the Friedmann equation is $K = 1 - \Psi^2 - \Phi^2$. For a flat RW background $K = 0$, and the constraint equation reads

$$\Phi^2 = 1 - \Psi^2. \tag{7}$$

Substituting into (5) we get

$$\Psi' = 3\Psi^3 + \frac{\sqrt{6}}{2}\lambda\Psi^2 - 3\Psi - \frac{\sqrt{6}}{2}\lambda. \tag{8}$$

The equilibrium points for (8) are $\Psi^2 = 1$ and $\Psi = -\frac{\lambda}{\sqrt{6}}$. From (7), we see that the points $\Psi^2 = 1$ correspond to the massless scalar-field case, which is also the initial axis. Now, for inflation to occur we must have $q < 0$, then, at the equilibrium point the solution will inflate if $\lambda^2 < 2$ [7]. Notice that the inflationary solution exists for positive Ψ if λ is negative and vice-versa. In view of this, and since we are interested in seeing how perturbations evolve as they approach the inflationary solution, in the next section we will restrict the analysis in the phase plane to $0 \leq \Psi \leq 1$ for $-\sqrt{2} < \lambda \leq 0$.

3. The evolution of inhomogeneities: qualitative analysis

We will use the covariant and gauge invariant (CGI) approach of [2]. This has also been applied to minimally coupled scalar-fields by Bruni et al [8]. But while [8] considers the scalar-field as a fluid with an equation of state (depending on the time), here the evolution equation for the scalar variable defined by

$$\Delta := a^2(\tau)\frac{D^2\psi}{\psi}, \quad \psi \equiv \dot{\phi} \tag{9}$$

is constructed without assuming any equation of state and leaving the form of the potential explicit. Then, for the flat case, and with an exponential potential we get [9]:

$$\mathcal{A}(\Psi) = -\left[1 + \frac{6\lambda}{\sqrt{6}}\frac{1-\Psi^2}{\Psi}\right] \tag{10}$$

$$\mathcal{B}(\Psi) = -\left[6(1-\Psi^2) + \frac{6\lambda}{\sqrt{6}}\frac{1-\Psi^2}{\Psi}\left(5+3\lambda\frac{1-\Psi^2}{\Psi}\right) + 3\lambda^2(1-\Psi^2) - \frac{n^2}{H^2a^2}\right] \tag{11}$$

Thus, equation (3), with the coefficients given by (10) and (11), together with the background equation (8), form an autonomous system of first-order nonlinear ODEs for $(\Psi,\mathcal{U}_{(n)})$, which describes the evolution of the phase of momentum-density perturbations for a scalar-field with exponential potential on a flat RW background.
At this point one should stress an important difference between the scalar-field and the fluid cases. In the former the equilibrium points in the $\Psi = 1$ axis are not independent from n while for the fluid cases $(\Omega = 1)$ they are. This means that we have two critical values for n, that we shall designate by $n^{(1)}_{crit}$ and $n^{(\lambda)}_{crit}$. At these values the fixed points of equation (3) coincide, being non-hyperbolic in nature, and the analysis must be carried to second-order in the Taylor expansion. In the $\Psi = 1$ line, $\mathcal{U}$ has real roots if

$$n^2 \leq \left(n^{(1)}_{crit}\right)^2 = \left(\frac{aH}{2}\right)^2. \tag{12}$$

- **$\lambda = 0$: massless scalar-field with a positive cosmological constant**

We observe that for $\lambda = 0$ we have a massless scalar-field with a positive cosmological constant. In this case, $\Psi = 0$ is the DeSitter solution (late-time attractor), the cosmological constant dominates and $\mathcal{U}$ has real roots if

$$n^2 \leq \left(n^{(0)}_{crit}\right)^2 = \left(\frac{5aH}{2}\right)^2. \tag{13}$$

It is easy to check that the asymptotic stationary state will always be on $\mathcal{U} > 0$, since for $n^2 = \left(\frac{5aH}{2}\right)^2$ gives $\mathcal{U} = \frac{1}{2}$. Above this value the roots are complex and the orbits are periodic and the perturbations have a sound wave character.

- **$\lambda \in]-\sqrt{2}, 0[$: inflationary solutions**

Here, we will restrict the study to $\lambda < 0$ such that the inflationary solution will be at $\Psi > 0$. On the axis $\Psi = -\frac{\lambda}{\sqrt{6}}$ we have

$$\mathcal{A}_{-\frac{\lambda}{\sqrt{6}}} = 5 - \lambda^2 \quad , \quad \mathcal{B}_{-\frac{\lambda}{\sqrt{6}}} = \left(24 - 18\sqrt{6}\right) + \left(6\sqrt{6} - 7\right)\lambda^2 + \left(\frac{1}{2} - \frac{\sqrt{6}}{2}\right)\lambda^4 + \frac{n^2}{a^2H^2} \tag{14}$$

and $\mathcal{U}$ has real roots if the condition

$$n^2 \leq \left(n^{(\lambda)}_{crit}\right)^2 = a^2H^2\left[\left(18\sqrt{6} - \frac{71}{4}\right) - \left(6\sqrt{6} - \frac{18}{4}\right)\lambda^2 - \left(\frac{1}{4} - \frac{\sqrt{6}}{2}\right)\lambda^4\right] \tag{15}$$

is satisfied. Moreover, the phase plane is structuraly stable if $n^2 \neq \left(n^{(\lambda)}_{crit}\right)^2$. Now, in order to ensure that the perturbations decay, the future atractor which corresponds to the solution $\mathcal{U}_+ = \frac{1}{2}\left(-\mathcal{A} + \sqrt{\mathcal{A}^2 - 4\mathcal{B}}\right)$ c.f. [4], must live in the half plane $\mathcal{U} < 0$, and then, the following conditions must be satisfied

$$\mathcal{B} > 0 \quad if \quad \mathcal{A} \geq 1 \quad or \quad \mathcal{B} < 0 \quad if \quad 0 < \mathcal{A} < 1 \tag{16}$$

Now, since the values of λ are in the open interval between 0 and $-\sqrt{2}$, we have that $\mathcal{A} > 1$ and the second condition in (16) is never satisfied. Thus we arrive at the following result:
Proposition: *First order scalar perturbations of a scalar-field with exponential potential in an*

expanding flat RW background decay, when approaching the inflationary solution, if the wave-number satisfies the following two conditions:

(i) $n^2 < a^2H^2\left[\left(18\sqrt{6}-\frac{71}{4}\right)+\left(\frac{18}{4}-6\sqrt{6}\right)\lambda^2+\left(\frac{\sqrt{6}}{2}-\frac{1}{4}\right)\lambda^4\right]$

(ii) $n^2 > a^2H^2\left[\left(18\sqrt{6}-24\right)+\left(7-6\sqrt{6}\right)\lambda^2+\left(\frac{\sqrt{6}}{2}-\frac{1}{2}\right)\lambda^4\right]$

where $\lambda \in]-\sqrt{2},0[$.

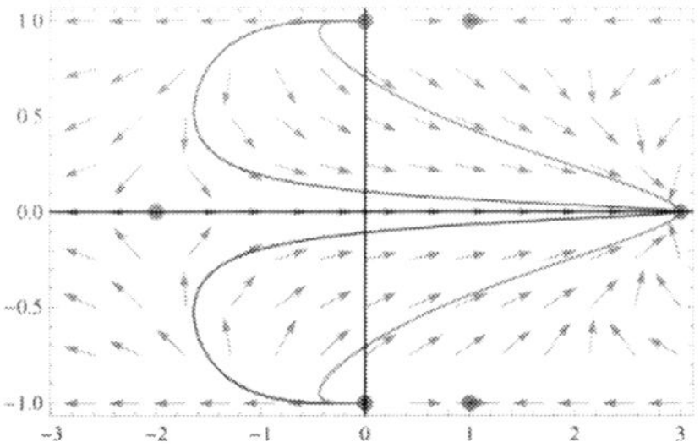

Figure 1. Phase plane for $\lambda = 0$ and in the long-wavelenght limit.

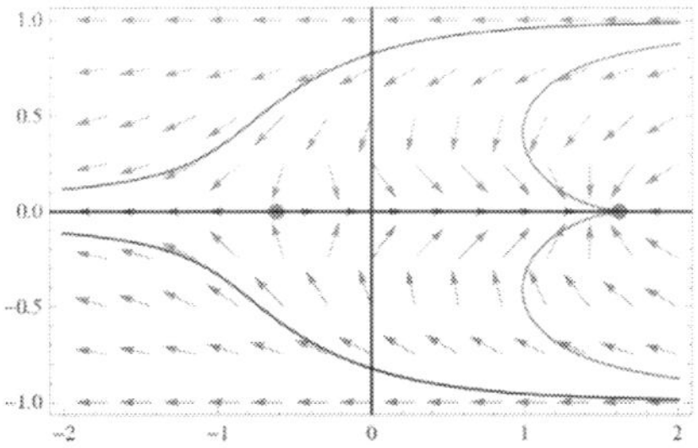

Figure 2. Phase plane for $\lambda = 0$ and $\left(n^{(1)}_{crit}\right)^2 < n^2 < \left(n^{(0)}_{crit}\right)^2$.

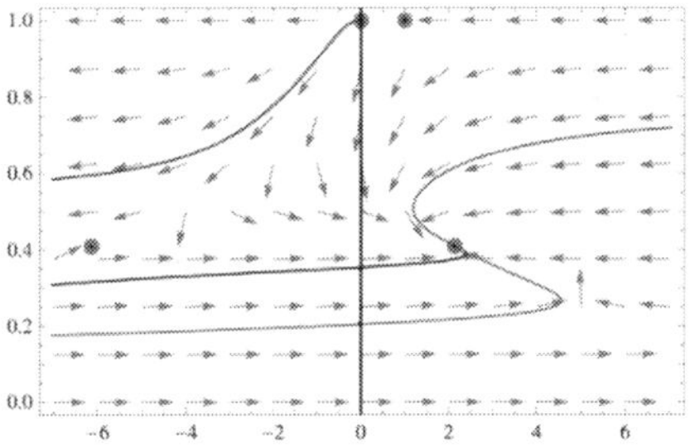

Figure 3. Phase plane for $\lambda = -1$ and $\Psi > 0$ in the long-wavelenght limit.

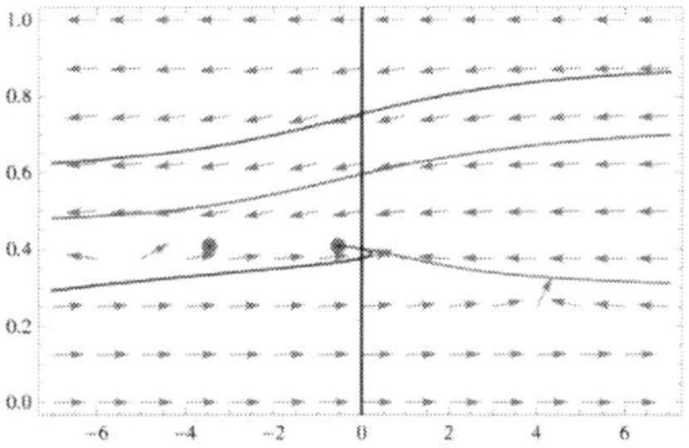

Figure 4. Phase plane for $\lambda = -1$ and $\left(n^{(1)}_{crit}\right)^2 < n^2 < \left(n^{(-1)}_{crit}\right)^2$.

Acknowledgments
The author thanks Filipe C. Mena for reading the manuscript, CMAT, Univ. Minho, and ERE09 organizing committee for support and FCT (Portugal) for grant SFRH/BD/48658/2008.

References
[1] Stewart J M 1990 *Class. Quantum Grav.* **7** 1169-80
[2] Ellis G F R and Bruni M 1989 *Phys. Rev.* D **40** 1804-18
[3] Ellis G F R and van Elst H 1999 *Proc. of the NATO ASIC (Cargese)* vol 541 (Dordrecht: Kluwer) pp 1-116
[4] Woszczyna A 1992 *Phys. Rev.* D **45** 1982-88
[5] Bruni M 1993 *Phys. Rev.* D **47** 738-42
[6] Coley A A 2003 *Dynamical Systems and Cosmology* (ASSL vol 291) ed Burton W B (Dordrecht: Kluwer)
[7] Halliwell J J 1987 *Physics Letters* B **185** 341-44
[8] Bruni M, Ellis G F R and Dunsby P K S 1992 *Class. Quantum Grav.* **9** 921-945
[9] Alho A and Mena F C On the linear stability of scalar-fields cosmologies *in preparation*

Spanish Relativity Meeting (ERE 2009) IOP Publishing
Journal of Physics: Conference Series **229** (2010) 012016 doi:10.1088/1742-6596/229/1/012016

Relating the Newman–Penrose constants to the Geroch–Hansen multipole moments

Thomas Bäckdahl

School of Mathematical Sciences, Queen Mary, University of London

E-mail: t.backdahl@qmul.ac.uk

Abstract. In these proceedings we investigate the relation between the Newman-Penrose constants and the Geroch-Hansen multipole moments for stationary spacetimes. This is a short description of the results of the paper [1] published in Classical and Quantum Gravity. This relation gives a simple characterization of which stationary spacetimes have vanishing Newman-Penrose constants.

1. Introduction

The Newman–Penrose (NP) constants were defined by Newman and Penrose in [2]. They are quantities of quadrupolar order defined on the null-infinities. Furthermore, they turn out to be conserved under time translations. Even though they have been studied for a long time, their meaning is still not fully understood. Lately it has been disputed whether the NP-constants are zero for stationary spacetimes or not. It has been shown that they are zero for all algebraically special stationary spacetimes including the Kerr solution [3]. The original paper [2] by Newman and Penrose, gives expressions of the NP-constants in terms of multipole moments. It is unclear however, how these moments were defined, if they are coordinate independent and if different moments can be specified independently. The Geroch–Hansen multipole moments have these properties, but were defined later [4, 5]. Therefore, this work is intended to clearly settle the matter by expressing the NP-constants in terms of the Geroch–Hansen multipole moments. These multipole moments also give a possibility of physical interpretation.

The Geroch–Hansen multipole moments can be freely specified under a simple convergence condition. That is, for any given choice of multipoles, satisfying the convergence condition, there is a unique stationary spacetime with these multipole moments. This was shown in [1] for the stationary axisymmetric case. Recently Herberthson [6] showed this for the general static case using results of Friedrich [7]. The result of Frierdich states that for static spacetimes one can freely specify null data under a convergence condition. These null data are related to the multipole moments, but the relation is fairly complicated. The results by Friedrich have been extended to the stationary case by Aceña [8]. Hopefully the results by Herberthson can also be extended to the stationary case, but for now it is still an open problem.

The aim of this work is to express the NP-constants in terms of well studied quantities, namely the Geroch-Hansen multipole moments, and investigate if they vanish in general. These proceedings are not intended to give a complete argument, only to introduce the necessary notation and present the result. For a more complete exposition, see the full paper [1].

2. Notation

2.1. Asymptotic expansion

We will use Bondi–Sachs coordinates $(u, r, \zeta, \bar{\zeta})$, $\zeta = e^{i\phi}\cot\frac{\theta}{2}$ and make expansions in terms of r^{-1}. The Weyl-curvature is then expressed as

$$\Psi_0 = \frac{\Psi_0^0}{r^5} + \frac{\Psi_0^1}{r^6} + \mathcal{O}(r^{-7}), \qquad \Psi_3 = \frac{\Psi_3^2}{r^4} + \frac{\Psi_3^3}{r^5} + \frac{\Psi_3^4}{r^6} + \mathcal{O}(r^{-7}),$$

$$\Psi_1 = \frac{\Psi_1^0}{r^4} + \frac{\Psi_1^1}{r^5} + \frac{\Psi_1^2}{r^6} + \mathcal{O}(r^{-7}), \qquad \Psi_4 = \frac{\Psi_4^4}{r^5} + \frac{\Psi_4^5}{r^6} + \mathcal{O}(r^{-7}),$$

$$\Psi_2 = \frac{\Psi_2^0}{r^3} + \frac{\Psi_2^1}{r^4} + \frac{\Psi_2^2}{r^5} + \frac{\Psi_2^3}{r^6} + \mathcal{O}(r^{-7}).$$

2.2. Newman–Penrose constants

The NP constants $\{G_m\}$ are defined as

$$G_m \equiv \int_0^{2\pi}\int_0^{\pi} \Psi_0^1\, \overline{{}_2Y_{2,m}} \sin\theta d\theta d\phi = \int_0^{2\pi}\int_0^{\pi} (\tfrac{10}{3}\Psi_1^0 - 5\Psi_2^0\Psi_0^0)\, \overline{{}_2Y_{2,m}} \sin\theta d\theta d\phi, \tag{1}$$

where ${}_sY_{j,m}$ are spin-weighted spherical harmonics. The second expression is useful for the actual computation.

2.3. Asymptotic flatness

Now consider a conformal compactification V of the 3-manifold of trajectories of t^a (the time-like Killing vector) with metric $h_{ab} = \Omega^2(-\lambda g_{ab} + t_a t_b)$. We want to choose Ω such that we can add a point Λ (the infinity point) such that h_{ab} extends smoothly to Λ. We also demand

$$\Omega = 0, \quad D_a\Omega = 0, \quad D_a D_b \Omega = 2h_{ab} \quad \text{at } \Lambda, \tag{2}$$

where D_a is the covariant derivative on h_{ab}.

2.4. Relativistic multipole moments

Define the complex potential

$$P = \frac{1 - \lambda - i\omega}{(1 + \lambda + i\omega)\sqrt{\Omega}}. \tag{3}$$

The Geroch-Hansen multipole moments are given by the limits of

$$P_{a_1\dots a_n} = C\left[D_{a_1} P_{a_2\dots a_n} - \frac{(n-1)(2n-3)}{2} R_{a_1a_2} P_{a_3\dots a_n}\right], \tag{4}$$

as one approaches Λ. Here $C[\cdot]$ represents the totally symmetric and trace-free part.

3. Development of results

We expand Ψ_0^0, Ψ_1^0 and Ψ_2^0 in terms of spin-weighted spherical harmonics:

$$\Psi_0^0 = \sum_{m=-2}^{2} A_m\, {}_2Y_{2,m} = \sqrt{5}\frac{A_{-2} + 2\zeta A_{-1} + \sqrt{6}\zeta^2 A_0 + 2\zeta^3 A_1 + \zeta^4 A_2}{2\sqrt{\pi}(1+\zeta\bar{\zeta})^2},$$

$$\Psi_1^0 = \sum_{m=-1}^{1} B_m\, {}_1Y_{1,m} = -\sqrt{3}\frac{B_{-1} + \sqrt{2}\zeta B_0 + \zeta^2 B_1}{2\sqrt{\pi}(1+\zeta\bar{\zeta})},$$

$$\Psi_2^0 = C.$$

Here C is real, B_m and A_m are complex.

The time-like Killing vector is computed to

$$t^a = \frac{\partial}{\partial u} + \mathcal{O}(r^{-5})\frac{\partial}{\partial r} + \mathcal{O}(r^{-5})\frac{\partial}{\partial \zeta} + \mathcal{O}(r^{-5})\frac{\partial}{\partial \bar{\zeta}}. \tag{5}$$

The following choice of conformal factor turns out to adequately simplify the computations

$$\Omega = (r^{-1} - Cr^{-2} + \tfrac{11}{8}C^2 r^{-3})^2. \tag{6}$$

We need good coordinates to verify the smoothness conditions. To this end we choose asymptotically Euclidian harmonic coordinates (x, y, z), $\rho^2 = x^2 + y^2 + z^2$. The conformal metric and the conformal factor are then found to be

$$h_{ab} = dx^2 + dy^2 + dz^2 + \mathcal{O}(\rho^3), \qquad \Omega = \rho^2 + \tfrac{1}{4}C^2\rho^4 + \mathcal{O}(\rho^5). \tag{7}$$

The limits in the definition of the multipole moments are found to be

$$\begin{aligned}
\lim_{\rho\to 0} P &= -C\\
\lim_{\rho\to 0} P_a &= \tfrac{\sqrt{6}}{12\sqrt{\pi}}(B_{-1} - B_1)dx - \tfrac{i\sqrt{6}}{12\sqrt{\pi}}(B_{-1} + B_1)dy + \tfrac{\sqrt{3}}{6\sqrt{\pi}}B_0 dz\\
\lim_{\rho\to 0} P_{ab} &= \tfrac{\sqrt{5}}{24\sqrt{\pi}}(\sqrt{6}A_0 - 3A_2 - 3A_{-2})dx^2 + \tfrac{i\sqrt{5}}{4\sqrt{\pi}}(-A_2 + A_{-2})dxdy\\
&\quad + \tfrac{\sqrt{5}}{4\sqrt{\pi}}(A_1 - A_{-1})dxdz + \tfrac{\sqrt{5}}{24\sqrt{\pi}}(\sqrt{6}A_0 + 3A_2 + 3A_{-2})dy^2\\
&\quad + \tfrac{i\sqrt{5}}{4\sqrt{\pi}}(A_1 + A_{-1})dydz - \tfrac{\sqrt{30}}{12\sqrt{\pi}}A_0 dz^2
\end{aligned} \tag{8}$$

By comparing the limits (8) with the definition of the moments (M, C_a, Q_{ab}), one can express A_m, B_m and C in terms of the multipole moments. Using this knowledge, one can compute the integrals (1).

4. Conclusion

The NP constants are

$$\begin{aligned}
G_{-2} &= -2\sqrt{5\pi}(3C_y^2 - 3C_x^2 + MQ_{xx} - MQ_{yy} + 2iMQ_{xy} - 6iC_xC_y),\\
G_{-1} &= -4\sqrt{5\pi}(iMQ_{yz} - 3C_xC_z - 3iC_yC_z + MQ_{xz}),\\
G_0 &= 2\sqrt{30\pi}(-C_x^2 - C_y^2 + 2C_z^2 + MQ_{xx} + MQ_{yy}),\\
G_1 &= -4\sqrt{5\pi}(iMQ_{yz} + 3C_xC_z - 3iC_yC_z - MQ_{xz}),\\
G_2 &= -2\sqrt{5\pi}(3C_y^2 - 3C_x^2 + MQ_{xx} - MQ_{yy} - 2iMQ_{xy} + 6iC_xC_y).
\end{aligned}$$

As expected, this is the same form as in the original paper by Newman and Penrose [2], i.e., linear combinations of dipole squared and monopole times quadrupole. It is easy to see that the NP-constants are independent of the choice of conformal factor, hence invariant under translations. As the NP-constants are expansion coefficients for spin-weighted spherical harmonics, they will depend on the spin-frame though.

For the axisymmetric case we see that $G_{-2} = G_{-1} = G_1 = G_2 = 0$ and $G_0 = 2\sqrt{30\pi}(2C_z^2 - MQ_{zz})$, where $Q_{zz} = -2Q_{xx} = -2Q_{yy}$ is the zz component of the quadrupole.

We can conclude that the NP-constants are in general not zero, but for some important solutions they are. For instance the Kerr solution has $C_z = iMa$, $Q_{zz} = -2Q_{xx} = -2Q_{yy} = -2Ma^2$, and all other components of C_a and Q_{ab} are zero. This yields the well known fact that all NP-constants are zero for the Kerr solution. In fact they are zero for all stationary, algebraically special solutions [3].

5. Acknowledgements
This work was supported by the Wenner-Gren foundations.

References
[1] Bäckdahl T 2009 *Class. Quantum Grav.* **26** 175021
[2] Newman E T and Penrose R 1968 *Proc. Roy. Soc. Lond. A* **305** 175
[3] Wu X and Shang Y 2007 *Class. Quantum Grav.* **24** 679
[4] Geroch R 1970 *J. Math. Phys.* **11** 2580
[5] Hansen R O 1974 *J. Math. Phys.* **15** 46
[6] Herberthson M 2009 *Class. Quantum Grav.* **26** 215009
[7] Friedrich H 2007 *Ann. Henri Poincaré* **8** 817
[8] Aceña A E 2009 *Ann. Henri Poincaré* **10** 275

Spacetime Brout–Englert–Higgs effect in General Relativity interacting with p-brane matter

Igor A Bandos

Ikerbasque, the Basque Foundation for Sciences, and Department of Theoretical Physics and History of Science, the Basque Country University, P.O. Box 644, 48080 Bilbao, Spain

E-mail: igor_bandos@ehu.es

Abstract. We review the manifestation of the Brout-Englert-Higgs effect in general relativity interacting with point-like and extended objects (p–branes including string for p=1 and membrane for $p = 2$), which manifests itself in the appearance of the brane source in the Einstein equation while the graviton remains massless [4]–[8], and discuss briefly its relation and differences with the model for massive spin 2 field proposed recently by G. t'Hooft [3].

Brout-Englert-Higgs effect [1], which was also known as Higgs effect, consists in that the gauge fields of an internal symmetry group acquires the mass when the gauge symmetry is spontaneously broken. The increasing of the number of polarizations is explained by that the gauge fields "eat" Goldstone fields of the spontaneously broken gauge symmetry and incorporates their degrees of freedom as additional polarizations.

To be more specific, let us consider, following [3], the Lagrangian $\mathcal{L} = -tr\left(\frac{1}{4}F_{\mu\nu}F^{\mu\nu} + \frac{1}{2}D_\mu\varphi^s D^\mu\varphi^s + V(\varphi^2)\right)$ for the gauge field $\mathcal{A}_\mu = \mathcal{A}_\mu^s T_s$ and the scalars $\phi = \psi^s T_s$ in the adjoint representation of the gauge group. If the potential $V(\phi)$ forces the scalar field to have a non-vanishing vacuum expectation value $< \phi >$, one can fix the 'preference gauge' [3] $\phi =< \phi >$. In this gauge the kinetic term for the scalar field, Goldstone field for the spontaneously broken gauge symmetry, gives rise to the mass term of the gauge field, $\frac{1}{2}D_\mu\varphi^s D^\mu\varphi^s \longmapsto \propto m^2\mathcal{A}_\mu\mathcal{A}^\mu$ with $m^2 \propto < \phi^2 >$, and the scalar field equations are reduced to $\partial^\nu \mathcal{A}_\nu = 0$. These are dependent: they can be obtained from selfconsistency conditions for the gauge field equations $D^\mu F_{\mu\nu} = m\mathcal{A}_\nu$.

General Relativity is invariant under the spacetime *diffeomorphisms* (general coordinate transformations) $Diff_4$ ($Diff_D$ in D–dimensional case)

$$\delta x^\mu := x^{\mu\prime} - x^\mu = -a^\mu(x)\, , \qquad \delta g_{\mu\nu} := g'_{\mu\nu}(x) - g_{\mu\nu}(x) = \partial_\mu a^\rho\, g_{\rho\nu} + \partial_\nu a^\rho\, g_{\mu\rho} + a^\rho\partial_\rho g_{\mu\nu}\, . \quad (1)$$

and graviton can be described by gauge field for the $Diff_4$ gauge symmetry, which is identified with vielbein $e^a = dx^\mu e_\mu{}^a(x)$ or with metric $g_{\mu\nu} = e_\mu{}^a(x)\eta_{ab}e_\nu{}^b(x)$. Then it was natural to expect that the spontaneous breaking of diffeomorphism symmetry should result in that the graviton becomes massive, see [2] and refs. in [3, 4].

On the other hand, the spontaneous breaking of diffeomorphism symmetry occurs in the presence of material objects: particles (0-branes), strings (1-branes), membranes (2-branes) and p-branes with $p > 2$ for higher dimensional cases $D \geq 4$. The corresponding spacetime Higgs effect, its supersymmetric generalization and consequences were the subjects of study in [4, 5, 6, 7] and in [8] (where some statements of previous papers on brane degrees of freedom were refined/improved). This spacetime Higgs effect manifests itself in the modification of the 'free' Einstein equations by the p-brane sources, *i.e.* by the energy-momentum tensor localized on the p-brane worldvolume W^{p+1},

$$\mathcal{G}^{\mu\nu}(g) := \sqrt{|g|}(\mathcal{R}_{\mu\nu} - \tfrac{1}{2}g_{\mu\nu}\mathcal{R}) = \tfrac{T_p}{4}\sqrt{|g^{[p+1]}|}\, g^{[p+1]mn}(\xi,\vec{0})\delta^{\mu}_{m}\delta^{\nu}_{n}\delta^{(D-p-1)}(x^I) \qquad (2)$$

(see below for the notation). However, although such an energy-momentum tensor is a counterpart of the mass term in $D^{\mu}F_{\mu\nu} = m\mathcal{A}_{\nu}$, it cannot be considered as a mass term for graviton but rather as a counterpart of the cosmological constant contribution [4]-[7].

The above statements of [4]-[7] might seem to be in contradiction with [2] and particularly with [3], where a model looking similar to brane interacting with gravity is discussed. The aims of the present contribution is to review the pure bosonic results of the study in [4]-[7] and to point out the differences between gravity interacting with branes and the model considered in [3] which result in different properties of spin 2 fields in these dynamical systems.

1. Spacetime Higgs effect in the presence of p-brane matter.

Material objects, p-branes (particles for p=0, strings for p=1) can be described by the coordinate functions $\hat{x}^{\mu}(\xi^m)$ defining parametrically their worldvolume W^{p+1} as a surface in spacetime M^D,

$$W^{(p+1)} \subset M^D \quad : \quad x^{\mu} = \hat{x}^{\mu}(\xi^m) = \hat{x}^{\mu}(\tau,\vec{\sigma}) , \qquad \{ {}^{\mu=0,1,\ldots,(D-1)\,,}_{m=0,1,\ldots,p\,.} \qquad (3)$$

Their transformations under $SDiff_D$

$$\delta\hat{x}^{\mu}(\xi) := \hat{x}^{\mu\prime}(\xi) - \hat{x}^{\mu}(\xi) = a^{\mu}(\hat{x}(\xi)) - \delta\xi^m\partial_m\hat{x}^{\mu}(\xi) \qquad (4)$$

indicate their Goldstone nature (see [9] for the flat spacetime and flat superspace).

As for any Goldstone fields of the gauge symmetry, the degrees of freedom in coordinate functions can be gauged away by imposing the 'static gauge' conditions (*cf.* $\phi =<\phi>$ above)

$$\hat{x}^{\mu}(\xi) = (\xi^m, 0, \ldots, 0) \qquad (\Leftrightarrow \qquad \phi =<\phi>) , \qquad \{ {}^{\mu=0,1,\ldots,(D-1)\,,}_{m=0,1,\ldots,p\,,} \qquad (5)$$

this is to say by choosing a local frame where $W^{(p+1)}$ looks locally as a flat hyperplane. In this gauge the p-brane degrees of freedom are carried (are 'eaten') by the metric gauge field.

The action describing the interaction of the gauge and Goldstone fields reads

$$S = \tfrac{1}{2\kappa}\int d^Dx\,\sqrt{|g|}\mathcal{R} + S_p \qquad \Big(\Longleftrightarrow \qquad \mathcal{L} = -\tfrac{1}{4}F_{\mu\nu}F^{\mu\nu} - \tfrac{1}{2}D_{\mu}\varphi^s D^{\mu}\varphi^s - V(\varphi^2)\Big) , \qquad (6)$$

where $\int d^Dx\,\sqrt{|g|}\mathcal{R}$ is the standard (D-dimensional) Einstein-Hilbert action and S_p is the p-brane action. In the simplest case of absence of the spacetime and worldvolume gauge fields (bulk and worldvolume fluxes) this reads ($S_0 = m\int d\tau\sqrt{|det(\dot{\hat{x}}^{\mu}\dot{\hat{x}}^{\nu}g_{\mu\nu}(\hat{x}(\tau)))|}$ for $p=0$, $T_0 = m$)

$$S_p = T_p\int d^{p+1}\xi\sqrt{|det(\partial_m\hat{x}^{\mu}\partial_n\hat{x}^{\nu}g_{\mu\nu}(\hat{x}))|} \qquad \Big(\Longleftrightarrow \qquad -\tfrac{1}{2}D_{\mu}\varphi^s D^{\mu}\varphi^s - V(\varphi^2)\Big) \qquad (7)$$

Varying the action with respect to the spacetime metric we arrive at the Einstein equation

$$\mathcal{G}_{\mu\nu}(g) := \sqrt{|g|}(\mathcal{R}_{\mu\nu} - \tfrac{1}{2}g_{\mu\nu}\mathcal{R}) = \kappa T_{\mu\nu} , \quad T^{\mu\nu} = \tfrac{T_p}{4}\int d^{p+1}\xi\sqrt{\gamma}\gamma^{mn}\partial_m\hat{x}^{\mu}\partial_n\hat{x}^{\nu}\delta^D(x-\hat{x}(\xi)) \quad (8)$$

with $\mathcal{R} = g^{\mu\nu}\mathcal{R}_{\mu\nu}$, $\mathcal{R}_{\mu\nu} = \mathcal{R}_{\mu\rho\nu}{}^{\rho}$ and the energy momentum tensor $T^{\mu\nu}$ having support on the p-brane worldvolume W^{p+1} (3). Varying (6) with respect to $\delta\hat{x}^{\mu}(\xi^m)$ one obtains the so-called minimal surface equation. This is dependent: as it was shown in [10] for the bosonic string and in [4] for super-p-brane cases it can be obtained as a consequence of the Einstein equation.

Thus the p-brane source term given by energy momentum tensor in the $r.h.s.$ of the Einstein eq. (2) is the counterpart of the gauge field mass term in $D^\mu F_{\mu\nu} = m\mathcal{A}_\nu$. However [4]–[7], this p-brane source term cannot be considered as a mass term for graviton, but rather as a counterpart of the cosmological constant term which cannot be identified with the graviton mass [7].

2. Spacetime filling brane (p–brane with $p = D - 1$) as cosmological constant.

In the p=D-1 case the $S_{p=D-1} =\propto T_{D-1}\int d^D\xi\sqrt{|det(\partial_m\hat{x}^\mu\partial_n\hat{x}^\nu g_{\mu\nu}(\hat{x}))|}$ term in the interacting action (6), can be reduced (by gauge fixing $\xi^m = \delta^m_\mu x^\mu$) to cosmological constant term

$$S_{p=(D-1)} = \frac{\Lambda}{\kappa}\int d^Dx\sqrt{|g_{\mu\nu}(x))|}\,. \tag{9}$$

We have to stress that the notion of spacetime filling brane becomes nontrivial in the String/M-theory context (the references can be found in [4]–[7]) in particular, because such branes ($e.g.$ D9–branes of type IIB string theory) are carriers of the gauge fields and fermionic Goldstone fields corresponding to spontaneous breaking of bulk supersymmetry.

With this in mind, *it is tempting to speculate on* that this simple observation can be useful as a basis to resolve the *cosmological constant problem*. Namely, the small value of cosmological constant with respect to a straightforward QFT estimation for vacuum energy, as well as the fact that its value is nonzero and positive while supersymmetry and supergravity prefer zero or negative cosmological constant, might be explained just by that the cosmological constant is determined by the tension of a spacetime filling brane, $\Lambda \propto T_{D-1}$. However, developing such a conjecture goes beyond the score of this contribution.

A simple proof of that the cosmological constant cannot be considered as graviton mass has been presented in [7]. Schematically, the arguments are as follows. In the case of small but non-vanishing cosmological constant, $\Lambda \to 0$ but $\Lambda \neq 0$, one may consider decomposition over the flat spacetime $g_{\mu\nu} = \eta_{\mu\nu} + h_{\mu\nu}$ of the Einstein equation with cosmological constant,

$$\mathcal{G}_{\mu\nu}(\eta + h) = \mathcal{G}^\eta_{\mu\nu}(h) + \mathcal{O}(hh) = \Lambda\eta_{\mu\nu} + \Lambda h_{\mu\nu}\,. \tag{10}$$

Here $\mathcal{G}^\eta_{\mu\nu}(h)$ contains the contributions of the first order in weak field h and the terms of higher order in h are denoted by $\mathcal{O}(hh)$ (the Einstein tensor calculated with the flat metric vanishes $\mathcal{G}_{\mu\nu}(\eta) = 0$). The first impression might be that $\Lambda\eta_{\mu\nu}$ provides the mass term for graviton $h_{\mu\nu}$. This, however, is not the case. Indeed, if the first order approximation to Eq. (10) is given by $\mathcal{G}_{\mu\nu}(h) = \Lambda h_{\mu\nu}$ then zero order approximation should be $\Lambda\eta_{\mu\nu} = 0$ which implies the vanishing of the cosmological constant $\Lambda = 0$ (in contradiction with the original assumption). Thus the selfconsistent weak (h-) field approximation to (10) with $\Lambda \neq 0$ can be formulated only assuming that the cosmological constant is of the same order as h, $h \propto \Lambda$, so that the first order approximation is given by $\mathcal{G}^\eta_{\mu\nu}(h) - \Lambda\eta_{\mu\nu} = 0$; this equation does not contain mass term for h.

3. Discussion and conclusion

The deep reason beyond the fact that cosmological constant contribution to field equations cannot be considered as a mass term is that the equation possesses the gauge symmetry with the same number of parameters as in the case of vanishing cosmological constant, and, hence, the graviton maintains the same number of polarizations $D(D-3)/2$ (which is 2 for $D = 4$) as in the case of massless spin 2 field. So the AdS graviton is also massless (like the Minkowski one).

Indeed, despite fixing the gauge (5), the functional (9) and the corresponding $p = D$ interacting action (6) is invariant under the spacetime diffeomorphism symmetry $Diff_D$. The reason for this is that the complete group of the gauge invariance of the interacting action (6) is the direct product $Diff_D \otimes Diff_{p+1}$ of the spacetime and worldvolume diffeomorphism transformations. These correspond to the parametric functions $a^\mu(x)$ and $\delta\xi^m(\xi)$ in Eq. (4) which can be used to fix (on the worldvolume W^{p+1}) the gauge (5). This remains invariant under combined (spacetime and worldvolume) $Diff_{p+1}$ diffeomorphisms.

In the case of spacetime filling brane p=D-1 this residual $Diff_{p+1}$ invariance is the spacetime diffeomorphism invariance $Diff_D$ of the Einstein action with cosmological constant. Then the number of polarization of graviton remains the same as in the case of absence of cosmological constants. In the case of not-spacetime–filling brane, p<D-1, the gauge symmetry is reduced to $Diff_{p+1}$, but only on the worldvolume W^{p+1}, while outside W^{p+1} the $Diff_D$ still acts on graviton. Thus after gauge fixing graviton acquires some additional degrees of freedom (counterparts of additional polarizations), but only on $W^{p+1} \in M^D$, the points of which for p<(D-1) form the set of measure zero with respect to the set of all M^D points. This is why one cannot identify the p–brane energy momentum tensor in (2) with a mass term for graviton.

The above discussion allows us to understand the difference of the gravity plus brane system with the model for massive spin 2 fields proposed not so long ago by G. 't Hooft [3],

$$S_{'t\ Hooft} = \frac{1}{2\kappa}\int d^Dx\,\sqrt{|g|}\mathcal{R} + \int d^4x\sqrt{|g|}g^{\mu\nu}(x)\partial_\mu X^a(x)\partial_\nu X^a(x)\,, \qquad a = 1,\ldots,D \quad (11)$$

Although seemingly similar to the spacetime filling brane action ((7) of (6)), the second term of (11) differs from that by that it involves not one but two spacetime metrics: $g_{\mu\nu}(x)$ and the constant δ_{ab} used to contract the indices of the Goldstone fields $X^a(x)$. As a result, the (spontaneous) breaking of the diffeomorphism symmetry by the dynamical system described by the action (11) is complete, the gauge fixed action ($X^a = \delta^a_\mu x^\mu \cdot \frac{m}{2\kappa}$)

$$S_{'t\ Hooft}|_{gauge} = \frac{1}{2\kappa}\int d^Dx\,\sqrt{|g|}\mathcal{R} + \frac{m^2}{2\kappa}\int d^4x\sqrt{|g(x)|}g^{\mu\mu}(x) \quad (12)$$

does not possess gauge symmetry and hence its 'graviton' acquires additional polarizations [3].

As far as the p-brane plus gravity interacting system described by Eqs. (6) and (7) is concerned, although the p-brane source terms play in it the same role as the mass terms of the gauge fields in the standard Higgs effect, they cannot be identified with the mass term, but rather with the counterpart of the cosmological constant term localized on the p-brane worldvolume [4]–[7]. The cosmological term itself can be associated with a spacetime feeling brane ($p = (D-1)$) and, as we have discussed above, cannot be considered as a mass term.

Acknowledgements. The author thanks Yury Zinoviev for useful communication. This work was partially supported by the research grants FIS2008-1980 from the Spanish MICINN and 38/50–2008 from Ukrainian National Academy of Sciences and Russian Federation RFFI.

References

[1] Englert F and Brout R 1964 *Phys. Rev. Lett.* **13** 321
Higgs P W 1964 *Phys. Lett.* **12** 132
[2] Duff M J 1975 *Phys. Rev.* D **12** 3969-3971
Omero C and Percacci R 1980 *Nucl. Phys.* B **165** 351
Percacci R 1991 *Nucl. Phys.* B **353** 271
Porrati M 2002 *J. High Energy Phys.* JHEP04(2002)058
Chamseddine A H 2003 *Phys. Lett.* B **557** 247
Zinoviev Yu M 2007 *Nucl. Phys.* B **770** 83-106
[3] 't Hooft G 2007 Unitarity in the Brout-Englert-Higgs Mechanism for Gravity *Preprint* arXiv:0708.3184
[4] Bandos I A, de Azcarraga J A, and Izquierdo J M 2002 *Phys. Rev.* D **65** 105010
[5] Bandos I A, de Azcarraga J A, Izquierdo J M and Lukierski J 2003 *Phys. Rev.* D **67** 065003
Bandos I A and Isidro J M 2004 *Phys. Rev.* D **69** 085009
[6] Bandos I A, de Azcarraga J A, Izquierdo J M and Lukierski J 2003 *Phys. Rev.* D **68** 046004
[7] Bandos I A 2003 Supergravity interacting with superbranes and spacetime Higgs effect in general relativity *Preprint* hep-th/0312095
[8] Bandos I A and de Azcarraga J A 2005 *J. High Energy Phys.* JHEP09(2005)064
[9] Hughes J and Polchinski J 1986 *Nucl. Phys.* B **278**, 147–169
Hughes J, Liu J and Polchinski J 1986 *Phys. Lett.* B **180** 370–374
[10] Gürses M and Gürsey F 1975 *Phys. Rev.* D **11** 967-969
Aragone C and Deser S 1975 *Nucl. Phys.* B **92** 327-333

Semiclassical instability of warp drives

C Barceló[1]*, S Finazzi[2] and S Liberati[3]
[1] Instituto de Astrofísica de Andalucía, IAA-CSIC, Glorieta de la Astronomía s/n, 18008 Granada, Spain
[2] [3] SISSA, via Beirut 2-4, Trieste 34151, Italy; INFN sezione di Trieste
E-mail: [1]`carlos@iaa.es`, [2]`finazzi@sissa.it`, [3]`liberati@sissa.it`

Abstract. Warp drives, at least theoretically, provide a way to travel at superluminal speeds. However, even if one succeeded in providing the necessary exotic matter to construct them, it would still be necessary to check whether they would survive to the switching on of quantum effects. In this contribution we will report on the behaviour of the Renormalized Stress-Energy Tensor (RSET) in the spacetimes associated with superluminal warp drives. We find that the RSET will exponentially grow in time close to the front wall of the superluminal bubble, hence strongly supporting the conclusion that the warp-drive geometries are unstable against semiclassical back-reaction.

1. Introduction

Since they were introduced by Alcubierre [1], warp-drive spacetimes have been certainly one of the most studied solutions of the Einstein equations among those requiring exotic matter [2]. They are not only an exciting theoretical test for our comprehension of general relativity and quantum field theory in curved spacetimes, but they might also be, at least theoretically, a way to travel at superluminal speed. The warp drive consists of a bubble containing an almost flat region, moving at arbitrary speed within an asymptotically flat spacetime.

After the proposal of this solution, its most investigated aspect has been the amount of exotic matter (*i.e.* energy-conditions violating matter) required to support such a spacetime [3, 4, 5, 6]. It has been found, using the so called *quantum inequalities* (QI), that such a matter must be confined in Planck-size regions at the edges of the bubble. This bound on the wall thickness turns into lower limits on the amount of exotic matter required to support the bubble (at least of the order of 1 solar mass).

Less effort has been devoted to other important issue regarding the feasibility of these spacetimes: the study of the warp-drive semiclassical stability. In particular, it was studied in [7] the case of an eternal superluminal warp drive by discussing its stability against quantum effects. It was there noticed that, to an observer within the warp-drive bubble, the backward and forward walls (along the direction of motion) look respectively as the future and past event horizon of a black hole. By imposing over the spacetime a quantum state which is vacuum at the null infinities (*i.e.* what one may call the analogue of the Boulware state for an eternal black hole) it was found that the renormalized stress-energy tensor (RSET) had to diverge on the horizons .

In this contribution we consider the more realistic case of a warp drive which is created with zero velocity at early times and then accelerated up to some superluminal speed in a finite time

Journal of Physics: Conference Series **229** (2010) 012018 doi:10.1088/1742-6596/229/1/012018

(a more detailed treatment can be found in [8]). We found, as expected, that in the centre of the bubble there is a thermal flux at the Hawking temperature corresponding to the surface gravity of the black horizon. However, this surface gravity is inversely proportional to the wall thickness, leading to a temperature of the order of the Planck temperature, for Planck-size walls. Even worse, we do show that the RSET does increase exponentially with time on the white horizon (while it is regular on the black one). This clearly implies that a warp drive becomes rapidly unstable once superluminal speed are reached.

2. Light-ray propagation

In the actual computation we shall restrict our attention to the $1+1$ dimensions case (since this is the only one for which one can carry out a complete analytic treatment as explained below).[1] Changing coordinates to those associated with an observer at the centre of the bubble, the warp-drive metric in [1] becomes

$$ds^2 = -c^2dt^2 + [dr - \bar{v}(r)dt]^2 , \qquad \bar{v} = v - v_c . \tag{1}$$

Here $v(r) = v_c f(r)$ with f a suitable smooth function satisfying $f(0) = 1$ and $f(r) \to 0$ for $r \to \infty$; v_c is the velocity of the centre of the bubble and r the signed distance from the centre of the bubble $r = 0$. In our dynamical situation the warp-drive geometry interpolates between an initial Minkowski spacetime [$\hat{v}(t,r) \to 0$, for $t \to -\infty$] and a final stationary superluminal ($v_c > c$) bubble [$\hat{v}(t,r) \to \bar{v}(r)$, for $t \to +\infty$]. To an observer living inside the bubble this geometry has two horizons, a *black horizon* $\mathscr{H}^+$ located at r_1 and a *white horizon* $\mathscr{H}^-$ located at r_2. For those interested, in [8] you can find the Penrose diagram of these spacetimes. Here let us just mention that from the point of view of the Cauchy development of $\mathscr{I}^-$ these spacetimes have Cauchy horizons.

Let us now consider light-ray propagation in the above described geometry. Only the behaviour of right-going rays determines the universal features of the RSET, just like outgoing modes do in the case of a black hole collapse [8, 9, 10]. Therefore, we need essentially the relation between the past and future null coordinates U and u, labelling right-going light rays. Following [9], this relation can be found by integrating the right-going-ray equation

$$\frac{dr}{dt} = c + \hat{v}(r,t) . \tag{2}$$

There are two special right-going rays defining, respectively, the asymptotic location of the black and white horizons. In terms of the right-going past null coordinate U let us denote these two rays by $U_{\rm BH}$ and $U_{\rm WH}$, respectively. The finite interval $U \in (U_{\rm WH}, U_{\rm BH})$ is mapped to the infinite interval $u \in (-\infty, +\infty)$ covering all the rays travelling inside the bubble. For rays which are close to the black horizon, in [8] the present authors proved that the relation between U and u can be approximated as a series of the form

$$U(u \to +\infty) \simeq U_{\rm BH} + A_1 e^{-\kappa_1 u} + \frac{A_2}{2} e^{-2\kappa_1 u} + \dots . \tag{3}$$

Here A_n are constants (with $A_1 < 0$) and $\kappa_1 > 0$ represents the surface gravity of the black horizon. This relation is the standard result for the formation of a black hole through gravitational collapse. As a consequence, the quantum state which is vacuum on $\mathscr{I}^-$ will show, for an observer inside the warp-drive bubble, Hawking radiation with temperature $T_H = \kappa_1/2\pi$.

[1] Indeed, we do expect that the salient features of our results would be maintained in a full 3+1 calculation, given that they will still be valid in a suitable open set of the horizons centred around the axis aligned with the direction of motion.

Equivalently, we find that the corresponding expansion in the proximity of the white horizon is

$$U(u \to -\infty) \simeq U_{\rm WH} + D_1 e^{\kappa_2 u} + \frac{D_2}{2} e^{2\kappa_2 u} + \ldots , \tag{4}$$

where $D_1 > 0$ and κ_2 is the white hole surface gravity and is also defined to be positive (κ_2 could be different from κ_1 in general, although it is expected, in principle, to be comparable with κ_1). The interpretation of this relation in terms of particle production is not as clear as in the black horizon case. For this reason, we shall consider now the RSET.

3. Renormalized stress-energy tensor

For the calculation of the RSET inside the warp-drive bubble we use the method proposed in [10]. In past null coordinates U and W the metric can be written as $ds^2 = -C(U,W)dUdW$. In the stationary region at late times, we use the previous future null coordinate u and $\tilde{w}$, defined as

$$\tilde{w}(t,r) = t + \int_0^r \frac{dr}{c - \bar{v}(r)} , \tag{5}$$

to write the metric as

$$ds^2 = -\bar{C}(u,\tilde{w}) du d\tilde{w} , \quad C(U,W) = \frac{\bar{C}(u,\tilde{w})}{\dot{p}(u)\dot{q}(\tilde{w})} , \tag{6}$$

where $U = p(u)$ and $W = q(\tilde{w})$. In this way, $\bar{C}$ depends only on r through $u, \tilde{w}$.

For concreteness, we refer to the RSET associated with a quantum massless scalar field living on the spacetime. When particularized to the stationary region, the RSET components [11] acquire the following form:

$$T_{UU} = -\frac{1}{12\pi} C^{1/2} \partial_U^2 C^{-1/2} = -\frac{1}{48\pi} \frac{1}{\dot{p}^2} \left[\bar{v}'^{\,2} + (1 - \bar{v}^2)\, \bar{v}\bar{v}'' - f(u) \right] , \tag{7}$$

$$T_{WW} = -\frac{1}{12\pi} C^{1/2} \partial_W^2 C^{-1/2} = -\frac{1}{48\pi} \frac{1}{\dot{q}^2} \left[\bar{v}'^{\,2} + (1 - \bar{v}^2)\, \bar{v}\bar{v}'' - g(\tilde{w}) \right] , \tag{8}$$

$$T_{UW} = T_{WU} = \frac{1}{96\pi} C\, R = -\frac{1}{48\pi} \frac{1}{\dot{p}\dot{q}} (1 - \bar{v}^2) \left[\bar{v}'^{\,2} + \bar{v}\bar{v}'' \right] . \tag{9}$$

Here we have put $c = 1$ and we have defined

$$f(u) \equiv \frac{3\ddot{p}^2(u) - 2\dot{p}(u)\, \dddot{p}(u)}{\dot{p}^2(u)} , \tag{10}$$

$$g(\tilde{w}) \equiv \frac{3\ddot{q}^2(\tilde{w}) - 2\dot{q}(\tilde{w})\, \dddot{q}(\tilde{w})}{\dot{q}^2(\tilde{w})} . \tag{11}$$

One can show [8] that $\dot{q}$ contains solely information associated with the dynamical details of the transition region. Moreover, for simple dynamical interpolations between Minkowski and the final warp drive, $\dot{q}(\tilde{w})$ goes to a constant at late times, such that $g(\tilde{w}) \to 0$. From now on, we will neglect this term.

We want to look at the energy density inside the bubble, in particular at the energy ρ as measured by a set of free-falling observers, whose four velocity is $u_c^\mu = (1, \bar{v})$ in (t,r) components. For these observers we obtain $\rho = T_{\mu\nu} u_c^\mu u_c^\nu = \rho_{\rm st} + \rho_{\rm dyn}$ where we define a static term $\rho_{\rm st}$, which depends only on the r coordinate through $\bar{v}(r)$, and a dynamic term $\rho_{\rm dyn}$ as

$$\rho_{\rm st} \equiv -\frac{1}{24\pi} \left[\frac{(\bar{v}^4 - \bar{v}^2 + 2)}{(1 - \bar{v}^2)^2} \bar{v}'^{\,2} + \frac{2\bar{v}}{1 - \bar{v}^2} \bar{v}'' \right] , \qquad \rho_{\rm dyn} \equiv \frac{1}{48\pi} \frac{f(u)}{(1 + \bar{v})^2} . \tag{12}$$

This latter term, depending also on u, corresponds to energy travelling on right-going rays, eventually red/blue-shifted by a term depending on r.

By studying now the behaviour of ρ at late times in the centre of the bubble and at the horizons $\mathscr{H}^+$ and $\mathscr{H}^-$, we reach the following conclusions.

4. Summary of results

(1) We found that the central region of the warp drive behaves like the asymptotic region of a black hole: In both of these regions the static term ρ_{st} vanishes and the whole energy density is due to the Hawking radiation generated at the black horizon. If one trusts the QI [3, 4], the wall thickness for a warp drive with $v_c \approx c$ would be $\Delta \lesssim 10^2 L_P$, and its surface gravity $\kappa_1 \gtrsim 10^{-2} t_P^{-1}$, where $t_P \approx 10^{-43}$ s is the Planck time. Hence, the Hawking temperature of this radiation would be unacceptably large: $T_H \sim \kappa_1 \gtrsim 10^{-2} T_P$.

(2) The formation of a white horizon produces a transient radiation which accumulates on the white horizon itself. This causes the energy density ρ, as seen by a free-falling observer, to grow unboundedly with time on this horizon. The semiclassical backreaction of the RSET will make the superluminal warp drive to become rapidly unstable, in a time scale of the order of $1/\kappa_2$ (i.e. of the inverse of the surface gravity of the white horizon). In fact, in order to get even a time scale $\tau \sim 1$ s for the growing rate of the RSET, one would need a wall as large as 3×10^8 m. Thus, most probably, one would be able to maintain a superluminal speed for just a very short interval of time.

(3) The formation of a Cauchy horizon gives rise to an instability, similar to the inner horizon instability in black holes. This is due to the blue-shift suffered by the Hawking radiation produced by the black horizon when it gets accumulated in the white horizon.

On the base of these evidences, we think that this work is convincingly ruling out the semiclassical stability of superluminal warp drives.

Acknowledgments

The authors wish to thank S. Sonego and M. Visser for illuminating discussions. C.B. has been supported by the Spanish MICINN under project FIS2008-06078-C03-01 and Junta de Andalucia under project FQM2288.

References

[1] Alcubierre M 1994 *Class. Quantum Grav.* **11** L73–L77
[2] Lobo F S N 2007 *Preprint* gr–qc/0710.4474
[3] Pfenning M J and Ford L H 1997 *Class. Quantum Grav.* **14** 1743–1751
[4] Van Den Broeck C 1999 *Class. Quantum Grav.* **16** 3973–3979
[5] Lobo F S N and Visser M 2004 *Class. Quantum Grav.* **21** 5871–5892
[6] Lobo F S N and Visser M 2004 *Preprint* gr–qc/0412065
[7] Hiscock W A 1997 *Class. Quantum Grav.* **14** L183–L188
[8] Finazzi S, Liberati S and Barcelo C 2009 *Phys. Rev.* D **79** 124017
[9] Barceló C, Liberati S, Sonego S and Visser M 2006 *Class. Quantum Grav.* **23** 5341–5366
[10] Barcelo C, Liberati S, Sonego S and Visser M 2008 *Phys. Rev.* D **77** 044032
[11] Birrell N D and Davies P C W 1984 *Quantum Fields in Curved Space* (Cambridge: Cambridge University Press)

Dark energy in vector-tensor theories of gravity

Jose Beltrán Jiménez[1*] **and Antonio L Maroto**[2]
Dpto. Física Teórica, Universidad Complutense de Madrid, 28040, Madrid, Spain.
E-mail: [1] `jobeltra@fis.ucm.es`; [2] `maroto@fis.ucm.es`

Abstract. We consider a general class of vector-tensor theories of gravity and show that solutions with accelerated expansion and a future type III singularity are a common feature in these models. We also show that there are only six vector-tensor theories with the same small scales behaviour as General Relativity and, in addition, only two of them can be made completely free from instabilities. Finally, two particular models as candidates for dark energy are proposed: on one hand, a cosmic vector that allows to alleviate the usual naturalness and coincidence problems and, on the other hand, the electromagnetic field is shown to give rise to an effective cosmological constant on large scales whose value can be explained in terms of inflation at the electroweak scale.

1. Introduction

The standard model of Cosmology accounts for the phase of accelerated expansion that the universe is currently undergoing by introducing a cosmological constant term in the gravitational action. According to observations, the value of this cosmological constant should be (10^{-3} eV)^4, which is very far away from the natural scale of gravitation set by the Planck mass $M_P = 1/\sqrt{G} \sim 10^{19}$ GeV. This fact poses the so-called naturalness problem associated to the cosmological constant. However, this could be signalling that the cosmological constant causing the accelerated expansion is not a true constant, but an effective description arising from some underlying physical theory. On the other hand, the accelerated expansion could also be caused by some other physical mechanism like a cosmological scalar field or a modification of the gravitational interaction [1]. Nonetheless, in most of these models, the original naturalness problem still remains because new dimensional scales are needed in order to produce the correct acceleration. Furthermore, they usually have some other problems related to the presence of instabilities or local gravity constrains. In this work we shall consider vector-tensor theories of gravity and show how vector fields can help to alleviate the naturalness problem. Indeed, we shall show that the very familiar electromagnetic field can play the role of dark energy without any of the aforementioned problems.

2. Vector-tensor theories of gravity

We shall consider the most general action for a vector-tensor theory leading to second order linear equations of motion for the vector field:

$$S[g_{\mu\nu}, A_\mu] = \int d^4x\sqrt{-g}\left[-\frac{1}{16\pi G}R + \omega R A_\mu A^\mu + \sigma R_{\mu\nu}A^\mu A^\nu + \lambda(\nabla_\mu A^\mu)^2 + \epsilon F_{\mu\nu}F^{\mu\nu}\right] \quad (1)$$

with ω, σ, λ and ϵ dimensionless parameters. The interest of this theory is that G is the only dimensional constant and we do not have potential terms for the vector field so that no new

Journal of Physics: Conference Series **229** (2010) 012019 doi:10.1088/1742-6596/229/1/012019

scales are introduced in the action. A general study of the homogeneous cosmological evolution for this action is given in [2] where a classification attending to the behaviour of the vector field throughout the different epochs of the universe evolution is performed. Also, the case in which the vector field dominates the universe is considered and the corresponding autonomous system is analysed. For this system, it is shown the presence of both attractors and repellers in which accelerated expansion can take place. These repellers are interesting for inflaton models because we have an initial state with accelerated expansion that can eventually end as the trajectory in the phase space goes away from the critical point. On the other hand, attractors with accelerated expansion can give rise to dark energy candidates since they provide late-time accelerated expansion. Finally, when a matter component is introduced in addition to the vector field, we have that, for quite general conditions, there are solutions with a transition from a matter dominated universe to a phase of accelerated expansion with a phantom behaviour for the vector field that leads to a future type III singularity in which the scale factor remains finite but the expansion rate diverges.

The small scales behaviour of the vector-tensor theory given by (1) is well described by means of its PPN parameters. Since GR is in excellent agreement with Solar System experiments, an alternative gravitational theory whose PPN parameters are the same as those of GR will agree with small scales viability conditions. The PPN parameters are obtained by performing perturbations around a Minkowski spacetime with a constant background vector field of the form $A_\mu = (A, 0, 0, 0)$. When we impose the PPN parameters of the vector-tensor theory to be identical to those of GR we find that there are six models which satisfy such a condition for any value of the background vector field [3]. These six models have $\omega = 0$ and can be classified as: Gauge non-invariant models: *i)* $\sigma = -4\lambda = -4\epsilon$, *ii)* $\sigma = -3\lambda = -2\epsilon$, *iii)* $\sigma = 0$; Gauge invariant models: $\lambda = 0$ and $\sigma = m\epsilon$ with $m = 0, -2, -4$. For these particular vector-tensor theories we study the presence of instabilities both at the classical and the quantum level in the inhomogeneous perturbations[1]. The classical stability can be achieved by imposing the absence of modes whose propagation speed is imaginary so that we do not have exponentially growing modes. On the other hand, the quantum stability will be guaranteed by eliminating the modes with negative energy (ghosts) in order to have a stable quantum vacuum (see Table 1). We find that the only two theories which can be made completely free from instabilities are the pure Maxwell action and Maxwell action supplemented with a gauge fixing term.

3. Scaling vector dark energy

The cosmic vector model having a scaling behaviour in the early universe proposed in [4] is described by the following action:

$$S = \int d^4x \left[-\frac{R}{16\pi G} - \frac{1}{4} F_{\mu\nu} F^{\mu\nu} - \frac{1}{2} (\nabla_\mu A^\mu)^2 + R_{\mu\nu} A^\mu A^\nu \right]. \tag{2}$$

In this model, the fraction of energy density stored in the vector field remains constant throughout the radiation dominated epoch and, in addition, the value of the vector field during such an epoch also remains constant. This prevents any dependence on the moment at which we set the initial conditions because the subsequent evolution of the universe will be the same irrespective of the initial redshift (as long as it is well inside the radiation dominated epoch). Moreover, the initial fraction of energy density corresponding to the vector field needed to explain the SNIa measurements turns out to be $\sim 10^{-6}$ and the vector field must take a value $\sim 10^{-4} M_P$, which are values that can arise *naturally* in the early universe, as quantum fluctuations for instance. That way, this model allows to alleviate the naturalness problem and it does not require any fine-tunings.

[1] We consider perturbations in both the vector field and the metric tensor.

Table 1. In this table we summarize the conditions obtained in order to have both classical and quantum stability for the models with the same set of PPN parameters as GR. The $m = 0$ gauge invariant model satisfies all the viability conditions.

	Model I	Model II	Model III	Gauge invariant models $m = -2, -4$
Classical stability	$-32\epsilon A^2 < 1$	$-15\lambda A^2 < 1$	Always	$\epsilon A^2 \notin \left[\frac{1}{(2+\frac{1}{2}m)m}, \frac{1}{2m}\right]$
Gravitational waves	$-16\epsilon A^2 < 1$	$-12\lambda A^2 < 1$	Unaffected	$4m\epsilon A^2 < 1$
Quantum stability	$1 < -64\epsilon A^2 < 4$	$\lambda A^2 \in (-0.098, -0.033)$	$\partial_\mu A^\mu = 0$ and $\epsilon < 0$	$\varepsilon A^2 \in (a_-, \frac{1}{2m}) \cup (b_+, 0)$
Viability condition	$1 < -64\epsilon A^2 < 2$	$\lambda A^2 \in (-\frac{1}{15}, -0.033)$	$\partial_\mu A^\mu = 0$ and $\epsilon < 0$	$\epsilon A^2 \in (b_+, \frac{1}{4m})$

Finally, we have confronted the predictions of the model to SNIa, CMB and BAO measurements. An interesting property of this model is that it provides the best fit to date to the SN Gold dataset with the same number of parameters as the standard ΛCDM. Moreover, CMB measurements predicts a closed universe for this model, being the flat case strongly disfavoured. Finally, we find some tension with BAO observations, which could be due to the dependence of such data on the fiducial model (ΛCDM) whose evolution is very different from that given by the cosmic vector model.

4. Electromagnetic dark energy

The second vector dark energy model we are going to discuss is nothing but the very familiar electromagnetic field. It is well-known [5] that the canonical quantization of the electromagnetic field needs to impose the subsidiary Lorenz condition $\nabla_\mu A^\mu = 0$ in order to reduce the corresponding Hilbert space so that we end up with the two transverse states for the photon. However, when we consider the covariant quantization in an expanding universe, the Lorenz condition turns out to be difficult to implement. This is due to the fact that $\nabla_\mu A^\mu$ behaves as a free scalar field so that it can be excited by means of gravitational fields. For these reasons, we consider the action including a gauge fixing term[6]:

$$S = \int d^4x\sqrt{-g}\left[-\frac{1}{4}F_{\mu\nu}F^{\mu\nu} + \frac{\xi}{2}(\nabla_\mu A^\mu)^2 + A_\mu J^\mu\right] \quad (3)$$

as the fundamental action for the electromagnetic field. In this action the gauge fixing term is a physical term, rather than being a mathematical trick, so that we break the $U(1)$ invariance, although a residual gauge symmetry $A_\mu \rightarrow A_\mu + \partial_\mu\vartheta$ with $\Box\vartheta = 0$ remains. Notice that this action is also the starting action in the path integral formalism once we have fixed the element of the orbits of the group. The corresponding modified Maxwell equations deduced from (3) are:

$$\nabla_\nu F^{\mu\nu} + \xi\nabla^\mu(\nabla_\nu A^\nu) = J^\mu. \quad (4)$$

Now, if we take the four divergence of this equation we obtain: $\Box(\nabla_\nu A^\nu) = 0$ where we have used the conservation of the electromagnetic current. Thus, as we said before, $\nabla_\mu A^\mu$ behaves as a free scalar field non-minimally coupled to gravity. Moreover, in an expanding universe, a free scalar field gets frozen for super-Hubble modes so that the gauge fixing term appearing in (3) will give an effective cosmological constant on large scales. On the other hand, we also know that a free scalar field decays as $1/a$ for sub-Hubble modes which implies that the gauge fixing term will be negligible at small scales and we effectively recover the usual Maxwell equations.

Since we are breaking the $U(1)$ invariance and we do not impose any extra subsidiary condition, the theory contains one extra degree of freedom so that the field can be decomposed as follows: $A_\mu = A_\mu^{(1)} + A_\mu^{(2)} + A_\mu^{(s)} + \partial_\mu\vartheta$ where $A_\mu^{(i)}$ with $i = 1, 2$ are the two transverse modes of the massless photon, $A_\mu^{(s)}$ is a new scalar state that represents the mode that would have been eliminated if we had imposed the Lorenz condition and, finally, $\partial_\mu\theta$ is a purely residual gauge mode, which can be eliminated by means of a residual gauge transformation in the asymptotically free regions, in a completely analogous way to the elimination of the A_0 component in the Coulomb gauge. The complete quantization procedure in an inflationary phase with De-Sitter expansion can be found in [6] where the primordial power spectrum defined by $\langle 0|(\nabla_\mu A^\mu)^2|0\rangle = \int \frac{dk}{k} P_A(k)$ for super-Hubble modes is computed and given by $P_A(k) = \frac{9H_I^4}{16\pi^2}$ with H_I the constant Hubble parameter during inflation. This result implies that the effective cosmological constant given by the gauge fixing term is $\sim H_I^4$ so that the measured value of the cosmological constant requires $H_I \sim 10^{-3}$ eV, which corresponds to an inflationary scale $M_I \sim 1$ TeV. Thus, we see that the cosmological constant scale can be naturally explained in terms of physics at the electroweak scale.

Although the homogeneous evolution is the same as in the standard ΛCDM, we have a crucial difference because, in this case, we can have inhomogeneous perturbations, unlike in the standard model where the cosmological constant is a true constant. In [7] we have considered the effects of the inhomogeneous perturbations on the CMB anisotropies and large scale structures and found that only at the largest scales we can have discriminatory consequences. If the primordial amplitude of the perturbations are not extremely large, the CMB temperature and matter power spectra are perfectly compatible with observations.

5. Conclusions

We have shown that vector fields are compelling candidates to drive an accelerated phase and they can alleviate or even solve some of the usual naturalness or coincidence problems arising in most of dark energy models. More interestingly, we have shown that the electromagnetic field can account for the present phase of accelerated expansion of the universe and provides a natural explanation for the observed value of the cosmological constant. This shows that it is possible to establish the fundamental nature of dark energy without resorting to new physics. *Acknowledgements*: This work has been supported by Ministerio de Ciencia e Innovación (Spain) project numbers FIS 2008-01323 and FPA 2008-00592, UCM-Santander PR34/07-15875, CAM/UCM 910309 and MEC grant BES- 2006-12059.

References

[1] Wetterich C 1988 *Nucl. Phys* B **302** 668
Caldwell R R, Dave R and Steinhardt P J 1998 *Phys. Rev. Lett.* 80 1582
Armendariz-Picon C, Damour T and Mukhanov V 1999 *Phys. Lett.* B **458** 209
Koivisto T and Mota D F 2007 *Phys. Rev.* D **75** 023518
Carroll S M, Duvvuri V, Trodden M and Turner M S 2004 *Phys. Rev.* D **70** 043528
Dvali G, Gabadadze G and Porrati M 2000 *Phys. Lett.* B **485** 208
[2] Beltrán Jiménez J and Maroto A L 2009 *Phys. Rev.* D **80** 063512
[3] Beltrán Jiménez J and Maroto A L 2009 *J. Cosmol. Astropart.* JCAP02(2009)025
[4] Beltrán Jiménez J and Maroto A L 2008 *Phys. Rev.* D **78** 063005
Beltrán Jiménez J, Lazkoz R and Maroto A L 2009 *Phys. Rev.* D **80** 023004
[5] Itzykson C and Zuber J B 1980 *Quantum field theory* (Dover Publications, INC.)
[6] Beltrán Jiménez J and Maroto A L 2009 *J. Cosmol. Astropart.* JCAP03(2009)016
Beltrán Jiménez J and Maroto A L 2009 *Preprint* arXiv:0903.4672
Beltrán Jiménez J and Maroto A L 2009 *Preprint* arXiv:0905.2589
Beltrán Jiménez J and Maroto A L 2009 *Preprint* arXiv:0911.3066
[7] Beltrán Jiménez J, Koivisto T S, Maroto A L and Mota D F 2009 *J. Cosmol. Astropart.* JCAP10(2009)029

Journal of Physics: Conference Series **229** (2010) 012020 doi:10.1088/1742-6596/229/1/012020

Racing a quantum computer through Minkowski spacetime

Jacob D Biamonte

Oxford University Computing Laboratory, Wolfson Building, Parks Road, Oxford, OX1 3QD, United Kingdom

Abstract. The Lorentzian length of a timelike curve connecting both endpoints of a computation in Minkowski spacetime is smaller than the Lorentzian length of the corresponding geodesic. In this talk, I will point out some properties of spacetime that allow an inertial classical computer to outperform a quantum one, at the completion of a long journey. We will focus on a comparison between the optimal quadratic Grover speed up from quantum computing and an n=2 speedup using classical computers and relativistic effects. These results are not practical as a new model of computation, but allow us to probe the ultimate limits physics places on computers.

1. Introduction

Consider a classical computation requiring N operations each taking time Δt, then $\Delta t N$ becomes the *total runtime* in a local inertial frame. Indeed, relativistic effects don't change complexity results inside an inertial frame, and the number N is agreed on by all observers. However, the total computational time experienced by observers in motion is relative. We will state this as a relation between a polynomial reduction inside the black box model and relativistic mass — this is accomplished in Section 2.2 where Theorem 1.1 is proven.

Theorem 1.1. *The minimal relativistic energy* E $(= mc^2)$ *required to perform a computation within time* N *in an inertial frame and time* $N^{1/n}$ *in the frame of an observer* O *is given as*

$$E = N^{1-1/n} m_0 c^2, \tag{1}$$

where m_0 *is the rest mass of* O, $\Delta t := 1$, n *is the order of the sought polynomial reduction and* $N \in \mathbb{N}^*, n \geq 1$.

We have found a non-linear tradeoff (1) between relativistic energy and reduction of computational runtime $\Delta t N$. This effect was experimentally observed in 1938 by Ives and Stilwell [1], observation in macroscopic clocks occurred in 1972 [2] with the current state of the art found in [3].

1.1. Computational Complexity Theory

The resources consumed by an *efficient algorithm* scale polynomially in the problem size — necessarily from the class P (of problems known to be efficiently solvable). The most famous open question in Computer Science concerns proving if it is impossible to efficiently solve a complete problem from the class NP [4, 5, 6].

Although the P≠NP question regards general algorithmic complexity, one can consider its physical analogue by asking if the laws of physics allow, even in principle, the existence of a physical process that can be harnessed to speed up the solution to an NP-complete problem [7, 8]. Deterministic query complexity in the black-box model is the ideal framework to address this question. To date, research connecting complexity theory and relativity theory has been focused on the implications the existence of closed time-like curves would have on computation [9, 10, 11, 12, 13, 14, 15].

Consider a classical device computing $f : \{0,1\}^n \rightarrow \{0,1\} :: x \mapsto f(x)$, for a given function defined on its range of inputs for positive integer n. In the case of an unstructured database, one is given f in a black-box with a promise that f outputs 1 for a single input x' and 0 otherwise. Using a classical computer, to determine x' in the worst case requires $N := 2^n$ *queries* of the search space — each taking time Δt. We will consider solving such an NP-complete problem.

2. Einstein's Computer of 1905

Let I denote an observer travelling along a geodesic that is the common origin of the coordinate system and let O denote an observer in relative motion. Consider an event **0** at which (i) O passes I; (ii) they synchronize their temporal and spatial orientations and (iii) I begins the computation.

In the following, we are working in units of ticks (and not roots or powers of seconds, as the slight abuse of notation might suggest). Now consider the general question of relating the proper time interval experienced in O's frame as the nth root of the time interval experienced in I's frame. This is done by relating the clocks of I and O — we insist that $t(\tau) = T^n$ ticks and $\tau(t) = T$ ticks, where O reaches d at the event $(cT^n, d, 0, 0)$ in I's frame, and so T^n and T are dimensionless constants. This is made possible by the Lorentz factor $\gamma(u)$ through the temporal relation:

$$T^n \text{ ticks} = \gamma(u) T \text{ ticks}, \tag{2}$$

where u is the velocity of O measured in I's frame. For analysis purposes, consider constant $T > 1, n \geq 1$ and let T^n and T have units of seconds only where applicable, where T^{n-1} is dimensionless. The velocity u is now expressed as $u(T,n) = c\sqrt{1 - T^{2-2n}}$, where we choose an origin $u \geq 0$ and note that the large positive constant T^{2-2n} is the ration of two clocks.

2.1. The Twins get computers

In 1911, Paul Langevin made Einstein's 1905 prediction of time-dilation vivid by noting asymmetry in a thought experiment involving twins (O and I) — both measure events on O's worldcurve. This became known as the *twin* or *clock paradox* and was a subject of debate during the first half of the last century [16, 17] and remains a research area today [16, 17, 18, 19, 20, 21, 22]. For completeness, let us then state this paradox in terms of our framework.

The twins calculate the time ($T^n = \Delta t N$) needed in I's frame to perform a computation with the understanding that O wishes to have the solution in the nth root of this time ($T = (\Delta t N)^{1/n}$

sec), upon returning from a journey. The time of the total trip measured by I is $T^n = 2d/u$ and by O is $\gamma(u)T = 2d/u$, and the spacetime path is given as (2.1):

Definition 2.1. (Spacetime path 2.1) *Consider three events:* (i) *O starts from rest, reaching a constant velocity within a negligibly short time leaving her twin I to perform a computation for time T^n;* (ii) *after journeying for time $T/2$ in O's frame, and some distance $d/2$ in I's frame, O suddenly reverses velocity;* (iii) *O arrives back at her starting point, stops, and recovers the result of the computation.*

Consider the Lorentz transform of the event where O reaches d, and hence recover the respective temporal and spatial relations: $cT^n = \gamma(u)Tc$ and $d = \gamma(u)Tu$. It is often stated that, "it is possible to travel as far as you like in as short a time as you like, provided the distance (d) is measured before you set off and the time (T) is measured along your world-line" (see [23, 24] for instance). The expression (1) now gives the preceding statement computational meaning in terms of a polynomial reduction n inside the black-box model.

2.2. Equivalence among the Polynomial Reduction and Relativistic Mass

We let E represent energy measured in the frame I, and use m_0 to denote the rest mass of O. We stated the equivalence among the polynomial reduction and relativistic mass in Theorem 1.1 in Section 1 — an outline of the proof follows:

Proof. (Theorem 1.1) The proof relies on the results of Sections 2 and 2.1. From the relation $E = mc^2$ it can be established that $E = mc^2 = T^{n-1}m_0c^2$ and Theorem 1.1 follows. □

Corollary 2.2. *From* (1) *it follows that* $n(E,T) = 1 + \log\left(E/(m_0c^2)\right)/\log(T)$.

The 4-momentum of O measured in the frame I can now be expressed as $P = T^{n-1}(E_0/c, p_1, p_2, p_3)$, where $p_i = T^{n-1}m_0u_i$ is the ith component of the 3-momentum. When $E = E_0 = m_0c^2$ one recovers the classical limit $n = 1$. In Minkowski spacetime, it is when the relativistic energy of an observer O increases past their rest energy in a frame I, that computational gains of order $n > 1$ become possible.

3. Racing a quantum computer through Minkowski Spacetime

Quantum query complexity broke classical lower bounds on the required number of queries and hence the total time interval (ΔtN) to solve certain black-box problems including database search [25, 26]. Let us examine the related speedup using classical computers together with relativistic effects. To recover a Grover speedup [26] for the case of constant velocity requires energy $E = \sqrt{N}m_0c^2$, in I's frame, where m_0 is the rest mass of O, Δt ($:= 1$) is the single query time and N gives the total number of items in a search space.

4. Conclusion

This study was aimed at understanding what class of computations are made possible or ruled out by the laws of physics [27, 28, 29, 7] in a relativistic setting. We have shown that finite nth root polynomial reductions in algorithmic run-time are made possible by relativistic effects. The runtime improvement is predicted by Einstein's theory of relativity and the connection to computation was explored by considering polynomial reductions inside the black-box model.

Acknowledgments
I thank Ettore Minguzzi, Paul Tod, Peter Love, John D. Barrow, Piotr Chrusciel, James Whitfield, Stephen Jordan, Raymond Lal, Mark Williamson and acknowledge financial support from EPSRC grant EP/G003017/1.

References
[1] Ives H E and Stilwell G R 1938 *J. Opt. Soc. Am.* **28** 215–226
[2] Hafele J C and Keating R E 1972 *Science* 166–168
[3] Reinhardt S *et al* 2007 *Nature Physics* **3** 861–864
[4] Papadimitriou C 1994 *Computational Complexity (1st ed.)* (Addison Wesley)
[5] Sipser M 1996 *Introduction to the Theory of Computation* (International Thomson Publishing)
[6] Garey M and Johnson D 1979 *Computers and Intractability: A Guide to the Theory of NP-Completeness* (W.H. Freeman and Co. New York, NY, USA)
[7] Aaronson S 2005 *SIGACT News* **36** 30–52 ISSN 0163-5700
[8] Kitaev A, Shen A and Vyalyi M 2002 *AMS, Graduate Studies in Mathematics* **47**
[9] Geroch R P and Hartle J B 1986 *Found. Phys.* **16** 533–550
[10] Deutsch D 1991 *Phys. Rev.* D **44** 3197–3217
[11] Etesi G and Nemeti I 2002 *Int. J. Theo. Phys.* **41** 341–370
[12] Nemeti I and David G 2006 *App. Math. Comp.* **178** 118–142
[13] Brun T 2003 *Foundations of Physics Letters* **16** 245–253
[14] Bacon D 2004 *Phys. Rev.* A **70** 032309
[15] Aaronson S and Watrous J 2008 Closed timelike curves make quantum and classical computing equivalent (*Preprint* `quant-ph/0808.2669`)
[16] Darwin C G 1957 *Nature* **180** 976–977
[17] McCrea W H 1957 *Nature* **179** 909–910
[18] Jarett K and Cover T 1981 *IEEE Trans. Inf. Theory* **27** 152–159
[19] Barrow J D and Levin J 2001 *Phys.Rev.* A **63** 044104
[20] Minguzzi E 2005 *Am. J. Phys* **73** 876
[21] Iorio L 2005 *Found. Phys. Lett.* **18** 1
[22] Minguzzi E 2006 *Found. Phys. Lett.* **19** 353
[23] Taylor E and Wheeler J 1992 *Spacetime Physics* (W.H.Freeman and Co Ltd)
[24] Woodhouse N 2007 *Special Relativity, 2nd ed* (Springer)
[25] Bennett C, Bernstein E, Brassard G and Vazirani U 1997 *SIAM J. Comp.* **5** 1510–1523
[26] Grover L K 1997 *Phys. Rev. Lett.* **79** 325
[27] Deutsch D 1985 *Proc. Roy. Soc. Lon. A* **400** 97–117
[28] Lloyd S 2000 *Nature* **406** 1047–1054
[29] Yao A C 2003 *J. ACM* **50** 100–105

Journal of Physics: Conference Series **229** (2010) 012021 doi:10.1088/1742-6596/229/1/012021

Complete classification of second-order symmetric spacetimes

Oihane F Blanco[1]*, Miguel Sánchez[1] and José M M Senovilla[2]

[1] Departamento de Geometría y Topología , Facultad de Ciencias, Universidad de Granada Campus Fuentenueva s/n, 18071 Granada, Spain

[2] Física Teórica, Universidad del País Vasco, Apartado 644, 48080 Bilbao, Spain

E-mail: `oihane@ugr.es`, `sanchezm@ugr.es`, `josemm.senovilla@ehu.es`

Abstract. We present the explicit local form of the metric of the second-order symmetric non-symmetric 4-dimensional Lorentzian manifolds. They turn out to be a specific subclass of plane waves.

1. Introduction

The historical roots of the subject under investigation go back to the classification of the locally symmetric spaces in the (proper) Riemannian case by Cartan [2, 3]. *Locally symmetric* pseudo-Riemannian manifolds are characterized by the condition $\nabla R = 0$, where R denotes the curvature tensor of the manifold. The pseudo-Riemannian manifolds satisfying $\nabla^2 R = 0$ constitute a logical generalization of the locally symmetric ones. But in the Riemannian case these two conditions are equivalent. Moreover, in the Riemannian case:

$$\nabla^m R = 0, m \geq 2 \Longrightarrow \nabla R = 0, \tag{1}$$

a result based on the orthogonal de Rham decomposition [10, 15]. In this Riemannian case the natural generalization are the semi-symmetric spaces (introduced in [4], studied in [13, 14]) defined by the relation $\mathcal{R}_{XY}R = 0$ for all vector fields X, Y, where $\mathcal{R}_{XY}$ is the curvature operator $\nabla_X \nabla_Y - \nabla_Y \nabla_X - \nabla_{[X,Y]}$.

In the Lorentzian case (1) does not hold, because of the failure of the orthogonal de Rham decomposition [16, 17, 18]. In fact,

$$\nabla^m R = 0, m \geq 2 \Longrightarrow g(\nabla^{m-1}R, \nabla^{m-1}R) = 0,$$

where the metric g is extended to the inner product between tensor fields in the standard fashion. So in the Lorentzian case we can introduce the *second-order symmetric* (also called *2-symmetric*) spaces [11], i.e., the spaces that satisfy the condition $\nabla^2 R = 0$.

The following relations hold: $\{\nabla R = 0\} \subset \{\nabla^2 R = 0\} \subset \{\mathcal{R}_{XY} R = 0\}$.

In this contribution we present a specific result from [1]. Namely, we find the 2-symmetric non-symmetric 4-dimensional spacetimes: the explicit local form of the metric, classification and properties. The main result is summarized in the following theorem:

Theorem 1.1 *A 2-symmetric non-symmetric 4-dimensional spacetime (M, g) is locally isometric to $\mathbb{R}^4$ endowed with the metric*

$$ds^2 = -2du(dv + Hdu) + dx^2 + dy^2,$$

where $H(u, x, y) = (\alpha_1 u + \beta_1)x^2 + (\alpha_2 u + \beta_2)y^2 + (\alpha_3 u + \beta_3)xy$ for some constants $\{\alpha_A, \beta_A\}_{A=1,2,3}$ with $\alpha_1^2 + \alpha_2^2 + \alpha_3^2 \neq 0$.

2. Sketch of the proof of theorem 1.1

The starting point for this proof is theorem 4.2 in [11]. From this theorem follows that a 2-symmetric non-symmetric spacetime must have a null parallel vector field and therefore its metric is of Brinkmann type:

$$ds^2 = -2du(dv + Hdu + W_i dx^i) + g_{ij}dx^i dx^j, \quad i, j \in \{2, 3\},$$

where H, W_i and $g_{ij} = g_{ji}$ are functions independent of v, otherwise arbitrary.

Using the 2-symmetry condition, one can prove that g_{ij} must be locally symmetric on each slice $u = u_0$ [1]. Since the only locally symmetric 2-dimensional Riemannian spaces are the constant curvature ones, we have that in appropriate local coordinates

$$g_{ij}dx^i dx^j = \frac{1}{\left(1 + \frac{k(u)}{4}(x^2 + y^2)\right)^2}\left(dx^2 + dy^2\right).$$

In fact, $k(u)$ must vanish. To prove this, we use the Petrov classification [12]. By [7] we know that all the semi-symmetric spacetimes (of dimension 4) are of type D, N or O. Moreover, as a consequence of this result, it was also proven that the 2-symmetric spacetimes are of Petrov type N, or its degenerate case, O. And this only happens when $k(u) = 0$.

Hence, the line element of a 2-symmetric non-symmetric 4-dimensional spacetime (M, g) simplifies to

$$ds^2 = -2du(dv + Hdu + W_2dx + W_3dy) + dx^2 + dy^2,$$

From now on, an overdot means derivative with respect to u, $f_{,x} = \partial f/\partial x$ and $f_{,y} = \partial f/\partial y$.

Using the Cartan method (see, for example, [5]) to calculate the connection 1-forms and the curvature 2-forms for the null coframe $\{\theta^0, \theta^1, \theta^2, \theta^3\} = \{du, dv + Hdu + W_2dx + W_3dy, dx, dy\}$ and solving the 2-symmetry equations in this frame, we conclude that there exist functions $\{a_{rs}(u), b_{rs}(u)\}_{\{r,s\}\in\{0,1\}, r+s<2}$, $\omega(u, x, y)$, $w_2(u)$, $w_3(u)$ and $r(u)$ such that

$$W_2(u, x, y) = \omega_{,x}(u, x, y) + w_2(u)y, \tag{2}$$

$$W_3(u, x, y) = \omega_{,y}(u, x, y) + w_3(u)x, \tag{3}$$

$$H(u, x, y) = \frac{a_{10}}{2}x^2 + \frac{b_{01}}{2}y^2 + b_{10}xy + a_{00}x + b_{00}y + r(u) + Q(u, x, y) \tag{4}$$

where $Q_{,y}(u, x, y) = \dot{W}_3(u, x, y)$, and the following equations must be satisfied:

$$-\ddot{a}_{10} + 2g\ddot{g} + 2\dot{g}a_{01} - 2g^2(b_{01} - a_{10}) + 4g\dot{a}_{01} = 0, \tag{5}$$

$$\ddot{b}_{01} - 2g\ddot{g} + 2\dot{g}b_{10} + 4g\dot{b}_{10} - 2g^2(b_{01} - a_{10}) = 0, \tag{6}$$

$$\ddot{a}_{01} + \dddot{g} - 4\dot{g}g^2 + \dot{g}(a_{10} - b_{01}) - 2g^2a_{01} - 2g(\dot{a}_{10} - \dot{b}_{01}) = 0, \tag{7}$$

$$b_{10} - a_{01} = 2\dot{g}, \tag{8}$$

$$g(u) = \frac{1}{2}(W_{2,y} - W_{3,x}). \tag{9}$$

The proof can then be completed by noticing that Eqs.(2), (3) and (4) together with (5)–(9) permit the coordinate change needed to arrive at the final form of the metric. This change of coordinates is given by

$$u' = u + u_0,$$

$$v' = v + F(u, x, y),$$

$$x' = \cos(\theta(u))\, x - \sin(\theta(u))\, y + B_2(u),$$

$$y' = \sin(\theta(u))x + \cos(\theta(u))\, y + B_3(u),$$

It is easily verified that the functions θ, B_2, B_3 and F can be chosen such that the line-element becomes

$$ds'^2 = -2du'(dv' + H'du') + (dx')^2 + (dy')^2, \tag{10}$$

where

$$H'(u', x', y') = p_1(u')(x')^2 + p_2(u')(y')^2 + p_3(u')x'y' \tag{11}$$

for some new functions $\{p_A(u')\}_{A=1,2,3}$, because the integrability conditions of the resulting differential equations are always met by virtue of equations (2)–(9).

To finalize the proof of theorem 1.1, one simply has to solve the equations (5)-(9) appropriately restricted to the new form of the metric (10) with (11). This readily implies that $p_1(u')$, $p_2(u')$ and $p_3(u')$ must be linear functions:

$$p_A(u') = \alpha_A u' + \beta_A, \ \alpha_A, \beta_A \in \mathbb{R}, \ A \in \{1, 2, 3\}.$$

The condition $\alpha_1^2 + \alpha_2^2 + \alpha_3^2 \neq 0$ must be enforced as otherwise $\nabla R = 0$ and the spacetime would be locally symmetric.

3. Concluding remarks

We would like to stress that, even though the family of metrics obtained in Theorem 1.1 depend on six parameters, a further analysis shows that only four of them are essential parameters, and the other two can be removed by the remaining freedom of coordinates.

Similarly, we want to remark that our result is local. A global classification is also feasible, though some global hypotheses on the spacetime seem unavoidable to that end. Such kind of hypotheses may involve the topology of the manifold, geodesics or causality (see [6, 8, 9] for a study of the last two issues in some wave-type spacetimes). In particular, it seems that the local solutions obtained in Theorem 1.1 are also the unique global 4-dimensional solutions under the requirements of simple connectedness and geodesic completeness.

An open question is whether or not the above pattern is maintained for higher-order symmetric 4-dimensional spacetimes. Specifically, we wonder if the m-symmetric spacetimes (those with $\nabla^m R = 0$) are given, for general m, precisely by the line-element (10) with (11) where the functions $p_1(u')$, $p_2(u')$ and $p_3(u')$ are polynomials of degree m.

A more important open question is the extension of the result to higher dimensions. Of course, many more possibilities appear in this general case, which are being analyzed in [1].

Acknowledgments

JMMS is supported by grants FIS2004-01626 (MICINN) and GIU06/37 (UPV/EHU). OFB and MS are partially supported by grants P06-FQM-01951 (J. Andalucía) and MTM2007-60731 (Spanish MEC with FEDER funds).

References

[1] Blanco O F, Sánchez M and Senovilla J M M 2010 2nd-order symmetric Lorentzian manifolds II: classification and explicit local form *in preparation*
[2] Cartan É 1926 Sur une classe remarquable d'espaces de Riemann *Bull. Soc. Math. France* **54** 214-64
[3] Cartan É 1927 Sur une classe remarquable d'espaces de Riemann *Bull. Soc. Math. France* **55** 114-34
[4] Cartan É *Leçons sur la Géométrie des Espaces de Riemann* 2nd edn (Paris: Gauthier-Villars)
[5] Chandrasekhar S *The Mathematical Theory of Black Holes* (Oxford University Press, Oxford, 1983)
[6] Candela A M, Flores J L and Sánchez M 2003 On general plane fronted waves. Geodesics *Gen. Relat. Gravit.* **35** 631-649.
[7] Eriksson I and Senovilla J M M 2010 Note on (conformally) semi-symmetric spacetimes *Class. Quantum Grav.* **27** 027001 (*Preprint* arXiv:0908.3246)
[8] Flores J L and Sánchez M 2003 Causality and conjugate points in general plane waves *Class. Quantum Gravit.* **20** 2275-2291
[9] Flores J L and Sánchez M 2008 The causal boundary of wave-type spacetimes *J. High Energy Phys.* no. 3 **036** 43
[10] Nomizu K and Ozeki H 1962 A theorem on curvature tensor fields *Proc. Nat. Acad. Sci. USA* **48** 206-7
[11] Senovilla J M M 2008 2nd-order symmetric Lorentzian manifolds I: characterization and general results, *Class. Quantum Grav.* **25** 245011
[12] Stephani H, Kramer D, MacCallum M, Hoenselaers C and Herlt E, *Exact Solutions of Einstein's Field Equations*, 2nd ed. (Cambrigde University Press., Cambrigde, 2003)
[13] Szabó Z I 1982 Structure theorems on Riemannian spaces satisfying $R(X,Y)\cdot R = 0$: I. The local version *J. Differ. Geom.* **17** 531-82
[14] Szabó Z I 1985 Structure theorems on Riemannian spaces satisfying $R(X,Y)\cdot R = 0$: I. The global versions *Geom. Dedicata* **19** 65-108
[15] Tanno S 1972 Curvature tensors and covariant derivatives *Ann. Mat. Pura Appl.* **96** 233-41
[16] Wu H 1964 On the de Rham decomposition theorem *Illinois J. Math.* **8** 291-311
[17] Wu H 1964 Decompositon of Riemannian manifolds *Bull. Am. Math. Soc.* **70** 610-7
[18] Wu H 1967 Holonomy groups of indefinite metrics *Pacific J. Math.* **20** 351-92

Gowdy waves as a test-bed for constraint-preserving boundary conditions

Carles Bona and Carles Bona-Casas*

Departament de Fisica, Universitat de les Illes Balears, Palma de Mallorca, Spain.
Institute for Applied Computation with Community Code (IAC [3])

E-mail: cbona@uib.es

Abstract. Gowdy waves, one of the standard 'apples with apples' tests, is proposed as a test-bed for constraint-preserving boundary conditions in the non-linear regime. As an illustration, energy-constraint preservation is separately tested in the Z4 framework. Both algebraic conditions, derived from energy estimates, and derivative conditions, deduced from the constraint-propagation system, are considered. The numerical errors at the boundary are of the same order than those at the interior points.

Constraint-preserving boundary conditions is a very active research topic in Numerical Relativity [1, 2, 3]. During this decade, many conditions have been proposed, adapted in each case to some specific evolution formalism: Fritelli-Reula [4], Friedrich-Nagy [5], KST [6, 7], Z4 [8], Generalized-Harmonic [9, 10, 1, 2], or BSSN [3]. Cross-comparison among different evolution formalisms has been carried out ('apples with apples' initiative [11, 12]). But only periodic boundary conditions have been considered up to now.

We endorse some recent claims (by Winicour and others) that the cross-comparison effort should be extended to the boundaries treatment. In this paper, we show that Gowdy waves [13], one of the 'apples with apples' tests, is suitable for boundary conditions cross-comparison in the non-linear regime. As an illustration, we test separately the energy-constraint preservation in the Z4 framework. We compare algebraic conditions, derived from energy estimates, with derivative conditions, deduced from the constraint-propagation system. The resulting numerical errors at the boundary are of the same order-of-magnitude than those at interior points.

1. The Gowdy waves metric

Let us consider the Gowdy solution [13], which describes a space-time containing plane polarized gravitational waves. The line element can be written as

$$\mathrm{d}s^2 = t^{-1/2}\, e^{Q/2}\, (-\mathrm{d}t^2 + \mathrm{d}z^2) + t\, (e^P\, \mathrm{d}x^2 + e^{-P}\, \mathrm{d}y^2) \tag{1}$$

where the quantities Q and P are functions of t and z only, and periodic in z. The initial slice $t = t_0$ is usually chosen so that the simulations can start with an homogeneous lapse.

Let us now perform the following time coordinate transformation

$$t \;=\; t_0\, e^{-\tau/\tau_0}\,, \tag{2}$$

so that the expanding line element (1) is seen in the new time coordinate τ as collapsing towards the $t = 0$ singularity, which is approached only in the limit $\tau \to \infty$. This "singularity avoidance" property of the τ coordinate is due to the fact that the resulting slicing by $\tau = constant$ surfaces is harmonic [14]. We will run our simulations in normal coordinates, starting with a constant lapse $\alpha_0 = 1$ at $\tau = 0$ $(t = t_0)$.

Standard cross-comparison tests [11, 12] are currently done with periodic boundary conditions. But one gets basically the same results by setting up algebraic boundary conditions, which take advantage of the symmetries of the Gowdy line element. For a rectangular grid, planar symmetry allows trivial boundary conditions along the x and y directions. Also, allowing for the fact that the z dependence in (1) is only through $\cos(2\pi z)$, one can set reflecting boundary conditions for the interval $0 \leq z \leq 1$. In this way, the Gowdy waves metric is obtained as a sort of stationary gravitational wave in a cavity with perfectly reflecting walls.

We can then set up a full set of algebraic boundary conditions, which are consistent with the Gowdy line element (1) for all times. This opens the door to a selective testing procedure, where one could for instance try some constraint-preserving condition for the longitudinal and transverse-trace modes, while keeping the exact condition for the transverse traceless ones. Or, as we will do below, testing just some energy-constraint preserving boundary conditions while dealing with all the remaining modes in an exact way.

2. Characteristic decomposition

We will consider here the the first-order version in normal coordinates, as described in refs. [15, 16]. For further convenience, we will recombine the basic first-order fields (K_{ij}, D_{ijk}, A_i, Θ, Z_i) in the following way:

$$\Pi_{ij} = K_{ij} - (\,trK - \Theta)\,\gamma_{ij} \qquad V_i = \gamma^{rs}(D_{irs} - D_{ris}) - Z_k \tag{3}$$

$$\mu_{ijk} = D_{ijk} - (\gamma^{rs} D_{irs} - V_i)\,\gamma_{jk} \qquad W_i = A_i - \gamma^{rs} D_{irs} + 2\,V_i \tag{4}$$

so that the new basis is (Π_{ij}, μ_{ijk}, W_i, Θ, V_i). Note that the vector Z_i can be recovered easily from this new basis as

$$Z_i = -\mu^k{}_{ik}. \tag{5}$$

In order to compute the characteristic matrix, we will consider the standard form of (the principal part of) the evolution system as follows

$$\partial_t\, \mathbf{u} + \alpha\, \partial_n\, \mathbf{F}^n(\mathbf{u}) = \cdots\;, \tag{6}$$

where $\mathbf{u}$ stands for the array of dynamical fields and $\mathbf{F}^n$ is the array of fluxes along the direction given by the unit vector $\mathbf{n}$. With this choice of basic dynamical fields, the principal part of the evolution system gets a very simple form in the harmonic slicing case:

$$F^n(W_i) = 0 \qquad F^n(\Theta) = V^n \qquad F^n(V_i) = n_i\,\Theta \tag{7}$$

$$F^n(\Pi_{ij}) = \lambda_{nij} \qquad F^n(\mu_{kij}) = n_k\,\Pi_{ij} \tag{8}$$

where the index n means a projection along n_i, and we have noted for short

$$\lambda_{nij} = \mu_{nij} + n_{(i}W_{j)} - W_n\,\gamma_{ij}\,, \tag{9}$$

where round brackets denote index symmetrization.

We can now identify the constraint modes, by looking at the Fluxes of Θ and Z_i in the array (7-8). It follows from (7) that the energy-constraint modes are given by the pair

$$E^{\pm} = \Theta\, \pm\, V_n \tag{10}$$

with propagation speed $\pm\alpha$. Also, allowing for (5,8), we can easily recover the flux of Z_i :

$$F^n(Z_i) = -\Pi^n{}_i \tag{11}$$

so that we can identify the momentum-constraint modes with the three pairs

$$M_i^{\pm} = \Pi_{ni} \,\pm\, \lambda_{nni} \,, \tag{12}$$

with propagation speed $\pm\alpha$. Note that, allowing for (3), the longitudinal component Π_{nn} does correspond with the transverse-trace component of the extrinsic curvature K_{ij}. We give now the remaining modes: the fully transverse ones, with propagation speed $\pm\alpha$,

$$T_{AB}^{\pm} = \Pi_{AB} \,\pm\, \lambda_{nAB} \tag{13}$$

(the capital indices denote a projection orthogonal to n_i), and the standing modes (zero propagation speed):

$$W_i \,, \qquad V_A \,, \qquad \mu_{Aij} \,. \tag{14}$$

Note that the standing modes (14), the energy modes (10) and the transverse momentum modes $M_A^{\pm}$ actually vanish for the Gowdy line element (1).

3. Energy-constraint preserving boundary conditions

In refs. [15, 16], the system above was shown to be symmetric hyperbolic, by providing a suitable energy estimate. We can rewrite it here as

$$\Pi_{ij}\Pi^{ij} + \lambda_{kij}\lambda^{kij} + \Theta^2 + V_k V^k + W^k W^k \tag{15}$$

This leads to the following sufficient condition for stability

$$(\Pi^{ij}\ \lambda_{nij} + \Theta\ V_n)\,|_\Sigma\ \geq\ 0 \tag{16}$$

where Σ stands for the boundary surface (**n** being here the outward normal).

Let us consider for instance the boundary at $z = 1$. We can enforce there the partial set of exact (reflection) boundary conditions $\lambda_{nij} = 0$, so that the requirement (16) reduces to

$$(\Theta\ V_n)\,|_\Sigma\ \geq\ 0 \tag{17}$$

which can be used for a separate test of energy-constraint preserving boundary conditions.

As an illustration, we will test two such conditions. The first one is given in the form of a logical gate:

$$(\Theta\ V_n)\,|_\Sigma\ < 0 \qquad \Rightarrow \qquad \Theta\,|_\Sigma\ = 0 \tag{18}$$

(Θ-gate), so that it only acts when condition (17) is violated. The second one is an advection equation

$$\partial_t\ \Theta\,|_\Sigma\ = -\alpha\,\partial_n\,\Theta - \eta\ \Theta\,, \tag{19}$$

where we have included a suitable damping term. Note that the principal part of the constraint-preserving condition (19) coincides with the 'maximal dissipation' one, $\partial_t\ E^- = 0$, which is not constraint-preserving in the generic case. Condition (19) can be understood as a sort of maximal dissipation condition for the energy-constraint evolution equation

$$\partial^2_{tt}\ \Theta - \alpha^2 \triangle\ \Theta = \cdots\ , \tag{20}$$

which follows from (the time component of) the covariant divergence of the Z4 field equations [15].

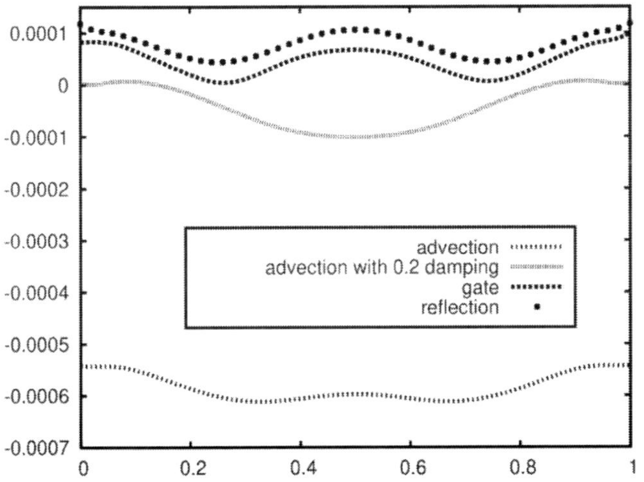

Figure 1. Θ profiles along the z direction for a Gowdy waves simulation ending at $\tau = 100$. From bottom to top, results for the pure advection condition (19), damped advection (with $\eta = 0.2$), Θ-gate (18), and exact reflection (included here for comparison).

We show in Fig. 1 our results for some numerical simulations ending at $\tau = 100$. We have included in the plot the exact (reflection) results for comparison. It is clear that the Θ-gate condition gets very close to (actually slightly better than) the exact result in this case. The pure advection condition (19) is off by half-an-order of magnitude. However, a suitable damping term (we have taken $\eta = 2$) greatly improves this, leading in this case to even less error than the pure reflection condition. Note that we are showing here the Θ profiles, giving the cumulated energy-constraint deviation. The average energy-constraint error in these strong-field simulations is actually smaller.

Acknowledgments
This work has been jointly supported by European Union FEDER funds and by the Spanish Ministry of Science and Education (projects FPA2007-60220, CSD2007-00042 and ECI2007-29029-E). C. Bona-Casas acknowledges the support of the Spanish Ministry of Science, under the FPU/2006-02226 fellowship.

References

[1] Winicour J 2009 *Gen. Rel. Grav.* **41** 1909–1926
[2] Rinne O, Buchman L, Scheel M and Pfeiffer H 2009 *Class. Quantum Grav.* **26** 075009
[3] Núñez D and Sarbach O 2009 Boundary conditions for the Baumgarte-Shapiro-Shibata-Nakamura formulation of Einstein's field equations *Preprint* arXiv:0910.5763
[4] Stewart JM 1998 *Class. Quantum Grav.* **15** 2865
[5] Friedrich H and Nagy G 1999 *Commun. Math. Phys.* **201** 619
[6] Calabrese G et al 2003 *Commun. Math. Phys.* **240** 377
[7] Sarbach O and Tiglio M 2005 *J. Hyperbol. Diff. Equat.* **2** 839
[8] Bona C, Ledvinka T, Palenzuela C and Žáček M 2005 *Class. Quantum Grav.* **22** 2615
[9] Babiuc M, Kreiss H–O and Winicour J 2007 *Phys. Rev.* D**75** 044002
[10] Kreiss H–O, Reula O, Sarbach O and Winicour J 2007 *Class. Quantum Grav.* **24** 5973
[11] Alcubierre M et al 2004 *Class. Quantum Grav.* **21** 589.
[12] Babiuc M et al 2008 *Class. Quant. Grav.* **25** 125012
[13] Gowdy R H 1971 *Phys. Rev.* D **27** 826
[14] Bona C and Massó J 1983 *Phys. Rev.* D **38** 2419
[15] Bona C, Ledvinka T, Palenzuela C and Žáček M 2003 *Phys. Rev.* D **67** 104005
[16] Bona C and Palenzuela C 2005 *Elements of Numerical Relativity* (Lect. Notes Phys. vol 673) (Berlin–Heidelberg–New York: Springer) pp 1–24

Black hole gas in TeV-gravity models

Mónica Borunda* **and Manuel Masip**

CAFPE and Departamento de Física Teórica y del Cosmos, Universidad de Granada, E-18071 Granada, Spain

E-mail: `mborunda@ugr.es, masip@ugr.es`

Abstract. In a plasma at temperature close to the fundamental scale a small fraction of particles will experience transplanckian collisions that may result in microscopic black holes (BHs). We study the dynamics of a system (a *black hole gas*) defined by radiation at a given temperature coupled to a distribution of BHs of different mass. Our analysis includes the production of BHs in photon-photon collisions, BH evaporation, the absorption of radiation, collisions of two BHs to give a larger one, and the effects of the expansion. We find that the system may follow two different generic paths depending on the initial temperature of the plasma.

1. Introduction

The presence of extra dimensions may accommodate a fundamental scale of gravity M_D much lower than the Planck mass $M_P \sim 10^{19}$ GeV [1, 2], with $M_D \approx 1$ TeV solving the so-called hierarchy problem. If that were the case, collisions of two particles at center of mass energies above M_D would be dominated by gravity. In particular, one expects that at small impact parameters gravity *bounds* the system and the particles collapse into a microscopic BH. This process, extensively studied as a possibility for the LHC, would be also relevant in the early universe if the initial temperature T was ever close to the TeV scale.

At such temperatures two particles in the high-energy tail of the distribution may experience a transplanckian ($\sqrt{s} > M_D$) collision and produce a mini-BH. If the BH is colder than the environment ($T_{BH} < T$) it will gain mass, and as it grows its temperature drops further. This would distinguish BHs from other massive particles or string excitations that may also be produced at high T: the heavier the BH, the colder and more stable it becomes, whereas heavier elementary particles have a shorter lifetime. BHs seem indeed a key ingredient in the dynamics of a plasma at temperatures close to the fundamental scale. Notice that the system is peculiar in the sense that the heat flows from the hot plasma to the colder BHs, but the effect is to cool the BHs.

In the usual TeV-gravity models with flat extra dimensions (ADD) a high T seems inconsistent with the standard cosmology [3]. There the KK excitations of the graviton have an effective 4-dim mass proportional to $m_c \equiv L^{-1}$ and very weak couplings ($\approx \sqrt{s}/M_P$) to ordinary matter. Even if the initial configuration consists of particles only in the 4-dim brane, at $T \gg m_c$ these bulk gravitons will be abundantly produced (due to their large multiplicity) in annihilation of brane particles. If T is large these (long-lived) gravitons will change the expansion rate at the time of primordial nucleosynthesis. Even if T is initially as low as 1 MeV, their late decay will distort the cosmic background radiation in an unacceptable way. Obviously, a temperature close to M_D would bring too many massive gravitons. In contrast, warped (RS) models predict KK gravitons just below M_D with unsuppressed couplings, so a few resonances are enough to give an order one gravitational coupling at M_D. In RS models the bulk is basically empty at $T < m_c \approx 0.1 M_D$.

Here we will consider *hybrid* models [4] with n extra dimensions and two uncorrelated parameters: the effective compactification scale m_c and the fundamental scale M_D. We will assume a mechanism that *pushes* the KK gravitons towards the 4-dim brane and increase their coupling to matter:

$$\frac{s}{M_P^2} \to \frac{s}{M_D^2}\left(\frac{m_c}{M_D}\right)^n . \tag{1}$$

In this way a smaller number of KK modes will imply an order one gravitational interaction at the same scale $\sqrt{s} = M_D$. If the *free* parameter m_c takes the value $m_c \approx M_D (M_D/M_P)^{2/n}$ we recover ADD, whereas for m_c approaching M_D we obtain RS. At distances below $1/m_c$ gravity would be $(4+n)$-dim (similar to ADD) whereas at larger distances it becomes 4-dim (like in the usual RS). The KK gravitons are not produced at $T < m_c$, while their larger couplings to matter decouple them from the plasma at T below their mass and make their lifetime shorter. This framework with $m_c > 1$ GeV avoids all astrophysical [5] and cosmological [3] bounds for any value of n.

There are two basic points that distinguish the TeV-gravity scenario from the standard 4-dim one. First, BHs are colder and live longer here than in 4-dim. Second, in these models the Hubble time $H^{-1} \approx 1/\sqrt{G_N \rho}$ at the temperature where the BHs are produced is much longer (in terms of the fundamental time scale) than in four dimensions:

$$H_{(4)}^{-1} \sim \frac{1}{M_{Pl}}, \qquad H_{(4+n)}^{-1} \sim \frac{M_{Pl}}{M_D}\left(\frac{m_c}{M_D}\right)^{n/2}\frac{1}{M_D} \tag{2}$$

where we have assumed a bulk in thermal equilibrium with the 4-dim brane and $\rho \approx T^{4+n}/m_c^n$. As a consequence, in a $(4+n)$-dim universe there is plenty of time for BHs to be produced and grow, changing the dynamics of the early universe. In a 4-dim universe the temperature drops much faster, the BHs of mass near the threshold M_P get then hotter than the environment and evaporate in a time $\approx 1/M_P$.

2. One black hole

Let us first consider a single BH in an expanding universe. We take $M_D = 1$ TeV, and $m_c = 10$ GeV, with initially the brane and the bulk in thermal equilibrium at a temperature $T = T_0$. Our hybrid model at energies above M_D and distances smaller than $1/m_c$ is strongly coupled and $(4+n)$-dim, so the BH is initially a regular $(4+n)$-dim one. If it is colder than the plasma ($T_{BH} < T_0$) it grows, and when it reaches a radius $r_H = 1/m_c$ it fills up the whole compact space. At this point the BH keeps gaining mass, but its 4-dim size does not increase because gravity is strong only up to distances of order $1/m_c$ (the inverse mass of the lightest KK graviton). Beyond this distance gravity is 4-dim, since only the massless graviton is effective. Therefore, the BH grows at constant radius and temperature $T_{BH} \approx m_c$ until its mass reaches $M \sim M_P^2/m_c$, when it becomes purely 4-dim. At $T > m_c$ the BH mass changes according to

$$\frac{dM}{dt} = \sigma_4 A_4 (T^4 - T_{BH}^4) + \sigma_{4+n} A_{4+n} (T^{4+n} - T_{BH}^{4+n}) \tag{3}$$

where the negative terms account for the evaporation into brane and bulk species and in A_{4+n} (the BH area) we take a maximum radius of $1/m_c$ along the extra dimensions.

The expansion rate (we assume that the extra dimensions are frozen) of the universe is dictated by the radiation energy density in the bulk and the brane,

$$\rho_{rad} \sim \frac{\pi^2}{30} T^4 \left(g_\star + g_b c_n \frac{T^n}{m_c^n} \right) . \tag{4}$$

At $T > m_c$ the bulk energy dominates and the temperature decreases as $T \sim t^{-2/(4+n)}$ with time, whereas at $T < m_c$ all the bulk energy is transferred to the brane and the universe expands like in the standard 4-dim cosmology.

In figure 1 we show the evolution of a BH of mass $M = 100$ TeV with $M_D = 1$ TeV, $n = 1$, just photons ($g_\star = 2$) and gravitons ($g_b = 5$) in the bulk at $T_0 = 100, 200$ GeV.

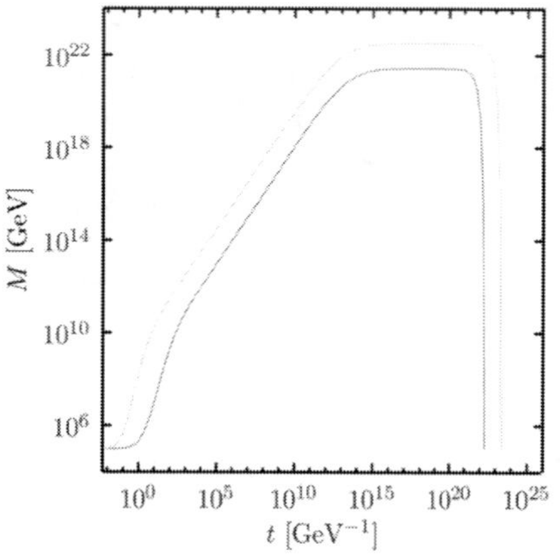

Figure 1. Mass of a single BH.

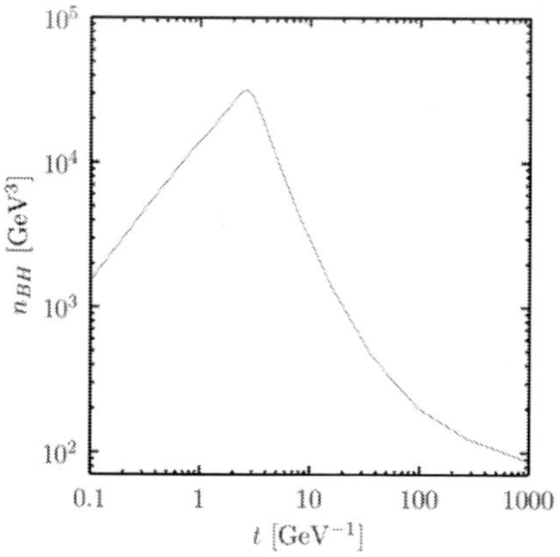

Figure 2. n_{BH} at $T_0 = 200$ GeV.

3. Black hole gas

Let us now simulate the evolution of the system, a *black hole gas*, including the different processes that may take place:

- Black hole formation in collisions of two particles.
- Collisions of two BHs into a larger one.
- Absorption and emission of radiation.
- Effect of the expansion of the universe.

We will consider an initial configuration with only radiation at T_0 both in the brane and the n dimensional bulk. Particles with enough energy will collide to form BHs that will grow if $T_{BH} < T_0$. If T_0 is smaller than M_D then the BHs are always non-relativistic matter, with a kinetic energy ($\approx T$) much smaller than the mass. We take $E_{BH} \sim M_{BH}$ and $v_{BH} = \sqrt{2T/M_{BH}}$.

We will assume that the BH gas is a two-component thermodynamic system formed by radiation with a time dependent temperature $T(t)$ and BHs described by a distribution $f(M,t)$. $f(M,t)$ describes the number of BHs of mass M at a given time t per unit mass and volume. In this way, the energy density of the universe is the addition of ρ_{rad} (for brane and bulk radiation) plus ρ_{BH} (the matter density in the BHs).

As the radiation cools down due to the absorption and the expansion, lighter BHs (of higher temperature) become hotter than the plasma and evaporate. This reheats the universe and provides energy to be absorbed by the larger BHs. We are interested in the evolution of the system. Does ρ_{BH} ever dominate the energy density of the universe? How big the BHs grow? How much the final configuration depends on the initial temperature of the universe?

4. Results

We have identified 2 generic scenarios depending on T_0. Larger values of T_0 (above $\approx 0.2M_D$) produce *many* BHs, which grow fast and dominate ρ. In this case the BH gas goes through four phases [6]:

- BHs of mass above a critical mass corresponding to $T_{BH} = T_0$ are produced (see Fig. 2) and absorb radiation. As their number grows collisions between two BHs become important. BHs dominate the energy density very fast ($t \approx 10^{-12}H^{-1}$), reducing the temperature of the radiation.
- As T drops BH production stops. Light BHs become hotter than the plasma and evaporate. This reduces the number of BHs but the average BH mass keeps growing, since the colder (bigger) BHs keep absorbing radiation. Once a BH reaches a mass around 10^7 GeV its radius $r \approx 1/m_c$ stops growing.
- When T drops to $T_{BH} \approx m_c$, the slow absorption of radiation by the heavier BHs is compensated with the evaporation of the lighter ones, keeping T basically constant. The energy density is still dominated by BHs.
- At times of order H^{-1} the expansion cools the radiation. The BHs decay fast and the universe becomes radiation dominated. Likewise, the lightest KK modes also decay fast and only 4-dim photons survive below $T \sim 0.1m_c$.

In the case with a lower initial temperature (T_0 below $\approx 0.1M_D$) the basic difference is that the BHs do not dominate the energy density at any time. In this second case we distinguish three phases:

- BHs are produced at a constant rate but since their number is so small collisions between them are negligible. These BHs grow like the single BH described in Fig. 1.
- At a Hubble time the expansion cools the radiation and stops BH production. The average BH mass keeps growing, but the universe is always radiation dominated.
- Due to the expansion the temperature of the universe drops below the temperature of the BHs, which evaporate in a time scale that depends on the ratio M_D/m_c (it is around $t \sim 10^{22}$ GeV^{-1} for $M_D/m_c = 10^2$).

Both scenarios seem consistent with primordial nucleosynthesis. In the second case the BHs are a small contribution to ρ that would not change the expansion rate of the universe. For larger values of M_D/m_c one may obtain BHs whose late decay may introduce distortions in the diffuse gamma-ray background, or even BHs with a lifetime larger than the present age of the universe. These ones could account for additional dark matter or become seeds for macroscopic primordial BHs.

We think that the work presented here is a first step in the search for possible observable effects of these BHs and on the understanding of the initial conditions in the early universe.

References

[1] Arkani-Hamed N Dimopoulos S and Dvali G R 1998 *Phys. Lett.* B **429** 263
[2] Randall L and Sundrum R 1999 *Phys. Rev. Lett.* **83** 3370
[3] Hannestad S 2001 *Phys. Rev.* D **64** 023515
Hall L J and Tucker-Smith D 1999 *Phys. Rev.* D **60** 085008
[4] Giudice G F, Plehn T and Strumia A 2005 *Nucl. Phys.* B **706** 455
[5] Hannestad S and Raffelt G G 2003 *Phys. Rev.* D **67** 125008
Hannestad S and Raffelt G G 2004 *Erratum-ibid.* D **69** 029901
[6] Borunda M and Masip M *Preprint* arXiv:0910.4532

$f(R)$ brane cosmology

Mariam Bouhmadi-López

Centro Multidisciplinar de Astrofísica - CENTRA, Departamento de Física, Instituto Superior Técnico, Av. Rovisco Pais 1, 1049-001 Lisboa, Portugal

E-mail: `mariam.bouhmadi@ist.utl.pt`

Abstract. Despite the nice features of the Dvali, Gabadadze and Porrati (DGP) model to explain the late-time acceleration of the universe, it suffers from some theoretical problems like the ghost issue. We present a way to self-accelerate the normal DGP branch, which is known to be free of the ghost problem, by means of an f(R) term on the brane action. We obtain the de Sitter self-accelerating solutions of the model and study their stability under homogeneous perturbations.

1. Introduction

One of the most puzzling problems nowadays in physics is the issue of the late-time acceleration of the universe [1]. A possible approach to tackle this problem is within the frame-work of self-accelerating universes [2, 3, 4, 5]; i.e. could a modification of gravity at late-time and on large scale be the cause of the current inflationary phase of the universe? In other words, could this modification of gravity on large scales provide an *effective negative pressure* that would fuel the late-time acceleration of the universe?

A possible approach to tackle these questions is the Dvali, Gabadadze and Porrati (DGP) scenario [2], which corresponds to a five-dimensional (5D) model. In this model, our universe is a brane; i.e. a 4D hyper-surface, embedded in a flat space-time. The DGP model has two types of solutions: the self-accelerating branch and the normal one. The self-accelerating brane is asymptotically de Sitter. This feature takes place without invoking any unknown dark energy component. On the other hand, the normal branch requires a dark energy component to accommodate the current observations [1]. Despite the nice features of the self-accelerating DGP branch, it suffers from serious theoretical problems like the ghost issue [6]. In this paper we propose a mechanism to self-accelerate the normal branch which is known to be free from the ghost issue [6]. This mechanism will be based on a modified Hilbert-Einstein action on the brane [5] and the simplest gravitational option is an $f(R)$ term [4].

2. The model

We consider a brane, described by a 4D hyper-surface (h, metric g), embedded in a 5D bulk space-time ($\mathcal{B}$, metric $g^{(5)}$), whose action is given by

$$\mathcal{S} = \int_{\mathcal{B}} d^5X\,\sqrt{-g^{(5)}}\,\left\{\frac{1}{2\kappa_5^2}R[g^{(5)}]\right\} + \int_h d^4X\,\sqrt{-g}\,\left\{\frac{1}{\kappa_5^2}K \;+\alpha f(R) + \mathcal{L}_m\right\}, \qquad (1)$$

where κ_5^2 is the 5D gravitational constant, $R[g^{(5)}]$ is the scalar curvature in the bulk and K the extrinsic curvature of the brane in the higher dimensional bulk. On the other hand, R is the scalar curvature of the induced metric on the brane, g, and α is a constant that measures the strength of the generalised induced gravity term $f(R)$ and has mass square units. Notice that therefore the function $f(R)$ has mass square units. Finally, $\mathcal{L}_m$ corresponds to the matter Lagrangian of the brane. We will assume that the brane splits the bulk in two symmetric pieces. The previous action, includes as a particular case the DGP model [2] for $f(R) = R$ and $\alpha = 1/2\kappa_4^2$ where κ_4^2 is proportional to the 4D gravitational constant. It can be shown that the total energy density of the brane is conserved (we refer the reader to [5] for more details). In particular, the energy density of matter on the brane is conserved.

In what follows, we consider a homogeneous and isotropic brane. The matter sector on the brane can be described by a perfect fluid with energy density $\rho^{(m)}$ and pressure $p^{(m)}$, where $\rho^{(m)}$ is conserved as we have pointed above. On the other hand, an effective energy density and an effective pressure associated to the energy momentum tensor coming from the $f(R)$ term on action can be defined as follows [5]

$$\rho^{(f)} = -2\alpha\left[3\left(H^2+\frac{k}{a^2}\right)f' - \frac{1}{2}(Rf'-f) + 3H\dot{R}f''\right], \tag{2}$$

$$p^{(f)} = 2\alpha\left\{\left(2\dot{H}+3H^2+\frac{k}{a^2}\right)f' - \frac{1}{2}(Rf'-f)\left[\ddot{R}f''+(\dot{R})^2f'''+2H\dot{R}f''\right]\right\}, \tag{3}$$

Notice that the definition of $\rho^{(f)}$ and $p^{(f)}$ is different from the standard 4D definition in $f(R)$ models [5]. On the other hand, the energy density is conserved on the brane.

The modified Friedmann equation on the brane can be written as

$$3H^2 = \frac{\kappa_5^4}{12}\rho^2. \tag{4}$$

While, the spatial component of Einstein equation can be expressed as

$$2\dot{H} + 3H^2 + \frac{k}{a^2} = -\frac{\kappa_5^4}{12}\rho(\rho+2p), \tag{5}$$

where the energy density ρ and the pressure p are defined as

$$\rho = \rho^{(m)} + \rho^{(f)}, \quad p = p^{(m)} + p^{(f)}. \tag{6}$$

For simplicity, on equations (4) and (6) we have used the spatially flat chart of the brane.

3. Self-accelerating branes

A de Sitter universe is the simplest cosmological solution that exhibits acceleration and therefore it is worthwhile to prove the existence of this solution in our model and study its stability. This would be a first step towards describing in a realistic way the late-time acceleration of the universe in an $f(R)$ brane-world model. This approach will also enable us to look for self-accelerating solutions on the modified normal DGP branch. So, in this section, we first obtain the fixed points of the model corresponding to a de Sitter space-time and then we study their stability under homogeneous perturbations.

3.1. Background solutions

In our model, the Hubble parameter for de Sitter solutions can be expressed as[1]

[1] For a maximally symmetric brane in our model, the matter content of the brane behaves like a cosmological constant. As such a term can always be reabsorbed in the $f(R)$ term we will disregard the matter content in our analysis of de Sitter branes.

$$2\kappa_5^4\alpha^2F_0^2H_0^2 = 1 + \frac{1}{3}\kappa_5^4\alpha^2F_0(R_0F_0 - f_0) + \epsilon\sqrt{1 + \frac{2}{3}\kappa_5^4\alpha F_0\left[\alpha(R_0F_0 - f_0)\right]} \qquad (7)$$

where $\epsilon = \pm 1$, the subscript 0 stands for quantities evaluated at the de Sitter space-time, $R_0 = 12H_0^2$ and $F = df/dR$. We recover the DGP model for $f(R) = R$. In fact, in that case, the de Sitter self-accelerating DGP branch is obtained for $\epsilon = 1$ and the normal DGP branch or the non-self-accelerating solution for $\epsilon = -1$. When the brane action contains curvature corrections to the Hilbert-Einstein action given by the brane scalar curvature, the branch with $\epsilon = -1$ is no longer flat and accelerates (cf. Fig. 1). Therefore, an $f(R)$ term on the brane action induce in a natural way self-acceleration on the normal branch. Most importantly, it is known that such a branch is free from the ghost problem (see [6] and references therein). The reason behind the self-acceleration of the generalised normal brane is the presence of the effective energy density

$$\rho_0^{(c)} = \alpha(F_0R_0 - f_0) \qquad (8)$$

on the modified Friedmann equation on the brane. This can be easily shown by comparing the Friedmann equation (7) with that of modified gravity on brane world-models [7]

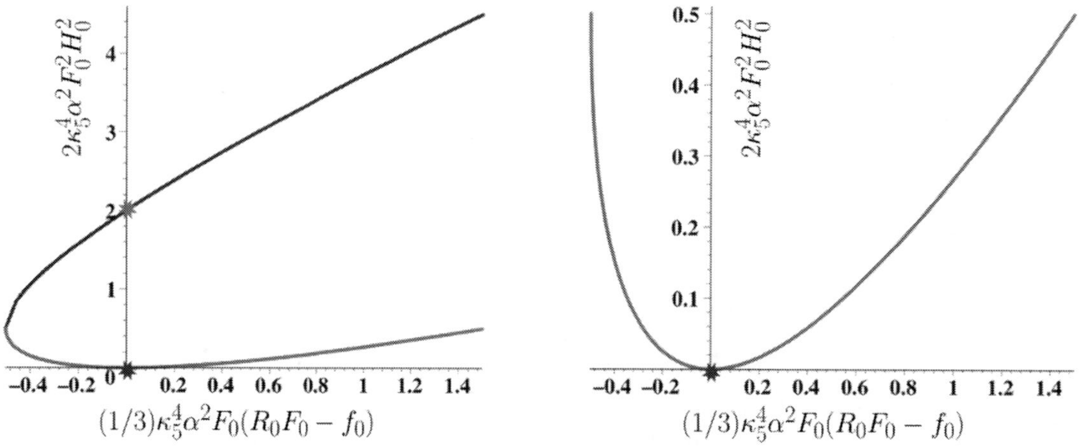

Figure 1. The figure on the left shows the behaviour of the rescaled squared Hubble rate $2\kappa_5^4\alpha^2F_0^2H_0^2$ for the two branches that generalise the DGP solution versus the rescaled energy density $\rho^{(c)}$ defined as $\frac{1}{3}\kappa_5^4\alpha^2F_0(R_0F_0 - f_0)$. The blue star corresponds to the normal DGP branch which is flat. The red star corresponds to the self-accelerating DGP branch. On the other hand, the blue curve corresponds to the generalised (by the inclusion of the $f(R)$ term) self-accelerating branch, while the red curve corresponds to the generalised (by the inclusion of the $f(R)$ term) normal branch. The figure on the right corresponds to a zoom of the normal branch as it appears on the figure of the left.

3.2. Stability of the self-accelerating solutions

We next analyse the stability of de Sitter solutions under homogeneous perturbations up to first order on $\delta H = H(t) - H_0$. We will follow the method used in [8].

The perturbed Friedmann equation (4) implies an evolution equation for δH:

$$\delta\ddot{H} + 3H_0\delta\dot{H} + m_{\text{eff}}^2\delta H = 0, \qquad (9)$$

where $m^2_{\rm eff}$ is defined as

$$m^2_{\rm eff} = m^2_{(4)} + m^2_{\rm shift} + m^2_{\rm pert}. \quad (10)$$

where

$$\begin{aligned} m^2_{(4)} &= \frac{1}{3}\left(\frac{F_0}{f_{RR}} - 2\frac{f_0}{F_0}\right), \quad (11)\\ m^2_{\rm back} &= -\frac{2}{\alpha^2\kappa_5^4 F_0^2}\left[1-\sqrt{1+\frac{2}{3}\alpha^2\kappa_5^4 F_0(f_0-\kappa_5^2 U F_0)}\right],\\ m^2_{\rm pert} &= \frac{F_0}{3f_{RR}}\left[1-\sqrt{1+\frac{2}{3}\alpha^2\kappa_5^4 F_0(f_0-\kappa_5^2 U F_0)}\right]^{-1}. \end{aligned}$$

and $f_{RR} = d^2f/dR^2$. All this quantities are evaluated at the de Sitter background solution. Any de Sitter solution is stable as long as $m^2_{\rm eff}$ is positive.

The terms defined on Eq. (11) have the following physical meaning: (i) $m^2_{(4)}$ is the analogous quantity to $m^2_{\rm eff}$ in a 4D f(R) model [8], (ii) $m^2_{\rm back}$ is a purely background effect due to the shift on the Hubble parameter respect to the standard 4D Case and (iii) $m^2_{\rm pert}$ is a purely perturbative extra-dimensional effect.

If we assume that we are close to the 4D regime; i.e. the Hubble rate of the brane is close to its analogous quantity in a 4D $f(R)$ model, then $m^2_{\rm back} > 0$ and $m^2_{\rm pert} < 0$. Consequently, $m^2_{\rm back}$ tends to make the perturbation heavier. However, the perturbative effect encoded on $m^2_{\rm pert}$ would make the perturbation lighter. It can be shown that the extra-dimension has a *benigner* effect in the 4D f(R) model; i.e. $m^2_{\rm eff} > m^2_{(4)}$, as long as[2]

$$F_0^2 < 4f_0 f_{RR}. \quad (12)$$

4. Conclusions

We have presented a mechanism to self-accelerate the normal DGP branch which unlike the original self-accelerating DGP branch is known to be free of the ghost problem. The mechanism is based in including curvature modifications on the brane action. For simplicity, we choose those terms to correspond to an $f(R)$ contribution, which in addition is known to be the only higher order gravity theories that avoid the so called Ostrogradski instability in 4D models.

Acknowledgments

M.B.L. is supported by the Portuguese Agency Fundação para a Ciência e Tecnologia through the fellowship SFRH/BPD/26542/2006.

References

[1] Durrer R and Maartens R 2008 *Preprint* arXiv:0811.4132
[2] Dvali G R, Gabadadze G and Porrati M 2000 *Phys. Lett.* B **485** 208
[3] Nojiri S and Odintsov S D 2007 *Int. J. Geom. Meth. Mod. Phys.* **4** 115
[4] Capozziello S and Francaviglia M 2008 *Gen. Rel. Grav.* **40** 357
[5] Bouhmadi-López M 2009 *J. Cosmol. Astropart. Phys.* JCAP11(2009)011
[6] Koyama K 2007 *Class. Quantum Grav.* **24** R231
[7] Bouhmadi-López M and Wands D 2005 *Phys. Rev.* D **71** 024010
[8] Faraoni V and Nadeau S 2005 *Phys. Rev.* D **72** 124005

[2] We have assumed the natural condition $F_0 > 0$; i.e. the effective gravitational constant of the brane is positive. On the other hand, we have also assumed that we are slightly perturbing the Hilbert-Einstein action of the brane, i.e. $f_0 \sim R_0$. Therefore, f_0 is positive because $R_0 = 12H_0^2$.

The quadrupole moment of slowly rotating stars

M Bradley[1*] **and G Fodor**[2]
[1]Department of Physics, Umeå University, SE-901 87 Umeå, Sweden
[2]KFKI Research Institute for Particle and Nuclear Physics, H-1525, Budapest 114, P.O.B. 49, Hungary

E-mail: michael.bradley@physics.umu.se, gfodor@rmki.kfki.hu

Abstract. The second order field equations for the interior of slowly and rigidly rotating stars are solved numerically for different classes of possible equations of state and these solutions are then matched to the general asymptotically flat axisymmetric vacuum metric to second order in the rotational parameter. For these solutions we find that the quadrupole moment differs from that of the Kerr metric. Further we consider the post-Minkowskian limit analytically.

1. Introduction

In the present work we use the second order formalism for slowly and rigidly rotating stars, developed in [1], to study the quadrupole moment and its deviation from that of the Kerr metric for different classes of possible equations of state. For more details the reader is referred to a longer version of this paper, [2], and references therein. Some earlier numerical studies find quadrupole moments differing from that of the Kerr metric, see e.g. [3, 4, 5, 6]. In some recent papers [7, 8] it is shown that in the rigidly rotating case incompressible fluids and fluids with polytropic equation of state cannot be sources of the Kerr metric in the post-Minkowskian limit. These results are in accordance with the general expectation that the ellipsoidal shape of the rotating fluid produces an extra contribution to the quadrupole moment.

We also consider the post-Minkowskian limit analytically by expanding the field equations in the small parameter $\lambda \equiv GM/r_1c^2$ and make a comparison with the results of [7, 8].

2. Preliminaries

To second order the metric of a slowly rotating axisymmetric object, both in the interior fluid region and the outside vacuum region, can be written as

$$ds^2 = (1+2h)A^2dt^2 - (1+2m)\frac{1}{B^2}dr^2 - (1+2k)r^2\left[d\theta^2 + \sin^2\theta\,(d\varphi - \omega dt)^2\right] , \qquad (1)$$

where ω is first order and h, m are k are second order in the rotational parameter. The requirements of regularity at the centre and asymptotic flatness imply that the first order function ω depends on r only. The second order functions can be given as $h = h_0 + h_2P_2(\cos\theta)$, $m = m_0 + m_2P_2(\cos\theta)$ and $k = k_2P_2(\cos\theta)$ where h_0, m_0 and h_2, m_2, k_2 are functions of r only and P_2 is the second order Legendre polynomial. For more details see [1].

The matter content of the interior is modelled by a perfect fluid $T_{ab} = (\rho + p)u_au_b - pg_{ab}$. The coordinate system used in (1) is assumed to be comoving with the fluid which also implies that the shear of the fluid is zero, so it rotates rigidly.

Einstein's equations then give a subsystem for the functions A, B, ω, m_2, k_2 and h_2 that decouples from m_0 and h_0. To close this system one more equation, e.g. an equation of state, $\rho = \rho(p)$, has to be specified. In this paper we will consider equations of state in the form:

$$\rho = d_1 p + d_2 \left(\frac{p}{p_c}\right)^{1/\gamma} + d_3 \ . \tag{2}$$

Here p_c is the central pressure. Equation (2) includes some of the more used approximations for the equation of state of dense stars, like Newtonian polytropes, relativistic polytropes with $d_1 = 1/(\gamma - 1)$, linear ones as well as the incompressible case.

Due to the requirement of a regular centre the solutions will only depend on three constants of integration, that are the central pressure to zeroth order, p_{0c}, the magnitude of the vorticity, ω_0 and a second order quantity h_1.

In the exterior vacuum region a frame adapted to the asymptotically non-rotating observer is chosen. The solution to second order in the rotational parameter is characterized by the mass M, the first order rotation parameter a, and the second order constants c_1, c_2 and q_1. When q_1 takes the value zero the metric is the general asymptotically flat stationary and axisymmetric vacuum metric to second order. Using the method in [9], the Geroch-Hansen relativistic moments up to order two are found to be the mass to second order $M - c_2$, the angular momentum $J = Ma$ and the quadrupole moment $Q = -M\left(a^2 + \frac{16}{5}M^4 c_1\right)$. We will be interested in the relative deviation of the quadrupole moment from that of the Kerr metric

$$\frac{\Delta Q}{Q} \equiv \frac{Q - Q_{Kerr}}{Q_{Kerr}} = \frac{16M^4 c_1}{5a^2}. \tag{3}$$

When $q_1 \neq 0$ the metric cannot be asymptotically flat. It is important to keep in mind, however, that without the inclusion of this constant the matching conditions on the zero pressure surface are overdetermined in general.

In the fluid region the matching surface $\mathcal{S}$ is the surface of vanishing pressure. For slow rotation $\mathcal{S}$ is given by $r = r_1 + \xi \equiv r_1 + \xi_0 + \xi_2 P_2(\cos\vartheta)$ where $r = r_1$ is the zeroth order radius and the constants ξ_0 and ξ_2 are determined from the pressure. Similarly, in the vacuum exterior region suitable surfaces for matching are determined by $r = r_1 + \chi \equiv r_1 + \chi_0 + \chi_2 P_2(\cos\vartheta)$, where χ_0 and χ_2 are constants to be determined from the matching conditions.

The two regions are matched using the Darmois-Israel procedure, where the respective induced metrics $ds^2_{(v)}$, ds^2 ($_{(v)}$ for vacuum) and induced extrinsic curvatures $K_{(v)}$, K are equated with each other as

$$ds^2_{(v)}|_{\mathcal{S}} = ds^2|_{\mathcal{S}} \qquad K_{(v)}|_{\mathcal{S}} = K|_{\mathcal{S}} \ . \tag{4}$$

Here $K \equiv K_{ab}dx^a dx^b \equiv h_a{}^c h_b{}^d n_{(c;d)} dx^a dx^b$ in terms of the unit normal n_a of the matching surface $\mathcal{S}$ and the projection operator $h_a{}^b = n_a n^b + \delta_a^b$. To adjust the different coordinate systems with each other we apply a rigid rotation in the fluid region by setting $\varphi \to \varphi + \Omega t$ where Ω is a constant. (The t-coordinates in the two regions are also adjusted by rescaling one of them.) The constants M, r_1, a, Ω, c_1, c_2, q_1, χ_0 and χ_2 can now be solved for consistently in terms of the interior solution. Note that we did *not* impose the equation of state (2) when calculating the matching conditions. Hence the vacuum metric above is general enough for describing the exterior of any axisymmetric rigidly rotating perfect fluid ball up to second order.

3. Numerical solutions

In this section we consider solutions with equations of state (2), with special attention to the quadrupole moment and its deviation, $\Delta Q/Q$, from that of the Kerr metric. In scanning the parameter space ω_0 was fixed to 0.1 due to a scaling invariance and the requirement of

Journal of Physics: Conference Series **229** (2010) 012025 doi:10.1088/1742-6596/229/1/012025

asymptotically flatness indirectly fixes the value of h_1. Hence, for a given equation of state, we only need to vary the central pressure, p_{0c}. When considering different equations of state we use another scale invariance to fix one of the constants d_2 or d_3. When performing the corresponding rescaling $r \to \alpha r$ the quantity $\Delta Q/Q$ is invariant.

3.1. Incompressible fluids (interior Schwarzschild)
When the central pressure approaches infinity and the radius the Buchdahl limit $\frac{9M}{4}$, then $\Delta Q/Q$ approaches 0.0212. This value is independent of the density. This case was also considered in [4].

3.2. Linear equations of state $\rho = d_1 p + d_3$
First consider the case $d_1 < -1$. For all these $\rho_{0c} = -p_{0c}$, corresponding to the anti-de Sitter spacetime, when $p_{0c} = d_3/(-d_1 - 1)$. Since the zero-pressure surface is further and further pushed outward when approaching this limit, all these configurations also become infinite in extension. The ratio $r_1/2M$ tends towards one and the quadrupole moment approaches that for Kerr. Even if these spacetimes are quite unphysical, a study of them still is helpful in understanding which conditions are needed for a successful matching to the Kerr metric.

In the numerical runs, due to scaling invariance, we put $d_3 = 1$. Negative values of d_3 are excluded if the surface density should be larger than zero, and configurations with $d_3 = 0$ are not finite in size. The quantity $\Delta Q/Q$ for a sequence of values for d_1 are given in Figure 1. As seen, for more realistic configurations with $d_1 \geq 1$ it differs significantly from zero.

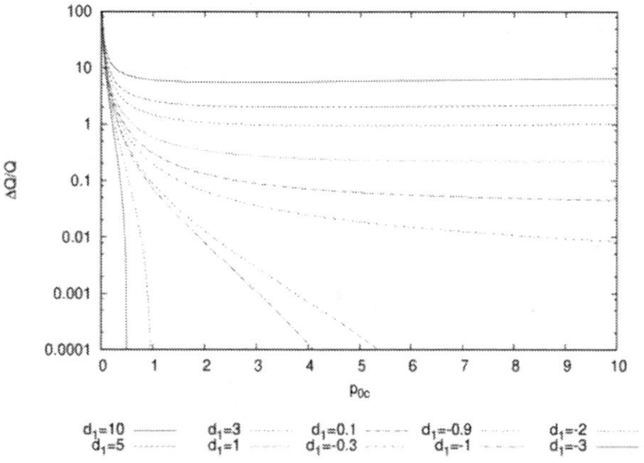

Figure 1. The quantity $\Delta Q/Q$ as a function of central pressure for a sequence of fluid balls with linear equation of state $\rho = d_1 p + 1$.

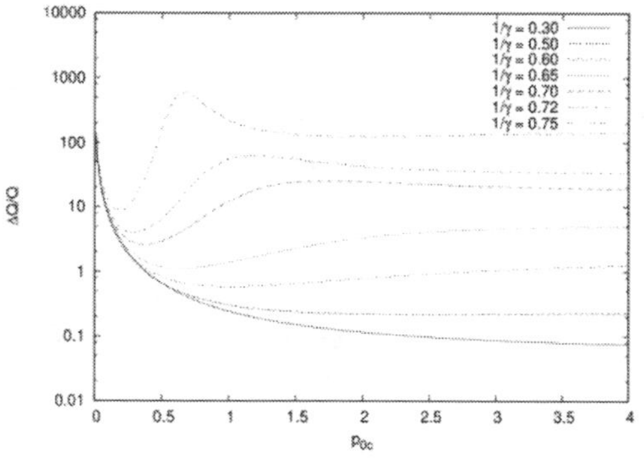

Figure 2. The quantity $\Delta Q/Q$ as a function of central pressure for a sequence of Newtonian polytopes with equation of state $\rho = \left(\frac{p}{p_c}\right)^{1/\gamma}$.

3.3. Polytropes
3.3.1. Newtonian polytropes In a recent paper [8] it was shown that slowly and rigidly rotating polytropes cannot be sources of the Kerr metric in the post-Minkowskian limit. Here we numerically find that slowly and rigidly rotating polytropes with arbitrary strength of the gravitational field cannot be matched to Kerr either.

We use the equation of state $\rho = d_2 \left(\frac{p}{p_c}\right)^{1/\gamma}$ and use scaling invariance to put $d_2 = 1$. In Figure 2 $\Delta Q/Q$ is shown as a function of the central pressure for a sequence of values of γ, including the physically interesting cases $\gamma = 4/3$ and $\gamma = 5/3$. As seen $\Delta Q/Q$ never seems to approach zero. Furthermore, for physically reasonable configurations with $p_c \leq \rho_c = 1$, the quantity differs significantly from zero.

3.3.2. Relativistic polytropes For a fluid with one type of constituent particles, a relativistic polytrope is given by $p = Cn^{\gamma}$, where C is a constant, in terms of the particle density n and the polytropic index γ. This equation is suitable, e.g., to describe an ideal degenerate neutron gas. Using the energy conservation equation for a perfect fluid it is then easy to show that the equation of state becomes $\rho = p/(\gamma-1)+d_2 p^{1/\gamma}$ in terms of the pressure p. As seen the equation of state approaches a linear equation of state for large pressures and a Newtonian polytrope for low pressures. The results of the numerical runs are similar to the previous ones.

4. The post-Minkowskian limit

Here the weak gravity limit is considered by making an expansion in the small parameter $\lambda = M/r_1$, or in SI units GM/r_1c^2, where M is the mass of the fluid ball and r_1 is its radius. A similar study, using global harmonic coordinates and the Lichnerowicz matching conditions, was performed in [7] and [8].

Using the condition that q_1 should vanish (to first order in λ) for an asymptotically flat solution one obtains

$$c_1 = -\frac{5\omega_{1/2}^2}{48 r_1^5 \left(\frac{dA_1}{dr}\right)\Big|_{r=r_1}^5} \frac{\frac{d}{dr}\log\left(\left(h_{21}+\omega_{1/2}^2 r^2/3\right)/\left(r^4\frac{dA_1}{dr}\right)\right)\Big|_{r=r_1}}{\frac{d}{dr}\log\left(r\left(h_{21}+\omega_{1/2}^2 r^2/3\right)/\frac{dA_1}{dr}\right)\Big|_{r=r_1}}, \tag{5}$$

that should be zero for Kerr. Here h_{21} is h_2 to order λ and $\omega_{1/2}$ =constant is ω to order $\sqrt{\lambda}$. Note that generically c_1 diverges as $1/\lambda^4$, whereas $\Delta Q/Q$ diverges as $1/\lambda$.

For the incompressible case ($\rho = \rho_1$ =constant) the equations can be integrated completely and the relative difference to the Kerr quadrupole moment in terms of the central pressure is $\Delta Q/Q = 25\rho_1/(16p_c)$.

For Newtonian polytropes, $p = p_c\left(\frac{\rho}{\rho_c}\right)^{\gamma}$, where ρ_c and p_c are the central density and pressure respectively, we can use a theorem in [8] to show that $c_1 \neq 0$, in agreement with their result. Furthermore, in [10] we show that the rotating fluid ball is oblate in shape iff $k_{2|r=r_1} + \frac{\xi_2}{r_1} < 0$. To first order in λ this is equivalent to $\left(h_{21}+\omega_{1/2}^2 r^2/3\right)_{|r=r_1} > 0$. For Newtonian polytropes this implies that c_1 as given by (5) is positive (see [2] for details). Hence $\Delta Q/Q = 16M^4c_1/5a^2 > 0$ for oblate Newtonian polytropes.

References

[1] Hartle J B 1967 *Astrophys. J.* **150** 1005
[2] Bradley M and Fodor G 2009 *Phys. Rev.* D **79** 044018
[3] Hartle J B and Thorne K S 1968 *Astrophys. J.* **153** 807
[4] Chandrasekhar S and Miller J C 1974 *Mon. Not. R. Astr. Soc.* **167** 63
[5] Berti E, White F, Maniopoulou A and Bruni M 2005 *Mon. Not. R. Astr. Soc.* **358** 923
[6] Laarakkers W G and Poisson E 1999 *Astrophys. J.* **512** 282
[7] Cabezas J A, Martín J, Molina A and Ruiz E 2007 *Gen. Rel. Grav.* **39** 707
[8] Martín J, Molina A and Ruiz E 2008 *Class. Quantum Grav* **25** 105019
[9] Fodor G, Hoenselaers C A and Perjés Z 1989 *J. Math. Phys.* **30** 2252
[10] Bradley M, Eriksson D, Fodor G and Rácz I 2007 *Phys. Rev.* D **75** 024013

Journal of Physics: Conference Series **229** (2010) 012026 doi:10.1088/1742-6596/229/1/012026

Spherically symmetric elasticity in Relativity

J Carot[1*], I Brito[2] and E G L R Vaz[3]

[1] Departament de Física, Universitat de les Illes Balears, Cra Valldemossa pk 7.5, E-07122 Palma, Spain

[2,3] Departamento de Matemática para a Ciência e Tecnologia, Universidade do Minho, Guimarães, Portugal

E-mail: [1]jcarot@uib.cat, [2]ireneb@mct.uminho.pt, [3]evaz@mct.uminho.pt

Abstract. The relativistic theory of elasticity is reviewed within the spherically symmetric context with a view towards the modelling of star interiors possessing elastic properties such as the ones expected in neutron stars. Emphasis is placed on generality in the main sections of the paper, and the results are then applied to specific examples. Along the way, a few general results for spacetimes admitting isometries are deduced, and their consequences are fully exploited in the case of spherical symmetry relating them next to the the case in which the material content of the spacetime is some elastic material. This paper extends and generalizes the pioneering work by Magli and Kijowski [1], Magli [2] and [3], and complements, in a sense, that by Karlovini and Samuelsson in their interesting series of papers [4], [5] and [6].

1. Introduction. Relativistic elasticity revisited

Let (M, g) be a spacetime, M then being a 4-dimensional Hausdorff, simply connected manifold of class C^2 at least, and g a Lorentz metric of signature $(-,+,+,+)$. The *material space* X is a 3-dimensional manifold endowed with a Riemannian metric γ, the *material metric*; points in X can then be thought of as the particles of which the material is made of. Coordinates in M will be denoted as x^a for $a = 0, 1, 2, 3$, and coordinates in X as y^A, $A = 1, 2, 3$. The material metric γ is not a dynamical quantity of the theory, but it is frozen in the material, and it roughly describes distances between neighbouring particles in the relaxed state of the material.

The spacetime configuration of the material is said to be completely specified whenever a submersion $\psi : M \to X$ is given; if one chooses coordinate charts in M and X as above, the coordinate representative of ψ is given by three fields $y^A = y^A(x^b)$, $A = 1, 2, 3$ and the physical laws describing the mechanical properties of the material can then be expressed in terms of a hyperbolic second order system of PDE. The differential map $\psi_* : T_pM \to T_{\psi(p)}X$ is then represented in the above charts by the rank 3 matrix $\left(y^A_b\right)_p$, $y^A_b = \partial_b y^A$ $A = 1, 2, 3$, $b = 0, 1, 2, 3$ which is sometimes called *relativistic deformation gradient*. Since ψ_* has maximal rank 3, its kernel is spanned at each point by a single timelike vector which we may take as normalized to unity, the resulting vector field, say $\vec{u} = u^a \partial_a$, satisfies then $y^A_b u^b = 0$, $u^a u_a = -1$, $u^0 > 0$, the last condition stating that we choose it future oriented; $\vec{u}$ is called the *velocity field of the matter*, and in the above picture in which the points in X are material points, it turns out that the spacetime manifold M (or, to be precise, an open submanifold of it) is then made up by the worldlines of the material particles, whose tangent vector is precisely $\vec{u}$.

The material space is said to be in a *locally relaxed state* at $p \in M$ if, at p, it holds $k_{ab} \equiv (\psi^*\gamma)_{ab} = h_{ab}$ where $h_{ab} = g_{ab} + u_a u_b$. Otherwise, it is said to be *strained*, and a measurement of the difference between k_{ab} and h_{ab} is the *strain*, whose definition varies in the literature. We shall follow the convention in [3] and use

$$K_{ab} \equiv k_{ab} - u_a u_b \tag{1}$$

The strain tensor determines the elastic energy stored in an infinitesimal volume element of the material space (or energy per particle). That energy will be a scalar function of K_{ab} called *constitutive equation* of the material, and its specification amounts to specifying the material. We shall denote it as $v = v(I_1, I_2, I_3)$, where I_1, I_2, I_3 are any suitably chosen set of scalar invariants associated with and characterizing K_{ab} completely. Following [3] we choose

$$I_1 = \frac{1}{2}\left(\mathrm{Tr}K - 4\right),\ I_2 = \frac{1}{4}\left[\mathrm{Tr}K^2 - (\mathrm{Tr}K)^2\right] + 3,\ I_3 = \frac{1}{2}\left(\mathrm{det}K - 1\right) \tag{2}$$

Notice that for $K_{ab} = g_{ab}$ (equivalently $k_{ab} = h_{ab}$) the induced metric on the rest frame of an observer moving with four-velocity $\vec{u}$, h, coincides with the material metric γ (its pull-back by ψ) describing the relaxed state of the material; thus it makes sense to have zero elastic energy stored and it is immediate to see that $I_1 = I_2 = I_3 = 0$.

The energy density ρ will then be the particle number density ϵ times the constitutive equation, that is

$$\rho = \epsilon v(I_1, I_2, I_3) = \epsilon_0 \sqrt{\det K}\, v(I_1, I_2, I_3) \tag{3}$$

where ϵ_0 is the particle number density as measured in the material space, or rather, with respect to the volume form associated with $k_{ab} = (\psi^*\gamma)_{ab}$, and ϵ is that with respect to h_{ab}; see [7].

In the case of elastic matter, the energy-momentum tensor is obtained from the Lagrangian $\Lambda = \sqrt{-g}\rho$, which depends on y^A, y^A_a and x^a, and performing the standard decomposition w.r.t. $\vec{u}$, the velocity of the matter, it follows:

$$T_{ab} = \rho u_a u_b + p h_{ab} + P_{ab}, \tag{4}$$

where $h_{ab} = g_{ab} + u_a u_b$, $P_{ab} = h^m_a h^n_b (T_{mn} - 3p h_{mn})$, $\rho = T_{ab}u^a u^b$, $p = \frac{1}{3}h^{ab}T_{ab}$; and the heat flux vanishes: $q_a = -T_{ab}u^b + \rho u_a = 0$. The quantities ρ, p, P_{ab} are respectively the energy density, isotropic pressure and anisotropic pressure tensor that a family of observers comoving with the matter would measure at each point in the spacetime. The resulting tensor is of the diagonal Segre type $\{1, 111\}$ or any of its degeneracies, $\vec{u}$ being its (unit) timelike eigenvector (see [8]), and it then follows that an orthonormal tetrad exists, $\{u_a, x_a, y_a, z_a\}$ (with $-u_a u^a = x^a x_a = y^a y_a = z^a z_a = 1$ and the mixed products zero) with respect to which T_{ab} may be written as

$$T_{ab} = \rho u_a u_b + p_1 x_a x_b + p_2 y_a y_b + p_3 z_a z_b, \quad p = \frac{1}{3}(p_1 + p_2 + p_3),$$
$$h_{ab} = x_a x_b + y_a y_b + z_a z_b, \quad \text{etc.} \tag{5}$$

The Dominant Energy Condition (DEC), see for instance [8], is fulfilled if and only if $\rho \geq 0,\ |p_A| \leq \rho, \mathrm{for} A = 1, 2, 3$.

2. Elasticity in spherical symmetry

Let us now consider in more detail the problem of elasticity in a spherically symmetric spacetime $(M, \bar{g})$ with associated material space (X, γ). The results given in this section generalize those

Journal of Physics: Conference Series **229** (2010) 012026 doi:10.1088/1742-6596/229/1/012026

in [3] in the sense that here we consider a non flat material metric γ. The reader is also referred to [9] and [10] for other related, interesting developments.

We demand that the submersion $\psi : M \longrightarrow X$ preserves the KVs, that is: $\psi_*(\vec{\xi}_A) = \vec{\eta}_A$ are also KVs on X. This implies that the metric γ is also spherically symmetric and therefore coordinates $y^A = (y, \tilde{\theta}, \tilde{\phi})$ exist with $y = y(t,r)$, $\tilde{\theta} = \theta$ and $\tilde{\phi} = \phi$, and are such that $\vec{\eta}_A = \vec{\xi}_A$ are KVs of the metric $\bar{\gamma}$. Thus, the line elements of g and γ may be written as:

$$ds^2 = -a(t,r)dt^2 + b(t,r)dr^2 + r^2 d\theta^2 + r^2 sin^2\theta d\phi^2 \tag{6}$$

$$d\Sigma^2 = f^2(y)(dy^2 + y^2 d\theta^2 + y^2 sin^2\theta d\phi^2), \tag{7}$$

Notice that this last expression is completely general, as any 3-dimensional spherically symmetric metric is necessarily conformally flat. The results in [3] correspond to $f(y) = 1$.

The velocity field of the matter takes then the form $\vec{u} = u^t(t,r)\partial_t + u^r(t,r)\partial_r$, as follows from $u^a y^A_a = 0$, and u^t and u^r can be easily derived from $g_{ab}\bar{u}^a\bar{u}^b = -1$ and $\bar{u}^0 > 0$. Now, much clarity is gained by making use of the comoving coordinates adapted to $\vec{u}$: it is easy to show that it is always possible to perform a coordinate change in the t, r plane so that, in the new coordinates one has:

$$ds^2 = -a(r,t)dt^2 + b(r,t)dr^2 + Y^2(r,t)\left(d\theta^2 + \sin^2\theta d\phi^2\right), \quad u^a = \left(a^{-1/2}, 0, 0, 0\right) \tag{8}$$

hence, for the material space (M, γ) there exist coordinates $y^A = (y, \tilde{\theta}, \tilde{\phi})$ such that $y = y(r)$, $\tilde{\theta} = \theta$ and $\tilde{\phi} = \phi$, as follows from the condition $y^A_a u^a = 0$ and the requirement that $\psi_*(\vec{\xi}_A) = \vec{\eta}_A$ are KVs of the metric $\bar{\gamma}$. Further, and since the line element of the material space is $d\sigma^2 = f^2(y)\left[dy^2 + y^2\left(d\theta^2 + \sin^2\theta d\phi^2\right)\right]$, with $y = y(r)$, no generality is lost if we set $y = r$, as this amounts to a redefinition of the r coordinate in spacetime which leaves unchanged the form of the metric and that of $\vec{u}$ in (8). We shall do that in the sequel.

As an aside, it is interesting to notice that, in spherical symmetry (and in more general classes of spacetimes, namely type B warped ones), P_{ab} and σ_{ab} are always proportional (the latter being the shear tensor of $\vec{u}$); thus, whenever the velocity flow is non-zero, it is always possible to interpret (at least formally) elastic matter as a viscous fluid.

The pulled-back material metric k is

$$k^a_b = \text{diag}\left(0, f^2(r)b^{-1}, r^2 f^2(r)Y^{-2}, r^2 f^2(r)Y^{-2}\right). \tag{9}$$

The operator $K^a_b = g^{ac}k_{cb} - u^a u_b$, used to measure the state of strain of the material has one eigenvalue equal to 1 (corresponding to the eigenvector $\vec{u}$), while the other eigenvalues η and s (algebraic multiplicity two) are $f^2(r)b^{-1}$, $s = r^2 f^2(r)Y^{-2}$, and one can then calculate the three invariants I_1, I_2, I_3 of K introduced in (2). In [3], the energy-momentum tensor was calculated from these invariants for a flat material metric. A similar calculation shows that

$$T^a_b = \rho\,\delta^a_b - \frac{\partial\rho}{\partial I_3}\det K\, h^a_b + \left(\text{Tr}\, K\,\frac{\partial\rho}{\partial I_2} - \frac{\partial\rho}{\partial I_1}\right)k^a_b - \frac{\partial\rho}{\partial I_2}\, k^a_c\, k^c_b. \tag{10}$$

Thus, Einstein's Field Equations read

$$-\frac{\dot{Y}}{Y^2 a} - \frac{\dot{Y}}{Y}\frac{\dot{b}}{ab} + \frac{2Y''}{Yb} + \frac{Y'^2}{Y^2 b} - \frac{Y'}{Y}\frac{b'}{b^2} - \frac{1}{Y^2} = \epsilon v\, 8\pi, \qquad 2\dot{Y}' - \frac{a'}{a}\dot{Y} - \frac{\dot{b}}{b}Y' = 0, \tag{11}$$

$$-\frac{\dot{Y}^2}{Y^2 a} + \frac{\dot{Y}}{Y}\frac{\dot{a}}{a^2} + \frac{Y'}{Y}\frac{a'}{ab} + \frac{Y'^2}{Y^2 b} - \frac{2\ddot{Y}}{Ya} - \frac{1}{Y^2} = -\epsilon\, 2\eta\,\frac{\partial v}{\partial\eta}\, 8\pi, \tag{12}$$

$$\frac{1}{2}\frac{\dot{Y}}{aY}\left(\frac{\dot{a}}{a}-\frac{\dot{b}}{b}\right)-\frac{1}{4}\frac{a'}{ab}\left(\frac{a'}{a}+\frac{b'}{b}\right)+\frac{1}{2}\frac{Y'}{bY}\left(\frac{a'}{a}-\frac{b'}{b}\right)$$
$$+\frac{Y''}{Yb}-\frac{\ddot{Y}}{Ya}+\frac{1}{2ab}\left(a''-\ddot{b}+\frac{\dot{b}}{2}\left(\frac{\dot{a}}{a}+\frac{\dot{b}}{b}\right)\right)=-\epsilon\, s\,\frac{\partial v}{\partial s}\,8\pi. \tag{13}$$

3. Shearfree solutions. Examples

We next consider the case of spacetimes with an elastic material content such that the velocity of the matter is shearfree, in which case, the interpretation as a viscous fluid with kinematical viscosity is not possible, and therefore the anisotropy in the pressures must be a consequence of the elastic properties of the material. In the shear-free case, coordinates exist such that

$$ds^2=-a(r,t)dt^2+Y^2(r,t)\left(dr^2+d\theta^2+\sin^2\theta d\phi^2\right) \tag{14}$$

and two cases can be distinguished: static or non-static.

In the static case, it is easy to see that solutions with a well posed constitutive equation and satisfying the DEC do indeed exist, as the following example shows:

$$ds^2=-e^{-5r^2}dt^2+e^{5r^2}\left(dr^2+d\theta^2+\sin^2\theta d\phi^2\right). \tag{15}$$

One has

$$\rho=\epsilon v=\frac{e^{-5r^2}}{8\pi}(25r^2+9),\quad p_1=-2\epsilon\eta\frac{\partial v}{\partial\eta}=-\frac{e^{-5r^2}}{8\pi}(25r^2+1),\quad p_2=-\epsilon s\frac{\partial v}{\partial s}=\frac{e^{-5r^2}}{8\pi}25r^2$$

which is obviously well behaved: satisfies the dominant energy condition and is non-singular at the origin. After some tedious algebra, we get: $f(r)=\exp(\frac{5}{2}r^2)(75r^2+1)^{-\frac{1}{3}}$, whence expressions for η, s and $\epsilon=\epsilon_0 s\sqrt{\eta}$ can be easily derived, and also: $v=F(\Sigma)\left[(75r^2+1)(75r^2+27)^{12}\right]^{-\frac{1}{39}}$, where $F(\Sigma(r))=(25r^2+9)^{-\frac{12}{39}}(75r^2+1)^{-\frac{1}{39}}$, and r must be the only real solution of $r^3-75e^{\frac{3}{2}\Sigma}r^2-e^{\frac{3}{2}\Sigma}=0$.

Thus we have proven that a solution exists, which is regular at the origin $r=0$, satisfies the dominant energy condition and possesses a constitutive equation that can be given in closed form.

Acknowledgments

JC acknowledges financial support from MEC through grant No.: HP2006-0074. Partial financial support from (MEC) through grant FPA-2007-60220 and from the "Govern de les Illes Balears" is also acknowledged. IB and EGLRV acknowledge financial support from (CRUP) through grant No.: E-89/07 and from FCT and CMAT.

References

[1] Magli G and Kijowski J 1992 *Gen. Rel. Grav.* **24** 139
[2] Magli G 1995 *Gen. Rel. Grav.* **25** 441
[3] Magli G 1993 *Gen. Rel. Grav.* **25** 441
[4] Karlovini M and Samuelsson L 2003 *Class. Quantum Grav.* **20** 3613
[5] Karlovini M and Samuelsson L 2004 *Class. Quantum Grav.* **21** 1559
[6] Karlovini M and Samuelsso, L 2004 *Class. Quantum Grav.* **21** 4531
[7] Kijowski J and Magli G 1992 *J. Geom. Phys.* **9** 207
[8] Kramer D, Stephani H, Hoenselaers C, MacCallum, M A H and Herlt H 2002 *Exact Solutions of Einstein's Field Equations* 2nd ed (Cambridge University Press)
[9] Park J 2000 *Gen. Rel. Grav.* **32** 235
[10] Beig R and Schmidt B G 2003 *Class. Quantum Grav.* **20** 889

On a recent proposal of the Penrose inequality

Alberto Carrasco[1*] **and Marc Mars**[2]
Dept. Física Fundamental, Univ. de Salamanca, Pl. de la Merced s/n, 37008 Salamanca, Spain
E-mail: [1]`acf@usal.es`, [2]`marc@usal.es`

Abstract. Recently Bray and Khuri [1] have proposed a version of the Penrose inequality in terms of a new type of horizons, the so-called generalized apparent horizons. In [2] we find counter-examples to this conjecture in the Kruskal spacetime. The main technical tool for this result is the implicit function theorem. In this short note we explain in detail how this theorem can be applied in our situation.

The standard version of the Penrose inequality bounds the total ADM mass of an asymptotically flat initial data set in terms of the minimal area needed to enclose the boundary of the outer trapped region. The heuristic argument supporting its validity [3] is based on a number of expected properties of gravitational collapse, most notably the weak cosmic censorship conjecture. The Penrose inequality, however, is logically independent of these hypotheses and proving (or disproving) its validity would therefore give rather convincing support for the standard viewpoint of gravitational collapse.

At present, the Penrose inequality has been proven only in spherical symmetry and when the second fundamental form of the initial data set vanishes (see [4] for a recent review). In a very interesting proposal to address the general case of the Penrose inequality, Bray and Khuri [1] have proposed a new version of the inequality which involves a novel type of surfaces called generalized trapped horizons, defined as follows. On an asymptotically flat initial data set $(\Sigma, \gamma_{ij}, A_{ij})$, consider *bounding* surfaces, i.e. compact and embedded surfaces S without boundary such that $\Sigma \setminus S$ consists of two disjoint open regions Ω^+ and Ω^-. Let Ω^+ denote the region containing the asymptotically flat end. Let p be the mean curvature of $S \subset \Sigma$ with respect to the unit normal pointing towards Ω^+, and let q be the trace of the pull-back of the second fundamental form A_{ij} of Σ onto S. A bounding surface S is a **generalized apparent horizon** if $p = |q|$. Eichmair [5] has proven that, if Σ contains at least one bounding surface satisfying $p \leq |q|$, then the initial data set contains a unique outermost generalized apparent horizon S_{out}. This surface is $C^{2,\alpha}$ and has the property of having less area than any other surface enclosing it. The Bray and Khuri proposal of the Penrose inequality is

$$M \geq \sqrt{\frac{|S_{\text{out}}|}{16\pi}}, \qquad (1)$$

where M is the ADM mass of the initial data and $|S_{\text{out}}|$ the area of the outermost generalized apparent horizon.

As discussed in [4], there exist slices in the Kruskal spacetime such that the outermost generalized apparent horizon penetrates, at least partially, inside the domain of outer communications. It becomes natural to consider whether the conjecture (1) holds true for such slices. In a recent paper [2] we have argued that, in fact, there exist slices of the Kruskal spacetime where the inequality (1) is violated. The main idea exploited in [2] is to consider slices Σ_ϵ which are small (but finite) perturbations of the time symmetric slice of the Kruskal spacetime and to use the implicit function theorem to locate a generalized apparent horizon S_ϵ near the bifurcation surface on each one of the perturbed slices Σ_ϵ. An explicit calculation of the area of S_ϵ shows $|S_\epsilon| > 16\pi M^2$ for $\epsilon \neq 0$ small enough. In a second step, we prove that the outermost generalized apparent horizon S^ϵ_{out} either coincides with S_ϵ or else satisfies $|S^\epsilon_{\text{out}}| \geq |S_\epsilon|$. In either case, the inequality (1) is violated in the slice Σ_ϵ. An important issue that has been recently put forward by M. Khuri [6] is whether the implicit function theorem can be applied to the functional $p - |q|$, since the appearance of an absolute value in $|q|$ makes the derivative potentially discontinuous. The aim of this contribution is to describe in detail how the implicit function theorem can be applied in our situation.

Consider the Kruskal spacetime of mass $M > 0$ with metric $ds^2 = \frac{32M^3}{r}e^{-r/2M}dudv + r^2\left(d\theta^2 + \sin\theta^2 d\phi^2\right)$, where $r(uv)$ solves $uv = e^{r/2M}(r-2M)/2M$. The null vectors ∂_u and $-\partial_v$ are future directed. Consider the one-parameter family of axially symmetric spacelike hypersurfaces $\Sigma_\epsilon = \mathbb{R} \times S^2$ defined by $\Sigma_\epsilon = \{u = \hat{y} - \epsilon x, v = \hat{y} + \epsilon x, \cos\theta = x, \phi = \phi\}$, where $\hat{y} \in \mathbb{R}$, $x \in [-1,1]$, $\phi \in [0, 2\pi]$. The discrete isometry of the Kruskal spacetime defined by $\{u, v\} \to \{v, u\}$ implies that, under reflection with respect to the equatorial plane, i.e. $(\hat{y}, x, \phi) \to (\hat{y}, -x, \phi)$, the induced metric of Σ_ϵ remains invariant, while the second fundamental form of Σ_ϵ changes sign. Uniqueness of the outermost generalized apparent horizon implies that S^ϵ_{out} must be axially symmetric and have equatorial symmetry. We therefore restrict ourselves to the subclass of surfaces $\mathcal{S}_\epsilon \subset \Sigma_\epsilon$ defined by $\mathcal{S}_\epsilon = \{\hat{y} = y(x,\epsilon), x = x, \phi = \phi\}$, with $y(-x,\epsilon) = y(x,\epsilon)$. We are seeking solutions of the equation $p - |q| = 0$. Thus, it becomes natural to restrict the space of functions to $U^{m,\alpha} \equiv \left\{y \in C^{m,\alpha}(S^2) : \partial_\phi y = 0, y(-x) = y(x)\right\}$, i.e. m-times differentiable functions on the unit sphere, with Hölder continuous m-th derivatives with exponent α and invariant under the axial Killing vector on S^2 and under reflection about the equatorial plane. $U^{m,\alpha}$ is a closed subset of the Banach space $C^{m,\alpha}(S^2)$ hence a Banach space itself. Let $I \subset \mathbb{R}$ be the closed interval where ϵ takes values. For each function $y \in U^{2,\alpha}$ the expression $p - |q|$ defines a non-linear map $f : U^{2,\alpha} \times I \to U^{0,\alpha}$. We are therefore looking for solutions of $f(y,\epsilon) = 0$. At $\epsilon = 0$, the hypersurface Σ_ϵ is totally geodesic, which implies that $q = 0$ for any surface on it. Consequently, all generalized apparent horizons satisfy $p = 0$ and are, in fact, minimal surfaces. The only closed minimal surface in $\Sigma_{\epsilon=0}$ is the bifurcation surface $\{u = 0, v = 0\}$. Thus, the equation $f(y,\epsilon) = 0$ has $y = 0$ as the unique solution when $\epsilon = 0$. We want to use the implicit function theorem to show that there exists a unique solution $y(x,\epsilon)$ of $f(y,\epsilon) = 0$ in a suitable neighbourhood of $y = 0$ and $\epsilon = 0$. This requires f to be $C^1(U^{2,\alpha} \times I)$ and the partial derivative $D_y f|_{(y=0,\epsilon=0)}$ to be an isomorphism between $U^{2,\alpha} \to U^{0,\alpha}$. As emphasized by Khuri, the appearance of an absolute value in $|q|$ makes it unlikely that the functional f is $C^1(U^{2,\alpha} \times I)$. To address this issue, we restrict ourselves to surfaces of the form $y = \epsilon Y$, where $Y \in U^{2,\alpha}$. An explicit computation of the mean curvature p on such surfaces gives $p = \epsilon\mathcal{P}(Y(x), \dot{Y}(x), \ddot{Y}(x), x, \epsilon)$, where dot denotes derivative with respect to x and where $\mathcal{P} : \mathbb{R}^3 \times [-1,1] \times I \to \mathbb{R}$ is a smooth (in fact, analytic) function. Similarly $q = \epsilon\mathcal{Q}(Y(x), \dot{Y}(x), x, \epsilon)$, where $\mathcal{Q} : \mathbb{R}^2 \times [-1,1] \times I \to \mathbb{R}$ is an analytic function. Moreover, the function $\mathcal{Q}$ has the symmetry $\mathcal{Q}(x_1, x_2, x_3, x_4) = -\mathcal{Q}(x_1, -x_2, -x_3, x_4)$, which reflects the fact that the extrinsic curvature of Σ_ϵ changes sign under a transformation $x \to -x$. Let us write $P(Y,\epsilon)(x) \equiv \mathcal{P}(Y(x), \dot{Y}(x), \ddot{Y}(x), x, \epsilon)$ and similarly $Q(Y,\epsilon)(x) \equiv \mathcal{Q}(Y(x), \dot{Y}(x), x, \epsilon)$.

Instead of analysing the functional f, we set out to consider the alternative functional $F : U^{2,\alpha} \times I \to U^{0,\alpha}$ defined by $F(Y,\epsilon) = P(Y,\epsilon) - |Q(Y,\epsilon)|$. This functional has the property that, for $\epsilon > 0$, the solutions of $F(Y,\epsilon) = 0$ correspond exactly to the solutions of $f(y,\epsilon) = 0$ via the relation $y = \epsilon Y$. However, the functional F is well-defined even at $\epsilon = 0$ (in fact, for all $\epsilon \in I$). By proving that $F = 0$ admits solutions in a neighbourhood of $\epsilon = 0$, we will conclude that $f = 0$ admits solutions for $\epsilon > 0$ and the solutions will in fact belong to a neighbourhood of $y = 0$ since $y = \epsilon Y$.

In order to show that F admits solutions we will use the implicit function theorem. A direct calculation yields $F(Y,0)(x) = G(L(Y)(x) - 3|x|)$, where G is a positive function of $(Y(x), \dot{Y}(x), x)$ and $L(Y) \equiv -(1-x^2)\ddot{Y} + 2x\dot{Y} + Y$. This operator is an isomorphism between $U^{2,\alpha}$ and $U^{0,\alpha}$. Let $Y_{sol} \in U^{2,\alpha}$ be the unique solution of the equation $L(Y) = 3|x|$. For later use, we note that $Q(Y_{sol}, 0) = -3x$. This vanishes *only* at $x = 0$. This is the key property that allows us to prove that F is $C^1(U^{2,\alpha} \times I)$.

The $C^1(U^{2,\alpha} \times I)$ property of the functional $P(Y,\epsilon)$ is standard. More subtle is to show that $|Q|$ is $C^1(U^{2,\alpha} \times I)$ in a suitable neighbourhood of $(Y_{sol}, \epsilon = 0)$. Let $r_0 > 0$ and define $\mathcal{V}_{r_0} = \{(Y,\epsilon) \in U^{2,\alpha} \times I : \|(Y - Y_{sol}, \epsilon)\|_{U^{2,\alpha}\times I} \leq r_0\}$. First of all we need to show that $|Q|$ is (Fréchet-)differentiable on $\mathcal{V}_{r_0}$, i.e. that for all $(Y,\epsilon) \in \mathcal{V}_{r_0}$ there exists a continuous linear mapping $D_{Y,\epsilon}|Q| : U^{2,\alpha} \times I \to U^{0,\alpha}$ such that, for all $(h,\delta) \in U^{2,\alpha} \times I$, $|Q(Y+h,\epsilon+\delta)| - |Q(Y,\epsilon)| = D_{Y,\epsilon}|Q|(h,\delta) + R_{Y,\epsilon}(h,\delta)$ where $\|R_{Y,\epsilon}(h,\delta)\|_{U^{0,\alpha}} = o(\|(h,\delta)\|_{U^{2,\alpha}\times I})$. The key observation is that

$$|Q(Y,\epsilon)(x)| = -\sigma(x) Q(Y,\epsilon)(x) \tag{2}$$

where $\sigma(x)$ is the *sign* function, (i.e. $\sigma(x) = +1$ for $x \geq 0$ and $\sigma(x) = -1$ for $x < 0$). For x away from a neighbourhood of 0, this follows from the fact that $Q(Y_{sol}, \epsilon) = -3x$, which is negative for $x > 0$ and positive for $x < 0$. Taking r_0 small enough, and using that $\mathcal{Q}$ is a smooth function of their arguments, the same holds for any $(Y,\epsilon) \in \mathcal{V}_{r_0}$. Moreover, the function $Q(Y,\epsilon)(x)$ is odd in x, so it passes through zero at $x = 0$. Hence, in a small enough neighbourhood of $x = 0$, the relation (2) holds provided we can prove that $Q(Y,\epsilon)$ is strictly decreasing at $x = 0$. But this follows easily from the fact that $\frac{dQ(Y_{sol},\epsilon=0)}{dx}|_{x=0} = -3$ and $\mathcal{Q}$ is a smooth function.

From its definition, it is immediate that $Q(Y,\epsilon)(x)$ is $C^{1,\alpha}$ and has derivative $D_{Y,\epsilon}Q(h,\delta) = A_{Y,\epsilon}(x)h(x) + B_{Y,\epsilon}(x)\dot{h}(x) + C_{Y,\epsilon}(x)\delta$, where $A_{Y,\epsilon}(x) \equiv \partial_1 \mathcal{Q}|_{(Y(x),\dot{Y}(x),x,\epsilon)}$, $B_{Y,\epsilon}(x) \equiv \partial_2 \mathcal{Q}|_{(Y(x),\dot{Y}(x),x,\epsilon)}$ and $C_{Y,\epsilon}(x) \equiv \partial_4 \mathcal{Q}|_{(Y(x),\dot{Y}(x),x,\epsilon)}$. We note that these three functions are $C^{1,\alpha}$ and that $A_{Y,\epsilon}$, $C_{Y,\epsilon}$ are odd, while $B_{Y,\epsilon}$ is even (as a consequence of the symmetries of $\mathcal{Q}$). Defining the linear map $D_{Y,\epsilon}|Q|(h,\delta) \equiv -\sigma(A_{Y,\epsilon}h + B_{Y,\epsilon}\dot{h} + C_{Y,\epsilon}\delta)$, it follows from (2) that $|Q(Y+h,\epsilon+\delta)| - |Q(Y,\epsilon)| = D_{Y,\epsilon}|Q|(h,\delta) + R_{Y,\epsilon}(h,\delta)$ with$\|R(h,\delta)\|_{U^{0,\alpha}} = o(\|(h,\delta)\|_{U^{2,\alpha}\times I})$. In order to conclude that $D_{Y,\epsilon}|Q|$ is the derivative of $|Q(Y,\epsilon)|$, we only need to check that, it is (i) well-defined, i.e. its image belongs to $U^{0,\alpha}$, and (ii) that it is continuous, i.e. $\|D_{Y,\epsilon}|Q|(h,\delta)\|_{U^{0,\alpha}} < C\|(h,\delta)\|_{U^{2,\alpha}\times I}$ for some constant C. To show (i), the most difficult term is $-\sigma B_{Y,\epsilon}\dot{h}$, because $B_{Y,\epsilon}(x)$ is even and need not vanish at $x = 0$. However $\dot{h}$ is an odd function, and hence $-\sigma B_{Y,\epsilon}\dot{h}$ is continuous. To show it is also Hölder continuous, we only need to consider points $x_1 = -a$ and $x_2 = b$ with $0 < a < b$ (if $x_1 \cdot x_2 \geq 0$, the sign function remains constant, so $-\sigma B_{Y,\epsilon}\dot{h}$ is in fact $C^{1,\alpha}$). Calling $w(x) \equiv -\sigma(x)B_{Y,\epsilon}(x)\dot{h}(x)$ and using that $w(x)$ is even, we find

$$|w(x_2) - w(x_1)| = |w(b) - w(-a)| = |(w(b) - w(a)| = \left|\frac{d(B_{Y,\epsilon}\dot{h})}{dx}\right|_{x=\xi}\right| |b - a| \leq$$
$$\leq \left|\frac{d(B_{Y,\epsilon}\dot{h})}{dx}\Big|_{x=\xi}\right| |b-a|^{1-\alpha}|x_2 - x_1|^{\alpha} \leq \left|\frac{d(B_{Y,\epsilon}\dot{h})}{dx}\Big|_{x=\xi}\right| |x_2 - x_1|^{\alpha}. \tag{3}$$

where $\xi \in (a,b)$ and we have used that $|b-a|^{\alpha} \leq |b+a|^{\alpha} = |x_2 - x_1|^{\alpha}$ and $|b - a| < 1$. This proves that $-\sigma B_{Y,\epsilon}\dot{h}$ is Hölder continuous with exponent α.

To check (ii), we first notice that $w(x)$ obviously satisfies $\sup_x |w| < C\|(h,\delta)\|_{U^{2,\alpha}\times I}$ because $B_{Y,\epsilon}(x)$ is $C^{1,\alpha}$. It remains to bound the Hölder constant $[w]_\alpha \equiv \sup_{x_1\neq x_2} \frac{|w(x_2)-w(x_1)|}{|x_2-x_1|^\alpha}$. Combining (3) with the fact that $B_{Y,\epsilon}(x)$ is $C^{1,\alpha}$, the bound $[w]_\alpha \leq C\|(h,\delta)\|_{U^{2,\alpha}\times I}$ follows at once. This proves (ii) for the term $-\sigma B_{Y,\epsilon}\dot{h}$. A similar argument applies to $-\sigma A_{Y,\epsilon}h$ and $-\sigma C_{Y,\epsilon}\delta$ and we conclude that $D_{Y,\epsilon}|Q|$ is indeed a continuous operator.

In order to apply the implicit function theorem, it is furthermore necessary that $|Q| \in C^1(U^{2,\alpha}\times I)$ (i.e. that $D_{Y,\epsilon}|Q|$ depends continuously on (Y,ϵ)). This means that given any convergent succession $(Y_n,\epsilon_n) \in \mathcal{V}_{r_0}$, the corresponding operators $D_{Y_n,\epsilon_n}|Q|$ also converge. Denoting by $(Y,\epsilon) \in \mathcal{V}_{r_0}$ the limit of the succession, we need to prove that $\|D_{Y_n,\epsilon_n}|Q| - D_{Y,\epsilon}|Q|\|_{\mathcal{L}(U^{2,\alpha}\times I, U^{0,\alpha})} \to 0$. It suffices to find a constant K (which may depend on (Y,ϵ)), such that

$$\|(D_{Y_n,\epsilon_n}|Q| - D_{Y,\epsilon}|Q|)(h,\delta)\|_{U^{0,\alpha}} < K\|(h,\delta)\|_{U^{2,\alpha}\times I}\|(Y_n - Y, \epsilon_n - \epsilon)\|_{U^{2,\alpha}\times I} \tag{4}$$

for all $(h,\delta) \in U^{2,\alpha}\times I$. Again, the most difficult case involves $\sigma(B_{Y,\epsilon} - B_{Y_n,\epsilon_n})\dot{h}$, so we concentrate on this term. Using the mean value theorem on the function $\mathcal{B} \equiv \partial_2\mathcal{Q}$ (recall that $B_{Y,\epsilon} = \mathcal{B}|_{(Y(x),\dot{Y}(x),x,\epsilon)}$) gives

$$\sup_x |\sigma(x)(B_{Y,\epsilon} - B_{Y_n,\epsilon_n})\dot{h}| \leq 2\sup_{\mathbb{K}}|\nabla\mathcal{B}|\sup_x|\dot{h}|\|(Y_n - Y,\epsilon_n - \epsilon)\|_{U^{2,\alpha}\times I}, \tag{5}$$

where $\nabla\mathcal{B}$ is the gradient of $\mathcal{B}$ and $\mathbb{K}\subset\mathbb{R}^4$ is a compact domain depending only on r_0 and Y_{sol} defined so that, for all $(Y,\epsilon) \in \mathcal{V}_{r_0}$, the quadruple $(Y(x),\dot{Y}(x),x,\epsilon) \in \mathbb{K}$, for all $x \in [-1,1]$. Inequality (5) is already of the form (4) (recall that $\mathcal{B}$ is smooth). It only remains to bound the Hölder constant of $z \equiv \sigma(B_{Y,\epsilon} - B_{Y_n,\epsilon_n})\dot{h}$ in a similar way. As before, this is done by distinguishing two cases, namely when $x_1\cdot x_2 \geq 0$ and when $x_1\cdot x_2 < 0$. Obtaining an inequality of the form $\sup_{x_1\neq x_2, x_1\cdot x_2\geq 0}\frac{|z(x_2)-z(x_1)|}{|x_2-x_1|^\alpha} \leq K_1\|(h,\delta)\|_{U^{2,\alpha}\times I}\|(Y_n - Y,\epsilon_n-\epsilon)\|_{U^{2,\alpha}\times I}$ is standard, because $\sigma(x)$ is a constant function. When $x_1\cdot x_2 < 0$, we exploit the parity of the functions as in (3) to get $|z(x_2) - z(x_1)| \leq \left|\left.\frac{d((B_{Y_n,\epsilon_n} - B_{Y,\epsilon})\dot{h})}{dx}\right|_{x=\xi}\right| |x_2 - x_1|^\alpha$, where $\xi \in (a,b)$ and we are assuming $x_1 = -a, x_2 = b, 0 < a < b$ without loss of generality. Bounding the right hand side in terms of $K_2\|(h,\delta)\|_{U^{2,\alpha}\times I}\|(Y_n - Y,\epsilon_n - \epsilon)\|_{U^{2,\alpha}\times I}|x_2 - x_1|^\alpha$ is again standard, since the sign function $\sigma(x)$ has already disappeared. This, combined with (5) gives (4) and hence continuity of the derivative of $D_{Y,\epsilon}|Q|$ with respect to $(Y,\epsilon) \in \mathcal{V}_{r_0}$.

The final requirement to apply the implicit function theorem to $F = P - |Q|$ is to check that $D_Y F|_{(Y_{\text{sol}},\epsilon=0)}$ is invertible. A simple computation gives $D_Y F|_{(Y_{\text{sol}},\epsilon=0)}(h) = cL(h)$, where c is a positive function of x and L is the elliptic operator defined above, which is an isomorphism between $U^{2,\alpha}$ and $U^{0,\alpha}$. Thus, the implicit function theorem can be used to conclude that there exists an open neighbourhood $\tilde{I}\subset I$ of $\epsilon = 0$ and a C^1 map $\tilde{Y} : \tilde{I} \to U^{2,\alpha}$ such that $\tilde{Y}(\epsilon = 0) = Y_{\text{sol}}$ and $y = \epsilon\tilde{Y}(\epsilon)$ defines a $C^{2,\alpha}$ generalized apparent horizon embedded in Σ_ϵ.

Acknowledgments
We acknowledge financial support from the projects FIS2009-07238 (MEC), Gr-234 (JCyL) and P09-FQM-4496 (JA and FEDER funds). AC acknowledges the PhD grant AP2005-1195 (MEC).

References

[1] Bray H L and Khuri M *Preprint* arXiv:0905.2622
[2] Carrasco A and Mars M *Preprint* arXiv:0911-0883v1
[3] Penrose R 1973 *Ann. N. Y. Acad. Sci.* **224** 125-134
[4] Mars M 2009 *Class. Quantum Grav.* **26** 193001
[5] Eichmair M *Preprint* arXiv:0805.4454
[6] Khuri M, private communication

Journal of Physics: Conference Series **229** (2010) 012028 doi:10.1088/1742-6596/229/1/012028

k-essence and extended tachyons in brane-worlds

Luis P Chimento[1,2*] **and Martín G Richarte**[2]

[1] Department of Theoretical Physics, University of the Basque Country, P.O. Box 644, 48080 Bilbao, Spain and IKERBASQUE, the Basque Foundation for Science, 48011, Bilbao, Spain.
[2] Departamento de Física, Facultad de Ciencias Exactas y Naturales, Universidad de Buenos Aires, Ciudad Universitaria, Pabellón I, 1428 Buenos Aires, Argentina

E-mail: `chimento@df.uba.ar`, `martin@df.uba.ar`

Abstract. We study a k-essence field evolving linearly with the cosmic time and the atypical k-essence model on a homogeneous and isotropic flat 3-brane. We show that the k-field is driven by an inverse quadratic polynomial potential. The solutions represent expanding, contracting or bouncing universes with a finite time span and some of them end in a big crunch or a big rip. Besides, by selecting the extended tachyonic kinetic functions we analyze the high and low energy limits of our model, obtaining the nearly power law solution. We introduce a tachyon field with negative energy density and show that the universe evolves between two singularities.

1. Introduction

Recent progresses in Superstring/M-theory have offered new perspectives to understand the cosmological evolution of the universe [1], [2]. In these theories our Universe is conceived as a 3-brane embedded in a higher dimensional spacetime usually referred as the bulk [3]. Among many interesting models coming from this new scenarios, there are two possibilities that seem to resolve different problems in the standard cosmology, the RS [2] and DGP [4] scenarios.

Several works have been investigated in cosmological scenarios with non-canonical kinetic term usually known as k-essence models [5]-[10]. In particular, some efforts in the framework of k-essence have been directed toward model building using power law solutions which preserve [6], [11],[12], [13] or violate the weak-energy condition [14]. In addition, a k-essence model with a divergent sound speed, called atypical k-essence, was carefully analyzed in Refs. [8], [15]. In the latter case it showed that the model fix the form of the Lagrangian for k-essence matter. Then it would be interesting to study brane-worlds models supported by k-essence.

Recently, power law solutions on a 3-dimensional brane coupled to the tachyonic field were obtained [16] by using a well known algorithm developed in [11]. Further the high and low energy limits for the tachyonic potential turns out to be $V = V_0\phi^{-1}$ and $V = V_0\phi^{-2}$ respectively.

Here, we present a model with k-essence localized on the Friedmann-Robertson-Walker (FRW) brane and obtain the scale factor and potential for a k-field evolving linearly with the time. Also, we examine the divergent sound speed model and apply these results to the extended tachyon field cases. Finally the conclusions are stated.

2. k-essence in brane-worlds

We will explore the evolution of a universe filled with a k-essence field ϕ in a flat FRW spacetime. Using the perfect fluid analogy, the energy density and pressure are given by

$$\rho_\phi = V(\phi)[F - 2xF_x], \qquad p_\phi = \mathcal{L} = -V(\phi)F(x). \tag{1}$$

where $F(x)$ is a function of the kinetic energy $x = -\dot{\phi}^2$, $F_x = dF/dx$ and $V(\phi)$ is a positive potential. We shall focus on cosmological branes with the induced metric $g_{\mu\nu} = \mathrm{diag}[-1, a^2\delta j_i]$, where a is the scale factor. Then the modified Einstein equations on the brane are [17]

$$3H^2 = \rho_\phi + \frac{3}{\lambda^2}\rho_\phi^2, \tag{2}$$

$$(F_x + 2xF_{xx})\ddot{\phi} + 3HF_x\dot{\phi} + \frac{V'}{2V}(F - 2xF_x) = 0, \tag{3}$$

with $H = \dot{a}/a$ the Hubble expansion rate and $' \equiv d/d\phi$. Also, by writing the equation of state for the k-essence as $p_\phi = (\gamma_\phi - 1)\rho_\phi$, the barotropic index reads $\gamma_\phi = -2xF_x/(F - 2xF_x)$. The quadratic term (2) modifies the standard cosmology dominating at the early times, $H \propto \rho_\phi/\lambda$, whereas in the limit $\lambda \to \infty$ we recover standard relativity $H \propto \sqrt{\rho_\phi}$.

For $x = x_0 = const$ and $F_x(x_0) = 0$, we get $\gamma_\phi = 0$. Hence, from Eqs. (2) and (3) the energy density and the expansion rate become constants and the brane exhibits a de Sitter phase.

Other kind of solutions can be found by re-writing the Eqs. (2)-(3)

$$3H^2 = V\alpha + \frac{3}{\lambda^2}\alpha^2 V^2, \qquad \frac{\dot{\alpha}}{\alpha} + \frac{\dot{V}}{V} + 3H\gamma_\phi = 0, \tag{4}$$

with $\alpha = \rho_\phi/V = F - 2xF_x$ and assuming the constraint $\alpha = \alpha_0 = const$. It is satisfied for any kinetic function F when *i.* the k-field evolves linearly with the cosmological time, $\phi = \phi_0 t$ with $x_0 = -\phi_0^2 = const$ or *ii.* for the kinetic function fulfilling $F^\infty - 2xF_x^\infty = \alpha_0$ [6], [8], [13], hence

$$F^\infty = \alpha_0 + \beta\sqrt{-x}, \tag{5}$$

where β is an arbitrary constant. It is associated with a divergent sound velocity and with the extended tachyon model considered in [6]. The function F^∞ generates a divergent sound speed model called atypical k-essence [8], [15].

i. In this case the barotropic index $\gamma_\phi = \gamma_0 = const$ and one finds the potential $V = V_0 a^{-3\gamma_0}$, after integrating the conservation equation (4b). For this potential the Eq. (4a) reads

$$\dot{y}^2 = 3\gamma_0^2\left[\rho_0 y + \frac{3\rho_0^2}{\lambda^2}\right], \tag{6}$$

with $y = a^{3\gamma_0}$ and $\rho_0 = V_0\alpha_0$. Its general solution is given by

$$a(\tau) = (3\rho_0)^{1/3\gamma_0}\left[\frac{\tau^2}{4} \pm \frac{\tau}{\lambda}\right]^{1/3\gamma_0}, \tag{7}$$

where $\tau = \gamma_0 t$. This solution has singularities at $\tau_s = 0$ and/or $\tau_s = \mp 4/\lambda$. For $\rho_0 > 0$, we have four expanding universes, two of them with $\gamma_0 > 0$ evolve from an initial singularity at $\tau_s = 0$ or $\tau_s = 4/\lambda$. The scale factor begins as $a \propto (\tau - \tau_s)^{1/3\gamma_0}$ in the high energy regime and ends as $a \propto \tau^{2/3\gamma_0}$ in the low energy regime. The remaining two universes with $\gamma_0 < 0$ end in a final big rip at $\tau_s = 0$ or $\tau_s = -4/\lambda$ having the final behavior $(\tau_s - \tau)^{1/3\gamma_0}$. For $\rho_0 < 0$, the solutions have an extremum at $\tau_e = 2/\lambda$ where $a_e = (-3\rho_0/\lambda^2)^{1/3\gamma_0}$. They represent two universes with a finite time span, one of them begins from a singularity, reaches a maximum value a_e, and ends in a big crunch at $\tau_s = 4/\lambda$. The other begins from a singularity at $\tau_s = 0$, reaches a minimum at a_e, where it bounces, and ends in a final big rip at $\tau_s = 4/\lambda$. In addition we have the time reversal of these solutions.

In turn, by combining $\phi(t)$ with $a(t)$ and $V = V_0 a^{-3\gamma_0}$, we obtain the following potential

$$V = \frac{V_0}{3\rho_0}\left[\frac{\gamma_0^2\phi^2}{4\phi_0^2} \pm \frac{\gamma_0\phi}{\lambda\phi_0}\right]^{-1}. \tag{8}$$

In the early universe quadratic contributions in ρ_ϕ become important and the potential approaches $V \propto \pm\lambda\phi^{-1}$ when $\lambda \to 0$. It seems to be the counterpart of the solution for quintessence cosmology driven by an inverse square potential [18], [19]. In the limit $\lambda \to \infty$, the potential $V \propto \phi^{-2}$ and we recover the power law solution [11], [12] [13].

The Eq. (8) shows that the brane correction shifts the inverse square potential to the inverse linear one at high energy. It has a minimum at $\phi_e = 2\phi_0/\lambda\gamma_0$ or a maximum at $\phi_e = -2\phi_0/\lambda\gamma_0$. Finally, by associating the brane equations to an effective fluid description, we find a linear barotropic equation of state in the low energy limit. However, in the high energy regime the k-essence source becomes a modified Chaplygin gas, $P_{ef} \approx (2\gamma_0 - 1)\rho_{ef} + \pm\gamma_0\lambda\rho_{ef}^{-1/2}/\sqrt{3}$.

ii. By combining the Eq. (5) and its associated barotropic index $\gamma_\phi = -\beta\sqrt{-x}/\alpha_0$ with the Eq. (4b) we find $H = \alpha_0 V'/(3\beta V)$. So by using $\rho_a = \alpha_0 V(\phi)$ into the Eq. (2) one gets the potential

$$V^\infty = \frac{1}{3}\left[\frac{\beta}{4\alpha_0}\phi^2 \pm \frac{\phi\beta}{\lambda}\right]^{-1}. \tag{9}$$

The potential on the brane (9) generalizes the inverse square one of the Friedmann cosmology shifting the divergence to $\phi_\infty = \pm 4\alpha_0/\lambda\beta$. So, for asymptotic power-law expansions the linear k-field model driven by a nearly inverse square potential and the atypical model are isomorphic.

2.1. Extended Tachyons in brane-worlds

Recently, it was proposed that the tachyon Lagrangian could be extended in such a way to allow the barotropic index any value [6] generating phantom and complementary tachyons in addition to the ordinary one [6], [20]. This extended tachyons generalize the Chaplygin gas introducing the phantom and complementary Chaplygin gases. The kinetic functions for these tachyons are

$$F_t = (1 - \dot\phi_t^{2r})^{1/2r}, \qquad 0 < \gamma_t = \dot\phi_t^{2r} < 1 \tag{10}$$

$$F_{ph} = (1 + \dot\phi_{ph}^{2r})^{1/2r}, \quad -\infty < \gamma_{ph} = -\dot\phi_{ph}^{2r} < 0 \tag{11}$$

$$F_c = -(\dot\phi_c^{2r} - 1)^{1/2r}, \quad 1 < \gamma_c = \dot\phi_c^{2r} < \infty \tag{12}$$

where r is a real parameter. The scale factor and potential for the three sets of extended tachyon fields are given by Eqs. (7) and (8) with $\gamma_0 = \phi_{0t}^{2r}$, $\gamma_0 = -\phi_{0ph}^{2r}$, $\gamma_0 = \phi_{0c}^{2r}$ and $\alpha_0 = (1 - \phi_{0t}^{2r})^{-1/2}$, $\alpha_0 = (1 + \phi_{0ph}^{2r})^{-1/2}$, $\alpha_0 = -(\phi_{0c}^{2r} - 1)^{-1/2}$ respectively.

The ordinary and complementary tachyons lead to expanding scenarios while the former has an accelerated expansion for $\phi_{0t}^{2r} < 2/3$. A universe dominated by a phantom tachyon ends in a big rip at $\tau_s = 0$ or $\tau_s = -4/\lambda$ where the scale factor blows up as $a \propto \tau^{-1/3\phi_{0ph}^{2r}}$ and the potential diverges as $V \propto \phi_{ph}^{-1}$. The ordinary and the complementary tachyons satisfy the weak energy condition $\rho \geq 0$ and $\rho + p \geq 0$, and the null energy condition $\rho + p \geq 0$ while the phantom tachyon violates both conditions.

As Eq. (2) is quadratic in the energy density, we introduce a tachyon with negative energy density by making $F \to -F$ in Eqs. (10)-(12) and keeping $V > 0$. The universe evolves between two singularities having a finite time span $t_s = 4/\lambda|\gamma_0|$. The scale factor has a maximum for ordinary and complementary tachyons, ending in a big crunch or has a minimum where the phantom tachyon bounces before it blows up in a big rip.

3. Summary

We have found solutions for 3-dimensional cosmological brane with k-essence and extended tachyons, whose metric corresponds to a spatially flat FRW spacetime, when the ratio ρ/V is constrained to be constant for linear k-field or for the atypical k-essence model. For $\rho_0 > 0$ we have obtained power law behavior at the initial and final stages. However, in the case $\rho_0 < 0$ the universe bounces, has a finite time span and ends in a big crunch ($\gamma_0 > 0$) or a big rip ($\gamma_0 < 0$). The quadratic brane correction has shifted the inverse square potential to the inverse linear one at high energy, so that the k-field is driven by an inverse quadratic polynomial potential. Finally, we have analyzed the extended tachyons with negative energy density and showed that the scale factor evolves between two singularities having a finite time span which depends on the brane tension and the barotropic index. For the ordinary and complementary tachyons the scale factor ends in a big crunch while for the phantom one it bounces and ends in a big rip.

Acknowledgments

LPC thanks the UBA for the partial support of this work under Project X044, and the CONICET under Project PIP 114-200801-00328. MGR is supported by the CONICET.

References

[1] Hořava P and Witten E 1996 *Nucl. Phys.* B **460** 506
Hořava P and Witten E 1996 *Nucl. Phys.* B **475** 94
Arkani-Hamed N, Dimopoulus S and Dvali G 1998 *Phys. Lett.* B **429** 263
Antoniadis I, Arkani-Hamed N, Dimopoulus S and Dvali G 1998 *Phys. Lett.* B **436** 257
Kokorelis C 2004 *Nucl. Phys.* B **677** 115
[2] Randall L and Sundrum R 1999 *Phys. Rev. Lett.* **83** 3370
Randall L and Sundrum R 1999 *Phys. Rev. Lett.* **83** 4690
[3] Maartens R 2004 *Living Rev. Rel.* **7** 7
[4] Dvali G, Gabadadze G and Porrati M 2000 *Phys. Lett.* B **485** 208
[5] Armendáriz-Picón C, Damour T and Mukhanov V 1999 *Phys. Lett.* B **458** 209
Riess A G, Garriga J and Mukhanov V F 1999 *Phys. Lett.* B **458** 219
[6] Chimento L P 2004 *Phys. Rev.* D **69** 123517
[7] Scherrer R J 2004 *Phys. Rev. Lett.* **93** 011301
[8] Chimento L P and Lazkoz R 2005 *Phys. Rev.* D **71** 023505
[9] Vikman A 2005 *Phys. Rev.* D **71** 023515
[10] Chimento L P and Lazkoz R 2006 *Phys. Lett.* B **639** 591-595
[11] Padmanabhan T 2002 *Phys. Rev.* D **66** 021301
[12] Feinstein A 2002 *Phys. Rev.* D **66** 063511
[13] Chimento L P and Feinstein A 2004 *Mod. Phys. Lett.* A **19** 761-768
[14] Aguirregabiria J M, Chimento L P and Lazkoz R 2004 *Phys. Rev.* D **70** 023509
[15] Malquarti M, Copeland E J and Liddle A R 2003 *Phys. Rev.* D **68** 023512
[16] Sami M 2003 *Mod. Phys. Lett.* A **18** 691
[17] Mukohyama S, Shiromizu T and Maeda K I 2000 *Phys. Rev.* D **62** 024028
Binétruy P, Deffayet C and Langlois D 2000 *Nucl. Phys.* B **565** 269
Maartens R 2000 *Phys. Rev.* D **62** 084023
[18] Mizuno S, Lee S -J and Copeland E J 2004 *Phys. Rev.* D **70** 043525
Maeda K I 2001 *Phys. Rev.* D **64** 123525
Huey G and Lidsey J E 2001 *Phys. Lett.* B **514** 217-225
[19] Chimento L P, Forte M, Kremer G M and Richarte M G 2009 *Phys. Rev.* D **79** 083527
[20] Chimento L P, Forte M, Kremer G M and Ribas M O *Preprint* arXiv:0809.1919

Journal of Physics: Conference Series **229** (2010) 012029 doi:10.1088/1742-6596/229/1/012029

On the local existence of maximal slicings in spherically symmetric spacetimes

Isabel Cordero-Carrión[1]*, José María Ibáñez[1] and Juan Antonio Morales-Lladosa[1]

[1] Departamento de Astronomía y Astrofísica, Universidad de Valencia, C/ Dr. Moliner 50, E-46100 Burjassot, Valencia, Spain

E-mail: isabel.cordero@uv.es, jose.m.ibanez@uv.es, antonio.morales@uv.es

Abstract. In this talk we show that any spherically symmetric spacetime admits locally a maximal spacelike slicing. The above condition is reduced to solve a decoupled system of first order quasi-linear partial differential equations. The solution may be accomplished analytical or numerically. We provide a general procedure to construct such maximal slicings.

1. Introduction

A *maximal* hypersurface is one that has vanishing mean extrinsic curvature, $K = 0$, K being the trace of the extrinsic curvature of the hypersurface. The name comes from the fact that the induced volume functional reaches a local maximum with respect the variations that keep fixed a given boundary. Maximal hypersurfaces were considered by Lichnerowicz [1] to solve Einstein's constraint equations, giving motivation for subsequent studies on the subject (see, for example [2, 3, 4, 5]). In fact, the existence of maximal hypersurfaces is extensively used in Mathematical Relativity. This property is a very simple geometric assumption to establish general results for broad classes of spacetimes, for instance, local or asymptotically stationary or conformally flat spacetimes.

We will use the term *maximal slicing* when referring to a (non intersecting) family of spacelike maximal hypersurfaces which locally foliates a certain domain of spacetime. This type of slicing has very nice properties as, for example: i) the well-known singularity avoidance capability [6], ii) it is well adapted to the propagation of gravitational waves [7, 8], and, iii) it gives the natural Newtonian analogous when, in addition, conformal flatness is imposed on each slice [9]. Maximal slicing condition has been recently used in the Fully Constrained Formulation of Einstein equations derived by the Meudon group [10, 11].

In spite of their extended use, the existence of maximal slicings in spherically symmetric spacetimes (SSSTs) has been only established for vacuum and for some particular energy contents (see [9, 12, 13, 14, 15, 16, 17]). There is, as far as we know, no theorem stating that always it is possible to build a maximal slicings in a SSST.

In this work, we aim to prove the local existence of maximal slicings in any SSST. We will follow a purely geometrical approach, independent of Einstein equations, according to [14], complementary to the standard time evolution strategy [9, 12, 13, 15, 16, 17].

Although our study is independent of the field equations, one by-product of our approach, which could be of interest in the field of Numerical Relativity, is that it provides a means to assess complex and sophisticated 3D numerical codes built to solve Einstein equations.

2. Local existence of maximal slicings

In this section we establish the following result:

Theorem. *Any spherically symmetric spacetime can be locally sliced by a family of maximal spacelike hypersurfaces.*

In order to prove the previous theorem, we derive a decoupled system of three first order partial differential equations that proves the local existence of a maximal slicings, and provides a general procedure allowing its construction.

Let us start with the canonical form of the metric of a SSST,

$$ds^2 = A\,dt^2 + 2C\,dt\,dr + B\,dr^2 + D\,d\Omega^2, \tag{1}$$

where $d\Omega^2 = d\theta^2 + \sin^2\theta d\varphi^2$ is the metric of the 2-sphere, A, B, C, D are smooth functions of t and r, and $AB - C^2 < 0$ to ensure the Lorentzian character of the metric. In addition we choose the signature $(-,+,+,+)$, and accordingly $D > 0$. Partial derivatives with respect to r will be denoted as $\dfrac{\partial f}{\partial r} = f'$, and with respect to t as $\dfrac{\partial f}{\partial t} = \dot{f}$. The spatial metric γ_{ij} induced on the hypersurfaces Σ_t, defined by t = constant, is $\gamma_{ij} = \mathrm{diag}(B, D, D\sin^2\theta)$, where $B > 0$ since we are considering spacelike hypersurfaces. Let n be the future pointing timelike unit normal to the hypersurfaces Σ_t,

$$n = \frac{1}{\alpha}\left(\frac{\partial}{\partial t} - \frac{C}{B}\frac{\partial}{\partial r}\right), \quad \alpha = \sqrt{\frac{C^2}{B} - A}. \tag{2}$$

The mean extrinsic curvature K of Σ_t is related with the expansion of n, $K = -\nabla \cdot n$, where ∇ is the covariant derivative with respect to the spacetime metric. In the given metric in Eq. (1), this relation is

$$K = \frac{1}{2\alpha B}\left(-\dot{B} - 2B\frac{\dot{D}}{D} + 2C' - C\frac{B'}{B} + 2C\frac{D'}{D}\right). \tag{3}$$

In the following, we assume that A, B, C and D are known functions. We look for a change of coordinates $\{\tilde{t} = \tilde{t}(t,r), \tilde{r} = \tilde{r}(t,r), \theta, \varphi\}$ such that the hypersurfaces $\tilde{t}$ = constant are maximal. We introduce two fields, X and Y, satisfying the commutation relation $[X, Y] = 0$. This condition assures the existence of two coordinate parameters, namely $\tilde{t} = \tilde{t}(t,r)$ and $\tilde{r} = \tilde{r}(t,r)$, such that

$$X = \frac{\partial}{\partial \tilde{t}}, \; Y = \frac{\partial}{\partial \tilde{r}}. \tag{4}$$

Then, we decompose these fields as $Y = \lambda\bar{Y}$, $X = a\bar{Y} + b\bar{Y}^{\perp}$, with $\bar{Y}^2 = 1$, $\bar{Y}\cdot\bar{Y}^{\perp} = 0$, $b \neq 0$, and $\lambda > 0$. The condition $\bar{Y}^2 = 1$ is equivalent to

$$\bar{Y} = f\frac{\partial}{\partial t} + P\frac{\partial}{\partial r}, \quad P = B^{-1}\left(-fC + \epsilon\sqrt{f^2 l^2 + B}\right), \tag{5}$$

being f an unknown function to be determined, $\epsilon = \pm 1$ and $l^2 = -AB + C^2 > 0$. Fixing the coefficient of $\dfrac{\partial}{\partial t}$ in the decomposition of $\bar{Y}^{\perp}$, $\bar{Y}\cdot\bar{Y}^{\perp} = 0$ leads to

$$\bar{Y}^{\perp} = \frac{\partial}{\partial t} + Q\frac{\partial}{\partial r}, \quad Q = B^{-1}\left(-C + \frac{\epsilon f l^2}{\sqrt{f^2 l^2 + B}}\right). \tag{6}$$

Consequently, the resulting fields are

$$X = (af+b)\frac{\partial}{\partial t} + (aP+bQ)\frac{\partial}{\partial r}, \quad Y = \lambda\left(f\frac{\partial}{\partial t} + P\frac{\partial}{\partial r}\right) = \lambda\left(f\alpha\, n + \frac{\epsilon}{B}\sqrt{f^2l^2+B}\,\frac{\partial}{\partial r}\right), \tag{7}$$

where we have taken into account Eqs. (2) and (5). The condition $[X,Y]=0$ is then equivalent to

$$\left[\frac{P}{bp}\right]' = -\left[\frac{f}{bp}\right]^{\cdot}, \quad -\left[\frac{aP+bQ}{bp\lambda}\right]' = \left[\frac{af+b}{bp\lambda}\right]^{\cdot}, \tag{8}$$

where $p = P - fQ = \dfrac{\epsilon}{\sqrt{f^2l^2+B}} \neq 0$.

Now, we denote with $\widetilde{K}$ the trace of the extrinsic curvature of the new hypersurfaces $\tilde{t}$ = constant. The condition $\widetilde{K}=0$ and the commutation relation provide 3 equations for 4 unknown functions, a, b, f, λ. Taking into account that D is a scalar under the above change of coordinates, we can add, without loss of generality, the following coordinate condition

$$\tilde{r}^2\, Y^2 = D, \tag{9}$$

saying that the metric on the hypersurfaces $\tilde{t}$ = constant is written in isotropic conformally flat form. From Eqs. (3) and (9), the condition $\widetilde{K}=0$ is equivalent to

$$2Y\,(X\cdot Y) - 3X\left(Y^2\right) + \left[\frac{4}{\tilde{r}} + \frac{Y\left(Y^2\right)}{Y^2}\right] X\cdot Y = 0. \tag{10}$$

From the decompositions (7), Eqs. (9) and (10) are expressed as $\lambda = \sqrt{D}/\tilde{r}$, that can be viewed as a definition of $\tilde{r}$ in terms of λ, and as

$$f\dot{a} + Pa' - a\left(\frac{f\dot{\lambda}+P\lambda'}{\lambda} - \frac{2}{\sqrt{D}}\right) = 3b\,\frac{\dot{\lambda}+Q\lambda'}{\lambda}. \tag{11}$$

After some algebraic calculations, the previous definition of λ and Eqs. (8) and (11) are equivalent to

$$a = b\sqrt{D}\left(\frac{\dot{\lambda}+Q\lambda'}{\lambda} - \frac{\dot{D}+QD'}{2D}\right), \tag{12}$$

$$f\frac{\dot{\lambda}}{\lambda} + P\frac{\lambda'}{\lambda} = f\frac{\dot{D}}{2D} + P\frac{D'}{2D} - \frac{1}{\sqrt{D}}, \tag{13}$$

$$f\frac{\dot{b}}{b} + P\frac{b'}{b} = P' - P\frac{p'}{p} + \dot{f} - f\frac{\dot{p}}{p}, \tag{14}$$

and

$$\frac{\dot{p}}{p} - \frac{\dot{D}}{D} - Q' + Q\left[\frac{p'}{p} - \frac{D'}{D}\right] = 0. \tag{15}$$

Notice that Eq. (15) involves only f when p and Q are written explicitly in terms of f.

First, Eq. (15) can be solved for f. Second, Eqs. (13) and (14) can be solved for λ and b. Finally, a can be obtained from Eq. (12). Assuming that A, B, C, D are continuously differentiable functions, the initial value problem with respect to this set of equations has always local (both in space and time) solution [18] (which is also continuously differentiable). Therefore, we have proved the announced theorem.

Notice that in order to solve this set of equations, it can be useful to distinguish two different cases, $f = 0$ and $f \neq 0$. In the case of $f = 0$, Eq. (15) is reduced to $K = 0$, and the rest of equations can be integrated easily. In the case of $f \neq 0$, it can be defined the variable $F = \dfrac{\epsilon f}{\sqrt{f^2 l^2 + B}} \Leftrightarrow f = \epsilon F \sqrt{\dfrac{B}{1 - l^2 F^2}}$, and Eqs. (13), (14) and (15) can be rewritten as a hyperbolic system of equations for F, λ and b.

3. Conclusions

Two basic results have been displayed: i) A theorem ensuring the existence of maximal slicings in any SSST. ii) A geometrical method to build up such slices by solving three decoupled first order quasi-linear partial differential equations (13), (14) and (15). The first result aims to fill a theoretical gap in the scientific literature. The second one tries to achieve an algorithmic procedure to obtain maximal slicings. An interesting by-product for Numerical Relativity of the approach presented in this paper has to do with the assessment of 3D codes written, as customary, in Cartesian coordinates. Let us consider two codes NC1 and NC2 such that only NC1 uses a gauge which is maximal. Hence, the evolution with code NC2 of any initial data admitting a spherically symmetric limit could be compared to the evolution produced by code NC1, by simply using our procedure to generate a SSST satisfying the maximal slicing condition.

Acknowledgments

I. C.-C. acknowledges support from the Spanish Ministerio de Educación y Ciencia (MEC) (AP2005-2857). This work has been also supported by the MEC Grant No. AYA2007-67626-C03-01, the MEC-FEDER project No. FIS2006-06062 and the MICINN-FEDER project No. FIS2009-07705.

References

[1] Lichnerowicz A 1944 *J. Math. Pures Appl.* **23** 37
[2] Choquet-Bruhat Y 1976 *Annali della Scuola Normale Superiore di Pisa Classe di Scienze* 4^e *srie*, tome 3, n3 361
[3] Cantor M, Fisher A, Marsden J, Murchadha N Ō and York J 1976 *Commun. Math. Phys.* **49** 187
[4] Marsden J E and Tipler F J 1980 *Phys. Rep.* **66** 109
[5] Bartnik R 1984 *Commun. Math. Phys.* **94** 155
[6] Smarr L and York J W 1978 *Phys. Rev.* D **17** 2529
[7] Smarr L and York J W 1978 *Phys. Rev.* D **17** 1945
[8] Shibata M and Nakamura T 1995 *Phys. Rev.* D **52** 5428
[9] Isenberg J A 2008 *Int. J. Mod. Phys. D* **17** 265
[10] Bonazzola S, Gourgoulhon E, Grandclément P and Novak J 2004 *Phys. Rev.* D **70** 104007
[11] Lin L-M and Novak J 2006 *Class. Quantum Grav.* **23** 4545
[12] Estabrook F, Wahlquist H, Christensen S, DeWitt B, Smarr L and Tsiang E 1973 *Phys. Rev.* D **7** 2814
[13] Beig R and Murchadha N Ó 1998 *Phys. Rev.* D **57** 4728
[14] Reinhart B L 1973 *J. Math. Phys.* **14** 719
[15] Malec E 1994 *Phys. Rev.* D **49** 6475
[16] Petrich L I, Shapiro S L and Teukolsky S A 1985 *Phys. Rev.* D **31** 2459
[17] Eardley D M and Smarr L 1979 *Phys. Rev.* D **19** 2239
[18] Zauderer E 1989 *Partial differential equations of applied mathematics*, Second edition (New York: John Wiley & Sons)

Spanish Relativity Meeting (ERE 2009) IOP Publishing
Journal of Physics: Conference Series **229** (2010) 012030 doi:10.1088/1742-6596/229/1/012030

Maximal analytic extensions of the Emparan-Reall black ring

P T Chruściel[1] **and J Cortier**[2⋆]

[1] LMPT, Fédération Denis Poisson, Tours, France
Mathematical Institute and Hertford College, Oxford, UK
[2] Institut de Mathématiques et de Modélisation de Montpellier, Université Montpellier 2, France

E-mail: [1] chrusciel@univ-tours.fr , [2] jcortier@math.univ-montp2.fr

Abstract. The Emparan-Reall black rings are a two-parameter family of black hole solutions of the vacuum Einstein's equation in dimension 4+1, the word "ring" refers to the event horizon, which has topology $\mathbb{S}^2 \times \mathbb{S}^1 \times \mathbb{R}$. We construct an analytic extension, whose global structure is similar to the Kruskal-Szekeres extension of the Schwarzschild space-time, and show that it is maximal, globally hyperbolic, and unique within a natural class of extensions.

1. Introduction

The Emparan-Reall [6] metrics form a remarkable class of vacuum black hole solutions of Einstein equations in dimension $4+1$. Some aspects of their global properties have been studied in [6], where it was shown that the solution contains a *Killing horizon* with $\mathbb{S}^2 \times \mathbb{S}^1 \times \mathbb{R}$ topology. In fact, the *event horizon* coincides with the Killing horizon, and therefore also has this topology (see [4]). We show how to construct an analytic extension with a bifurcate Killing horizon; and we establish some global properties of the extended space-time. The extension resembles closely the Kruskal-Szekeres extension of the Schwarzschild space-time. We show the maximality of this extension, as well as global hyperbolicity, establish uniqueness of our extension within a natural class, and verify existence of a conformal completion at null infinity.

2. The Emparan-Reall space-time

In local coordinates (t, ψ, z, x, φ), the Emparan-Reall metric can be written in the form

$$\begin{aligned} g \;=\; & -\frac{F(x)}{F(z)}\left(dt+\sqrt{\frac{\nu}{\xi_F}}\frac{\xi_1-z}{A}d\psi\right)^2+\frac{F(z)}{A^2(x-z)^2}\times \qquad (1)\\ & \left[-F(x)\left(\frac{dz^2}{G(z)}+\frac{G(z)}{F(z)}d\psi^2\right)+F(z)\left(\frac{dx^2}{G(x)}+\frac{G(x)}{F(x)}d\varphi^2\right)\right], \end{aligned}$$

where $A>0$, ν et ξ_F are constants, and

$$F(\xi)=1-\frac{\xi}{\xi_F}\,, \qquad (2)$$

$$G(\xi)=\nu\xi^3-\xi^2+1=\nu(\xi-\xi_1)(\xi-\xi_2)(\xi-\xi_3)\,, \qquad (3)$$

are polynomials, with ν chosen so that $\xi_1 < 0 < \xi_2 < \xi_3$). The study of the coordinate singularities at $x = \xi_1$ and $x = \xi_2$ leads to the determination of ξ_F as:

$$\xi_F = \frac{\xi_1\xi_2 - \xi_3^2}{\xi_1 + \xi_2 - 2\xi_3} \in (\xi_2, \xi_3) . \tag{4}$$

Emparan and Reall have established the asymptotically flat character of (1), as well as existence of an analytic extension across an analytic Killing horizon at $z = \xi_3$. The extension given in [6] is somewhat similar of the extension of the Schwarzschild metric that one obtains by going to Eddington-Finkelstein coordinates, and is *not* maximal. Now, existence of an analytic extension with a bifurcate horizon, *à la* Kruskal-Szekeres, is guaranteed from this by an analytic version of the analysis of Rácz and Wald in [9]. But the global properties of an extension so constructed are not clear. It is therefore of interest to present an explicit extension with good properties. This extension is outlined in Section 3, and its global properties are studied in Section 4.

The coordinate z can be seen as a radial coordinate, and runs in $]\xi_3, +\infty[\cup]-\infty, \xi_1[$, whereas ψ and φ are angular coordinates, and t is a time coordinate define in $\mathbb{R}$. As in [6] we assume throughout that

$$\xi_1 \le x \le \xi_2 . \tag{5}$$

As discussed in [6], the extremities correspond to a north and south pole of S^2, with a function θ defined by $d\theta = dx/\sqrt{G(x)}$ providing a latitude on S^2, except for the limit $x - z \to 0$, $x \to \xi_1$, which corresponds to an asymptotically flat region, see [6]; a detailed proof of asymptotic flatness can be found in [3]. The coordinate φ provides the corresponding azimuth on S^2. The surface "$\{z = \infty\}$" can be identified with "$\{z = -\infty\}$" by introducing a coordinate $Y = -1/z$, with the metric extending analytically across $\{Y = 0\}$, see [6] for details. The symmetries are given by the Killing vector fields ∂_t, ∂_ψ and ∂_φ, and one can check that ∂_t is timelike in the asymptotic flat region, and becomes null at $\{Y = 0\}$. This gives the location of the *ergosurface*, and the space-time is *stationary*.

3. The extension

We start by working in the range $z \in (\xi_3, \infty)$; there we define new coordinates w, v by the formulae

$$dv = dt + \frac{bdz}{(z - \xi_3)(z - \xi_2)} , \tag{6}$$

$$dw = dt - \frac{bdz}{(z - \xi_3)(z - \xi_2)} , \tag{7}$$

where b is a constant. (Our coordinates v and w are closely related to, but not identical, to the coordinates v and w used in [6] when extending the metric through the Killing horizon $z = \xi_3$). Similarly to the construction of the extension of the Kerr metric in [1,2], we define a new angular coordinate $\hat{\psi}$ by:

$$d\hat{\psi} = d\psi - adt , \tag{8}$$

where a is a constant. Then, similarly to what is done in [1, 2], and to remove a signature degeneracy at $z = \xi_3$, one introduces

$$\hat{v} = \exp(cv) , \quad \hat{w} = -\exp(-cw) , \tag{9}$$

where c is some constant to be chosen. With a suitable choice of the constants a, b and c (see [4] for details), one may finally construct an analytic extension $(\widehat{\mathcal{M}}, \hat{g})$ of the original Emparan-Reall spacetime $(\mathcal{M}_I, g)$ (see Figure 1), with a bifurcate Killing horizon $\{\hat{w}\hat{v} = 0\}$ extended from the Killing horizon $\{z = \xi_3\}$.

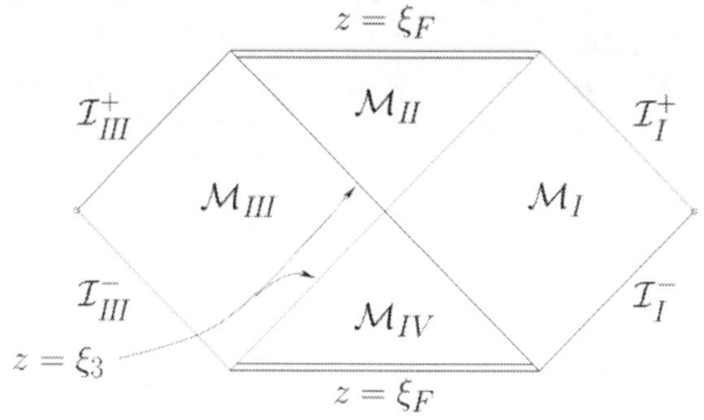

Figure 1. $\widehat{\mathcal{M}}$ with its various subsets. For example, $\mathcal{M}_{I\cup II}$ is the union of $\mathcal{M}_I$ and of $\mathcal{M}_{II}$ and of that part of $\{z=\xi_3\}$ which lies in the intersection of their closures; this is the manifold constructed in [6]. Very roughly speaking, the various $\mathcal{I}$'s correspond to $x=z=\xi_1$. It should be stressed that this is neither a *conformal* diagram, nor is the space-time a product of the figure times $\mathbb{S}^2\times\mathbb{S}^1$. But the diagram represents accurately the causal relations between the various $\mathcal{M}_N$'s, as well as the geometry near the bifurcate horizon $z=\xi_3$, as the manifold *does have* a product structure there.

4. Global properties

4.1. Maximality

The space-time $(\widehat{\mathcal{M}},\hat{g})$ just constructed is a maximal analytic extension of the original Emparan-Reall space-time $(\mathcal{M}_I,g)$. One way to show this is to prove that $(\widehat{\mathcal{M}},\hat{g})$ is *timelike s-complete*: by this we mean that any timelike, maximally extended geodesic of the extended space-time is either complete, or reaches a curvature singularity in finite affine time. In fact, the metric is singular on the set $\{z=\xi_F\}$, which follows from the asymptotic expansion, derived by Alfonso Garcia-Parrado and José María Martín García using the symbolic algebra package xAct [8], for the Kretschmann scalar:

$$R_{\alpha\beta\gamma\delta}R^{\alpha\beta\gamma\delta}=\frac{12A^4\xi_F^4G(\xi_F)^2(x-z)^4\left(1+O(z-\xi_F)\right)}{(\xi_F-x)^2(z-\xi_F)^6}\,. \tag{10}$$

This shows that the Emparan-Reall metric is C^2–inextendible across $\{z=\xi_F\}$. Then, to prove the *s-timelike completeness* of $(\widehat{\mathcal{M}},\hat{g})$, we study first the timelike geodesic segments of finite affine length which lie in the domain of outer communication $\langle\langle\mathcal{M}_{\rm ext}\rangle\rangle$. To do so, we use various constants of motion, such as $g(\dot\gamma,\dot\gamma)$ and $g(\dot\gamma,X)$ where X is a Killing vector field, to study the evolution equations for x and z obtained from the Lagrangian $L=g(\dot\gamma,\dot\gamma)$. One finds that the geodesics which are maximally extended within $\langle\langle\mathcal{M}_{\rm ext}\rangle\rangle$ are either complete, or can be smoothly extended through the bifurcate Killing horizon $\{\hat w\hat v=0\}$. Then, one shows that causal geodesics in the region $\{\xi_F<z<\xi_3\}$ reach the horizon $\{\hat w\hat v=0\}$ and are smoothly extendible there in one time direction, and reach the singular set $\{z=\xi_F\}$ in finite affine time in the other direction. A detailed proof of these facts is provided in [4]. Note that restricted classes of geodesics in the Emparan-Reall metrics have been analysed by different methods in Durkee, using constants of motion as well as the separability of the Hamilton-Jacobi equation for these geodesics (see [5]).

4.2. Global hyperbolicity

The maximal extension $(\widehat{\mathcal{M}},\hat{g})$ is globally hyperbolic. Indeed, the surface $\mathcal{S}:=\{\hat v+\hat w=0\}$ is a *Cauchy surface* of $\widehat{\mathcal{M}}$; note that $\mathcal{S}$ is a smooth extension of the hypersurface $\{t=0\}$ $(=\mathcal{S}\cap\{\hat v>0\})$ of $\mathcal{M}_I$ across the bifurcation surface $\{\hat w=\hat v=0\}$, to its image in $\mathcal{M}_{III}$

under the map $(\hat{w}, \hat{v}, \hat{\psi}, x, \varphi) \mapsto (-\hat{w}, -\hat{v}, -\hat{\psi}, x, -\varphi)$. The proof of this uses the s-geodesic completeness of the metric.

4.3. A uniqueness theorem
We have the following uniqueness result:

THEOREM 4.1 *The space-time* $(\widehat{\mathcal{M}}, \hat{g})$ *is unique within the class of simply connected analytic extensions of the Emparan-Reall space-time* $(\mathcal{M}_I, g)$ *which have the property that all maximally extended causal geodesics on which* $R_{\alpha\beta\gamma\delta}R^{\alpha\beta\gamma\delta}$ *is bounded are s-complete.*

Uniqueness is understood up to isometry. Theorem 4.1 follows from the analysis of geodesics just described and from Theorem 4.2, which is a Lorentzian version of [7, Theorem 6.3, p. 255] where geodesic completeness is weakened to *timelike s-completeness* (see [4] for details).

THEOREM 4.2 *Let* $(\mathcal{M}, g)$, $(\mathcal{M}', g')$ *be analytic Lorentzian manifolds of dimension* $n+1$, $n \geq 1$, *with* $\mathcal{M}$ *connected and simply connected, and* $\mathcal{M}'$ *timelike s-complete. Then every isometric immersion* $f_U : U \subset \mathcal{M} \hookrightarrow \mathcal{M}'$, *where* U *is an open subset of* $\mathcal{M}$, *extends uniquely to an isometric immersion* $f : \mathcal{M} \hookrightarrow \mathcal{M}'$.

Acknowledgments

The authors are grateful to the Mittag-Leffler Institute, Djursholm, Sweden, for hospitality and financial support during a significant part of work on this subject.

References

[1] Boyer R H and Lindquist R W 1967 Maximal analytic extension of the Kerr metric *Jour. Math. Phys.* **8** 265-81
[2] Carter B 1968 Global structure of the Kerr family of gravitational fields *Phys. Rev.* **174** 1559-71
[3] Chruściel P T 2006 Beijing lecture notes on mathematical relativity `www.phys.univ-tours.fr/~piotr/papers/BeijingAll.pdf`
[4] Chruściel P T and Cortier J 2008 Maximal analytic extensions of the Emparan-Reall black ring *Preprint* gr-qc/0807.2309v2
[5] Durkee M 2009 Geodesics and symmetries of the doubly-spinning black rings *Class. Quantum Grav.* **26** 085016 (*Preprint* gr-qc/0812.0235v2)
[6] Emparan R and Reall H S 2002 A rotating black ring solution in five dimensions *Phys. Rev. Lett.* **88** 101101 (*Preprint* hep-th/0110260)
[7] Kobayashi S and Nomizu K 1963 *Foundations of differential geometry* (New York: Interscience Publishers)
[8] Martín-García J M xAct: Efficient Tensor Computer Algebra `http://metric.iem.csic.es/Martin-Garcia/xAct`.
[9] Rácz I and Wald R M 1996 Global extensions of space-times describing asymptotic final states of black holes *Class. Quantum Grav.* **13** 539-52 (*Preprint* gr-qc/9507055)

Spanish Relativity Meeting (ERE 2009) IOP Publishing
Journal of Physics: Conference Series **229** (2010) 012031 doi:10.1088/1742-6596/229/1/012031

The gravitational force on a gyroscope and the electromagnetic force on a magnetic dipole as analogous tidal effects

L Filipe O Costa* and Carlos A R Herdeiro

Centro de Física do Porto, Faculdade de Ciências Universidade do Porto, Rua do Campo Alegre 687, 4169-007 Porto, Portugal

E-mail: filipezola@fc.up.pt, crherdei@fc.up.pt

Abstract. We compare the covariant expression of the electromagnetic force exerted on a magnetic dipole with Papapetrou's equation for the gravitational force exerted on a spinning test particle. We show that if Pirani's supplementary spin condition holds, there is an exact, covariant, and fully general analogy relating these two forces: both are determined by a contraction of the spin 4-vector with a magnetic-type tidal tensor. Moreover, these tidal tensors obey strikingly analogous equations which are covariant forms for (some of) Maxwell's and Einstein's field equations. These equations allow for an insightful comparison between the two interactions. It is shown that, in the special case that the gyroscope/dipole are "at rest" and far away from a stationary source, the two forces are similar (in accordance with the results known from linearized theory); but that for generic dynamics key differences arise. In particular we show that the time projection of the force on a dipole is the power transferred to it by Faraday's induction, whereas the fact that the force on a gyroscope is spatial signals the absence of an analogous gravitational effect; that whereas the total work done on a magnetic dipole by a stationary magnetic field is zero, a stationary gravitomagnetic field, by contrast, does work on mass currents, which quantitatively explains the Hawking-Wald spin interaction energy.

1. Introduction

It is known, since the works of Mathisson and Papapetrou that spinning particles follow worldlines which are not geodesics. Wald [1] later showed that, in linearized theory, the gravitational force exerted on a spinning pole-dipole test particle (hereafter a gyroscope) *whose center of mass is at rest in a stationary field*, takes a form $\vec{F}_G = -\nabla(\vec{B}_G.\vec{S})$ similar to the electromagnetic force on a magnetic dipole. The analogy was cast in an exact form by Natário [2], who, in the framework of the "quasi-Maxwell" formalism, showed that in *stationary* spacetimes the force exerted on a gyroscope *whose worldline is tangent to a time-like Killing vector* (i.e., its center of mass is at rest relative to some stationary observer) consists of an electromagnetic-like term plus a term interpreted therein as being "the weight of the energy of the gravitomagnetic dipole". Recently it was shown that there is actually an exact, covariant and fully general analogy relating the two forces, which is made explicit in the tidal tensor formalism introduced in [3]. Herein we will exemplify how this analogy provides new intuition for the understanding of spin curvature coupling.

2. Force on a magnetic dipole vs. Force on a gyroscope

The force exerted on a magnetic dipole under the action of an electromagnetic field is given (see e.g. [4]) by Eq. (1a), where $S^{\mu\nu}$ is the spin tensor and σ the gyromagnetic ratio:

$$F^{\alpha}_{EM} \equiv \frac{DP^{\alpha}}{d\tau} = \frac{\sigma}{2}F^{\mu\nu;\alpha}S_{\mu\nu} \quad \text{(a)}; \qquad F^{\alpha}_{G} \equiv \frac{DP^{\alpha}}{d\tau} = -\frac{1}{2}R^{\alpha}{}_{\beta\mu\nu}U^{\beta}S^{\mu\nu} \quad \text{(b)}. \tag{1}$$

This equation needs to be supplemented by a spin supplementary condition; two covariant conditions are found in literature: the Moller-Tulczyjew condition $S^{\alpha\beta}P_{\beta} = 0$, and the Pirani condition $S^{\alpha\beta}U_{\beta} = 0$. We choose the latter, which means that $S^{\mu\nu} = \epsilon^{\mu\nu\tau\lambda}S_{\tau}U_{\lambda}$, where S^{α} is the spin 4-vector, defined as being the 4-vector with components $(0, \vec{S})$ in the dipole's rest frame. In this case (1a) becomes Eq. (2a) of Table 1. F^{α}_{EM} is the covariant form of the familiar textbook

Table 1. Analogy between the EM force on a magnetic dipole and the GR force on a gyroscope

Electromagnetic Force on a Magnetic Dipole		Gravitational Force on a Spinning Particle	
$F^{\beta}_{EM} = \sigma B_{\alpha}{}^{\beta}S^{\alpha}; \quad B^{\alpha}{}_{\beta} \equiv \star F^{\alpha}{}_{\mu;\beta}U^{\mu}$	(2a)	$F^{\beta}_{G} = -\mathbb{H}_{\alpha}{}^{\beta}S^{\alpha}; \quad \mathbb{H}^{\alpha}{}_{\beta} \equiv \star R^{\alpha}{}_{\mu\beta\nu}U^{\mu}U^{\nu}$	(2b)
Eqs. Magnetic Tidal Tensor		Eqs. Gravito-magnetic Tidal Tensor	
$B^{\alpha}{}_{\alpha} = 0$	(3a)	$\mathbb{H}^{\alpha}{}_{\alpha} = 0$	(3b)
$B_{[\alpha\beta]} = \frac{1}{2}\star F_{\alpha\beta;\gamma}U^{\gamma} - 2\pi\epsilon_{\alpha\beta\sigma\gamma}j^{\sigma}U^{\gamma}$	(4a)	$\mathbb{H}_{[\alpha\beta]} = -4\pi\epsilon_{\alpha\beta\sigma\gamma}J^{\sigma}U^{\gamma}$	(4b)

3-D expression $\vec{F}_{EM} = d\vec{P}/dt = \nabla(\vec{S}.\vec{B})q/2m$ (which is valid only in the dipole's frame).

The equation of motion for the center of mass of a spinning test particle in a gravitational field is given (e.g. [5]) by the Papapetrou equation (1b). Again, if Pirani's supplementary condition $S^{\alpha\beta}U_{\beta} = 0$ holds, then Eq. (1b) becomes Eq. (2b) of Table 1, revealing the physical analogy $B_{\alpha\beta} \leftrightarrow \mathbb{H}_{\alpha\beta}$. These tidal tensors are dubbed *magnetic* because $B_{\alpha\beta}$ is the covariant derivative of the magnetic field $B^{\alpha} = \star F^{\alpha}{}_{\beta}U^{\beta}$ measured by the observer of (fixed) 4-velocity U^{β}; and it turns out that they obey the formally similar equations (3) and (4), which in one case are Maxwell equations, and in the other are *exact* Einstein equations. That is: Eqs. (4a) and (3a) are, respectively, the space projection of Maxwell equations $F^{\alpha\beta}{}_{;\beta} = 4\pi j^{\alpha}$ and the time projection of Bianchi identity $\star F^{\alpha\beta}{}_{;\beta} = 0$; Eqs. (4b) and (3b) are, respectively the time-space projection of Einstein equations $R_{\mu\nu} = 8\pi(T_{\mu\nu} - \frac{1}{2}g_{\mu\nu}T^{\alpha}{}_{\alpha})$ and the time-time projection of the algebraic Bianchi identities $\star R^{\gamma\alpha}{}_{\gamma\beta} = 0$. There is an electric counterpart to this analogy which is manifest in the wordline deviations of both theories, and together they form the gravito-electromagnetic analogy based on tidal tensors [3].

Note that P^{α} in Eqs. (1) is *not* parallel to U^{α} (i.e., $P^{\alpha} \neq mU^{\alpha}$); the momentum of the spinning particle will be discussed in detail in [7]; for now we refer to definitions given in [4, 5].

Comparing (1a) to (1b) we see that in gravity the gyromagnetic ratio is 1 (i.e., the gravitational analogue of the magnetic dipole moment is the angular momentum); the relative minus sign manifests the fact that parallel charge (mass) currents attract (repel) one another, which that can be traced back to the fact that masses (charges of the same sign) attract (repel).

Apart from that, Eqs. (1) are formally similar; hence the analogy is ideally suited to compare the two interactions, because it amounts to compare $B_{\alpha\beta}$ and $\mathbb{H}_{\alpha\beta}$, which is crystal clear from equations (3) and (4). The most important differences between them are: i) $B_{\alpha\beta}$ is linear in the fields whereas $\mathbb{H}_{\alpha\beta}$ is not; ii) *in vacuum* $\mathbb{H}_{[\alpha\beta]} = 0$ (symmetric tensor), whereas $B_{[\alpha\beta]} = \frac{1}{2}\star F_{\alpha\beta;\gamma}U^{\gamma} \neq 0$ (generically non symmetric, even in vacuum); iii) time components: $\mathbb{H}_{\alpha\beta}$ is a spatial tensor, whereas $B_{\alpha\beta}$ is not. In what follows we explore some physical consequences.

2.1. Symmetries of tidal tensors

According to Eqs. (2), it is the magnetic tidal tensor, *as seen by the test particle* of 4-velocity U^α, that determines the force exerted upon it. Thus, if the fields do not vary along the test particle's worldline (so that $\star F_{\alpha\beta;\gamma}U^\gamma = 0$), Eqs. (4) allow for a similarity between F^α_{EM} and F^α_G. Consider the simple example of analogous physical apparatus: a dipole in the electromagnetic field of a spinning charge (charge Q, magnetic moment μ), and a gyroscope in the gravitational field of a spinning mass (mass M, angular momentum J), asymptotically described by the Kerr solution. For large r, the force exerted on a gyroscope *whose center of mass is at rest* relative to the central mass is similar (identifying μ with J) to its electromagnetic counterpart:

$$F^i_G \simeq -\frac{3}{c}\left[\frac{(\vec{r}\cdot\vec{J})}{r^5}\delta^{ij} + 2\frac{r^{(i}J^{j)}}{r^5} - 5\frac{(\vec{r}\cdot\vec{J})r^i r^j}{r^7}\right]S_j \overset{J\leftrightarrow\mu}{=} -F^i_{EM} \tag{5}$$

Thus we have a gravitational spin-spin force in analogy with the situation in electromagnetism [1]. However, in the general case that the dipole/gyroscope are allowed to move, then Eqs. (4) make clear that the two forces differ significantly (even in the weak field and slow motion approximation; see [6]), due to the different symmetries of the tidal tensors. Its implications in the dynamics will be discussed in detail in [7].

2.2. Time Components

Time components in test particle's frame — The electromagnetic force exerted on a magnetic dipole has a non-vanishing time projection $F^\alpha_{EM}U_\alpha$ in the dipole's proper frame; it is physically interpreted as *minus* the rate of work done by the induced electric field $\vec{E}$ due to the time varying magnetic field $\vec{B}$ (as measured by the inertial observer momentarily comoving with the dipole). This is more easily seen if we think of the magnetic dipole as a small current loop:

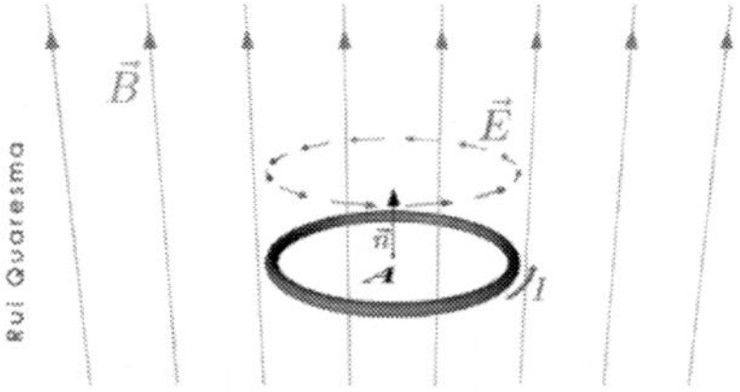

$$F^\alpha_{EM}U_\alpha = \frac{\partial\vec{B}}{\partial t}.\vec{\mu} = \frac{\partial\Phi}{\partial t}I = -I\oint_{loop}\vec{E}.d\vec{s}$$

$\vec{\mu} = IA\vec{n}$ (magnetic moment), $\Phi \equiv \vec{B}.\vec{n}A$ (magnetic flux),
$\vec{E} \equiv$ induced electric field, $A \equiv$ area of the loop.

This effect has no counterpart in gravity. Since $\mathbb{H}_{\alpha\beta}$ is a spatial tensor, we *always have* $F^\alpha_G U_\alpha = 0$ which means that the energy of the gyroscope, *as measured in its proper frame*, is constant. Hence, the spatial character of gravitational tidal tensors *precludes* induction effects analogous to the electromagnetic ones.

Time components as measured by *static observers* — in a given frame, the time component $P_0 = -E$ yields minus the energy of the test particle as measured in that frame; thus its variation $DP_0/d\tau$ is minus the rate of work done on it. For an arbitrary frame, in which the dipole has 4-velocity $U^\beta = \gamma(1,\vec{v})$, the time component of the force exerted on a magnetic dipole is:

$$(F_{EM})_0 = \frac{DP_0}{d\tau} = -\frac{DE}{d\tau} = \frac{F^\beta_{EM}U_\beta}{\gamma} - F^i_{EM}v_i = -(\mathcal{P}_{mech} + \mathcal{P}_{ind}) \tag{6}$$

where, in accordance with discussion above, we identify $\mathcal{P}_{ind} = -F^\beta_{EM}U_\beta/\gamma$ as the power transferred to the dipole by Faraday's induction, which is reflected in a variation of its internal energy, and $\mathcal{P}_{mech} \equiv F^i_{EM}v_i$ as the familiar "mechanical" power transferred to the dipole by the 3-force F^i_{EM} exerted on it, which is reflected in the kinetic energy of translation of the center of mass. Consider now the situation in Fig. 1a): a magnetic dipole moving in the field generated by a strong magnet. An observer at rest relative to the magnet sees a time-constant field, thus, for this observer, the time component of the force on the dipole vanishes:

$$(F_{EM})_0 \equiv \frac{DP_0}{d\tau} = \sigma\star F_{\alpha\beta;0}U^\beta S^\alpha = 0.$$

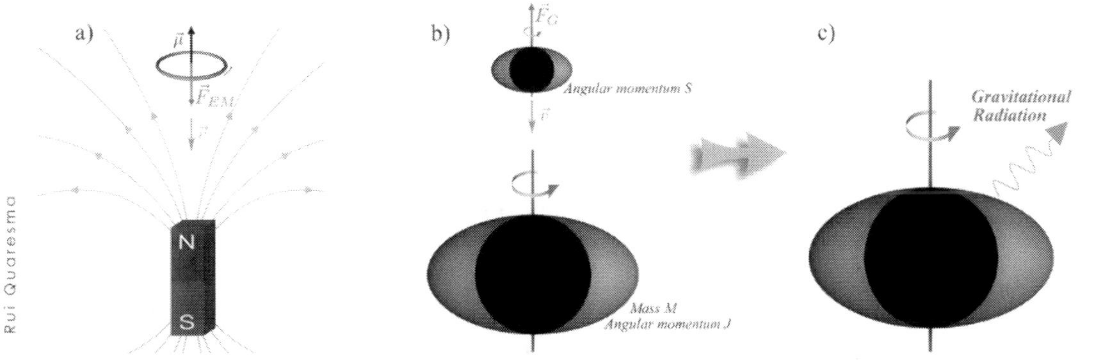

Figure 1. a) Magnetic dipole in the field of a strong magnet; b) gyroscope (small Kerr black hole) in the field of a large Kerr black hole; c) black hole merging.

That tells us that for this observer the total work done on the dipole is zero; which is related to a basic principle from electromagnetism: that a (stationary) magnetic field does no work. From the viewpoint of Eq. (6) this is due to an exact cancellation between $\mathcal{P}_{mech}$ and $\mathcal{P}_{ind}$: on the one hand the dipole gains translational kinetic energy; on the other hand there is a variation is of its internal energy by induction, which allows for the total work to vanish (this is in agreement with discussion in [8]).

In gravity, since those induction effects are absent, such cancellation does not occur: $(F_G)_0 = -DE/d\tau = -F_G^i v_i \neq 0$; in other words, a stationary "gravitomagnetic field", *unlike its electromagnetic counterpart, does work* on mass currents. And there is a known consequence of this fact: the spin dependent upper bound for the energy released by gravitational radiation (Fig. 1c) when two black holes collide, obtained by Hawking from the area law. For the case with spins aligned, from Hawking's expression one can infer a gravitational spin-spin interaction energy [1], which indeed accounts for the rate of work $(F_G)_0$. In order to see that, consider the apparatus in Fig. 1b): two Kerr black holes with spins aligned, a large one (mass m, spin $J = am$) which is our source, and small one (4-velocity U^α, spin $S \equiv \sqrt{S^\alpha S_\alpha}$) which we take to be the gyroscope, falling into the former along the symmetry axis (that ensures that F_G^α arises from the spin-spin interaction only). The observer is *at rest relative to the source.* The time component of the force acting on the small black hole is given by expression (7a):

$$(F_G)_0 = -F_G^i v_i = \frac{2ma\left(a^2 - 3r^2\right)U^r S}{\left(r^2 + a^2\right)^3} \quad \text{(a)}; \quad E_s = \frac{aS}{2m\left[m + \sqrt{m^2 - a^2}\right]} = \int_\infty^{r_+} (F_G)_0 \, d\tau \quad \text{(b)}. \quad (7)$$

The Hawking-Wald spin-spin interaction energy for this particular setup [1] is (7b) which is precisely what one obtains integrating (7a) from infinity to the horizon.

References

[1] Wald R 1976 *Phys. Rev.* D **6** 406
[2] Natário J 2007 *Gen. Rel. Grav.* **39** 1477 (*Preprint* gr-qc/0701067)
[3] Costa L F O and Herdeiro C A R 2008 *Phys. Rev.* D **78** 024021
[4] Corben H C 1968 *Classical and Quantum Theories of Spinning Particles* (San Francisco: Holden Day)
[5] Semerák O 1999 *Mont. Not. R. Astron. Soc.* **308** 863
[6] Costa L F O and Herdeiro C A R 2009 *Proc. IAU* vol 5 (Cambridge University Press) p 31 (*Preprint* arXiv:0912.2146)
[7] Costa L F O, Herdeiro C A R, Natário J and Zilhão M, in preparation.
[8] Gralla S, Hartle A I and Wald R M 2009 *Phys. Rev.* D **80** 024031
[9] Hawking S W 1971 *Phys Rev. Lett.* **26**

Journal of Physics: Conference Series **229** (2010) 012032 doi:10.1088/1742-6596/229/1/012032

Getting Wahlquist's metric from a perturbation theory solution

J E Cuchí[1]*, J Martín[1], A Molina[2] and E Ruiz[1]

[1]Departamento de Física Fundamental. Universidad de Salamanca, 37008 Salamanca
[2]Departament de Física Fonamental. Universitat de Barcelona, Diagonal 647, 08028 Barcelona

E-mail: jecuchi@usal.es, chmm@usal.es, alfred.molina@ub.edu, eruiz@usal.es

Abstract. Perturbation theory is a very convenient and powerful tool to study solutions of Einstein's equations, and its use is extensive nowadays. Nevertheless from the formal point of view, the fact of working with an infinite dimensional manifold results in that it can not be guaranteed in general that a particular perturbation solution corresponds to the Taylor expansion of a supposed exact solution with the desired properties. Here we will present a result obtained from a solution of the perturbed Einstein's system following the CMMR scheme for a perfect fluid region with equation of state $\mu + (1-n)p = \mu_0$, being μ density and p pressure. Then we will show how it can be reached from a Post-Minkowskian and slow-rotation approximation of the Wahlquist metric. Additionally we show how conditions we got for it to be Petrov type D are automatically satisfied when we try to mimic Wahlquist's solution, and also identify the Wahlquist parameter r_0 in terms of physical quantities.

1. Previous work

We work within the analytical approximation scheme developed in [1]. Let $(\mathcal{M}, g)$ be a stationary and axisymmetric spacetime, corresponding to a compact source interior. Let also $\boldsymbol{\xi}$ be the time-like Killing and $\boldsymbol{\zeta}$ the closed-orbits space-like one.

Adapting t and φ to the Killings and r and θ to the two-dimensional surfaces orthogonal to the Killing ones, [2] we can write the inner metric with the structure

$$\begin{aligned}\mathbf{g} = {} & \gamma_{tt}\, \boldsymbol{\omega}^t \otimes \boldsymbol{\omega}^t + \gamma_{t\varphi}(\boldsymbol{\omega}^t \otimes \boldsymbol{\omega}^\varphi + \boldsymbol{\omega}^\varphi \otimes \boldsymbol{\omega}^t) + \gamma_{\varphi\varphi}\, \boldsymbol{\omega}^\varphi \otimes \boldsymbol{\omega}^\varphi \\ & + \gamma_{rr}\, \boldsymbol{\omega}^r \otimes \boldsymbol{\omega}^r + \gamma_{r\theta}(\boldsymbol{\omega}^r \otimes \boldsymbol{\omega}^\theta + \boldsymbol{\omega}^\theta \otimes \boldsymbol{\omega}^r) + \gamma_{\theta\theta}\, \boldsymbol{\omega}^\theta \otimes \boldsymbol{\omega}^\theta \end{aligned} \quad (1)$$

in the cobasis $\boldsymbol{\omega}^t = dt$, $\boldsymbol{\omega}^r = dr$, $\boldsymbol{\omega}^\theta = r\, d\theta$, $\boldsymbol{\omega}^\varphi = r \sin\theta\, d\varphi$. We require these coordinates to be spherical-like associated to harmonic Cartesian-like ones (in the sense that they are Cartesian in the Minkowskian limit).

$(\mathcal{M}, g)$ is filled with a compact perfect fluid without convective motions, so that its 4-velocity can be written as $\mathbf{u} = \psi(\boldsymbol{\xi} + \omega\boldsymbol{\zeta})$ with ψ a normalization factor and ω the module of the rotation speed of the fluid. Its equation of state (EOS) is $\mu + (1-n)p = \mu_0$, where μ is energy density, p pressure, n a non zero real and μ_0 a positive constant. Expressions for μ and p in the source can be found integrating Euler's equation $dp = (\mu + p) d \ln \psi$ and obtaining ψ from the normalization condition $u^a u_a = -1$. We expand the $p = 0$ surface embedding in Legendre polynomials.

As already mentioned we follow an analytical two-parameter perturbation method. Thus we have a two-parameter family of spacetimes $(\mathcal{M}_{\lambda,\Omega}, g_{\lambda,\Omega})$. The post-Minkowskian parameter is chosen to be $\lambda \propto \mu_0$ and the slow rotation one as $\Omega^2 = r_s^3\omega^2/m$, with r_s the radius of the sphere with density μ_0 and mass m. Having said that, we can proceed to solve iteratively in λ the system of Einstein's equations plus harmonic condition

$$\begin{cases} \triangle h^{(n)}_{\alpha\beta} = -16\pi(T_{\alpha\beta} - \tfrac{1}{2}g_{\alpha\beta}T)^{(n)} + 2N^{(n)}_{\alpha\beta} - \partial_\alpha P^{(n)}_\beta - \partial_\beta P^{(n)}_\alpha \,, \\ \partial^k\left[h^{(n)}_{k\alpha} - \tfrac{1}{2}h^{(n)}\,\eta_{k\alpha}\right] = -P^{(n)}_\alpha \,, \end{cases} \tag{2}$$

where $\triangle$ is the flat Laplacian, $T_{\alpha\beta}$ the energy–momentum tensor $N_{\alpha\beta}$ and $P_{\alpha\beta}$ are the second and higher order terms of the Ricci tensor and the harmonic condition respectively. Finally, we make use of the slow rotation approximation to expand the shape of the surface as a power series in Legendre polynomials

$$r_\Sigma(\cos\theta) \to r_\Sigma = r_s\left(1 + S\Omega^2 P_2(\cos\theta)\right) + O(\Omega^4) \tag{3}$$

with S a constant (that could be determined through the matching with a certain exterior metric) and r_s the radius of the fluid in the static limit.The full structure of the general solution can be found in [1] and contains the following free constants: m_0, m_2, j_1, j_3, b_0, a_0, a_2.

This scheme has already been applied to a pack of EOS. In [1], $\mu = \mu_0$ was used and in [3] the fluid was a polytrope. In both cases, as well as in the present one [4], the results pointed out that such source configurations could not create a Kerr exterior already in the linear solution. Here we are interested in comparing with the Wahlquist metric, so we will make us of a theorem stating that for a stationary perfect–fluid solution with rigid rotation, the metric is given by the Wahlquist family iff the solution is type D, axisymmetric and with equation of state $\mu + 3p =$ const. [5]. This translates into the conditions

$$\begin{gathered} 18 + 5m_2(n-1) = 0 \\ m_2\left(\frac{23}{70} - \frac{n}{5}\right) + j_3(n-1) + \frac{3(n-1)}{35} = 0\,. \end{gathered} \tag{4}$$

The free parameters in g can be chosen to satisfy them, and in that case our perturbed metric corresponds to Wahlquist.

2. Wahlquist metric and its coordinates

Our aim now is to proceed in the opposite direction: take the singularity free Wahlquist metric, expand it in the appropriate parameters and try to use every freedom available to recover a particular case of the CMMR metric.

The singularity free Wahlquist metric is

$$ds^2 = -f(dt + A\,d\varphi)^2 + r_0^2(\xi^2+\eta^2)\left[\frac{c^2h_1h_2}{h_1-h_2}\,d\varphi^2 + \frac{d\xi^2}{(1-k^2\xi^2)h_1} + \frac{d\eta^2}{(1+k^2\eta^2)h_2}\right] \tag{5}$$

with $\mathbf{u} = f^{-1/2}\partial_t$, so it is written in comoving coordinates, and the energy density and pressure are given in terms of the free parameters μ_0 and b as

$$\mu = \frac{1}{2}\mu_0(3b^2f - 1)\,, \qquad p = \frac{1}{2}\mu_0(1 - b^2 f) \tag{6}$$

and where

$$f(\xi,\eta) = \frac{h_1-h_2}{\xi^2+\eta^2}\,, \qquad A = c\,r_0\left(\frac{\xi^2h_2+\eta^2h_1}{h_1-h_2} - \eta_0^2\right),$$

$$h_1(\xi) = 1 + \xi^2 + \frac{\xi}{b^2}\left[\xi - \frac{1}{k}(1-k^2\xi^2)^{1/2}\arcsin(k\,\xi)\right], \qquad \text{and } k^2 \equiv \frac{1}{2}\,\mu_0\,r_0^2b^2\,.$$

$$h_2(\eta) = 1 - \eta^2 - \frac{\eta}{b^2}\left[\eta - \frac{1}{k}(1+k^2\eta^2)^{1/2}\text{arcsinh}(k\,\eta)\right],$$

The parameter μ_0 is the energy density in the static limit and r_0 is related with the rotation speed of the fluid although, to the best of our knowledge, there is no proved explicit expression for it. This will be the natural choice of parameters to expand Wahlquist for us, but first we must find the change to spherical-like coordinates. If we study the static limit[6], making $\eta = \cos\chi$, $\xi = \frac{R}{r_0}$ and letting $r_0 \to 0$, one obtains the Whittaker metric. In Minkoskian limit, the same change gives us the flat metric in Boyer-Lindquist coordinates. This is interesting because the coordinate change from these to spherical coordinates is known, and then, what we should look for seems to be two new functions such that in the mentioned limits $\eta = \cos\chi$, $\xi = \frac{R}{r_0}$ holds. Once written in these coordinates, the change from Boyer-Lindquist to spherical should give us Wahlquist in the coordinates we need. The functions we choose are

$$\{\xi,\eta\} \to \{R_1, R_2\} : \quad \begin{cases} h_1(\xi) = 1 + \dfrac{R_1^2}{r_0^2} \\ 1 - h_2(\eta) = \cos^2\chi = \dfrac{R_2^2}{r_0^2} \end{cases} \tag{7}$$

As we are dealing with a perturbed calculation, we only need the approximate expressions of $\xi = \xi(R_1)$ and $\eta = \eta(R_2)$. Expanding in μ_0 and r_0 all the constants in $h_1(\xi)$ and $h_2(\eta)$, we get

$$\xi^2 = \frac{R_1^2}{r_0^2} - \frac{1}{6}\mu_0\frac{R_1^4}{r_0^2}\left\{1 - \frac{1}{15}\mu_0 R_1^2\left[2 - \mu_0\left(\sigma_1 + \frac{37}{84}R_1^2\right)\right]\right\} + \mathcal{O}(\mu_0)^4 \tag{8}$$

$$\eta^2 = \frac{R_2^2}{r_0^2} + \frac{1}{6}\mu_0\frac{R_2^4}{r_0^2}\left\{1 + \frac{1}{15}\mu_0 R_2^2\left[2 - \mu_0\left(\sigma_1 - \frac{37}{84}R_2^2\right)\right]\right\} + \mathcal{O}(\mu_0)^4 \tag{9}$$

And now, the change

$$\begin{aligned} r^2\sin^2\theta &= \sin^2\chi(r_0^2 + R_1^2) \\ r\cos\theta &= R_1\cos\chi \end{aligned} \tag{10}$$

should give us the spherical-like coordinates desired.

3. From Wahlquist to CMMR

Now we face the problem of the identification of the approximation parameters and making every term in Wahlquist and CMMR equal.

To get an idea of the problems arising in the general case, we analyze the static limit. Prior to make any comparison, we stop the CMMR metric making $\varphi' = \varphi - \omega t \longrightarrow \mathbf{u} = \psi\partial_t$. Now, we use the clear relation between mass parameters: $\lambda = \frac{3}{4}\pi\mu_0 r_s^2$ and make $r_0 \to 0$. The g_{rr} term of the expression of Whittaker we get is

$$dr^2\left(1 - \frac{r^2\lambda}{{r_s}^2} + \frac{2r^2\lambda^2\left(7r^2 + 15\sigma\right)}{5{r_s}^4}\right)$$

and upon comparison with the corresponding static limit of CMMR

$$dr^2\left(1 + m_0\lambda - \frac{r^2\lambda}{{r_s}^2} - \frac{2r^4\lambda^2}{35{r_s}^4} + \frac{2m_0 r^2\lambda^2}{5{r_s}^2} - \frac{24r^2 S\lambda^2}{5{r_s}^2}\right)$$

we can see that there are discrepancies among r^4 terms. To some extent this was to be expected since CMMR is written in coordinates associated to harmonic ones and no such a condition has been imposed on Wahlquist yet. In this particular case, the two metrics can be rendered exactly equal with a change of the form $r \to r(1 + \lambda^2 f(r))$

In the general case we keep the relation between mass parameters. From comparing both $g_{t\varphi}$ terms, we can guess that $r_0 = w\sqrt{\lambda}\Omega r_s$, with w a proportionality factor. This leads to a new kind of discrepancy, giving rise to $const \times \Omega^2$ and $const \times \Omega^4$ terms. This issue can be solved using the freedom in the time scale. Eventually, every "bad term" is eliminated through the changes

$$t = t'(1 + \mu_0 F_1 + \mu_0^2 F_2) + \cdots \tag{11}$$

$$\begin{aligned} r &= r'(1 + \mu_0 G_1[r,\theta] + \mu_0^2 G_2[r,\theta]) + \cdots \\ \theta &= \theta' + \mu_0 \sin\theta H_1[r,\theta] + \mu_0^2 \sin\theta H_2[r,\theta] + \cdots \end{aligned} \tag{12}$$

3.1. Adjusting terms

Once the dependencies in r^n and Ω^n are corrected, only the adjustment of the constant factors remains. Here we can still carefully make use of a change with the form (12), and finally, we find that both metrics exactly coincide up to order (λ^1, Ω^2) if the free constants of CMMR take the values

$$\begin{aligned} m_0 &= 0 \qquad & m_2 &= \frac{6}{5}(1 + 2\lambda S) \\ j_1 &= \frac{9}{5}\Omega^2 S \qquad & j_3 &= \frac{36}{175}\,. \end{aligned} \tag{13}$$

These values satisfy the Petrov type D condition (4), as was to be expected. The final adjustment also fixes the proportionality factor w and the Wahlquist surface potential b to be

$$w(r_0 = w\sqrt{\lambda}\Omega r_s) \longrightarrow w = 1 - \frac{\Omega^2}{10} \tag{14}$$

$$b^2 = (1 + \Omega^2)(1 + 2\lambda S) \tag{15}$$

4. Conclusions

In this work we have arrived at the same metric through two different ways: direct construction on perturbation theory and perturbing a known exact solution. Thus, we have checked the reliability of the method at least when applied to the singularity free Wahlquist case. Additionally, we have identified the uncertain Wahlquist parameter r_0 in terms of physical quantities.

Acknowledgments

This work was supported by projects FIS2006-05319 and FIS2007-63034 from the Spanish Ministerio de Educación y Ciencia. JEC thanks Junta de Castilla y León for PhD grant EDU/1165/2007.

References

[1] Cabezas J A, Martín J, Molina A and Ruiz E 2007 *Gen. Rel. Grav.* **39** 707–736
[2] Papapetrou A 1966 *Ann. Inst. H. Poincaré A* **4** 83–105
[3] Martín J, Molina A and Ruiz E 2008 *Class. Quantum Grav.* **25** 105019
[4] Cuchí J E, Gil-Rivero A, Molina A and Ruiz E 2008 *Spanish Relativity Meeting 2007. Relativistic Astrophysics and Cosmology* vol 30 ed Oscoz A, Mediavilla E and Serra-Ricart M (EAS Publications Series) pp 315–318 URL `http://dx.doi.org/10.1051/eas:0830052`
[5] Senovilla J M M 1987 *Phys. Lett.* A **123** 211–214
[6] Stephani H, Kramer D, MacCallum M, Hoenselaers C and Herlt E 2003 *Exact solutions of Einstein's field equations* (Cambridge University Press)

Spanish Relativity Meeting (ERE 2009) IOP Publishing
Journal of Physics: Conference Series **229** (2010) 012033 doi:10.1088/1742-6596/229/1/012033

Black holes in modified gravity theories

A de la Cruz-Dombriz*, A Dobado and A L Maroto
Departamento de Física Teórica I, Universidad Complutense de Madrid, 28040 Madrid, Spain.
E-mail: dombriz@fis.ucm.es

Abstract. In the context of $f(R)$ gravity theories, the issue of finding static and spherically symmetric black hole solutions is addressed. Two approaches to study the existence of such solutions are considered: first, constant curvature solutions, and second, the general case (without imposing constant curvature) is also studied. Performing a perturbative expansion around the Einstein-Hilbert action, it is found that only solutions of the Schwarzschild-(Anti-) de Sitter type are present (up to second order in perturbations) and the explicit expressions for these solutions are provided in terms of the $f(R)$ function. Finally we consider the thermodynamics of black holes in Anti-de Sitter space-time and study their local and global stability.

1. Introduction

In the last years, increasing attention has been paid to modified theories of gravity in order to understand several open cosmological issues such as the present accelerated expansion of the universe [1]. Some of those theories modify general relativity by adding higher powers of the scalar curvature R, i.e. $f(R)$ theories, the Riemann and Ricci tensors or their derivatives [2]. It is therefore quite natural to ask about black hole (BH) features in those gravitational theories since, on the one hand, some BH signatures may be peculiar to Einstein's gravity and others may be robust features of all generally covariant theories of gravity. On the other hand, the obtained results may lead to rule out some models which will be in disagreement with expected physical results. For those purposes, research on thermodynamical quantities of BH is of particular interest.

In this work we will restrict ourselves to the so called $f(R)$ gravity theories in metric formalism in Jordan's frame where the gravitational Lagrangian is given by $R + f(R)$ and modified Einstein's equations are usually fourth order in the metric. Using the euclidean action method [3, 4] in order to determine different thermodynamical quantities, Anti-de Sitter (AdS) BHs in $f(R)$ models and their stability will be studied [5].

2. Black Holes in $f(R)$ gravities

The most general static and spherically symmetric $D \geq 4$ dimensional external metric for the gravitational field produced by a nonrotating object in $f(R)$ gravity theories can be written as:

$$ds^2 = e^{-2\Phi(r)}A(r)dt^2 - A^{-1}(r)dr^2 - r^2 d\Omega_{D-2}^2 \equiv \lambda(r)dt^2 - \mu^{-1}(r)dr^2 - r^2 d\Omega_{D-2}^2 \qquad (1)$$

where $d\Omega_{D-2}^2$ is the metric on the S^{D-2} sphere. In this case the scalar curvature R in D dimensions depends only on r and for the first parametrization in equation (1) is given by

$$R = \frac{1}{r^2}\left\{(D-2)[(D-3)(1-A) + 2r(-A' + A\Phi')] + r^2[3A'\Phi' - A'' - 2A(\Phi'^2 - \Phi'')]\right\}. \qquad (2)$$

where the prime denotes derivative with respect to r from now on. Therefore the most general static and spherically symmetric metrics with constant scalar curvature R_0 can be found solving the equation $R = R_0$. For a constant $\Phi(r) = \Phi_0$ it is straightforward to see that the solution is:

$$A(r) = 1 + a_1 r^{3-D} + a_2 r^{2-D} - \frac{R_0}{D(D-1)} r^2 \tag{3}$$

with $a_{1,2}$ being arbitrary integration constants. For the particular case $D = 4$, $R_0 = 0$ and $\Phi_0 = 0$, the metric can be written exclusively in terms of the function $A(r) = 1 + a_1 r^{-1} + a_2 r^{-2}$ where by establishing the identifications $a_1 = -2G_N M$ and $a_2 = Q^2$, this solution corresponds to a charged massive BH solution with mass M and charge Q (Reissner-Nordström solution).

On the other hand, if the constant curvature case $R = R_0$ is considered, the modified Einstein's equation from $f(R)$ theories for the function $A(r)$ reduces to:

$$R + f(R) + (1 + f_R)\left[A'' + \frac{(D-2)}{r}(A' - 2A\Phi') - 3A'\Phi' + 2A\left(\Phi'^2 - \Phi''\right)\right] = 0 \tag{4}$$

where f_R holds for $df(R)/dR$. As it is explained in [5], the vacuum constant curvature solutions of $f(R)$ gravities are given by:

$$R_0 = \frac{D\, f(R_0)}{2(1 + f_{R_0}) - D} \tag{5}$$

If the hypothesis $\Phi'(r) = 0$ is made, then equation (4) , by using (5), reduces in this case to

$$A'' + (D-2)\frac{A'}{r} = -\frac{2}{D}R_0 \;\rightarrow\; A(r) = c_1 + c_2 r^{3-D} - \frac{R_0}{D(D-1)} r^2 \tag{6}$$

where $c_{1,2}$ are two arbitrary constants. However this solution has no constant curvature in the general case since, as it can be seen in expression (3), the constant curvature requirement demands $c_1 = 1$. Then, for negative R_0, this solution is basically the D dimensional generalization obtained by Witten [4] of the BH in AdS space-time solution considered by Hawking and Page [3]. With the natural choice $\Phi_0 = 0$ the solution can be written as:

$$A(r) = 1 - (R_S/r)^{D-3} + r^2/l^2 \tag{7}$$

where R_S and l have the usual interpretations found in the literature.

Thus we have concluded that the only static and spherically symmetric vacuum solutions with constant (negative) curvature of any $f(R)$ gravity is just the Hawking-Page BH in AdS space. However this kind of solution is not the most general static and spherically symmetric metric with constant curvature as can be seen by comparison with the solutions given by expression (3).

3. Perturbative results

In Einstein-Hilbert (EH) theory the most general static solution with spherical symmetry is the one with constant curvature. However, this is not guaranteed to be the case in $f(R)$ theories. The problem of finding the general static spherically symmetric solution in arbitrary $f(R)$ theories without imposing the constant curvature condition is in principle too complicated. For that reason a perturbative analysis of the problem is performed here, assuming that the modified action is a small perturbation around EH theory of the form $f(R) = -(D-2)\Lambda_D + \alpha\, g(R)$ where $\alpha \ll 1$ is a dimensionless parameter and $g(R)$ is assumed to be analytic. Assuming that the $\lambda(r)$ and $\mu(r)$ functions appearing in the second parametrization of the metric (1) are also analytical in α, they can be written as follows

$$\lambda(r) = \lambda_0(r) + \sum_{i=1}^{\infty} \alpha^i \lambda_i(r) \quad ; \quad \mu(r) = \mu_0(r) + \sum_{i=1}^{\infty} \alpha^i \mu_i(r) \tag{8}$$

where $\{\lambda_0(r), \mu_0(r)\}$ are the unperturbed solutions for the EH action with cosmological constant Λ_D given by $\mu_0(r) = 1 + \frac{C_1}{r^{D-3}} - \frac{\Lambda_D}{(D-1)} r^2$ and $\lambda_0(r) = -C_2(D-2)(D-1)\,\mu_0(r)$ which are the standard BH solutions in a D dimensional AdS spacetime with constant curvature $R_0 \equiv D\Lambda_D$.

By inserting the proposed $f(R)$ expansion and equation (8) in modified Einstein's equations the first and second order equations in α are obtained, whose solutions are

$$\begin{aligned}
\lambda_1(r) &= (D-1)(D-2)\left\{C_4 + \frac{C_1C_4 - C_2C_3}{r^{D-3}} - \left[\frac{C_4\Lambda_D}{D-1} + \frac{C_2 A(g;\, D,\, \Lambda_D)}{(D-1)(D-2)}\right] r^2\right\} \\
\mu_1(r) &= \frac{C_3}{r^{D-3}} + \frac{A(g;\, D,\, \Lambda_D)}{(D-2)(D-1)} r^2 \\
\lambda_2(r) &= C_6 + \frac{C_5 + C_6C_1}{r^{D-3}} - \left[\frac{C_6\Lambda_D}{D-1} - A(g;\, D\,\Lambda_D)\left(C_4 + C_2 g_{R_0} - \frac{2C_2R_0 g_{R_0R_0}}{D-2}\right)\right] r^2 \\
\mu_2(r) &= \left(C_3C_4 + \frac{-C_5}{(D-2)(D-1)}\right)\frac{C_2^{-1}}{r^{D-3}} + \frac{A(g;\, D,\, \Lambda_D)\,(2R_0 g_{R_0R_0} - (D-2)g_{R_0})}{(D-2)^2(D-1)} r^2
\end{aligned} \tag{9}$$

where $g_{R_0} \equiv dg(R)/dR|_{R=R_0}$, $g_{R_0R_0} \equiv d^2g(R)/dR^2|_{R=R_0}$, and $A(g;\, D,\, \Lambda_D) \equiv g(R_0) - 2\Lambda_D g_{R_0}$. Higher order solutions $\{\lambda_{3,4,...}(r), \mu_{3,4,...}(r)\}$ can be obtained by inserting the previous results in the corresponding equations order $\alpha^{3,4,...}$ which become increasingly complicated.

Notice that from the obtained results up to second order in α, the corresponding metric has constant scalar curvature for any value of the parameters $C_{1,...,6}$. As a matter of fact, this metric is nothing but the standard Schwarzschild-AdS geometry, and can be easily rewritten in the usual form by making a trivial time reparametrization. Therefore, at least up to second order, the only static, spherically symmetric solutions which are analytical in α are the standard Schwarzschild-AdS space-times.

4. Black-hole thermodynamics

For simplicity we will concentrate only on constant curvature AdS BH solutions given by expression (7) with $\Phi(r) \equiv 0$ as a natural choice. Then, usual definitions of temperature lead to $\beta \equiv 1/T = 4\pi l^2 r_H/[(D-1)r_H^2 + (D-3)l^2]$ where r_H satisfies $A(r_H) = 0$. Notice that the temperature is a function of r_H only, i.e. it depends only on the BH size. It is easy to prove that T has a minimum T_0 at r_{H0} and whose existence was established in [3] for $D = 4$. Notice that for any $T > T_0$, there are two possible BH sizes: one corresponding to the small BH phase with $r_H < r_{H0}$ and the other corresponding to the large BH phase with $r_H > r_{H0}$. The remaining thermodynamic quantities are determined by using the Euclidean method so that the free energy F is given by

$$F = -\frac{(R_0 + f(R_0))\mu_{D-2}}{36\pi(D-1)G_D}(l^2 r_H^{D-3} - r_H^{D-1}) \tag{10}$$

where provided $-(R_0 + f(R_0)) > 0$, which is the usual case in EH gravity, $F > 0$ for $r_H < l$ and $F < 0$ for $r_H > l$. On the other hand, the heat capacity C can be written as:

$$C = \frac{\partial E}{\partial T} = \frac{\partial E}{\partial r_H}\frac{\partial r_H}{\partial T} = \frac{-(R_0 + f(R_0))(D-2)\mu_{D-2} r_H^{D-2} l^2}{8G_D(D-1)} \frac{(D-1)r_H^2 + (D-3)l^2}{(D-1)r_H^2 - (D-3)l^2}. \tag{11}$$

Let us assume from now on that, like in the EH case, the condition $(R_0 + f(R_0)) < 0$ holds. This is the required condition, as explicitly shown in [5] to get both positive energy and entropy in these theories and according to (5) it implies $1 + f_{R_0} > 0$ in agreement with cosmological requirements [6]. With this assumption, we find $C > 0$ for $r_H > r_{H0}$ (the large BH region) and $C < 0$ for $r_H < r_{H0}$ (the small BH region). For $r_H \sim r_{H0}$ (T close to T_0) C is divergent. In any case, for $f(R)$ theories with $R_0 + f(R_0) < 0$, we have found a similar scenario to the one described in full detail in [3] for the EH case: for $T > T_0$ two possible BH solutions are present, the small BH and the large BH: The small one has $C < 0$ and $F > 0$ as the standard Schwarzschild BH. For the large BH two possibilities appear; if $T_0 < T < T(r_H = l)$ then $C > 0$ and $F > 0$ and the BH will decay by tunnelling into radiation, but if $T > T(r_H = l)$ then $C > 0$ and $F < 0$. In this case F of the large BH will be less than that of pure radiation.

To illustrate the previous calculations, a model $f(R) = \alpha(-R)^{\beta}$ is considered where the heat capacity C and the free energy F signs will be studied to know the local and global stability of BHs. This model has a nonvanishing constant curvature solution $R_0 = -\left[(2 - D)/\alpha(2\beta - D)\right]^{1/(\beta-1)}$ calculated by using (5) and therefore two separated regions, where $R_0 < 0$ holds, have to be studied: Region 1 $\{\alpha < 0, \beta > D/2\}$ and Region 2 $\{\alpha > 0, \beta < D/2\}$. Further models following an analogous procedure were presented in [5].

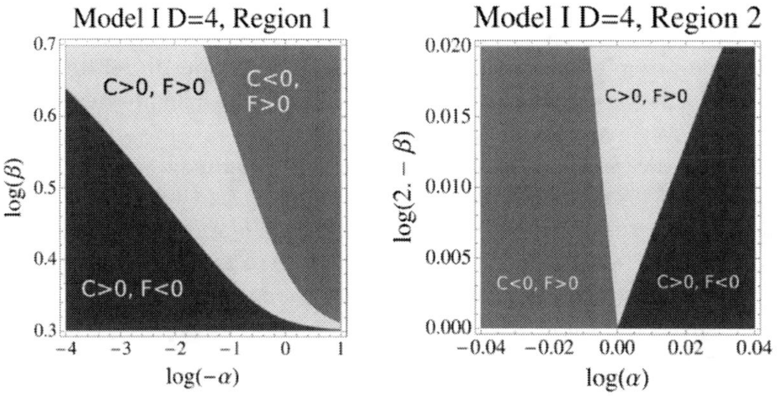

Figure 1. Regions for C and F signs have been compared in the (α, β) plane for Model I in $D = 4$. Region 1(left), Region 2 (right). Schwarzschild radius $R_S = 2$ was assumed.

Acknowledgments
This work has been supported by the MICINN (Spain) project numbers FIS 2008-01323 and FPA 2008-00592, UCM-Santander PR34/07-15875 and UCM-BSCH GR58/08 910309.

References
[1] Carroll S M *et al.* 2004 *Phys. Rev.* D **70** 043528
Dvali G, Gabadadze G and Porrati M 2000 *Phys. Lett.* B **485** 208
de la Cruz-Dombriz A and Dobado A 2006 *Phys. Rev.* D **74** 087501
Cembranos J A R 2006 *Phys. Rev.* D **73** 064029
[2] Dobado A and Maroto A L 1993 *Phys. Lett.* B **316** 250 [Erratum-ibid. 1994 B **321** 435]
[3] Hawking S W and Page D N 1983 *Commun. Math. Phys.* **87** 577
[4] Witten E 1998 *Adv. Theor. Math. Phys.* **2** 505
[5] de la Cruz-Dombriz A, Dobado A and Maroto A L 2009 *Phys. Rev.* D **80** 124011
[6] Pogosian L and Silvestri A 2008 *Phys. Rev.* D **77** 023503

Unleashing the aether

Alberto Diez-Tejedor
Instituto de Ciencias Nucleares, Universidad Nacional Autónoma de México, México D.F. 04510, México

E-mail: alberto.diez@nucleares.unam.mx

Abstract. We follow an effective field theory approach to identify the general class of theories that describes a real vector field coupled to gravity at low energies, large distances and long timescales. Here we restrict ourselves to the case in which Lorentz invariance is spontaneously broken at low energies (i.e. at energies bellow the characteristic scale of the effective theory). We find the constraints on the parameter space of the theory coming from the absence of classical and quantum instabilities. We apply our results to the flat case before closing.

The symmetries of a spacetime severely constrain the transformation properties of its matter content. In flat spacetime for instance Lorentz invariance only allows scalar fields to have a non-vanishing expectation value. However, we do *not* live in a Lorentz invariant universe: at large scales it is homogeneous and isotropic, but non-static. In an expanding universe no symmetry principle prevents the appearance of non-vanishing tensor fields, as long as they were constant in space and invariant under spatial rotations. Since only bosonic fields can acquire a large occupation number, and if we restrict ourselves to fields of spin less than two, this leaves vector fields as essentially the only additional type of matter field that may be *classically* relevant in in a cosmological context. (By classical we mean that at the quantum level the state of the system is described by a coherent excitation.)

Massless vector-tensor theories were considered long time ago to explore different modifications of gravity [1, 2, 3]. More recently, theories with vectors have experienced a modest revival in connection with inflation [4], the current accelerated expansion of the universe [5] and a possible breakdown of Lorentz invariance [6]. Contrary to other proposals we do not assume here any specific model and consider the most general vector-tensor theory expected to describe our universe at large distances and long timescales. We thus follow a low energy effective action approach and retain all possible terms compatible with the symmetries of the theory, which we assume to consist just of locality, general covariance and a $\mathbb{Z}_2$ symmetry for the vector field (i.e. $A^\mu \to -A^\mu$). If we are interested in low energies however, we expect only those terms with the least number of field derivatives to appear into the action. In principle, such an action would also contain arbitrary powers of the vector field, but since a vector has dimensions of mass, terms with too many powers of A^μ must be suppressed by corresponding powers of an energy scale Λ. If the background value of the field is smaller than Λ, the dominant terms are hence given by those with no suppression, that is, by the operators of dimension four. Therefore, we just have to consider the most general action with two derivatives acting on the metric and the vector field, but with no vector field operators of dimension higher than four:

$$S = \int d^4x\sqrt{-g}\left[\frac{R}{16\pi G} - \frac{\beta_1}{2}F_{\mu\nu}F^{\mu\nu} - \beta(\nabla_\mu A^\mu)^2 + \beta_{13}R_{\mu\nu}A^\mu A^\nu + \beta_4 R A_\mu A^\mu - V\right]. \quad (1)$$

Here β_i are some dimensionless coefficients, $F_{\mu\nu} \equiv \partial_\mu A_\nu - \partial_\nu A_\mu$ is the electromagnetic tensor and V a potential for the vector field which, under the previously mentioned assumptions and up to a constant term, is given by

$$V(A_\mu A^\mu) = \lambda(A_\mu A^\mu + M^2)^2. \tag{2}$$

Other low energy terms not appearing into the expression (1) can be reabsorbed by the appearing ones after integrating by parts. Notice that, since we deal with massive vectors, a consistent quantum description at low energies does not require gauge invariance, so many different kinetic terms for the vector field are possible. It is also interesting to point out here that for generic values of the potential parameters different from zero Lorentz invariance will be spontaneously broken by the vacuum expectation value of the theory. The action (1) is to some extent a generalization of previously considered theories. Setting $V \equiv 0$ one recovers the massless vector-tensor theories studied in [3], while replacing the potential (2) by a fixed-norm constraint leads to the Einstein-aether models considered in [6]. Although a fixed normed constraint seems to be difficult to justify from a particle physics perspective, it can be naively recovered from a potential of the form (2), at least classically, by taking the limit $\lambda \to \infty$. This is the reason for which we call the theories considered here *aether unleashed.* (Similar kind of actions have been considered for instance in [7] under the name of "bumblebee models".)

However, in order to consider the aether unleashed as a sensible model to describe the observable universe, we should be sure that at least it is self-consistent in a cosmological setting. With this in mind, let us analyze the conditions that classical and quantum-mechanical stability place on the parameters of the theory.[1] The stability of the system will be determined by the dynamics of the small perturbations b_μ around the background value $\bar{A}_\mu$ of the vector field,

$$A_\mu = \bar{A}_\mu + b_\mu. \tag{3}$$

We neglect here metric perturbations, which relies on the assumption that $\Lambda \ll M_G$. In conformal time coordinates, the metric and background field are hence given by

$$ds^2 = a^2(\eta)(-d\eta^2 + d\vec{x}^{\,2}), \quad \bar{A}^\mu = a^{-1} \cdot (\bar{A}(\eta), 0, 0, 0). \tag{4}$$

To obtain the Lagrangian for the field fluctuations b^μ we insert the expansion (3) and the background (4) into the action (1). Expanding to quadratic order (free level) in the small perturbations we arrive at

$$\begin{aligned}\mathcal{L}_b = & - \beta_1 \partial_\mu b_\nu \partial^\mu b^\nu - (\beta - \beta_1)(\partial_\mu b^\mu)^2 - 4\beta(\mathcal{H}^2 b_0^2 - \mathcal{H} b_0 \partial_\mu b^\mu) + \\ & + \beta_{13} R_{\mu\nu} b^\mu b^\nu + a^2 \beta_4 R\, b_\mu b^\mu - a^2 V' b_\mu b^\mu - 2a^4 V'' \cdot (\bar{A}^\mu b_\mu)^2,\end{aligned} \tag{5}$$

where indexes are raised with the Minkowski metric $\eta^{\mu\nu}$, $\mathcal{H} = H/a \equiv d\log a/d\eta$, and a prime denotes a derivative with respect to $A_\mu A^\mu$. For arbitrary parameters β_1 and β the determinant of the Hessian matrix is non-zero and the four components of the vector field are dynamical. For $\beta = 0$ or $\beta_1 = 0$ the determinant of the Hessian vanishes, signalling that some of the vector field components are constrained. This is what happens for instance for the gauge invariant Maxwell theory ($\beta = 0$). These two degenerate cases should be analyzed separately. In aether models the fixed norm constrain requires $b^0 = 0$, eliminating one of the four dynamical fields. This will be no longer true for the case in which the aether is unleashed.

To proceed any further it is convenient to decompose the field b^μ in irreducible representations of the unbroken symmetry group: the Euclidean group in three dimensions $E(3)$. Expanding as usual

$$b_0 \equiv -u \quad \text{and} \quad b_i \equiv \partial_i r + v_i, \quad \text{with} \quad \partial_i v^i = 0, \tag{6}$$

[1] For more details about the possible viability of the unleashed aether models please see the reference [8].

Table 1. Stability conditions in the vector sector. The checkmark means that the condition is automatically satisfied.

Vector	Low p	High p
Classical	$\beta_1 m_v^2 \geq 0$	✓
Quantum	$\beta_1 \geq 0$	$\beta_1 \geq 0$

decouples the two scalar modes u and r from the two vector modes in $\vec{v}$. (Here we are talking about scalar and vector modes in a 3-dimensional sense). The Lagrangian of the vector sector is then given by

$$\mathcal{L}_v = -\beta_1 \partial_\mu \vec{v} \cdot \partial^\mu \vec{v} - a^2 m_v^2 \vec{v} \cdot \vec{v}, \quad \text{where} \quad m_v^2 \equiv V' - 3(\beta_{13} + 4\beta_4)H^2 - (\beta_{13} + 6\beta_4)\dot{H}. \tag{7}$$

Notice that the quantity $a^2 m_v^2/\beta_1$ one would naively identify as the mass of the vector excitations is time-dependent. We can nevertheless quantize the theory under the approximation that the frequency of each mode is constant if the relative change in $\omega^2 = p^2 + a^2 m_v^2/\beta_1$ during the characteristic time $\Delta\eta \approx \omega^{-1}$ is small [9]. Under this assumption the vector sector is ghost free for $\beta_1 \geq 0$, and free of classical instabilities for $m_v^2/\beta_1 \geq 0$. These conditions are summarized in Table 1. Note that if $\beta_1 = 0$, the vector sector is not dynamical.

Regarding the scalar sector, it is more convenient to switch to Fourier space. Assuming again that all explicitly time-dependent quantities are constant we arrive at the action

$$S_s = -\int \frac{d^4p}{(2\pi)^4} \begin{pmatrix} u \\ pr \end{pmatrix}^\dagger D \begin{pmatrix} u \\ pr \end{pmatrix}, \tag{8}$$

where the matrix D is given by

$$D = \begin{pmatrix} \beta\omega^2 - \beta_1 p^2 + m_u^2 & -i(\beta - \beta_1)\omega p - 2\beta H p \\ i(\beta - \beta_1)\omega p - 2\beta H p & -\beta_1 \omega^2 + \beta p^2 + m_v^2 \end{pmatrix} \tag{9}$$

and we have also defined

$$m_u^2 \equiv 2V''\bar{A}^2 - V' + (2\beta + 3\beta_{13} + 12\beta_4)H^2 - (2\beta - 3\beta_{13} - 6\beta_4)\dot{H}. \tag{10}$$

Here $p \equiv |\vec{p}|$. The inverse of the matrix D is just the field propagator. In order to find the propagating modes we just have to find the values of ω^2 at which its eigenvalues have poles, or, equivalently, the values of ω^2 at which the eigenvalues of D have zeros. Requiring that $\det D$ vanish we thus arrive at the frequencies of the *two* propagating scalar modes. Furthermore, demanding them to be real guarantees the classical stability of the system. On the other hand, its quantum character will be determined by the residuals at the poles of the eigenvalues of D^{-1}. They should be positive in order to avoid ghosts. The conditions obtained for the stability of the scalar sector are summarized in Table 2. Notice that ghosts are inevitable at high energies. However, we are following an effective field theory approach so it should not represent a serious problem for the self-consistency of the model: for appropriate choices of the free parameters it is still possible to avoid the existence of low-momentum instabilities.

Our conclusions above do not apply to the two degenerate limits $\beta = 0$ and $\beta_1 = 0$. For $\beta = 0$ the conditions for the stability of the vector sector coincide with those summarized in Table 1. However, only one scalar mode propagates now. Its stability conditions can be found in Table 3. Notice that some of them are the same as in the vector sector, and that if the

Table 2. Stability conditions in the scalar sector for $\beta_1 \neq 0$ and $\beta \neq 0$. The cross means that the condition cannot be satisfied.

Scalar ($\beta_1 \neq 0$, $\beta \neq 0$)	Low p	High p
Classical	$\beta_1 \cdot m_v^2 > 0,\ \beta \cdot m_u^2 < 0$	$(\beta_1^{-1} - \beta^{-1})(m_u^2 + m_v^2) - 4\beta\beta_1^{-1}H^2 > 0$
Quantum	$\beta_1 > 0, \beta < 0$	X

Table 3. Stability conditions in the scalar sector for the cases $\beta = 0$ and $\beta_1 = 0$ respectively.

Scalar	Low p	High p
Classical	$\beta_1 m_v^2 > 0$	$m_v^2 m_u^2 < 0$
Quantum	$\beta_1 > 0$	$m_v^2 - m_u^2 > 0$

Scalar	Low p	High p
Classical	$\beta m_u^2 < 0$	$m_v^2(m_u^2 - 4\beta H^2) < 0$
Quantum	$\beta < 0$	$m_u^2 - m_v^2 - 4\beta H^2 < 0$

theory is stable both in the low and high momentum limits it will be stable for all momenta. If $m_u^2 = 0$ the theory does not have any propagating scalar degrees of freedom, as it happens in the electromagnetic case. For $\beta_1 = 0$ the vector sector is not dynamical. Once again, only one scalar mode propagates; the corresponding stability conditions are given in Table 3. If $m_v^2 = 0$ there are no propagating modes.

As an application of the results presented here, let us briefly discuss the stability conditions obtained for the simplest possible case in which the vector field sits at the minimum of the potential. This is generically possible only in flat spacetime (i.e. $H = 0$, $m_v^2 = 0$ and $m_u^2 = 4\lambda M^2$). If $\beta, \beta_1 \neq 0$ the spectrum contains two vector and two scalar degrees of freedom, three of them (the two vectors plus a scalar) massless: the Nambu-Goldstone bosons associated with the spontaneous breaking of the three generators of Lorentz boosts. However, at low momenta, the theory is either classically or quantum-mechanically stable, but not both. The only way to scape this conclusion is to choose a well-behaved set of massless modes ($\beta > 0$ and $\beta_1 > 0$) and push the massive mode out the domain of validity of the effective theory (taking $\beta(\Lambda/M)^4 \lesssim 1$). In this sense, the aether unleashed models with $\beta, \beta_1 \neq 0$ can be seen as a high-momentum completion of the Einstein-aether theories. The instability can be also averted by setting $\beta = 0$ or $\beta_1 = 0$. For $\beta = 0$, $\beta_1 > 0$ two well-behaved massless vector modes propagate. On the other hand, if $\beta_1 = 0$ there are not propagating modes.

References

[1] Will C M and Nordtvedt K J 1972 Conservation laws and preferred frames in relativistic gravity. I. Preferred-frame theories and an extended PPN formalism *Astrophys. J.* **177** 757
[2] Hellings R W and Nordtvedt K 1973 Vector-metric theory of gravity *Phys. Rev.* D **7** 3593
[3] Will C M 1981 *Theory And Experiment In Gravitational Physics* (UK: Cambridge University Press)
[4] Ford L H 1989 Inflation driven by a vector field *Phys. Rev.* D **40** 967
[5] Armendariz-Picon C 2004 Could dark energy be vector-like? *J. Cosmol. Astropart. Phys.* JCAP07(2004)007
[6] Jacobson T and Mattingly D 2001 Gravity with a dynamical preferred frame *Phys. Rev.* D **64** 024028
[7] Kostelecky V A and Potting R 2009 Gravity from spontaneous Lorentz violation *Phys. Rev.* D **79** 065018
[8] Armendariz-Picon C and Diez-Tejedor A 2009 Aether Unleashed *J. Cosmol. Astropart. Phys.* JCAP12(2009)018
[9] Mukhanov V and Winitzki S 2007 *Introduction to quantum effects in gravity* (UK: Cambridge University Press)

Petrov D vacuum spaces revisited: complete tables and invariant classification by GHP analysis

S Brian Edgar[1*], Alfonso García-Parrado Gómez-Lobo[1,4] and José M Martín-García[1,2,3]

[1] Matematiska institutionen, Linköpings Universitet, SE-581 83 Linköping, Sweden.
[2] Laboratoire Univers et Théories, Observatoire de Paris, CNRS, Univ. Paris Diderot, 5 place Jules Janssen, 92190 Meudon, France.
[3] Institut d'Astrophysique de Paris, Univ. Pierre et Marie Curie, CNRS, 98bis boulevard Arago, 75014 Paris, France.
[4] Current address: Centro de Matemática, Universidade do Minho. 4710-057 Braga, Portugal

E-mail: bredg@mai.liu.se, alfonso@math.uminho.pt, Jose.Martin-Garcia@obspm.fr

Abstract. The efficiency of the GHP formalism combined with the power of the computer system *xAcT* enables an exhaustive systematic analysis to be made of Petrov D vacuum spaces. This gives a complete involutive set of tables for the four GHP derivatives on each of the four GHP spin coefficients and the one Weyl tensor component. It follows directly from these results that the theoretical upper bound on the order of covariant differentiation of the Riemann tensor required for a Karlhede invariant classification of these spaces is reduced to two.

1. Introduction

Kinnersley's work [16] is not just the definitive work on the Type D vacuum spacetimes, but it was also the calculation which established the power of the NP formalism [18]. So it would appear rather presumptuous, 40 years later, to draw attention to some related results which had not been presented by Kinnersley, and also to suggest that the GHP formalism [12] (the lesser-known descendant of the NP formalism) is an even more powerful formalism for such calculations. However, the difference today — compared to 40 years ago — is the availability of powerful computer support: in particular, we use the system *xAct* [17]; this permits calculations which would not have been possible 40 years ago.
So although the results in this talk are important in their own right — for Petrov type D vacuum spaces — they also illustrate how methods and techniques in the GHP formalism [7] [8], [10], [13], [14] can be exploited even more fully when powerful computer support is available.

2. Calculations

<u>Nonvanishing weighted GHP spin coefficients and curvature components</u>

$$\mathcal{S} \equiv \{\rho, \quad \tau, \quad \rho', \quad \tau', \quad \Psi_2\}.$$

Beginning with the familiar Ricci and Bianchi equations in the GHP formalism [12], we use the GHP commutator equations and the introduction of *extension variables* in a systematic manner

to obtain complete and involutive tables for all non-zero spin coefficients and Ψ_2. In particular, all information is extracted by checking integrability conditions and consistency at all stages.
Ricci equations

$$\begin{aligned} þ\rho &= \rho^2, & ð\rho &= \tau(\rho-\bar\rho),\\ þ\tau &= \rho(\tau-\bar\tau'), & ð\tau &= \tau^2,\\ þ'\rho &= ð'\tau+\rho\bar\rho'-\tau\bar\tau-\Psi_2, \end{aligned}$$

and their $'$ counterparts;
Bianchi equations

$$þ\Psi_2 = 3\rho\Psi_2, \qquad ð\Psi_2 = 3\tau\Psi_2,$$

and their $'$ counterparts;
Commutator equations

$$\begin{aligned} [þþ'-þ'þ]\eta_{pq} &= \Big((\bar\tau-\tau')ð+(\tau-\bar\tau')ð'-p(\Psi_2-\tau\tau')-q(\bar\Psi_2-\bar\tau\bar\tau')\Big)\eta_{pq},\\ [þð-ðþ]\eta_{pq} &= \Big(\bar\rhoð-\bar\tau'þ+q\bar\rho\bar\tau'\Big)\eta_{pq},\\ [þð'-ð'þ]\eta_{pq} &= \Big(\rhoð'-\tau'þ+p\rho\tau'\Big)\eta_{pq},\\ [ðð'-ð'ð]\eta_{pq} &= \Big((\bar\rho'-\rho')þ+(\rho-\bar\rho)þ'+p(\Psi_2+\rho\rho')-q(\bar\Psi_2+\bar\rho\bar\rho')\Big)\eta_{pq}, \end{aligned}$$

together with their $'$ counterparts, acting on a scalar η_{pq} of spin and boost weight $(p-q)/2$, $(p+q)/2$ respectively.
When we substitute the Bianchi equations into the commutators for Ψ_2 and also make use of the Ricci equations, we obtain
Post-Bianchi equations

$$\begin{aligned} \Psi_2(-þ'\rho+þ\rho'-\tau\bar\tau+\tau'\bar\tau')=0 &\quad\Leftrightarrow\quad -þ'\rho+þ\rho'-\tau\bar\tau+\tau'\bar\tau'=0,\\ \Psi_2(ð'\rho-þ\tau')=0 &\quad\Leftrightarrow\quad ð'\rho-þ\tau'=0 \end{aligned}$$

and their $'$ counterparts.
To continue the analysis, it will be convenient to introduce the zero-weighted Z_1, the $\{0,2\}$-weighted Z_2 and the $\{2,0\}$-weighted Z_2' (using the notation of Carminati and Vu [1]),
Extension variables

$$Z_1 = ð'\tau\ (=Z_1'), \qquad Z_2 = ð'\rho, \qquad Z_2' = ð\rho'.$$

These variables enable us to write down
Complete tables for all spin coefficients

$$þ\rho=\rho^2,\quad ð\rho=\tau(\rho-\bar\rho),\quad ð'\rho=Z_2,\quad þ'\rho=Z_1-\tau\bar\tau+\tau'\bar\tau'$$

$$þ\tau=\rho(\tau-\bar\tau'),\quad ð\tau=\tau^2,\quad ð'\tau=Z_1,\quad þ'\tau=Z_2',$$

and their $'$ counterparts.
Next, by applying commutators to all spin coefficients, we obtain
Complete tables for Z_1, Z_2, Z_2'

$$þZ_1=\rho(2Z_1-\bar Z_1+\tau'\bar\tau'-\bar\rho\rho'+\bar\rho'\rho)+(\tau-\bar\tau')Z_2,$$

$$\begin{aligned}
\eth Z_1 &= \tau(2Z_1 + \Psi_2 + \bar\Psi_2 + \rho\bar\rho' + \bar\rho'\bar\rho) - (\bar\rho - \rho)Z_2' - \bar\tau'(\rho\bar\rho' - \rho\rho') \\
\text{þ} Z_2 &= 3\rho Z_2, \\
\eth' Z_2 &= 3\tau' Z_2, \\
\eth Z_2 &= \tau Z_2 + 2(\rho - \bar\rho)Z_1 + \bar\rho\Psi_2 - \rho\bar\Psi_2 + 2\rho\bar\rho'(\rho - \bar\rho), \\
\text{þ}' Z_2 &= \rho' Z_2 + \bar\Psi_2\tau' + \Psi_2(\bar\tau - 2\tau') - 2(\bar\tau - \tau')(Z_1 + \rho\bar\rho')
\end{aligned}$$

and their $'$ counterparts.

At this stage we have a complete and involutive set of tables: for the eight complex quantities $\{\rho, \tau, \rho', \tau', \Psi_2, Z_1, Z_2, Z_2'\}$.

Next, by applying commutators to Z_1, Z_2, Z_2', we obtain

Complex identity

$$\Psi_2(\tau\bar\tau - \tau'\bar\tau' + \rho\bar\rho' - \bar\rho\rho') = 0 \quad \Leftrightarrow \quad \tau\bar\tau - \tau'\bar\tau' + \rho\bar\rho' - \bar\rho\rho' = 0;$$

and

Explicit expressions for Z_2, Z_2' in terms of spin coefficients

$$\Psi_2(Z_2 + \rho\bar\tau - 2\rho\tau' + \tau'\bar\rho) = 0 \quad \Leftrightarrow \quad Z_2 = -\rho\bar\tau + 2\rho\tau' - \tau'\bar\rho,$$

together with the $'$ counterpart.

Substituting these expressions for Z_2, Z_2' into the above tables for Z_2, Z_2' respectively, we obtain

Consistency conditions for Z_2, Z_2'

$$\begin{aligned}
\bar\tau' Z_1 - \tau \bar Z_1 &= \tau'\bar\tau'(\tau - -\bar\tau'), \\
\bar\rho' Z_1 - \rho' \bar Z_1 &= 2\rho'\bar\rho'(\rho - \bar\rho) + \tau'\bar\tau'(\bar\rho' - \rho') + \bar\rho'\Psi_2 - \rho'\bar\Psi_2.
\end{aligned}$$

In this talk, we restrict attention to the *generic case*: $\tau\tau'\rho'\rho \neq 0 \neq \bar\rho'\tau - \rho'\bar\tau'$; we then obtain

Explicit expression for Z_1

$$Z_1 = \tau\Big(2\bar\rho\rho'^2 - \bar\Psi_2\rho' + (\bar\tau' - 2\tau)\bar\tau\rho' - 2\bar\rho'\bar\rho\rho' + \Psi_2\bar\rho' + \bar\rho'\tau\bar\tau\Big)\Big/\bar\rho'\tau - \rho'\bar\tau'.$$

Substitution into the above table for Z_1 of this value for Z_1 — together with the values for Z_2, Z_2' — gives only trivial identities.

So finally we have obtained

Complete and involutive tables for all spin coefficients

$$\begin{aligned}
&\text{þ}\rho = \rho^2, \quad \eth\rho = \tau(\rho - \bar\rho), \quad \eth'\rho = -\rho\bar\tau + 2\rho\tau' - \tau'\bar\rho, \\
&\text{þ}'\rho = \tau\Big(2\bar\rho\rho'^2 - \bar\Psi_2\rho' + (\bar\tau' - 2\tau)\bar\tau\rho' - 2\bar\rho'\bar\rho\rho' + \Psi_2\bar\rho' + \bar\rho'\tau\bar\tau\Big)\Big/\bar\rho'\tau - \rho'\bar\tau'; \\
&\text{þ}\tau = \rho(\tau - \bar\tau'), \quad \eth\tau = \tau^2, \quad \text{þ}'\tau = -\rho'\bar\tau' + 2\rho'\tau - \tau\bar\rho', \\
&\eth'\tau = \tau\Big(2\bar\rho\rho'^2 - \bar\Psi_2\rho' + (\bar\tau' - 2\tau)\bar\tau\rho' - 2\bar\rho'\bar\rho\rho' + \Psi_2\bar\rho' + \bar\rho'\tau\bar\tau\Big)\Big/\bar\rho'\tau - \rho'\bar\tau',
\end{aligned}$$

together with the $'$ counterparts.

The analogous calculations for the non-generic cases are easily carried out. Full details of all calculations are given in [9].

The system *xAct* was used to cross-check all the computations derived by hand and to derive others which due to their complication were difficult to obtain without the aid of the computer. A *Mathematica* notebook in which all the equations of the GHP formalism are obtained from scratch is available in [17] at [11].

Earlier relevant work on these spacetimes is given in [4], [5], [6], [1]. In particular, in [5] Czapor and Mclenaghan state that they have obtained the complete tables for all of the NP spin coefficients, but these tables, together with the intermediate calculations, were too long to print out in the NP formalism; in [1], Carminati and Vu presented *partial* tables (for *three* GHP operators acting on each GHP spin coefficient) in the GHP formalism.

3. Conclusions

The calculations in the previous section supply a *complete and involutive set of tables* for

$$\mathcal{S} \equiv \{\rho, \quad \tau, \quad \rho', \quad \tau', \quad \Psi_2\}.$$

There is an immediate application to the invariant classification scheme [15], [20]. All elements of $\mathcal{S}$ occur at the order of first derivative of the Weyl tensor, and no new quantities appear at second derivative order; moreover, the invariance group remains unchanged from first to second derivative order: therefore *the Karlhede algorithm for the invariant classification of these spacetimes terminates at second order.* This result strengthens earlier results on the invariant classification of Petrov type D vacuum spaces [2], [3], [21].

These results enable the classification scheme for vacuum Petrov type D spacetimes in [4] to be strengthened, and the full details are given in [9]; recently, a similar classification has been given in [19].

Finally we point out, that when a non-zero cosmological constant is introduced, it is very straightforward to repeat the calculations using *xAct*, and once again a complete and involutive set of tables is obtained, with the same result for the invariant classification.

References

[1] Carminati J and Vu K T 2001 *Gen. Rel. Grav.* **33** 295
[2] Collins J, d'Inverno R A and Vickers J A 1991 *Class. Quantum Grav.* **7** 2005
[3] Collins J, d'Inverno R A and Vickers J A 1991 *Class. Quantum Grav.* **8** L215
[4] Czapor S R and McLenaghan R G 1982 *J. Math. Phys.* **23** 2159
[5] Czapor S R and McLenaghan R G 1987 *Gen. Rel. Grav.* **19** 623
[6] Debever R and McLenaghan R G 1981 *J. Math. Phys.* **22** 1711
[7] Edgar S B 1980 *Gen. Rel. Grav.* **12** 347
[8] Edgar S B 1992 *Gen. Rel. Grav.* **24** 1267
[9] Edgar S B, García-Parrado Gómez-Lobo A and Martín-García J M 2009 *Class. Quantum Grav.* **26** 105022
[10] Edgar S B and Ludwig G 2000 *Gen. Rel. Grav.* **32** 637
[11] García-Parrado Gómez-Lobo A and Martín-García J M `http://metric.iem.csic.es/Martin-Garcia/xAct/Spinors/`
[12] Geroch R, Held A and Penrose R 1973 *J. Math. Phys.* **14** 874
[13] Held A 1974 *Gen. Rel. Grav.* **7** 177
[14] Held A 1975 *Comm. Math. Phys.* **44** 211
[15] Karlhede A 1980 *Gen. Rel. Grav.* **12** 693
[16] Kinnersley W 1969 *J. Math. Phys.* **10** 1195
[17] Martín-García J M `http://metric.iem.csic.es/Martin-Garcia/xAct/`
[18] Newman E and Penrose R 1962 *J. Math. Phys.* **3** 566
[19] Wylleman L 2008 *Ph. D. thesis*: Invariant classification of purely electric and magnetic perfect fluids (Universiteit Ghent)
[20] Åman J E and Karlhede A 1981 *Preprint* **46** Institute of Theoretical Physics, University of Stockholm
[21] Åman J E 1984 *Computer-aided classification of symmetries in general relativity; Example: the Petrov type D vacuum metrics*, Classical General Relativity: Bonnor W B, Islam J N and MacCallum M A H (Cambridge University Press)

Spanish Relativity Meeting (ERE 2009) IOP Publishing
Journal of Physics: Conference Series **229** (2010) 012036 doi:10.1088/1742-6596/229/1/012036

Using black hole physics to compute the energy loss of a quark in strongly coupled QCD

Daniel Fernández

Departament de Física Fonamental & Institut de Ciències del Cosmos, Diagonal 647, E-08028 Barcelona, Spain

E-mail: daniel@ffn.ub.es

Abstract. The AdS/CFT duality can be applied to study strongly coupled regimes of QCD, such as the quark-gluon plasma. Gravity is a key factor in the understanding of the duality and an intriguing correspondence between the gravitational properties of higher-dimensional black holes and quark/gluon physics can be established. In this context, I will show that the dual theory allows us to identify a new mechanism for the energy loss of a deconfined quark in this plasma, which can be described as Cherenkov radiation of vectorial mesons. The corresponding results might be observed in heavy-ion collision experiments.

1. Introduction

The purpose of this work is to study an experimental situation: The emission of meson Cherenkov radiation when a quark goes through a quark-gluon plasma. This is a phase of QCD which exists at extremely high temperature and density. It consists basically of free, deconfined quarks and gluons. Current heavy-ion collision experiments at the RHIC of Brookhaven are studying the properties of the quark/gluon plasma, and future experiments at the LHC will continue this effort.

2. New mechanism of energy loss

The classical phenomenon known as Cherenkov radiation is an electromagnetic radiation emitted when a charged particle, such as an electron, passes through an insulator faster than the speed of light in that medium. An analogous phenomenon is expected to happen in the context of quantum chromodynamics. The quark, in this case, would radiate mesons.

As a consequence of this radiation, the quark loses energy. Therefore, it represents a new mechanism of energy loss, "new" in the sense that it has not been studied before. But, because the plasma is a strongly-coupled phase of QCD, its theoretical study remains a challenge. There are no systematic methods with which we can analyze its non-perturbative properties. This is why the gauge/string duality, so-called AdS/CFT, is a useful tool to treat the problem, providing at the same time a theoretical understanding. In this case, it will allow us to identify the mechanism of this loss of energy.

3. String theory background

In the dual string theory, a deconfined quark corresponds to an open string stretching between D-branes, in the 10-dimensional spacetime. D-branes are solitonic solutions of the equations of

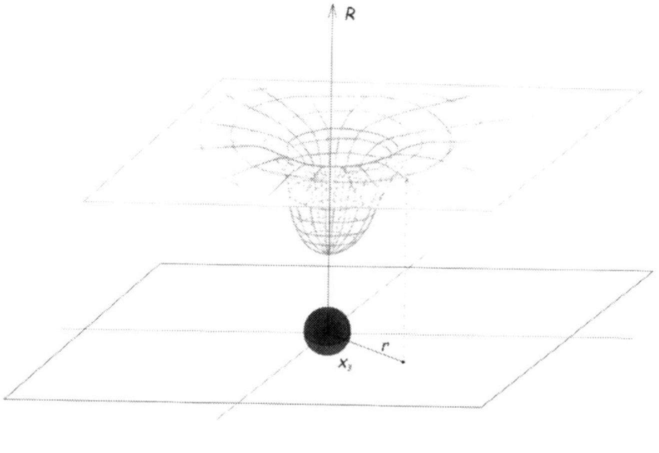

Figure 1. Affected by the black branes, the D7-branes are bent towards the horizon.

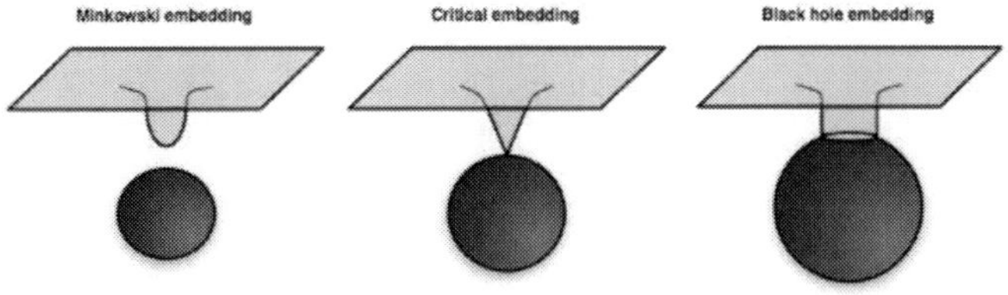

Figure 2. Various probe brane configurations.

supergravity. And to be precise, our framework involves N_c coincident black D3-branes and also N_f coincident D7-branes. The D3-branes are surrounded by a horizon, like a black hole. This is required because we are interested in studying the plasma at temperature different from zero, and the temperature of the dual gauge theory corresponds to the Hawking temperature of this black hole (By considering $T > 0$ we will extend the analysis considered in [1]).

The metric sourced by the black branes is asymptotically flat, but in the neighboring region it develops a throat geometry described by 5-dimensional Anti-DeSitter spacetime (plus a 5-sphere at each point: $AdS_5 \times S^5$). And, since the number of flavors is considered to be much smaller than the number of colors, $N_f << N_c$, the D7-branes can be considered as probes in this geometry, neglecting their gravitational effect. Besides, the D7-branes, so-called flavor branes, introduce new degrees of freedom in the theory, corresponding to strings stretching between them and the D3, which would be matter in the fundamental representation in the dual gauge theory (such as quarks). And also, there are open strings with both ends on the flavor branes, which correspond to bound states of quarks and antiquarks; mesons.

These open string excitations are represented by fields, which live on the worldvolume of the branes, and their spectrum can be found by computing the brane fluctuations. The full spectrum of meson excitations includes scalar and vector modes. These meson fields will couple to the external quark, that is why they are interesting for us.

I will not go into the details of the calculations here , but they involve the embedding of the D7-branes which bend towards the black hole (see figure 1), and the corresponding induced metric. Physically, the scalar fields have a geometrical interpretation, in terms of the

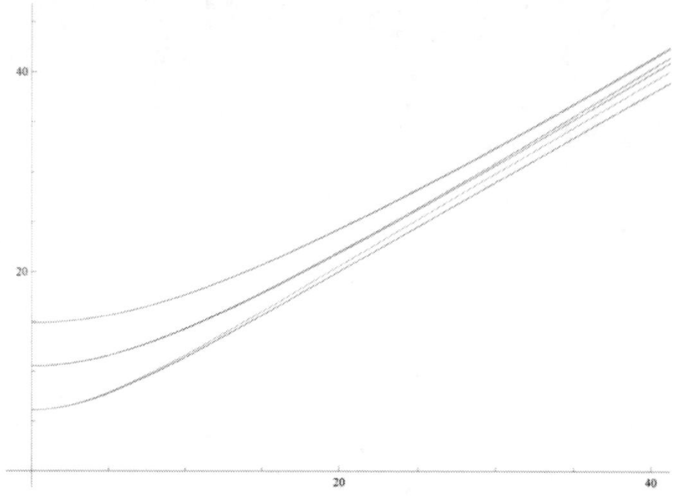

Figure 3. Dispersion relations $\omega(q)$.

deformations of the shape of the branes along the transverse directions.

4. Mesons on the brane

First, we extremize the action of the brane to find the solution for the embedding. The solutions fall into two classes: "Black hole" embeddings, if the brane falls into the horizon, and "Minkowski" embeddings, if the brane closes-off smoothly outside (see figure 2). The presence of one or the other depends on the size of the horizon, which is related to the temperature. In the dual theory, this corresponds to a phase transition exemplified by discontinuities in some observable quantities. For instance, in the Minkowski phase, the mesons are stable and the spectrum is discrete with a mass gap; whereas in the black hole phase, stable mesons cease to exist and instead we find a continuous gapless spectrum of excitations. Hence the phase transition is characterized by the dissociation or "melting" of the mesons (see the discussion in [2]).

In the next step, we consider fluctuations around this background embedding, and obtain the linearized equations of motion for the scalar fields. In the case of vector mesons, this is a little more complicated because we need to consider the contribution of non-Abelian gauge fields. For each choice of the three-momentum $\vec{q}$, we tune the frequency ω in order to find solutions that vanish asymptotically. Thus, by selecting normalizable solutions, we can find the **dispersion relations** for the modes of the meson fields (see figure 3).

For the vector fields, there is a decomposition into two transverse modes and one longitudinal mode. Each one has a different dispersion relation, although they converge to the same value at zero momentum, because in that case there is no distinction between these modes. There is a set of normalizable solutions, indexed by n. We can also solve numerically the equations and plot their **radial profiles** (see figure 4). Note that the profiles of $n > 0$ cross the axis, and the number of crossings is precisely n.

And notice also that the profiles corresponding to greater q are more peaked around the origin. The intuitive explanation is that greater frequency means greater energy, and the gravitational pull of the black hole has a greater effect over those more energetic mesons. The most important characteristic of these plots is that examining the regime of large q, we get a linear relation, and

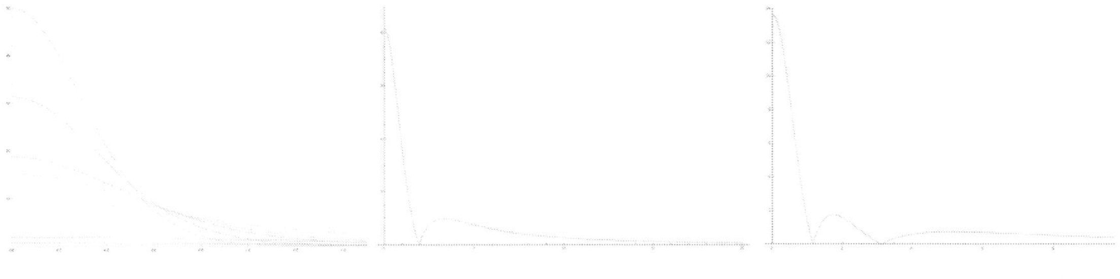

Figure 4. Radial profiles of the modes. From left to right, $n = 0, 1, 2$.

the slope is the same for all the modes. Its value is given by the local speed of light, which is affected by the redshift of the black hole:

$$c = \sqrt{-\frac{g_{tt}}{g_{xx}}} \tag{1}$$

The intuitive explanation is that the radial profiles of the mesonic fields are very peaked around the origin, which is the point of maximum curvature of the embedding of the D7-branes. So we can roughly think of them as excitations propagating along the bottom of the branes, and the speed of these signals is therefore set by the speed of light. Actually, this is a clear and beautiful result that supports the physics behind the duality: Basically, the phase velocity of the mesons is given by the redshift of a black hole.

5. Phenomenology

The mesonic Cherenkov radiation will be produced when the quark exceeds this limiting velocity of the mesons. Just in the same way as its classical counterpart were produced when the electron surpassed the limiting velocity of light.

Hopefully, we will be able to study the coupling between the quark and the meson fields and calculate the energy loss of the quark. This is work in progress, recent results can be found in [3]. The aim is to obtain results which might be contrasted with results from real experiments.

Although we should draw into attention that the correspondence we are working with relates the 10-dimensional physics to $\mathcal{N} = 4$ SYM gauge theory, not QCD. However, finite temperature breaks both the supersymmetry and the conformal invariance of the SYM theory, and this encourages us to think that some properties of the SYM plasma may be shared by the real-life QCD plasma.

References

[1] Kruczenski M, Mateos D, Myers R C and Winters D J 2003 Meson spectroscopy in AdS/CFT with flavour *J. High Energy Phys.* JHEP07(2003)049

[2] Mateos D, Myers R C and Thomson R M 2007 Thermodynamics of the Brane *J. High Energy Phys.* JHEP05(2007)067

[3] Casalderrey-Solana J, Fernández D and Mateos D 2010 A New Mechanism of Quark Energy Loss *Preprint* arXiv:0912.3717v2

Equation of state and singularities in FLRW cosmological models

Leonardo Fernández-Jambrina[1]* and Ruth Lazkoz[2]

[1]ETSI Navales, Universidad Politécnica de Madrid, Arco de la Victoria s/n, 28040-Madrid, Spain

[2]Física Teórica, Facultad de Ciencia y Tecnología, Universidad del País Vasco, Apdo. 644, E-48080 Bilbao, Spain

E-mail: leonardo.fernandez@upm.es, ruth.lazkoz@ehu.es

Abstract. We consider FLRW cosmological models with standard Friedmann equations, but leaving free the equation of state. We assume that the dark energy content of the universe is encoded in an equation of state $p = f(\rho)$, which is expressed with most generality in the form of a power expansion. The inclusion of this expansion in Friedmann equations allows us to construct a perturbative solution and to relate the coefficients of the equation of state with the formation of singularities of different types.

1. Introduction

Since the end of the 20th century there has been an increase of evidence for an accelerated expansion of the universe [1]. These puzzling observations do not fully fit in our present theoretical framework and therefore there are two trends of development attempting to amend it.

One trend is grounded on modifications of general relativity as the correct theory of gravity [2]. The other major trend assumes the validity of general relativity and postulates the existence of an exotic component in the content of the universe known as dark energy [3].

So far the only final state for our universe could be either infinite expansion or a final collapse, the big crunch. But these new scenarios open the possibility of a different final fate, since both violate some of the classical energy conditions [4]. New candidates for a final fate of our universe are the big rip [5], sudden singularities[6], big brake [7], big freeze [8], inaccessible singularities [9], directional singularities [10], w-singularities [11], in braneworld models [12], among others.

In [13] we dealt with such singular scenarios from the point of view of geodesic completeness in modified theories of gravity. In this talk we do it following the approach of a dark energy fluid. More details can be found in [14].

2. Equations of state

Instead of restricting to a specific model, we propose a framework which may comprise most of them. With this aim in mind, we assume the validity of standard Friedmann equations,

$$H^2 = \rho, \qquad \dot{\rho} + 3H(\rho + p) = 0, \tag{1}$$

assuming that pressure p has a power series in density ρ of the matter content around a value ρ_*, for which a qualitative change of behaviour is expected,

$$p = p_0(\rho - \rho_*)^{\alpha_0} + p_1(\rho - \rho_*)^{\alpha_1} + \cdots, \qquad (2)$$

where the exponents are real and ordered, $\alpha_0 < \alpha_1 < \cdots$.

Combining cosmological equations with our power expansion, we get

$$\dot\rho = -3\sqrt{\rho_*}\left(\rho_* + \frac{3}{2}(\rho - \rho_*) + \frac{3}{8}\frac{(\rho-\rho_*)^2}{\rho_*} + \cdots + p_0(\rho - \rho_*)^{\alpha_0} + p_1(\rho - \rho_*)^{\alpha_1} + \cdots\right). \qquad (3)$$

We may integrate it for ρ in powers of $(t_0 - t)$, where t_0 satisfies $\rho_* = \rho(t_0)$, if it exists,

$$\rho(t) = \rho_* + \rho_1(t_0 - t)^{\beta_1} + \rho_2(t_0 - t)^{\beta_2} + \cdots.$$

In order to compare our results with previous ones in terms of expansions of the scale factor of the universe [15],

$$a(t) = a(t) = c_0|t - t_0|^{\eta_0} + c_1|t - t_1|^{\eta_1} + \cdots, \qquad (4)$$

we are to relate both series through Friedmann equation,

$$\rho = \frac{\eta_0^2}{(t-t_0)^2} + \frac{2c_1}{c_0}\eta_0(\eta_1 - \eta_0)(t_0 - t)^{\eta_1 - \eta_0 - 2} + \frac{2c_2}{c_0}\eta_0(\eta_2 - \eta_0)(t_0 - t)^{\eta_2 - \eta_0 - 2} + \cdots,$$

These expansions cope with almost every cosmological model, except those with oscillatory behaviour as the ones in [16]. The results are consigned in table 1.

Table 1. Expansions of density and scale factor.

ρ_*	β_1	β_2	η_0	η_1	η_2
0	$(-\infty, -2)$	(β_1, ∞)	-	-	-
0	-2	$(-2, \infty)$	$\pm\sqrt{\rho_1}$	(η_0, ∞)	(η_1, ∞)
0	$(-2, 0) \cup (0, \infty)$	(β_1, ∞)	0	$\frac{\beta_1+2}{2}$	(η_1, ∞)
$\neq 0$	$(0, 1)$	(β_1, ∞)	0	1	$(1, 2)$
$\neq 0$	1	$(1, \infty)$	0	1	$[2, \infty)$
$\neq 0$	$(1, \infty)$	(β_1, ∞)	0	1	2

3. Classification of singularities

We follow the popular classification of singularities in [4]:

- Big bang / crunch: zero a, divergent H, density and pressure.
- Type I: "Big rip": divergent a. [5]
- Type II: "Sudden": finite a, H, density, divergent $\dot H$ and pressure. [6]
- Type III: "Big freeze": finite a, divergent H, density and pressure. [8]
- Type IV: "Big brake": finite a, H, $\dot H$, density, pressure, divergent higher derivatives. [7]

In [17] a detailed analysis of the strength of these singularities is performed, making use of the concepts of geodesic completeness [18] and the definitions of strong singularities due to Ellis, Schmidt, Tipler and Królak [19]. Sudden and big brake singularities cannot be considered as such [20], since the cosmological spacetime may be continued beyond the singular event. And particles travelling at the speed of light do not experience the big rip, since they need infinite proper time to reach it [17]. These results may be checked in table 2 in terms of the exponents of the expansion of the density. Conformal diagrams for these models can be found in [21].

Table 2. Singularities and density expansions.

ρ_*	β_1	**Tipler**	**Królak**	**N.O.T.**
No	$(-\infty, -2)$	Strong	Strong	Big crunch/rip
No	$(-2, 0)$	Weak	Strong	III
No	$(0, 2)$	Weak	Weak	II
No	$[2, \infty)$	Weak	Weak	IV
Yes	$(0, 1)$	Weak	Weak	II
Yes	$[1, \infty)$	Weak	Weak	IV

4. Singularities and equations of state

Now we have all the ingredients for checking the formation of singularities in cosmological models with an equation of state of the form (2). We may integrate (3) in terms of power series of time and get the relevant exponents and coefficients. Comparing them with the results in table 2, we are ready to obtain the types of singularities which may appear in these models. The results may be found in table 3.

Table 3. Singularities and equation of state.

ρ_*	p_0	α_0	β_1	**N.O.T.**
0	Any	$(-\infty, 0)$	$2/(1-2\alpha_0)$	II
0	Any	$[0, 1/2)$	$2/(1-2\alpha_0)$	IV
Yes	Any	$(-\infty, 0)$	$(1-\alpha_0)^{-1}$	II
Yes	$\neq -\rho_*$	0	1	IV
Yes	$-\rho_*$	0	$1/(1-\alpha_1)$	IV
Yes	Any	$(0, \infty)$	1	IV

5. Conclusions

In this talk a detailed classification of the future behaviour of FLRW cosmologies in terms of the equation of state is provided. It depends at most on the exponents of the first two terms of a power expansion of the equation of state. The scheme provides an easy route to conclude the sort of singular behaviour.

Acknowledgments

R.L. is supported by the University of the Basque Country through research grant GIU06/37 and by the Spanish Ministry of Education and Culture through research grant FIS2007-61800.

References

[1] Riess A G *et al* 1998 *Astron. J.* **116** 1009
Perlmutter S *et al* 1999 *Astrophys. J.* **517** 565
Davis T M *et al* 2007 *Astrophys. J.* **666** 716
Wood-Vasey W M *et al* 2007 *Astrophys. J.* **666** 694
Spergel D N *et al.* 2003 *Astrophys. J. Suppl.* **148** 175

Spergel D N *et al* 2007 *Astrophys. J. Suppl.* **170** 377
[2] Maartens R 2007 *J. Phys.: Conf. Ser.* **68** 012046
Durrer R and Maartens R 2008 *Gen. Rel. Grav.* **40** 301
Capozziello S and Francaviglia M 2008 *Gen. Rel. Grav.* **40** 357
Nojiri S and Odintsov S D 2007 *Int. J. Geom. Meth. Mod. Phys.* **4** 115
Starobinsky A A 2007 *JETP Lett.* **86** 157
Hu W and Sawicki I 2007 *Phys. Rev. D* **76** 064004
Appleby S A and Battye R A 2007 *Phys. Lett. B* **654** 7
Tsujikawa S 2008 *Phys. Rev. D* **77** 023507
Nojiri S and Odintsov S D 2007 *Phys. Lett.* B **652** 343
Nojiri S and Odintsov S D 2007 *Phys. Lett.* B **657** 238
Nojiri S and Odintsov S D 2008 *Phys. Rev.* D **77** 026007
Capozziello S, De Laurentis M, Nojiri S and Odintsov S D 2009 *Phys. Rev.* D **79** 124007
Bamba K, Geng C, Nojiri S and Odintsov S D 2009 *Phys. Rev.* D **79** 083014
Bamba K, Odintsov S D, Sebastiani L and Zerbini S *Preprint* arXiv:0911.4390
Bamba K *Preprint* arxiv:/0909.2991
Nojiri S and Odintsov S D *Preprint* arXiv:0911.2781
[3] Padmanabhan T 2006 *AIP Conf. Proc.* **861** 179
Albrecht A *et al Preprint* astro-ph/0609591
Sahni V and Starobinsky A 2006 *Int. J. Mod. Phys. D* **15** 2105
[4] Nojiri S, Odintsov S D and Tsujikawa S 2005 *Phys. Rev.* D **71** 063004
[5] Caldwell R R, Kamionkowski M, Weinberg N N 2003 *Phys. Rev. Lett.* **91** 071301
[6] Barrow J D 2004 *Class. Quantum Grav.* **21** L79
Nojiri S and Odintsov S D 2004 *Phys. Lett.* B **595** 1
Barrow J D 2004 *Class. Quantum Grav.* **21** 5619
Lake K 2004 *Class. Quantum Grav.* **21** L129
Nojiri S and Odintsov S D 2004 *Phys. Rev.* D **70** 103522
Dąbrowski M P 2005 *Phys. Rev.* D **71** 103505
Barrow J D and Tsagas C G 2005 *Class. Quantum Grav.* **22** 1563
Chimento L P and Lazkoz R 2004 *Mod. Phys. Lett.* A **19** 2479
Dąbrowski M P 2005 *Phys. Lett.* B **625** 184
Balcerzak A and Dąbrowski M P 2006 *Phys. Rev.* D **73** 101301
Barrow J D, Batista A B, Fabris J C and Houndjo S 2008 *Phys. Rev.* D **78** 123508
Barrow J D and Lip S Z W 2009 *Phys. Rev.* D **80** 043518
Nojiri S and Odintsov S D 2008 *Phys. Rev.* D **78** 046006
Setare M R and Saridakis E N 2009 *Phys. Lett.* B **671** 331
[7] Gorini V, Kamenshchik A Y, Moschella U and Pasquier V 2004 *Phys. Rev.* D **69**, 123512
[8] Bouhmadi-López M, Gonzalez-Díaz P F and Martín-Moruno P 2008 *Phys. Lett.* B **659** 1
[9] McInnes B 2007 *Class. Quantum Grav.* **24** 1605
[10] Fernández-Jambrina L 2007 *Phys. Lett.* B **656** 9 (*Preprint* arXiv:0704.3936)
[11] Dabrowski M P and Denkiewicz T 2009 *Phys. Rev.* D **79** 063521
Fernández-Jambrina L *(in preparation)*
[12] Shtanov Y and Sahni V 2002 *Class. Quantum Grav.* **19** L101
Brown R A, Maartens R, Papantonopoulos E and Zamarias V 2005 *J. Cosmol. Astropart. P.* JCAP11(2005)008
Tretyakov P, Toporensky A, Shtanov Y and Sahni V 2006 *Class. Quantum Grav.* **23**, 3259
[13] Fernández-Jambrina L and Lazkoz R 2009 *Phys. Lett.* B **670** 254-258 (*Preprint* arXiv:0805.2284)
[14] Fernández-Jambrina L and Lazkoz R *(in preparation)*
[15] Cattoën C and Visser M 2005 *Class. Quantum Grav.* **22** 4913
Cattoën C 2006 MSc thesis (Victoria University of Wellington)
[16] Cailleteau T, Cardoso A, Vandersloot K and Wands D *Preprint* arXiv: 0808.0190v2
[17] Fernández-Jambrina L and Lazkoz R 2004 *Phys. Rev.* D **70** 121503 (*Preprint* gr-qc/0410124)
[18] Hawking S W and Ellis G F R 1973 *The Large Scale Structure of Space-time* (Cambridge University Press)
[19] Ellis G F R and Schmidt B G 1977 *Gen. Rel. Grav.* **8** 915
Tipler F J 1977 *Phys. Lett.* A **64** 8
Królak A 1986 *Class. Quantum Grav.* **3** 267
Clarke C J S and Królak A 1985 *Journ. Geom. Phys.* **2** 127
[20] Fernández-Jambrina L and Lazkoz R 2006 *Phys. Rev.* D **74** 064030 (*Preprint* gr-qc/0607073)
[21] Hohn P A and Scott S M 2009 *Class. Quantum Grav.* **26** 035019

Spinor calculus on 5-dimensional spacetimes

Alfonso García-Parrado Gómez-Lobo[1]* **and José M Martín-García**[2,3]

[1] Centro de Matemática, Universidade do Minho. 4710-057 Braga, Portugal
[2] Laboratoire Univers et Théories, Observatoire de Paris, CNRS, Univ. Paris Diderot, 5 place Jules Janssen, 92190 Meudon, France
[3] Institut d'Astrophysique de Paris, Univ. Pierre et Marie Curie, CNRS 98bis boulevard Arago, 75014 Paris, France

E-mail: `alfonso@math.uminho.pt`, `jose.martin-garcia@obspm.fr`

Abstract. We explain how Penrose's spinor calculus of 4-dimensional Lorentzian geometry is implemented in a 5-dimensional Lorentzian manifold. A number of issues, such as the essential spin algebra, the spin covariant derivative and the algebro-differential properties of the curvature spinors are discussed .

1. Introduction

Recently, the study of General Relativity in dimensions higher than four has experienced much progress. Specific fields which have been the subject of interest are the discovery and classification of new exact solutions [3], the generalisation of the Petrov classification [1] and the Newman-Penrose formalism [6, 9] or the study of isolated bodies. When the spacetime dimension is five one may expect that the study of these subjects become simpler than when one works in generic dimension N. Indeed, there are important discoveries which took place first when the spacetime dimension was assumed to be five. Perhaps the best known example is the discovery of the *black ring* solution [2] which led to the realisation that the stationary black hole uniqueness theorems known in four dimensions are not generalisable to higher dimensions. In this work we contribute to this trend by showing how the spinor calculus of Penrose [7] which was developed on a four-dimensional spacetime is formulated in a spacetime of five dimensions.

2. Spinor algebra

Let $\mathbf{L}$ be a 5-dimensional real vector space endowed with a real scalar product $g(\ ,\ )$ of Lorentzian signature (signature convention $(+,-,-,-,-)$) and let $\mathbf{S}$ be a complex vector space whose dimension is for the moment left unspecified (complex conjugate of scalars will be denoted by an overbar). Using the vector space $\mathbf{L}$ and its dual $\mathbf{L}^*$ as the starting point one builds a tensor algebra in the standard fashion. Similarly a tensor algebra is built from $\mathbf{S}$ and its dual $\mathbf{S}^*$. We denote these algebras by $\mathfrak{T}(\mathbf{L})$ and $\mathfrak{T}(\mathbf{S})$ respectively [1]. In this work abstract indices will be used throughout to denote tensorial quantities: in this way small Latin indices $a, b, \ldots$ will denote abstract indices on elements of $\mathfrak{T}(\mathbf{L})$ and capital Latin indices $A, B, \ldots$ will be used for abstract

[1] Strictly speaking only the algebras $\mathfrak{T}^r_s(\mathbf{L})$ of tensors r-contravariant s-covariant can be defined (and the same applies to $\mathfrak{T}^r_s(\mathbf{S})$). To lessen the notation we will suppress the labels r, s in the notation and they will only be made explicit when confusion may arise.

indices of elements in $\mathfrak{T}(\mathbf{S})$. The tensor algebra $\mathfrak{T}(\mathbf{S})$ will be referred to as the *spin algebra* and its elements will be called spinors. One can also build tensor algebras by taking tensor products of elements in $\mathfrak{T}(\mathbf{L})$ and elements in $\mathfrak{T}(\mathbf{S})$. Quantities in these tensor algebras will be referred to as *mixed tensors* and they will carry abstract indices of both types. The algebras $\mathfrak{T}(\mathbf{L})$ and $\mathfrak{T}(\mathbf{S})$ shall be regarded as complex vector spaces.

Our departure point is a mixed tensor $\gamma_{aB}{}^{C}$ fulfilling the algebraic property

$$\gamma_{aA}{}^{B}\gamma_{bB}{}^{C} + \gamma_{bA}{}^{B}\gamma_{aB}{}^{C} = -\delta_{A}{}^{C} g_{ab}, \tag{1}$$

where $\delta_{A}{}^{C}$ is the identity tensor (also known as the *Kronecker delta*) on the vector space $\mathbf{S}$ (note that each tensorial quantity has a number of slots arranged sequentially on which the indices are filled). Eq. (1) means that $\gamma_{aA}{}^{B}$ belongs to a representation on the vector space $\mathbf{S}$ of the *Clifford algebra* $Cl(\mathbf{L}, g)$. When this representation turns out to be irreducible then there are strong constraints on the vector space $\mathbf{S}$ which we summarise in the following result.

Theorem 1. *If the quantity $\gamma_{aB}{}^{C}$ belongs to an irreducible representation of $Cl(\mathbf{L}, g)$, then the dimension of $\mathbf{S}$ is 4 and there exist two antisymmetric spinors ϵ_{AB}, $\widehat{\epsilon}^{AB}$, unique up to a constant, such that*

$$\epsilon_{AB}\widehat{\epsilon}^{CB} = \delta_{A}{}^{C}. \tag{2}$$

These antisymmetric spinors are related to $\gamma_{aB}{}^{C}$ by means of the algebraic property

$$\gamma_{aD}{}^{A}\gamma^{a}{}_{C}{}^{B} = \frac{1}{2}\delta_{D}{}^{A}\delta_{C}{}^{B} - \delta_{C}{}^{A}\delta_{D}{}^{B} + \epsilon_{CD}\widehat{\epsilon}^{AB}. \tag{3}$$

This theorem is a particular case of a more general result described in appendix B of [8]. Another important consequence of the irreducibility of the Clifford algebra representation is that the quantity $\gamma_{aA}{}^{A}$ must vanish, for otherwise the 1-form $\gamma_{aA}{}^{A}$ would be invariant under the action of any endomorphism of $\mathbf{L}$ keeping g_{ab} invariant (orthogonal group), and this can only happen for scalars and 5-forms [8].

Previous ideas lead naturally to the concept of *spin structure*.

Definition 1. *Under the conditions stated in theorem 1 we will refer to $\gamma_{aA}{}^{B}$ as a spin structure on $\mathbf{L}$. The complex vector space $\mathbf{S}$ is then called the spin space of the spin structure.*

The spinors ϵ_{AB} and $\widehat{\epsilon}^{AB}$ can be regarded as a metric tensor and its inverse in the vector space $\mathbf{S}$ and therefore they can be used to raise and lower spinorial indices. Since they are antisymmetric quantities we must pay close attention to the conventions introduced for these operations, which in our case are

$$\xi^{A}\epsilon_{AB} = \xi_{B}\,, \quad \xi^{A} = \widehat{\epsilon}^{AB}\xi_{B}.$$

In particular we can raise the indices of ϵ_{AB} getting $\epsilon^{AB} = \widehat{\epsilon}^{AB}$ and from now on, only the symbol ϵ will be used for the symplectic metric and its inverse. Note also the property

$$\delta^{A}{}_{B} = -\delta_{B}{}^{A}. \tag{4}$$

Here the quantity $\delta_{B}{}^{A}$ is the *Kronecker delta* on $\mathbf{S}$ and $\delta^{A}{}_{B}$ is a derived quantity obtained from it by the raising and lowering of indices It is possible to use the quantity $\gamma^{a}{}_{AB}$ to relate spinors to tensors and back. For example for a vector v^{a}, its spinor counterpart is given by

$$v^{AB} = \gamma_{a}{}^{AB}v^{a}.$$

A consequence of (3) is that $\gamma^{a}{}_{[AB]} = \gamma^{a}{}_{AB}$ which means that v^{AB} is anti-symmetric and also traceless, because $\gamma_{aA}{}^{A} = 0$. Reciprocally, any antisymmetric and traceless spinor ξ^{AB} has a unique vector counterpart given by $\xi^{AB}\gamma^{a}{}_{AB}$. It is straightforward to generalise this simple example to spinors and tensors of higher rank (see [4] for the precise statement).

3. Spinor calculus

In this section we explain how the algebraic considerations discussed before are translated to a 5-dimensional Lorentzian manifold This requires the introduction of the concepts of *spin bundle* and *spin structure*. Once these are defined we will go on to show how a calculus involving spinors is developed.

3.1. Spin structures on a 5-dimensional Lorentzian manifold

Suppose that $(\mathcal{M}, g)$ is a 5-dimensional Lorentzian manifold and let $T_p(\mathcal{M})$ be the tangent space at a point p. This is a vector space which can be endowed with the Lorentzian scalar product $g(\ ,\ )|_p$. Therefore the vector space $T_p(\mathcal{M})$ has properties similar to $\mathbf{L}$ and we can introduce a spin space $\mathbf{S}_p$ and a spin structure $\gamma_{aA}{}^{B}|_p$ at each point p.

Definition 2 (Spin bundle). *The union*

$$S(\mathcal{M}) \equiv \bigcup_{p\in\mathcal{M}} \mathbf{S}_p, \tag{5}$$

is a vector bundle with the manifold $\mathcal{M}$ as the base space and the group of linear transformations on $\mathbb{C}^4$ as the structure group. We will call this vector bundle the spin bundle *and the sections of $S(\mathcal{M})$ are the contravariant rank-1 spinor fields on $\mathcal{M}$.*

One can now formulate a similar definition starting from spin spaces of spinors of arbitrary rank and we will use the name "spin bundle" for any of the vector bundles constructed in this way. Also it is possible to construct new vector bundles by combining spin bundles and vector bundles of tensors, using the tensor product operation. We will use the symbol $\mathfrak{S}(\mathcal{M})$ to denote generically any of these vector bundles.

Definition 3 (Spin structure on a 5-dimensional manifold). *If the quantity $\gamma_{aA}{}^{B}|_p$ varies smoothly on the manifold $\mathcal{M}$, then one can define a smooth section on the bundle $\mathfrak{S}(\mathcal{M})$, denoted by $\gamma_{aA}{}^{B}$. We call this smooth section a smooth spin structure on the Lorentzian manifold $(\mathcal{M}, g)$.*

A necessary and sufficient condition for the existence of a spin structure is that the second Stiefel-Whitney class of $\mathcal{M}$ vanishes (see e.g. [5]).

3.2. Spin covariant derivative

One can introduce a covariant derivative in the bundle $\mathfrak{S}(\mathcal{M})$ by following the standard axioms which define a covariant derivative in any vector bundle.

Definition 4 (Spin covariant derivative). *Suppose that $\mathfrak{S}(\mathcal{M})$ admits a spin structure $\gamma_{aA}{}^{B}$. We say that a torsion-free covariant derivative D_a defined on $\mathfrak{S}(\mathcal{M})$ is compatible with the spin structure $\gamma_{aA}{}^{B}$ if it fulfils the property*

$$D_a \gamma_{bC}{}^{D} = 0. \tag{6}$$

The covariant derivative D_a is then called a spin covariant derivative with respect to the spin structure $\gamma_{aA}{}^{B}$.

Elementary properties of a spin covariant derivative are $D_a g_{bc} = 0$ and $D_a \epsilon_{AB} = \epsilon_{AB} D_a Y$, for some scalar function Y. The first property means that D_a is just the Levi-Civita covariant derivative when acting on tensors. From these properties is clear that a spin covariant derivative with respect to a given spin structure is not unique unless additional restrictions are imposed. The next result, proven in [4] spells out which these restrictions are.

Theorem 2. *There is one and only one spin covariant derivative ∇_a on $\mathfrak{S}(\mathcal{M})$ with respect to the spin structure $\gamma_{aA}{}^B$ which fulfils the property*

$$\nabla_a \epsilon_{AB} = 0. \tag{7}$$

From now on we will work with the spin covariant derivative described in theorem 2. Some spinor equations adopt a simpler form when written in terms of the differential operator $\nabla_{AB} \equiv \gamma^a{}_{AB}\nabla_a$.

3.3. The curvature spinors

The curvature of ∇_a has two contributions: one coming from its role as a connection in the spin bundle and another from its role as a connection in the tensor bundle. The latter is just the Riemann tensor of the Levi-Civita covariant derivative whereas the former is characterised by a certain set of *irreducible* spinors called *curvature spinors*. The Riemann tensor and the curvature spinors are related as shown in the following theorem, proven in [4].

Theorem 3. *Define the quantity $G^{ab}{}_{AC} \equiv -\gamma^a{}_{(A}{}^B\gamma^b{}_{C)B}$. Then the Riemann tensor R_{abcf} of the covariant derivative ∇_a can be decomposed in the form*

$$R_{abcf} = \Lambda(g_{af}g_{bc} - g_{ac}g_{bf}) - \frac{1}{2}G_{ab}{}^{AB}G_{cf}{}^{CD}\Psi_{ABCD} - G_{ab}{}^{AB}G_{cf}{}^{CD}\Omega_{ACBD}. \tag{8}$$

The quantities Λ, Ω_{ABCD} and Ψ_{ABCD} are known as the curvature spinors. They fulfil the algebraic properties

$$\Psi_{(ABCD)} = \Psi_{ABCD}\,, \quad \Omega_{ABCD} = \Omega_{[AB]CD} = \Omega_{CDAB}\,, \quad \Omega_{AB}{}^C{}_C = \Omega_A{}^C{}_{CD} = 0\,,$$
$$\Omega_{ABCD} + \Omega_{BCAD} + \Omega_{CABD} = 0\,, \tag{9}$$

which correspond to the first Bianchi identity of R_{abcd}, and the differential identity

$$\nabla_{(Z}{}^W\Psi_{V)BAW} - \nabla_{(Z}{}^W\Omega_{V)ABW} - \nabla_{(Z}{}^W\Omega_{V)BAW} - 2\epsilon_{(A|(V}\nabla_{Z)|B)}\Lambda = 0\,, \tag{10}$$

which corresponds to the second Bianchi identity of R_{abcd}.

References

[1] Coley A 2008 Classification of the Weyl tensor in higher dimensions and applications *Class. Quantum Grav.* **25** 033001
[2] Emparan R and Reall H S 2002 A Rotating Black Ring Solution in Five Dimensions *Phys. Rev. Lett.* **88** 101101
[3] Emparan R and Reall H S 2008 Black Holes in Higher Dimensions *Living Rev. Rel.* **11** 6
[4] García-Parrado A and Martín-García J M 2009 Spinor calculus on 5-dimensional spacetimes *J. Math. Phys.* in press
[5] Nakahara M 1990 *Geometry Topology and Physics* (Bristol: IOP Publishing Ltd.)
[6] Ortaggio M, Pravda V and Pravdová A 2007 Ricci identities in higher dimensions *Class. Quantum Grav.* **24** 1657-1664
[7] Penrose R 1960 A spinor Approach to General Relativity *Annals of Physics* **10** 171-201
[8] Penrose R and Rindler W 1986 *Spinors and Space-Time* vol 2 (Cambridge: Cambridge University Press)
[9] Pravda V, Pravdová A, Coley A and Milson R 2004 Bianchi identities in higher dimensions *Class. Quantum Grav.* **21** 2873-2897

Journal of Physics: Conference Series **229** (2010) 012039 doi:10.1088/1742-6596/229/1/012039

Some classical and quantum aspects of cosmological braneworlds with a time-dependent extra dimension

Suman Ghosh

Department of Physics and Meteorology and Centre for Theoretical Studies, Indian Institute of Technology, Kharagpur 721 302, India

E-mail: `suman@phy.iitkgp.ernet.in`

Abstract. We discuss certain classical and quantum features of warped braneworld models with a dynamic extra dimension (analytically, wherever possible, otherwise numerically). We are able to demonstrate, that it is indeed possible to construct such generalised braneworld models with different bulk matter sources satisfying the weak energy conditions. The effect of the metric functions - the cosmological scale factor, scale of the extra dimension and the warp factor on geodesic motion are studied in detail. Distinctive features of the geodesics in different braneworld scenarios are thus highlighted. Also the issue of geodesic confinement is addressed. Particle creation, which is essentially a quantum consequence of the dynamic nature of spacetime, is investigated and how the nature of particle production on the brane depends on the nature of warping, type of cosmological evolution as well as the temporal evolution of the extra dimension is illustrated. We, in fact, have found a class of models for which the back reaction due to particle production can be controlled.

1. Introduction

Beginning with Kaluza-Klein [1], a large variety of models with extra dimensions have been proposed over the years. In the recently proposed warped braneworld models [2, 5] our world is viewed as a four dimensional sub-manifold (3-brane) embedded in a curved five dimensional spacetime where the line element on the 3-brane is scaled by a warp factor. The braneworld models seem to provide a viable resolution of the long-standing hierarchy problem in high energy physics and also suggest a dynamic way of compactification [6]. In this article, we briefly discuss few classical and quantum aspects of a so called generalisation of such braneworld scenarios. In Sec. 1 we will investigate whether those generalised braneworld models are compatible with the framework of general relativity or not [7]. In Sec. 2, we analyse the geodesic motion in such universes and address the issue of confinement of test particles [8]. In Sec. 3, we briefly discuss particle creation phenomenon and effects of warped extra dimension on the same [9].

2. Bulk spacetime with a cosmological brane

We have essentially considered the following line element as a generalisation to originally proposed Randall-Sundrum (RS) Braneworld model [3, 4]

$$ds^2 = e^{2f(\sigma)} \left[-dt^2 + a^2(t) \left(dr^2 + r^2 d\theta^2 + r^2 \sin^2\theta d\phi^2\right)\right] + b^2(t) d\sigma^2 \tag{1}$$

where, we have chosen a cosmological line element on the $\sigma = constant$ hypersurface. $a(t)$ is the cosmological scale factor, $e^{2f(\sigma)}$ is the warp factor and we now have a time-dependent function

$b(t)$ associated with the extra dimension. In general, such a function, when it is dependent on all the four coordinates (t, r, θ, ϕ), is known as the *radion* field. It has been demonstrated that for typical choices for the various metric functions, such as a growing $a(t) = \sqrt{t}$ (radiative) or e^t (de-Sitter), a stabilisable $b(t) = c_1 + c_2 e^{-t}$ and $f(\sigma) = \pm \log(\cosh \sigma)$ which represents a thick brane located at $\sigma = 0$, the above geometry does satisfy the Weak Energy Condition for specific domains of the parameters involved. This suggests the viability of these choices for physical models.

Next we look for exact solutions of Einstein equations, which gives rise to the type of geometry given by Eq.1, with matter sources of various kinds. To begin with, we look at possible matter sources that may arise if we assume an exponential (in σ) warp factor and some other typical constraints on matter stress energy. The solution space for $[a(t), b(t)]$ is analysed using a dynamical systems approach. Subsequently, we look at solutions with various types of scalar fields as sources eg. bulk normal scalar, the Brans-Dicke scalar and the dilaton scalar. In our approach to obtaining solutions we exploit the known fact that with a conformally related metric and a redefined scalar one can convert BD theory into canonical Einstein Gravity coupled to a scalar. The following table shows some of the typical analytic solutions found.

Table 1. The above table shows few of the typical solutions found and the status of the energy conditions through the energy inequality functions F_1, F_2 and F_3 (see [7] for details).

Bulk Field	$f(\sigma) = const.+$	$a(t) \sim$	$b(t) \sim$	$F_1 \geq 0$	$F_2 \geq 0$	$F_3 \geq 0$
Ordinary Scalar	$0.25 \log(\sigma - \sigma_0)$	$(t - t_0)^{0.48}$	$(t - t_0)^{-0.43}$	$\surd$	$\times$	$\surd$
Brans-Dicke	$0.37 \log(\sigma - \sigma_0)$	$(t - t_0)^{0.43}$	$(t - t_0)^{-0.55}$	$\surd$	$\times$	$\surd$
(for $\omega = 1$)	$0.075 \log(\sigma - \sigma_0)$	$(t - t_0)^{0.515}$	$(t - t_0)^{-0.32}$	$\surd$	$\surd$	$\surd$
Dilaton Scalar	$0.5 \log(\sigma - \sigma_0)$	$(t - t_0)^{0.33}$	$(t - t_0)^{-0.82}$	$\times$	$\times$	$\surd$
($\omega = -1$)	$-0.5 \log(\sigma - \sigma_0)$	$(t - t_0)^{0.57}$	$(t - t_0)^{-0.18}$	$\surd$	$\surd$	$\surd$

In the case with a Brans-Dicke scalar and matter in the bulk, it is observed that, for traceless matter energy momentum, exact and viable analytical solutions can be found with decaying or growing warp factors as well as decelerating/accelerating $a(t)$. We note that in the case of the dilaton we can have decaying as well as growing warp factor solutions too. We are also able to find desirable solutions with growing $a(t)$ and decaying $b(t)$ in several of our examples. We check the energy conditions, the nature of the energy inequality functions F_1, F_2 and F_3 and figure out whether an obtained solution is desirable or not. The details are tabulated in one of the columns in the table commenting on the desirability of the solution. We do find several classes of desirable solutions, which satisfy our requirements and the energy conditions. The bulk singularities that appear in our solutions can be resolved by placing thin branes, where the singularities occur, using proper junction conditions. One can say that those bulk singularities are in fact providing us with places for branes, one of which can be chosen as the Standard Model brane we live in.

With three different functions appearing in the line element, it is always very difficult to find exact solutions. That we have found some is indeed encouraging. We hope to use these solutions in actual brane cosmological scenarios and arrive at relevant conclusions by making use of existing observational data on supernova and CMB anisotropies. We would also like to extend our results to other types of stress energy expressions. A disturbing aspect of our solutions is that they are singular in the bulk, though we do provide ways of resolving them. It remains to be seen whether we can find newer solutions (a) which are not necessarily of a power law type, (b) which have, in appropriate limits, both deceleration and acceleration, (c) which have a decaying extra dimension stabilisable to a finite value and (d) where the bulk warp factor gives rise to a non-singular bulk metric, similar to RS. In essence, an appropriate combination of different bulk fields, with some dominating over the others in specific time intervals, will then be able to generate the expansion history (with proper decelerating and accelerating phase), of our universe on the brane.

3. Geodesics and confinement of test particles

To investigate the effects of warp factor and the dynamic nature of the bulk on the motion of test particles, we study the geodesics in the geometry given by metric (1). We have assumed the metric functions as earlier. The geodesic equations in such a background can be rewritten as a first order autonomous dynamical system. Analytical insights as well as some specific solutions can therefore be obtained. In particular, the analysis of these equations as a dynamical system shows the role of the warp factor (growing/decaying) in controlling the nature of the solutions (oscillatory/exponential). Apart from a few specific cases, the geodesic equations, in general, can only be solved numerically since they are highly coupled differential equations. However, we are able to demonstrate some qualitative features of the dynamical system through numerical investigations. It is found that, a decaying warp factor results into runaway trajectories (for both timelike and null geodesics) but in presence of a growing warp factor massive particles are constrained to be confined near the brane, while massless particles like photons or gravitons can indeed access the bulk. This observation is important in the context of localisation of fields on the brane. The behaviour of localisation of test particles in a geometry with a growing warp factor is reminiscent of fermion localisation with a growing warp factor. Nature of the effective geodesic potential also clearly indicates existence of bound or confined trajectories. Even in the presence of a decaying warp factor, particles can also be localised, though very special initial conditions are required. The effect of a dynamical (time dependent) extra dimension on the trajectories is found to be largely quantitative. The issue of confinement also puts important constraints on possible dynamic behaviour of the extra dimension.

We mention that it would be useful to investigate geodesic deviation and the kinematics of geodesic flows in a bulk geometry with a thick brane. In the latter case, we need to solve the geodesic and Raychaudhuri equations simultaneously, in order to obtain the evolution and initial condition dependencies of the kinematical quantities: expansion, rotation and shear (or ESR). We believe that such work may improve our understanding about the nature of warped extra dimensions. We hope to report on related work on the above-mentioned aspects, at a later stage.

4. Particle creation in presence of a dynamic extra dimension

Particle creation can be viewed as quantum consequence of existence of a dynamic extra dimension. We have investigated the possible effects of the presence of a warped extra dimension (static or dynamic) on particle production in a 4d universe. Following the standard convention we used the conformal time $\eta = \int \frac{dt}{a(t)}$ for this study. From the numerical analysis, it is easy to point out the main characteristic features. The warp factor plays its part through the values of the allowed modes arising because of the warped extra dimension. These modes will have discrete values in the case of a two brane model, otherwise it will be a continuum. The location and heights of the peaks in the plots for the particle number density and energy density (as a function of 4D modes) decreases in the presence of an extra dimension. This is a feature of the breaking of conformal invariance. Here we took $a^2(\eta) = b_1^2 + b_2\eta^2$ (which asymptotically reaches a radiative universe) and for the scenario with a time-varying extra dimension, we took $b^2(\eta) = \frac{b_1^2+b_2^2\eta^2}{b_3^2+b_2^2\eta^2/b_1^2}$. We note an increase in the heights of the peaks for $b(\eta, b_3 < 1)$ and a decrease for $b(\eta, b_3 > 1)$. Though these features look to be case specific and model dependent, it is worth noting that we did not choose $b(\eta)$ arbitrarily. It is chosen to be equal to b_1 (or asymptotically reaching the same value in the dynamic cases) which is the minimum value of $a(\eta)$ during its evolution. In effect we are assuming some correlation between $a(\eta)$ and $b(\eta)$, i.e. there may be a single mechanism which drives both the scale factors and therefore their ought to be a connection between them. We have also found a mechanism (for $b(\eta, b_3 > 1)$) to have control over contributions of higher KK modes to the energy density of the universe. We can suppress the production of those higher modes at our will by choosing higher and higher value of b_3 . The back reaction thus remains negligible and the classical cosmological model which is our background for all these studies does not break down. To get an idea about the thermal/nonthermal nature of the spectrum of created particles we have tried to fit our numerical data with a thermal profile with a new set of parameters. Our analysis reveals that the spectrum could be assumed to be nearly thermal though marked deviations seem to appear for larger values of 4D modes. As a next step, one should study particle creation in the context of more realistic cosmological models and also further investigate the possibility of creation of other types of particles (like fermions, or spin one particles etc.) on the brane.

We admit, in conclusion, that our results are based on a toy scenario and can only be thought of as a build-up towards the study of more relevant and realistic situations in future. However, we do believe that some of the features observed (eg. possible avoidance of significant backreaction effects and non-thermal nature of the spectral profile) here will indeed be carried over in generically similar situations and can provide pointers towards a better understanding of quantum fields in the presence of a warped extra dimension.

Acknowledgements
SG thanks IIT Kharagpur for providing financial support and Centre for Theoretical Studies, IIT Kharagpur for its research facilities.

References

[1] Kaluza T 1921 *Sitzunober. Preuss. Akad. Wiss.* Berlin 966
Klein O 1926 *Z. Phys.* **37** 895
[2] Gogberashvili G 2002 *Int. J. of Mod. Phys.* D **11** 1635
[3] Randall L and Sundrum R 1999 *Phys. Rev. Lett.* **83** 3370
[4] Randall L and Sundrum R 1999 *Phys. Rev. Lett.* **83** 4690
[5] Arkani-Hamed N *et al.* 2000 *Phys. Rev. Lett.* **84** 586
[6] Bajc B and Gabadadze G 2000 *Phys. Lett.* B **474** 282
Randjbar-Daemi S and Shaposhnikov M 2000 *Phys. Lett.* B **492** 361
Dubovsky S L, Rubakov V A and Tinyakov P G 2000 *Phys. Rev.* D **62** 105011
Grossman Y and Neubert N 2000 *Phys. Lett.* B **474** 361
Ringeval C, Peter P and Uzan J P 2002 *Phys. Rev.* D **65** 044416
Ichinose S 2002 *Phys. Rev.* D **66** 104015
Koley R and Kar S 2005 *Class. Quantum Grav.* **22** 753
[7] Ghosh S and Kar S 2009 *Phys. Rev.* D **80** 064024
[8] Ghosh S, Nandan H and Kar S 2009 Confinement of test particles in warped spacetimes *Preprint* gr-qc/0904.2321
[9] Ghosh S and Kar S 2008 *J. Cosmol. Astropart. P.* JCAP08(2008)001

Journal of Physics: Conference Series **229** (2010) 012040 doi:10.1088/1742-6596/229/1/012040

Conformastat eletrovacuum spacetimes

Guillermo A González[1]* and Raül Vera[2]

[1]Escuela de Física, Universidad Industrial de Santander, A. A. 678, Bucaramanga, Colombia
[2]Dept. Física Teórica, Universidad del País Vasco/Euskal Herriko Unibertsitatea, 644 PK, 48080 Bilbao, Basque Country, Spain

E-mail: guillego@uis.edu.co, raul.vera@ehu.es

Abstract. We present the first step towards the complete solution of the Einstein-Maxwell equations outside the sources of static spacetimes in which the space of orbits is conformally flat. Under the assumption of functional dependence between the two potentials that define the geometry and the electromagnetic field, the solution consists only of the well known class which depends on a function satisfying the Laplace equation in flat 3-space plus four explicit bi-parametric families of line-elements.

1. Conformastationary spacetimes

The line element of a stationary spacetime $(\mathcal{M}, g_{\mu\nu})$ in adapted coordinates $\{t, x^a\}$ reads [1]

$$ds^2 = -e^{2U}(dt + A_a dx^a)^2 + e^{-2U}\widehat{h}_{ab}dx^a dx^b, \tag{1}$$

where U, A_a and $\widehat{h}_{ab}$ do not depend on t. For a static spacetime $A_a = 0$. A *conformastationary* spacetime is a stationary spacetime where $(\Sigma_3, \widehat{h}_{ab})$ is conformally flat, i.e. the York tensor density vanishes [1]

$$Y_a{}^e = \hat{\eta}^{bce}\left(2\hat{\nabla}_c\widehat{R}_{ba} - \frac{1}{2}\widehat{h}_{ab}\hat{\nabla}_c\widehat{R}\right) = 0. \tag{2}$$

Here $\widehat{R}_{ab}$ and $\hat{\nabla}$ are the Ricci tensor and covariant derivative relative to $\widehat{h}_{ab}$, and $\hat{\eta}_{abc}$ is the volume form of $(\Sigma_3, \widehat{h}_{ab})$. The York tensor density satisfies $Y_{ae} = Y_{ea}$ and $Y_a{}^a = 0$. *Conformastat* spacetimes are conformastationary spacetimes which are static.

We will restrict ourselves to Maxwell fields $F_{\alpha\beta}$ for which $\mathcal{L}_\xi \boldsymbol{F} = 0$ with $\vec{\xi} \equiv \partial_t$. The Einstein-Maxwell equations outside the sources reduce to the equations (see e.g. [1, 2])

$$\widehat{R}_{ab} = G_a\overline{G}_b + \overline{G}_a G_b - (H_a\overline{H}_b + \overline{H}_a H_b), \tag{3}$$

$$\hat{\nabla}^a H_a + \frac{1}{2}\overline{G}\cdot H - \frac{3}{2}G\cdot H = 0, \tag{4}$$

$$\hat{\nabla}^a G_a - \overline{H}\cdot H - (G - \overline{G})\cdot G = 0, \tag{5}$$

for $H_a \equiv (\Re\mathcal{E} + \Phi\overline{\Phi})^{-1/2}\Phi_{,a}$ and $G_a \equiv 1/2(\Re\mathcal{E} + \Phi\overline{\Phi})^{-1}(\mathcal{E}_{,a} + 2\overline{\Phi}\Phi_{,a})$, where $\Phi(x^a)$ is the electromagnetic potential and $\mathcal{E}(x^a)$ the Ernst potential. The integrability conditions for the two complex potentials are, in fact, $\mathrm{d}H = H\wedge\Re G$ and $\mathrm{d}G = G\wedge\overline{G} + \overline{H}\wedge H$.

The metric function U and the 1-form A_a are determined by

$$e^{2U} = \Re\mathcal{E} + \Phi\overline{\Phi}, \quad \mathrm{d}A_{ab} = 2e^{-4U}\hat{\eta}_{abc}\Im G^c.$$

The electromagnetic field, $\boldsymbol{\mathcal{F}} \equiv \boldsymbol{F} + i * \boldsymbol{F}$, is recovered by

$$\boldsymbol{\mathcal{F}} = -e^{-U}\left[\boldsymbol{H}\wedge\boldsymbol{\xi} + i * (\boldsymbol{H}\wedge\boldsymbol{\xi})\right],$$

where the 1-form H_μ in $(\mathcal{M}, g_{\mu\nu})$ is given by $H_\mu = (0, H_a)$ in the above coordinates adapted to $\vec{\xi}$. Note that $\xi_\mu = -e^{2U}(1, A_a)$. The real and imaginary parts of H_μ correspond to the electric and magnetic fields with respect to the observer defined by $\boldsymbol{u} \equiv e^{-U}\boldsymbol{\xi}$, this is $H_\mu = E_\mu + iB_\mu = \mathcal{F}_{\mu\nu}u^\nu$. The intrinsic definition of G_μ in $(\mathcal{M}, g_{\mu\nu})$ is given by $G_\mu \equiv u^\nu(\nabla_\nu u_\mu + i * \nabla_\nu u_\mu)$ and its real and imaginary parts $G_\mu = a_\mu + i\frac{1}{2}w_\mu$ correspond to the acceleration and twist vectors of the congruence $\boldsymbol{u}$.

For the ***Vacuum Case*** we have $\Phi = 0$, so that $H_a = 0$ and hence the integrability condition reads simply $\mathrm{d}G = G\wedge\overline{G}$, while (3)-(5) reduce to

$$\widehat{R} = G_a\overline{G}_b + \overline{G}_aG_b, \qquad \hat{\nabla}^aG_a - (G - \overline{G})\cdot G = 0.$$

On the other hand, for the ***Static Case*** we have $G - \overline{G} = 0$. Then $\mathrm{d}G = 0$ (in fact $G_a = U_{,a}$), $\mathrm{d}H = H\wedge G$ and $H\wedge\overline{H} = 0$. This leads eventually to $\overline{H} = e^{-2i\theta}H$, with θ constant. We define $X_a \equiv e^{-i\theta}H_a$, so that $E_a = \cos\theta X_a$ and $B_a = \sin\theta X_a$, and consider the real potential $\Psi = e^{-i\theta}\Phi$. Therefore $X_a = e^{-U}\Psi_{,a}$ and equations (3)-(5) reduce to

$$\widehat{R} = 2(G_aG_b - X_aX_b), \quad \hat{\nabla}^aX_a - G\cdot X = 0, \quad \hat{\nabla}^aG_a - X\cdot X = 0. \tag{6}$$

2. A common framework

We define j satisfying $j^2 = 1$ and a "conjugation" $\breve{j} = -j$. Given any two real objects f and g we define a j-object $F = f + jg$ and its conjugate $\breve{F} = f - jg$, so that $f = 1/2(F + \breve{F})$ and $g = 1/2(F - \breve{F})$. The equation $F = 0$ holds if and only if $\breve{F} = 0$ and therefore $F = 0 \Leftrightarrow f = 0 = g$. Considering the static case, we are now ready to define

$$\Sigma_a \equiv \frac{1}{2}(G_a + jX_a),$$

in terms of which equations (6) read

$$\widehat{R}_{ab} = 4(\Sigma_a\breve{\Sigma}_b + \breve{\Sigma}_a\Sigma_b), \qquad \hat{\nabla}^a\Sigma_a - (\Sigma - \breve{\Sigma})\cdot\Sigma = 0, \tag{7}$$

and the integrability conditions satisfied by G and X now simply reduce to $\mathrm{d}\Sigma = \Sigma\wedge\breve{\Sigma}$. The equations for the hyperbolic complex Σ_a in the static case and the complex G_a in the vacuum case are analogous except for the number appearing in the corresponding equations for the Ricci tensor. We can thus set a common problem by denoting by ι any of i and j, so that $\iota^2 = \pm 1$ accordingly, a general conjugation by $\tilde{\ }$, so that $\tilde{\iota} = -\iota$, and call $f + \iota g$ a composed object. Consider then a composed vector field $\mathcal{Y}^a$ which satisfies the system of equations

$$\hat{\nabla}^a\mathcal{Y}_a - (\mathcal{Y} - \widetilde{\mathcal{Y}})\cdot\mathcal{Y} = 0, \tag{8}$$

together with the relations

$$\mathrm{d}\mathcal{Y} = \mathcal{Y}\wedge\widetilde{\mathcal{Y}}, \tag{9}$$

and the metric $\widehat{h}_{ab}$ satisfying

$$\widehat{R}_{ab} = N(\mathcal{Y}_a\widetilde{\mathcal{Y}}_b + \widetilde{\mathcal{Y}}_a\mathcal{Y}_b). \tag{10}$$

The vacuum case is recovered by taking $N = 1$, $\mathcal{Y}_a = G_a$, a complex 1-form, and the conjugate being the complex conjugate. The static case corresponds to $N = 4$, $\mathcal{Y}_a = \Sigma_a$, a j-1-form and the conjugate being the j-conjugation.

Conformastationarity demands the vanishing of the York tensor density of $\widehat{h}_{ab}$. Let us introduce first the vector

$$L \equiv \star(\mathcal{Y} \wedge \widetilde{\mathcal{Y}}), \tag{11}$$

where $\star$ denotes the Hodge-dual in $(\Sigma_3, \widehat{h}_{ab})$, i.e. $L_a = \mathcal{Y}^b \widetilde{\mathcal{Y}}^c \hat{\eta}_{bca}$. By construction we have $\widetilde{L} = -L$ and $L \cdot \mathcal{Y} = L \cdot \widetilde{\mathcal{Y}} = 0$. Introducing (10) into (2) one obtains the real equation

$$(\mathcal{Y}_a - \widetilde{\mathcal{Y}}_a)L^e + \hat{\eta}^{bce}(\widetilde{\mathcal{Y}}_b \hat{\nabla}_c \mathcal{Y}_a + \mathcal{Y}_b \hat{\nabla}_c \widetilde{\mathcal{Y}}_a) - \frac{1}{2}\widehat{h}_{ab}\hat{\eta}^{bce}\hat{\nabla}_c(\mathcal{Y} \cdot \widetilde{\mathcal{Y}}) = 0. \tag{12}$$

Since $Y_a{}^a = 0$ this equation contains at most 5 independent components. Two very different situations arise in the study of this system of equations for $\mathcal{Y}$: the class of solutions for which $L \neq 0$ and those for which $L = 0$. Nevertheless, before entering into the study of the two classes one has to consider the case $\mathcal{Y} \cdot \mathcal{Y} = 0$. In the static case $\mathcal{Y} = \Sigma$ is j-composed and $\Sigma \cdot \Sigma = 0$ implies, in particular, $G \cdot G + X \cdot X = 0$, which leaves us only with the trivial case $G = X = 0$. However, in the vacuum case $\mathcal{Y} = G$ is complex and one can have, in principle, fields for which $G \cdot G = 0$. The study of these null fields was performed in [2], where it was proven that no null conformastationary vacuum spacetimes exist apart from the trivial case of flat spacetime. We can thus take $\mathcal{Y} \cdot \mathcal{Y} \neq 0$ without loss of generality.

3. The class $L = 0$

The vanishing of L means, first, that $\mathcal{Y}$ and $\widetilde{\mathcal{Y}}$ are parallel. Secondly, because of (9) $\mathcal{Y}$ is the gradient of some composed potential. As a result, the real and "imaginary" parts of the potential must be functionally dependent. In the *vacuum* case this means that $G \wedge \overline{G} = 0$, and therefore that $\mathcal{E}_{[,a}\overline{\mathcal{E}}_{,b]} = 0$, so that $\mathcal{E} = \mathcal{E}(\overline{\mathcal{E}})$. This case was studied in [2], and the solution consists only of three explicit bi-parametric families of line-elements. In the *static* case $\mathcal{Y} \wedge \widetilde{\mathcal{Y}} = 0$ stands for $\Sigma \wedge \widetilde{\Sigma} = 0$, this is $G \wedge X = 0$, which is in turn equivalent to a functional relationship $U = U(\Psi)$. The divergence equation (8) firstly fixes this relationship to be (see e.g. [1])

$$e^{2U} = 1 - 2k\Psi + \Psi^2,$$

with $k = const$, which can be rewritten in parametric form as

$$k^2 = 1: \quad \Psi = k - 1/V, \qquad e^{2U} = V^{-2}, \tag{13}$$

$$k^2 > 1: \quad \Psi = k - \sqrt{k^2 - 1}\coth V, \quad e^{2U} = (k^2 - 1)\sinh^{-2} V, \tag{14}$$

$$k^2 < 1: \quad \Psi = k - \sqrt{1 - k^2}\cot V, \quad e^{2U} = (1 - k^2)\sin^{-2} V, \tag{15}$$

and secondly implies $\hat{\nabla}^2 V = 0$ in all cases. The Ricci equations (10) reduce now to

$$k^2 = 1: \widehat{R}_{ab} = 0, \qquad k^2 > 1: \widehat{R}_{ab} = 2V_{,a}V_{,b}, \qquad k^2 < 1: \widehat{R}_{ab} = -2V_{,a}V_{,b},$$

and the remaining equation that V_a has to satisfy is encoded in (12).

The solutions for $k^2 = 1$ correspond to the well known Majumdar-Papapetrou class of solutions, which depend on functions satisfying the Laplace equation in flat $\widehat{h}_{ab}$ 3-space [1]. The important point to note for case $k^2 > 1$ is that in the *conformastat vacuum* case $\Phi = 0$ and $G_a = u_{,a}$, so that the Einstein-Maxwell equations reduce to

$$\widehat{R}_{ab} = 2u_{,a}u_{,b}, \qquad \hat{\nabla}^2 u = 0.$$

These equations together with equation (12) for u_a constitute a problem for u and $\widehat{h}_{ab}$ which is equivalent to that for V and $\widehat{h}_{ab}$ in the *conformastat electrovacuum* case for $k^2 > 1$. Accordingly, given a conformastat vacuum solution, a conformastat electrovacuum solution with $k^2 > 1$ is obtained by means of the substitution

$$V = u, \tag{16}$$

while keeping $\widehat{h}_{ab}$. Now, the general solution of the *conformastat vacuum* problem was found in [3, 4] (also included in [2]), and consists of three explicit one-parameter families of line-elements, plus the trivial solution, the flat case. This last case has to be considered in principle, but it does not produce a different spacetime eventually. By using those three "seeds" we can construct the *conformastat electrovacuum* line-elements with $k^2 > 1$ as

$$ds^2 = -e^{2U}dt^2 + e^{-2U}\left\{dr^2 + e^{2V}r^2\left[d\vartheta^2 + f^2(\vartheta)d\varphi^2\right]\right\}, \tag{17}$$

where e^{2U} and Ψ are computed by means of expression (14) from each one of the following four possibilities

$$\begin{aligned} V(r) &= \tfrac{1}{2}\ln\tfrac{b}{r}, & f(\vartheta) &= \vartheta, \\ V(r) &= \tfrac{1}{2}\ln\left[\tfrac{r-2b}{r}\right], & f(\vartheta) &= \sin\vartheta, \\ V(r) &= \tfrac{1}{2}\ln\left[\tfrac{2b-r}{r}\right], & f(\vartheta) &= \sinh\vartheta, \end{aligned}$$

where b is a real parameter. One thus obtains three explicit bi-parametric families of line-elements for the corresponding electromagnetic potentials computed from (14). Finally, one can show that the class $k^2 < 1$ contains only one bi-parametric family, whose explicit form and study will be presented elsewhere.

4. Concluding Remarks

Conformastat electrovacuum solutions with a functional relationship $U = U(\Psi)$ consist only of the well known class of solutions which depend on a function satisfying the Laplace equation in a flat 3-space, plus four explicit bi-parametric families of line elements, three of which are presented above. At present we are studying conformastat electrovacuum solutions without a functional relationship between the potentials U and Ψ. A preliminary result is that the spacetime $(\mathcal{M}, g_{\mu\nu})$ admits a further spacelike isometry.

Acknowledgments

GAG wants to thank the warm hospitality of Dept. de Física Teórica, UPV/EHU, where this work was made. RV is funded by project IT-221-07 from the Basque Government, and thanks support from project FIS2007-61800 (MICINN).

References

[1] Stephani H, Kramer D, MacCallum M A H, Hoenselaers C and Herlt E 2003 *Exact solutions of Einstein's field equations* Second Edition (Cambridge: Cambridge University Press)
[2] Lukács B, Perjés Z and Sebestyén Á 1983 *Gen. Rel. Grav.* **15** 511
[3] Das A 1971 *J. Math. Phys.* **12** 1136
[4] Lukács B and Perjés Z 1982 *Phys. Lett.* A **88** 267

Dark matter halos: consitency between rotational curves and gravitational lenses

Alma X González-Morales* **and Darío Núñez**

Instituto de Ciencias Nucleares. Universidad Nacional Autńoma de México. Apdo. 70-543, CU, 04510 México D.F., México

E-mail: `alma.gonzalez@nucleares.unam.mx`, `nunez@nucleares.unam.mx`

Abstract. The main point of this work is to emphasize the fact that the estimation of the mass content of a dark matter halo must be done in a consistent way with the nature of the dark matter. In general relativity, depending on how dark matter nature is modelled, different expressions arise for the computation of the mass and the deflection of light. Here we present the comparison of the mass estimated by assuming two different dark matter models, the first one in which dark matter is modelled by a perfect fluid and the second in which it is modelled by a scalar field, but assuming they both reproduce the same rotation curve and discuss the consequences for the lensing observations.

1. Introduction

It is well known that the ΛCDM cosmology has been very successful in describing the large scale universe but at small scales there is still discrepancies when comparing the results of N-body simulations with the observations. A related issue is the fact that the inferred mass from different approaches, as dynamical analysis and gravitational lensing, presents some discrepancies [1], possibly related with the nature of dark matter. The ΛCDM paradigm usually assumes that the dark matter is a perfect fluid of negligible pressure. As dark matter nature is not known yet this assumption could be not correct. For example, in some models of scalar field dark matter pressure can play an important role [2]. So that, the formalisms used to study dark matter, e.g. gravitational lensing modelling, could be incomplete. On the other hand, it has been proposed that combined observations of rotational curves and gravitational lensing could constraint the equation of state of the dark matter [3]. This is a good approach as far as complete expressions for the determination of the gravitational potential, mass function, and lensing observables quantities are used. In the following sections we review the relation between the metric functions with the observable quantities and the validity of the Newtonian limit for dark matter models. Finally we give an example for the calculation of the mass inferred if we consider that dark matter is a perfect fluid or a scalar field.

2. Static and spherical system in general relativity

The most general static and spherically symmetric metric that can be used to describe a dark matter halo is given by

$$ds^2 = -e^{2\phi/c^2}\, dt^2 + \frac{dr^2}{1 - \frac{2Gm}{c^2 r}} + r^2\, d\Omega^2 \tag{1}$$

where, $d\Omega = d\theta^2 + \sin^2\theta\, d\varphi^2$. This metric is completely determined by the two metric functions m and ϕ which depends only on r; a prime stands for derivative respect to r. The dynamics of a test particle moving in a space-time described by this metric is given by the geodesic equations. One of the equations leads to the relation

$$\phi' = \frac{v_c^2}{r}. \tag{2}$$

for the case of massive test particles in circular orbit [4]. It is important to note that this relation is exactly the same as in Newtonian dynamics [5], but at this point we have not made any assumption on the magnitude of these quantities. Because the potential is given in terms of the circular velocity of the halo, equation (2), and this is very small compared to the speed of light it is well justified to take the weak field limit from now on, that is $\phi(r)/c^2 << 1$ and $e^{2\phi/c^2} \approx (1 + 2\phi/c^2)$.

On the other hand, we can consider unbound trajectories for photons as test particles. These trajectories are given by

$$\frac{d\varphi}{dr} = \frac{L}{r^2\sqrt{\left(1 - \frac{2Gm}{rc^2}\right)\left[E^2 e^{-2\frac{\phi}{c^2}} - \frac{L^2}{r^2}\right]}}, \tag{3}$$

where L and E are constants of motion associated to the metric [6]. If the trajectory satisfy the conditions $\varphi = \varphi(r_m)$, $r = r_m$ and $dr/d\varphi = 0$ at the moment of the closest approach, the trajectory suffers a deflection. Using these conditions the equation for the trajectory can be simplified. In principle, if somehow we are able to identify the trajectory of photons, $d\varphi/dr$, directly to an observable quantity, then we can recover the mass function by solving equation (3). Unfortunately the trajectory of photons is not an observable quantity.

What is closely related to an observable quantity is the deflection angle of the light rays, which is given by the relation $\alpha(r) = 2(\varphi_m - \varphi_\infty) - \pi$. So, by integrating the simplified equation (3) we obtain the deflection angle, in terms of the mass function, m, and the impact parameter, r_m

$$\varphi_\infty - \varphi_m = -\int_\infty^{r_m} \frac{r_m\, dr}{r^2\sqrt{\left(1 - \frac{2Gm}{rc^2}\right)\left(1 + 2\frac{\phi(r_m)}{c^2} - 2\frac{\phi}{c^2} - \frac{r_m^2}{r^2}\right)}}. \tag{4}$$

The deflection angle has to satisfy the lens equation $\beta = \theta - \frac{D_{OL}}{D_{OS}}\alpha(D_{LS}\theta)$, where β is the actual position of the source, $\theta = r_m/D_{OL}$ is the observed apparent position of the image, D_{OL},D_{OS} and D_{LS} are the distances between the lens, the source and the observed are given by the lens configuration. Up to now we have just given the relations between the metric potentials in terms of two types of observations, the rotational curves and gravitational lensing. To see the influence of the dark matter on the determination of these quantities we will consider Einstein's equations for two types of dark matter modelling, the perfect fluid and the scalar field.

2.1. Einstein equations

For a perfect fluid (PF) the stress energy tensor is a function the density, ρ, and the pressure, p, of the fluid. While the scalar field (SF) stress energy tensor is a function of the potential, $V(\chi)$, and the scalar field χ. These stress energy tensors has been used to model the dark matter halos.

In terms of the component of the stress energy tensor Einstein's equations can be written as

$$\frac{4\pi G r}{c^4} T^r{}_r = \left(1 - \frac{2Gm}{c^2 r}\right)\frac{v_c^2}{rc^2} - \frac{Gm}{c^2 r^2} \tag{5}$$

$$T^r{}_r{}' + \left(\frac{1}{r}\left(\frac{v_c}{c}\right)^2 + \frac{2a}{r}\right)\left(T^r{}_r - T^t{}_t\right) = 0 \tag{6}$$

and $m'(r) = \frac{-4\pi}{c^2} r^2 T^t{}_t$ gives the density. The factor a differentiates between the two stress energy tensors, i.e. $a = 0$ for the perfect fluid and $a = 1$ for the scalar field. Note we have substituted the potential function in terms of the circular velocity.

In the weak field limit, equation (5) and (6) are combined to get the following relations between the mass and the circular velocity.

$$\left(\frac{m}{r^3}\right)' = \frac{c^2}{G}\left(\frac{v_c^2}{c^2 r^2}\right)' \qquad \text{PF} \tag{7}$$

$$(m\,r)' = -\frac{c^2}{G} r^2 \left(\frac{v_c^2}{c^2}\right)' \qquad \text{SF} \tag{8}$$

From equations (7) and (8) we see that the mass function is also determined by the circular velocity, but it has completely different expressions if the matter is modelled by a perfect fluid or by a scalar field. This means that although the gravitational potential is completely determined by the observation of the circular velocity, the mass inferred from this observation depends on the kind of matter we are assuming. In fact, equation (7), leads to the familiar Newtonian expression $v_c^2 = \frac{GM(r)}{r}$ [5], but equation (8) do not have a Newtonian counterpart.

3. Comparison between models

To get a better idea of what we are talking about, we will see an example. Rotational curves of spiral galaxies have been fitted from a wide variety of models, some based on the luminosity properties of the galaxies and some others based on the result of $N-$body simulations. We consider that there is a set of observations that are well fitted by the expression

$$v_c^2 = \frac{k}{r}\left(-\frac{r/r_0}{1 + r/r_0} + \ln\left[1 + r/r_0\right]\right) \tag{9}$$

where k is a constant associated to the mass of the galaxy and r_0 is some characteristic radius.

According to equations (7) and (8) the mass function is

$$m(r) = \frac{k}{G}\left(-\frac{r/r_0}{(1 + r/r_0)} + \ln\left[1 + r/r_0\right]\right) \qquad \text{PF} \tag{10}$$

$$\begin{aligned} m(r) &= \frac{k}{Gr/r_0}\left[\frac{1}{1 + r/r_0} - 3\left(1 + r/r_0\right)\right. \\ &+ \left. 3\ln\left[1 + r/r_0\right] + \left(1 + r/r_0\right)\ln\left[1 + r/r_0\right]\right] \qquad \text{SF} \end{aligned} \tag{11}$$

Note that equation (10) is the mass profile of the Navarro-Frenk-White model ($k = 4\pi G \rho_0 r_0^3$), not surprising because the rotational curve we chose is exactly the one predicted from the NFW density profile [7]. In the case of the scalar field dark matter, the expression for the mass results different, equation (11). In figure (1.a) we show a graphical comparison of the expressions given above where we can see that the mass can be over (under) estimated if we are not modelling dark matter correctly (dashed line for the perfect fluid and solid line for the scalar field dark matter).

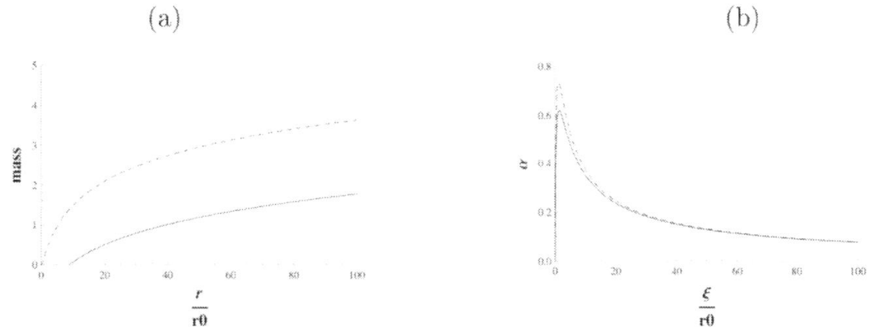

Figure 1. (a) Mass profile for the dark matter models (normalized to $k = 1$). (b) Angle of deflection for the NFW profile in the weak field and Newtonian limit. See text for explanation

4. Gravitational lensing as a consistency test

Observations of gravitational lensing are used to estimate the mass content of the dark matter distribution that causes the lens. The formalism used consider that the deflection of light caused by an extended mass distribution can be calculated from the superposition of the deflection caused by point masses, this assumptions implies that equation (4) reduces to

$$\alpha = \frac{4\,G\,M(\xi)}{c^2\,\xi} \tag{12}$$

where $M(\xi) = 2\pi \int_0^{\xi} d\xi'\xi'\Sigma(\xi')$ is the mass projected on to a plane defined by the point of closest approach [6], i.e. where de deflection occurs, being this expression the one used for study gravitational lensing. In the case of the dark matter is modelled by a perfect fluid there are no differences in using expression (4)(solid curve in figure (1.b)) or equation (12)(dashed line in figure (1.b)), except for the case where the deflection occurs at the inner regions of the halo.

When the dark matter is a scalar field then we have to look for a consistency on the mass estimated by expressions presented above and the ability of this mass distribution to reproduce the observed lensed system. For doing this it is important to compute the lensing quantities using the complete expression, equations (2),(4) and (8).

5. Conclusions

We have presented a simple way to test the nature of dark matter based on the observation of rotational curves of galaxies and its consistency with gravitational lensing observations. In Figure (1.a) should be noted that a simple scaling of the constant k for the scalar field is not enough to reconcile the results. To get good conclusions from this consistency test the most important task is to have observations of the circular velocity and gravitational lensing for the same system, and have a large number of systems so the statistics becomes relevant. Finally fig.(1.b) shows small differences for the deflection angle for deflections at the very inner regions, so that for future observations may be necessary to take into account this small differences.

References

[1] Wu X-P and Fang L-Z 1996 *Astrophys. J. L.* **467** L45
[2] Matos T, Vazquez A and Magaña J 2009 *MNRAS* **393** 1359-69
[3] Faber T and Visser M 2006 *MNRAS* **372** 136 (and references there in)
[4] Matos T, Guzmán F S and Núñez D 2000 *Phys. Rev.* D **62** 061301
[5] Binney J and Tremaine S 2008 *Galactic Dynamics: Second Edition* (Princeton University Press)
[6] Mollerach S and Rulet E 2002 *Gravitational Lensing and Microlensing* (World Scientific)
[7] Navarro J F, Frenk C S and White S D M 1997 *Astrophys. J.* **490** 493

The causal boundary and its relations with the conformal boundary

J Herrera

Departamento de Álgebra, Geometría y Topología, Facultad de Ciencias, Universidad de Málaga, Campus Teatinos, 29071 Málaga, Spain

E-mail: jherrera@agt.cie.uma.es

Abstract. Our aim in this note is to present the results (obtained in [2]) which ensure that, under certain regularity conditions, the conformal boundary becomes equal to the causal boundary, not only as a point set, but in a topological and chronological level. In particular, under these conditions the conformal boundary becomes a powerful tool to compute the causal one.

1. Preliminaries

In this section, we will introduce the basic concepts related with the causal and conformal boundaries.

1.1. *Causal boundary.*

The construction of the causal boundary (called *c-boundary for short*) will follow from the natural idea (introduced in [3]) of adding ideal points in order to: ensure that any timelike curve have an endpoint; and two timelike curves with the same past (or future) are attached the same ideal point. In order to formalize this idea, the following notions are introduced: A subset $P \subset M$ is called a *past set* if it coincides with its past; i.e, $P = I^-[P] := \{p \in M : p \ll q \text{ for some } q \in P\}$. If P cannot be expressed as the union of two proper past sets, it will be called *indecomposable past set*, IP. IPs can be classified in two different sets: $PIPs$, *proper indecomposable past sets*, which are IPs defined as the past of a point; $TIPs$, *terminal indecomposable past sets*, which are IPs defined as the past of a future-directed inextendible timelike curve. The *common past* of $S \subset M$ is defined by $\downarrow S := I^-[\{p \in M : p \ll q \quad \forall q \in S\}]$. The analogous concepts of future sets, IF, PIF, TIF and common future are obtained by interchanging the roles of past and future.

Thanks to the fact that the spacetime is strongly causal (and then, in particular, future and past distinguishing), M can be identified with the set of $PIPs$ ($PIFs$). Then, the *future causal boundary* $\hat{\partial} M$ of M will be defined as the set of all $TIPs$ in M. Finally, the *future causal completion* $\hat{M}$ will be defined as the set of all IPs:

$$M \equiv PIPs, \qquad \hat{\partial} M \equiv TIPs, \qquad \hat{M} \equiv IPs.$$

Analogously, the concepts of *past causal boundary* $\check{\partial} M$ and *past causal completion* $\check{M}$ rise naturally as:

$$M \equiv PIFs, \qquad \check{\partial} M \equiv TIFs, \qquad \check{M} \equiv IFs.$$

In order to define a (total) causal boundary, the union of future and past boundaries becomes pathological and make necessary to consider a different approach. After some unsatisfactory approaches, Marolf and Ross introduce in [4] an alternative procedure based on forming pairs with TIPs and TIFs. To formulate the construction, the following relation has to be introduced (see [5] for details): two terminal sets P, F are *S-related*, $P \sim_S F$, if P is maximal IP into $\downarrow F$ and F is maximal IF in $\uparrow P$. Then, the *c-completion* and *c-boundary* will be defined as

$$\overline{M} := \{(P,F) : P \sim_S F,\ P \text{ an IP},\ F \text{ an IF }\} \qquad \partial M := \overline{M} \setminus M$$

where M is identified with the pairs $(I^-(p), I^+(p))$ with $p \in M$ and if P (analogously for F) is not S-related with nobody, the pair $(P, \emptyset)$ (analogously $(\emptyset, F)$) is formed.

The natural way to define a chronological relation to the c-completion is the following: $(P,F),(P',F') \in \overline{M}$ are *chronologically related*, $(P,F) \ll (P',F')$, if $P' \cap F \neq \emptyset$. The topology considered will be the *chronological topology* (see [1] and [2] for details).

1.2. Conformal boundary.

In order to compute the conformal boundary of the spacetime M, we will consider a *(conformal) envelopment*, i.e, an open conformal embedding $i : M \to M_0$, where M_0 is a *("aphysical")* strongly causal spacetime. Then, we will define the conformal completion (associated to the envelopment i) as the topological adherence of the image of M, $\overline{M}_i := \overline{i(M)}$; and the conformal boundary will be defined in the natural way, $\partial_i M = \overline{M}_i \setminus i(M)$. The set of points in $\partial_i M$ which are endpoints of inextendible timelike curves of M are called the accessible part of the boundary, and it will be denoted by $\partial_i^* M$. Without lost of generality, we will suppose that $M \subset M_0$, leaving the sub-index i just in order to distinguish the conformal and causal boundaries (or completions). Finally, the topology on $\overline{M}_i$ will be the induced one from M_0 and the chronological and causal relations, denoted by $\ll_i$ and $\leq_i$ resp., will be defined in the following way: $p \ll_i q$ iff there exists some continuous curve $\gamma : [a,b] \to \overline{M}_i$ with $\gamma(a) = p, \gamma(b) = q$ such that $\gamma\mid_{(a,b)}$ is future-directed smooth timelike and contained in M; and $p \leq_i q$ iff there exist a continuous curve $\gamma : [a,b] \to \overline{M}_i$ with $\gamma(a) = p, \gamma(b) = q$ and causal in M_0.

In general, any envelopment will not generate a complete conformal completion, i.e, all inextendible timelike curve of M will not have necessarily an endpoint (in the natural sense) in M_0: for instance, consider M as the right semiplane of $\mathbb{L}^2$ in natural coordinates and, as envelopment, the identity over $\mathbb{L}^2$. In this sense, we will say that an envelopment is *chronologically complete* if any inextendible timelike curve of M (and so, a curve which generate a TIP or a TIF) have an endpoint in M_0.

Theorem 1.1 *An envelopment is chronologically complete if and only if the natural projections of the future and past causal preboundaries, $\hat{\partial} M \to \partial_i^* M, \check{\partial} M \to \partial_i^* M$, which map each TIP $P = I^-[\gamma]$ or TIF $F = I^+[\gamma]$ in the endpoint of γ, are well defined on all $\hat{\partial} M$ and $\check{\partial} M$. In this case, the natural projections of the precompletions*

$$\hat{\pi} : \hat{M} \to \overline{M}_i, \qquad \check{\pi} : \check{M} \to \overline{M}_i$$

are obviously defined.

2. Main result.

In general, we cannot expect a totally general result relating the conformal and the c-boundaries or completions (see appendix in [2] for details). In order to obtain relevant results, we have to impose some regular conditions for the envelopment. In this sense, we introduce the concept of *regular points*:

Definition 2.1 *Let $i : M \hookrightarrow M_0$ be an envelopment. We say that a continuous curve $\gamma : [a,b] \to \overline{M}_i$ such that $\gamma\mid_{[a,b)}$ is future-directed (resp. past-directed) smooth timelike curve, contained in M and such that $\gamma(b) \in \overline{M}_i$ is future (resp. past)* timelike deformable (in its endpoint), *if there exists a neighbourhood $U = U_0 \cap \overline{M}_i$ of $\gamma(b)$ (where U_0 is an open set of M_0) such that $\gamma(a) \ll_i w$ (resp. $w \ll_i \gamma(a)$) for all $w \in U$.*

A point $z \in \partial_i^* M$ is regularly accessible *if: any $TIPs$ and $TIFs$ associated to z is defined by a timelike deformable curve; and it admits a neighbourhood $V = \overline{M}_i \cap V_0$ (V_0 open in M_0) such that for any $x, x' \in \overline{M}_i$ with $x \ll_i z \leq_i x'$ (or $x \leq_i z \ll_i x'$) then $x \ll_i x'$.*

Its boundary is regularly accessible *if all the points in $\partial_i^* M$ are regularly accessible.*

Then, the following theorem is obtained.

Theorem 2.2 *Let $i : M \hookrightarrow M_0$ be a chronologically complete envelopment. If its boundary is regularly accessible then the conformal and c-completion are equivalent, i.e.:*

a) The map $\pi : \overline{M} \to \overline{M}_i^*$ is well defined and bijective.

b) π is an homeomorphism and a chronological isomorphism.

where

$$\pi : \overline{M} \to \overline{M}_i^*, \qquad \pi((P,F)) = \begin{cases} \hat{\pi}(P) & \text{if } P \neq \emptyset \\ \check{\pi}(F) & \text{if } F \neq \emptyset. \end{cases}$$

The following theorem yield sufficient conditions to ensure regular accessibility, which are very easy to check in practice. The condition of *strong accessibility* ensure that any point $z \in \partial_i^* M$ which is accessible by a future (or past) timelike curve, is accessible by a future (or past) timelike curve which is smoothly extendible to its endpoints with timelike velocity. The other condition, *causally tame* (see [2] for details), is imposed in order to avoid pathological cases which would not happen in physical examples.

Theorem 2.3 *An envelopment $i : M \hookrightarrow M_0$ with C^1 boundary, strong accessible and causally tame is regularly accessible. So, its conformal completion and c-completion are equivalent (in the sense of previous theorem).*

The following two corollaries are practical applications of previous theorem. The first result allow us to compute the c-boundary of any asymptotically flat spacetime (see [6] for definition).

Corollary 2.4 *The c-boundary of the asymptotic part of an asymptotically flat spacetime will be two lightlike cones (for the future and past resp.), which base is topologically a sphere and with vertexes i^+ and i^- respectively.*

About the second result, the implication to the right is well-known, but we are not aware of a rigorous proof for the implication to the left.

Corollary 2.5 *A spacetime which admits a chronologically complete envelopment with C^1 boundary $\partial_i M$ is globally hyperbolic if and only if $\partial_i M$ does not have timelike points.*

References

[1] Flores J L 2007 The causal boundary of spacetimes revisited *Commun. Math. Phys.* **276** 611-643
[2] Flores J L, Herrera J and Sánchez M On the final definition of the causal boundary and its relations with the conformal boundary, in progress
[3] Geroch R P, Kronheimer E H and Penrose R 1972 Ideal points in spacetime *Proc. Roy. Soc. Lond.* A **237** 545-67
[4] Marolf D, Ross S F 2003 A new recipe for causal completions *Class. Quantum Grav.* **20** 4085-4117
[5] Szabados L B 1988 Causal boundary for strongly causal spaces *Class. Quantum Grav.* **5** 121–34
[6] Wald R 1984 *General Relativity* (Chicago and London: The University of Chicago Press)

Journal of Physics: Conference Series **229** (2010) 012043 doi:10.1088/1742-6596/229/1/012043

Flat synchronizations in spherically symmetric space-times

Alicia Herrero[1]* **and Juan Antonio Morales-Lladosa**[2]

[1] Institut de Matemàtica Multidisciplinar, Universitat Politècnica de València, 46022 València, Spain

[2] Departament d'Astronomia i Astrofísica, Universitat de València, C/ Dr. Moliner 50, E-46100 Burjassot, València, Spain

E-mail: aherrero@mat.upv.es, antonio.morales@uv.es

Abstract. It is well known that the Schwarzschild space-time admits a spacelike slicing by flat instants and that the metric is regular at the horizon in the associated adapted coordinates (Painlevé-Gullstrand metric form). We consider this type of flat slicings in an arbitrary spherically symmetric space-time. The condition ensuring its existence is analyzed, and then, we prove that, for any spherically symmetric flat slicing, the densities of the Weinberg momenta vanish. Finally, we deduce the Schwarzschild solution in the extended Painlevé-Gullstrand-Lemaître metric form by considering the coordinate decomposition of the vacuum Einstein equations with respect to a flat spacelike slicing.

1. Introduction

The Schwarzschild solution was extended inside its horizon by Painlevé [1], Gullstrand [2] and Lemaître [3], who used a non-orthogonal curvature coordinate system $\{t, r, \theta, \varphi\}$ in which the metric takes the form

$$ds^2 = -\left(1 - \frac{2m}{r}\right) dt^2 + 2\sqrt{\frac{2m}{r}}\, dt\, dr + dr^2 + r^2 d\Omega^2 \tag{1}$$

with $d\Omega^2 = d\theta^2 + \sin^2\theta d\varphi^2$. In these Painlevé-Gullstrand (PG) coordinates, the metric is non diagonal, asymptotically flat, and regular at the horizon. So, the space-time appears foliated by a synchronization of flat instants.

This kind of foliations is being studied with renewed interest nowadays due to its applications in both, relativistic and non-relativistic, situations (see, for example, [4, 5, 6]).

In general, for a spherically symmetric space-time (SSST), the existence of flat synchronizations is usually taken for granted but, recently, an obstruction to it has been pointed out [7]. However, as far as the authors are aware, a definitive interpretation of this obstruction as well as the analysis of the domains where a flat synchronization exists have not been done up to now. The question is: does every SSST admit a region of physical interest where a synchronization by flat instants exists?

Here, in Section 2, we consider flat synchronizations in SSST analyzing their existence. In Section 3 we show that the energy and momenta densities of a flat slice vanish for any SSST. Next, Section 4 is devoted to recovering the Painlevé-Gullstrand-Lemaître (PGL) extension of the Schwarzschild solution by solving the vacuum Einstein equations for a SSST in PG coordinates.

2. Existence of flat synchronizations in SSST

Let us consider a SSST, whose metric can be written in the form [8, 9]

$$ds^2 = A(T,R)dT^2 + B(T,R)dR^2 + 2C(T,R)dTdR + D(T,R)d\Omega^2$$

being $d\Omega^2 = d\theta^2 + \sin^2\theta d\varphi^2$, $D(T,R) \neq 0$ an arbitrary function, and $A(T,R)$, $B(T,R)$ and $C(T,R)$ arbitrary functions satisfying the condition $\delta \equiv AB - C^2 < 0$ in order to guarantee the Lorentzian character of the metric.

In general, these coordinates are not PG coordinates. In fact, if we consider the induced metric γ on the 3-surfaces $T = constant$,

$$\gamma = B(T,R)dR \otimes dR + D(T,R)\sigma$$

($\sigma \equiv d\theta \otimes d\theta + \sin^2\theta\, d\varphi \otimes d\varphi$ being the metric on the unit 2-sphere) then we can calculate its Ricci tensor,

$$\mathcal{R}ic(\gamma) = \left(\frac{B}{2}\mathcal{R} - \frac{B}{D}F\right) dR \otimes dR + \left(\frac{D}{4}\mathcal{R} + \frac{F}{2}\right)\sigma$$

and its scalar curvature,

$$\mathcal{R} \equiv \mathcal{R}(\gamma) = \begin{cases} \dfrac{2F}{D} + \dfrac{4\partial_R F}{\partial_R D} & \partial_R D \neq 0 \\ \dfrac{2}{D} & \partial_R D = 0 \end{cases}$$

where

$$F = 1 - \frac{(\partial_R D)^2}{4BD}.$$

So, we have that γ is a flat metric ($\mathcal{R}ic(\gamma) = 0$) if and only if $F = 0$, that is, if and only if $(\partial_R D)^2 = 4BD$.

Consequently, in the general case, we can look for a coordinate transformation of the form $T(t,r)$ and $R(t,r)$, which leads to write the metric of the SSST in PG coordinates. This transformation will allow to put the metric in the form

$$ds^2 = \xi^2\, dt^2 + \chi^2 dr^2 + 2\xi \cdot \chi\, dt\, dr + \mathcal{D}(t,r)d\Omega^2 \tag{2}$$

with $\mathcal{D}(t,r) \equiv D(T(t,r), R(t,r))$, and the vector fields ξ and χ defined as

$$\xi \equiv \dot{T}\frac{\partial}{\partial T} + \dot{R}\frac{\partial}{\partial R}, \qquad \chi \equiv T'\frac{\partial}{\partial T} + R'\frac{\partial}{\partial R}$$

where $J \equiv \dot{T}R' - T'\dot{R} \neq 0$, in order to assure coordinate regularity (over-dot and prime stand for partial derivative with respect to t and r, respectively).

Given that the induced metric on the 3-surfaces $t = constant$ is flat if and only if

$$F = 1 - \frac{\mathcal{D}'^2}{4\xi^2\mathcal{D}} = 0 \Leftrightarrow 4\mathcal{D}\xi^2 = \mathcal{D}'^2 \Leftrightarrow (d\sqrt{\mathcal{D}})^2 \leq 1,$$

and r is a coordinate of curvature for the spherically symmetric form (2) if $\mathcal{D}(t,r) = r^2$, then we have that

$$\chi^2 = 1, \quad \mathcal{A} \equiv \xi^2 = J^2\delta(dr)^2, \quad \xi \cdot \chi = \pm J\sqrt{\delta[(dr)^2 - 1]}$$

So, the real function $\mathcal{B}(t,r) \equiv \xi \cdot \chi$ exists if $(dr)^2 \leq 1$ and in this case the metric results

$$ds^2 = \mathcal{A}(t,r)dt^2 + 2\mathcal{B}(t,r)dt\,dr + dr^2 + r^2 d\Omega^2$$

Then, we have proved the following result.

▷ *Let r be the radius of curvature of the orbits (2-spheres) of the isometry group of a spherically symmetric space-time with metric g. In the region defined by the condition*

$$(dr)^2 \equiv g^{\mu\nu}\partial_\mu r \partial_\nu r \leq 1 \tag{3}$$

a curvature coordinate system $\{t, r, \theta, \varphi\}$ exists in which the metric line element may be written as

$$ds^2 = \mathcal{A}(t,r)dt^2 + 2\mathcal{B}(t,r)dt\,dr + dr^2 + r^2 d\Omega^2 \tag{4}$$

where $d\Omega^2 = d\theta^2 + \sin^2\theta d\varphi^2$.

The above condition $(dr)^2 \leq 1$ is equivalent to say that the Misner-Sharp gravitational energy of a sphere of radius r is nonnegative [10].

3. Energy and momenta densities of a flat slice in SSST

In this section we consider a SSST in PG coordinates and write the metric (4) in the form $g = \eta + h$, with η the Minkowski metric and $h_{00} = 1 + \mathcal{A}$, $h_{0i} = \mathcal{B}x_i/r$, and $h_{ij} = 0$.

Moreover, we have that the Weinberg pseudotensor is defined by [11]

$$2Q^{i0\lambda} = \frac{\partial h^\mu_\mu}{\partial x_0}\eta^{i\lambda} - \frac{\partial h^\mu_\mu}{\partial x_i}\eta^{0\lambda} - \frac{\partial h^{\mu 0}}{\partial x^\mu}\eta^{i\lambda} + \frac{\partial h^{\mu i}}{\partial x^\mu}\eta^{0\lambda} + \frac{\partial h^{0\lambda}}{\partial x_i} - \frac{\partial h^{i\lambda}}{\partial x_0},$$

where Latin and Greek indexes go from 1 to 3 and from 0 to 3, respectively, and all indexes are raised and lowered with the flat metric η.

So, the Weinberg pseudotensor for a SSST in PG coordinates results

$$Q^{i00} = 0, \qquad 2Q^{i0j} = \left(\frac{\mathcal{B}}{r} + \mathcal{B}'\right)\delta_{ij} + \left(\frac{\mathcal{B}}{r} - \mathcal{B}'\right)\frac{x_i}{r}\frac{x_j}{r}.$$

By differentiating and contracting indexes in this expression we directly obtain

$$\tau^{0\lambda} \equiv -\frac{1}{8\pi G}\frac{\partial Q^{i0\lambda}}{\partial x^i} = 0$$

and then, the angular momentum densities, $\mathrm{j}^{i\lambda} = x^i\tau^{0\lambda} - x^\lambda\tau^{0i}$ also vanish.

So, we have proved the following result.

▷ *In any spherically symmetric space-time, the energy and momenta densities vanish for every Painlevé-Gullstrand synchronization.*

4. Painlevé-Gullstrand-Lemaître extension of the Schwarzschild solution

In this section we recover the Schwarzschild solution, including the horizon and its interior region. For that, we consider a SSST in PG coordinates whose metric expression is given by (4) and write the vacuum Einstein equations for this synchronization. According to the standard 3+1 splitting of the Einstein equations (see for example [12]), in this case the constraint equations are

$$\Phi(\Phi + 2\Psi) = 0, \qquad r\Phi' + \Phi - \Psi = 0 \tag{5}$$

and the evolution equations result

$$\dot{\Psi} = \Psi(2\Phi - \Psi) + \frac{1}{\mathcal{B}}(\mathcal{B}^2\Psi)', \qquad \dot{\Phi} = \Psi\Phi + \frac{\mathcal{B}}{r^2}(r^2\Phi)' \tag{6}$$

where

$$\Phi = \frac{1}{\alpha}\frac{\mathcal{B}}{r}, \qquad \Psi = \frac{\mathcal{B}'}{\alpha} \tag{7}$$

being $\alpha^2 = \mathcal{B}^2 - \mathcal{A}$ the lapse function.

The first constraint equation in (5) implies that, for $\Phi \neq 0$, $\Phi = -2\Psi$ and then, from the expressions in (7), we have that

$$\mathcal{B} = f(t)r^{-1/2}$$

with $f(t)$ and arbitrary function. Substituting this expression in the other constraint equation (5), it results that the α is a function of the sole variable t. So, by re-scaling the t coordinate parameter, we can take $\alpha = 1$. Then, we have that

$$\Phi = f(t)r^{-3/2} = -2\Psi$$

and taking into account the evolution equations (6) one has that $f(t)$ must be a constant function. So, if we take $f(t) = \sqrt{2m}$ we have that the metric can be written as in (1). Then, we have recovered the Schwarzschild metric and now the r coordinate can take any positive value, $0 < r < \infty$. In fact, from (1), we have $(dr)^2 = g^{rr} = 1 - \frac{2m}{r} < 1$, and the domain of the PG coordinates extends for every value of $r \neq 0$.

In the case $\Phi = 0$ the resolution of the equations (5) and (6) lead to the Minkowski space-time.

5. Conclusions

Summarizing, in this work we have interpreted the condition allowing the existence of a flat synchronization in any SSST in terms of the curvature radius of the 2-spheres, see Eq. (3). We have also shown that, for these flat slices, the energy and momenta densities vanish. And, finally, we have recovered the Painlevé-Gullstrand-Lemaître extension of the Schwarzschild solution by solving the vacuum Einstein equations for a SSST in a coordinate system adapted to a flat synchronization.

Acknowledgments

We would like to thank Dr. Marc Mars his comment during the presentation of this talk about the physical interpretation of the results. This work has been also supported by the MEC-FEDER project No. FIS2006-06062 and the MICINN-FEDER project No. FIS2009-07705.

References

[1] Painlevé P 1921 *C. R. Acad. Sc. Paris* **173** 677
[2] Gullstrand A 1922 *Archiv. Mat. Astron. Fys.* **16** 1
[3] Lemaître G 1933 *Ann. Soc. Sci. Bruxeles* **A 53** 51. Reprinted 1997 *Gen. Rel. Grav.* **29** 641
[4] Martel K and Poisson E 2001 *Am. J. Phys.* **69** 476
[5] Visser M 2005 *Int. J. Mod. Phys.* D **14** 2051
[6] Barceló C, Liberati S and Visser M 2005 Analogue Gravity *Living Rev. Relativity* **8** 12
[7] Lin C-Y and Soo C 2009 *Phys. Lett.* B **671** 493
[8] Takeno H 1966 The Theory of Spherically Symmetric Space-Times (Revised ed.) *Scientific Reports of the Research Institute for Theoretical Physics* (Hiroshima University) n. 5
[9] Plebański J and Krasiński A 2006 *An Introduction to General Relativity ans Cosmology* (Cambridge U. P.)
[10] Hayward S 1996 *Phys. Rev.* D **53** 1938
[11] Weinberg S 1972 *Gravitation and Cosmology* (John Wiley & Sons, New York)
[12] Gourgoulhon É 2007 3+1 formalism and bases of Numerical Relativity *Preprint* ArXiv:gr-qc/0703035

Journal of Physics: Conference Series **229** (2010) 012044 doi:10.1088/1742-6596/229/1/012044

Thermodynamic structure of gravitational field equations from near-horizon symmetries

Dawood Kothawala [1]

Inter-University Centre for Astronomy and Astrophysics, Post Bag 04, Ganeshkhind, Pune 411007, India

E-mail: dawood@iucaa.ernet.in

Abstract. There exists a strong mathematical resemblance between the on-horizon structure of the gravitational field equations and the *first law of thermodynamics*. In this talk, we discuss how such a structure arises and show that the field equations near any static horizon can be written as: $T\mathrm{d}S - \mathrm{d}E = P\mathrm{d}V$. Moreover, the result extends beyond Einstein theory and holds for Lanczos-Lovelock Lagrangians as well. The entropy S we obtain is precisely the Noether charge entropy of Wald, and E provides a natural generalization of *quasi-local energy* of the horizon. We comment on several implications of this result, particularly the notion of gravitational entropy [treated as the Noether charge of diffeomorphism invariance] associated with horizons and it's role in gravitational dynamics arising out of virtual displacements of the horizon.

1. Introduction

There is an intriguing analogy between the gravitational dynamics of horizons and thermodynamics, which is not yet understood at a deeper level. It was first noted by Padmanabhan [1] that gravitational field equations near a static spherically symmetric horizon has a intriguing structure; the differential form of this structure can be interpreted as virtual displacement of the horizon, and is, in fact, equivalent to the first law fo thermodynamics: $T\mathrm{d}S = \mathrm{d}E + P_{\perp}\mathrm{d}V$. In this talk, we describe how such a result arises out of near-horizon symmetries of the gravitational field tensor and holds not only for Einstein theory, but arbitrary Lanczos-Lovelock theory as well. In particular, our discussion brings out the key ingredients responsible for the thermodynamic structure of field equations as well as the generality of the result. Moreover, one can identify the expression for horizon entropy with the Noether charge entropy defined by Wald. Therefore, the question of how "gravity knows about thermodynamics at all", finds a natural explanation in the analysis we present. [2]

Notation: The metric signature is $(-,+,+,+)$; latin indices go from 0 to 3, and greek indices from 1 to 3. Capitalized latin indices go over the $(D-2)$ transverse coordinates.

2. Gravitational field equations as a thermodynamic identity

2.1. Background

The coordinate system best suited for our discussion is given by the metric:

$$\mathrm{d}s^2 = -N^2\mathrm{d}t^2 + \mathrm{d}n^2 + \sigma_{AB}\mathrm{d}y^A\mathrm{d}y^B \qquad (1)$$

[1] Work done in collaboration with T. Padmanabhan, IUCAA, India.

[2] For further details and an exhaustive list of references, see [2].

where $\sigma_{AB}(n, y^A)$ is the transverse metric, and the Killing horizon, generated by the timelike Killing vector field $\boldsymbol{\xi} = \boldsymbol{\partial}_t$, is approached as $N^2 \to 0$. Demanding the finiteness of curvature invariants on the horizon leads to the following Taylor series expansion for the lapse $N(n, x, y)$ and the transverse metric σ_{AB} [3]:

$$N(n, y) = \kappa n \left[1 - \frac{1}{2} R_\perp(y; n = 0) n^2 + O(n^3)\right] \tag{2}$$

$$\sigma_{AB} = [\sigma_H(y)]_{AB} + \frac{1}{2} [\sigma_2(y)]_{AB}\, n^2 + \mathcal{O}(n^3) \tag{3}$$

where we have collectively called the transverse coordinates as y, and $R_\perp$ is the Ricci scalar corresponding to y = constant surface. We shall be interested in a small region in the neighbourhood of the spacelike $(D-2)$-surface; more precisely, we shall assume $n \ll R_\perp^{-1/2}$, and would be interested in the $\kappa \to \infty$. That is, we shall require the length scale set by κ to be the smallest of all length scales. The t = constant part of the metric is written by employing *Gaussian normal coordinates* for the spatial part of the metric spanned by (n, y^A), n being the normal distance to the horizon. Consider the (y = constant) null vectors given by $l_i = \left(-1, +N^{-1}, 0, 0\right)$ and $k_i = \left(-1, -N^{-1}, 0, 0\right)$. In the limit we are interested in, it is easy to check that these vectors satisfy the geodesic equation in affinely parametrized form, that is; $\nabla_{\boldsymbol{l}} \boldsymbol{l} = 0 = \nabla_{\boldsymbol{k}} \boldsymbol{k}$ [3]. The affine parameter λ, defined by $\boldsymbol{l} \cdot \nabla\lambda = 1$ can be found by using the above form of $N(n, y)$; to the leading order, we find that, $\lambda \sim \lambda_H + (1/2)\kappa n^2$ where $\lambda = \lambda_H$ is the location of the horizon. Note that, $N^2 \boldsymbol{l} \to \boldsymbol{\xi}|_H$, which implies, $2\kappa\,(\lambda - \lambda_H)\,\boldsymbol{l} \to \boldsymbol{\xi}|_H$. In subsequent analysis, the differentials of various geometric quantities (such as entropy) defined on the horizon, which are directly involved in the statement of the first law of thermodynamics, are to be interpreted as variations with respect to the affine parameter along the outgoing null geodesics, i.e., λ. This, of course, is the most natural variation that can be chosen on a *null surface*. All throughout, we shall take the *on the horizon* limit by considering a foliation defined by n = constant surfaces and then taking the limit $n \to 0$.

2.2. Thermodynamic structure of Lanczos-Lovelock tensor

A natural generalization of the above result would be to look at Lanczos-Lovelock Lagrangians, which are the unique generalizations of Einstein tensor to higher dimensions, and yield equations of motion which are well behaved. We shall simply outline the derivation here; details can be found in [2].

The Lanczos-Lovelock (LL) Lagrangians are given by

$$\mathcal{L}_m^{(D)} = \left(\frac{1}{16\pi}\frac{1}{2^m}\right) \boldsymbol{\delta}^{a_1 b_1 \ldots a_m b_m}_{c_1 d_1 \ldots c_m d_m}\, \boldsymbol{R}^{c_1 d_1}_{\quad a_1 b_1} \cdots \boldsymbol{R}^{c_m d_m}_{\quad a_m b_m} \tag{4}$$

The corresponding equations of motion are $2E^i_j = T^i_j$, where

$$E^i_{j(m)} = \left(\frac{1}{16\pi}\frac{m}{2^m}\right) \boldsymbol{\delta}^{a_1 b_1 \ldots a_m b_m}_{j d_1 \ldots c_m d_m}\, \boldsymbol{R}^{i d_1}_{\quad a_1 b_1} \cdots \boldsymbol{R}^{c_m d_m}_{\quad a_m b_m} - \frac{1}{2}\delta^i_j \mathcal{L}_m \tag{5}$$

where m is an integer, and Einstein theory corresponds to $m = 1$. These Lagrangians have various special properties which have been discussed extensively in the literature. In particular, these Lagrangians satisfy: $\nabla_a \left(\partial\mathcal{L}/\partial R_{abcd}\right) = 0$.

[3] Explicit calculation shows that $\nabla_{\boldsymbol{l}} \boldsymbol{l}$ has only "y"-components, given by: $-\left(2\kappa^{-2}\right)\left[\sigma^{AB}\, \partial_B R_\perp\right]_{n=0} \boldsymbol{\partial}_A + O(n^2)$, which go to zero in the limit $\partial R_\perp/\kappa^2 \to 0$, which is the limit we are interested in.

A detailed analysis using Gauss-Codazzi decomposition and Combinatorics leads to

$$E^{\hat{a}}_{\hat{b}}|_H = \left[\begin{array}{cc|c} E_\perp & 0 & 0 \\ 0 & E_\perp & 0 \\ \hline 0 & 0 & E_{\|}{}_{[(D-2)\times(D-2)]} \end{array}\right] \tag{6}$$

which generalizes the result of Einstein theory. The object of interest, viz. $E_\perp$, turns out to be

$$E_\perp = \underbrace{\left(\frac{1}{16\pi}\frac{m}{2^m}\right)\sigma^{CA}\ \mathcal{E}^B_A\ [\sigma_2]_{CB}}_{\text{corr to variation of a lower dim. action!}} - \left(\frac{1}{2}\right)\mathcal{L}^{(D-2)}_m \tag{7}$$

Once again, some simple algebraic manipulations similar to those done in the Einstein case puts this in the form (valid on the horizon):

$$2E_\perp\ \sqrt{\sigma}\ \delta\lambda = \underbrace{\frac{\kappa}{2\pi}\left(\frac{1}{8}\frac{m}{2^{m-1}}\right)\mathcal{E}^{BC}\ \delta_\lambda\sigma_{BC}\ \sqrt{\sigma}}_{\delta_\lambda S} - \underbrace{\mathcal{L}^{(D-2)}_m\sqrt{\sigma}\ \delta\lambda}_{\delta_\lambda E} \tag{8}$$

where

$$S = 4\pi m\int \mathrm{d}\Sigma\ \mathcal{L}^{(D-2)}_{m-1} \qquad \mapsto \frac{1}{4}\int \mathrm{d}\Sigma \tag{9}$$

$$E = \int^\lambda \delta\lambda \int \mathrm{d}\Sigma\ \mathcal{L}^{(D-2)}_m \qquad \mapsto \int_\lambda \frac{1}{16\pi}\left[\int_y R_{\|}\sqrt{\sigma}\ \mathrm{d}^2y\right]\delta\lambda \tag{10}$$

where the arrows give corresponding expressions for $D = 4$ Einstein case. The thing to note is that both terms on RHS are expressible as "variations" of quantities locally defined on the horizon! In fact, S is precisely the Noether charge entropy as defined by Wald [4]. The quantity E as defined above gives an expression for energy which matches with known expressions for specific cases [for example, it gives $(1/2)r_H$ for spherically symmetric solutions in $D = 4$ Einstein theory, and reproduces the correct mass for spherically symmetric black hole solutions in Lovelock theory as calculated by others]. In fact, this expression for energy deserves a closer look, since it provides a very natural generalization of quasilocal energy for aspheric black holes in Einstein as well as Lovelock gravity.

3. Comments and Discussion

To put the result in appropriate context [5], begin by noting that what we have shown is that the following relations

$$\underbrace{\boldsymbol{E}(\hat{\boldsymbol{n}},\hat{\boldsymbol{n}}) = -\boldsymbol{E}(\boldsymbol{u},\boldsymbol{u})}_{near-horizon\ symmetry} \quad \& \quad \underbrace{\boldsymbol{E}(\boldsymbol{u},\boldsymbol{u}) = (1/2)\ \boldsymbol{T}(\boldsymbol{u},\boldsymbol{u})}_{field\ eq}$$

have a thermodynamic structure. Democracy of all observers then implies that $\boldsymbol{E} = (1/2)\boldsymbol{T}$ are thermodynamic:

$$\boldsymbol{E}(\boldsymbol{u},\boldsymbol{u}) = \frac{1}{2}\ \boldsymbol{T}(\boldsymbol{u},\boldsymbol{u}) \underset{all\ \boldsymbol{u}}{\longmapsto} \boldsymbol{E} = \frac{1}{2}\ \boldsymbol{T}$$

Further, note that, in the limit $N \to 0$, we have $N^2\boldsymbol{l} \to \boldsymbol{\xi}$. Using this and $\boldsymbol{u} = \boldsymbol{\xi}/N$, we see that $\boldsymbol{T}(\boldsymbol{u},\boldsymbol{u}) \to \boldsymbol{T}(\boldsymbol{\xi},\boldsymbol{l})$, which is precisely the flow of energy when the horizon undergoes a virtual displacement along the outgoing null geodesics. [4]

[4] Incidentally, there is another way to look at this which is worth mentioning: the object $g^\perp_{ab} = 2\xi_{(a}l_{b)}$ is the transverse part of the induced metric on the horizon, and hence $T_{ab}\xi^a l^b = (1/2)\ \mathrm{tr}_2[T_{ab}]$, where tr_2 is the trace with respect to $g^\perp_{ab}$.

Finally, let us briefly comment on Lagrangians which depend on metric and Riemann tensor (but not it's derivatives); i.e., $L = L\left[g_{ab}, R_{abcd}\right]$. Equations of motion are given by:

$$P_a^{\ cde} R_{bcde} - 2\nabla^c\nabla^d P_{acdb} - \frac{1}{2}Lg_{ab} = 8\pi T_{ab} \tag{11}$$

where P_{abcd} is defined as $\partial L/\partial R_{abcd}$, and inherit all the symmetries of the Riemann tensor. It is obvious that analyzing the near-horizon symmetries is not going to be easy, particularly due to the presence of the middle term on LHS; in fact, no such symmetries might exist at all for general Lagrangians without imposing additional restrictions. One might concentrate [taking hint from the Lovelock case] on $T_{ab}\xi^a l^b$ without worrying about the symmetries of the field tensor, but then one risks loosing some crucial link between such symmetries and black hole entropy. Indeed, that such a link exists is most clearly evident from the work of Carlip [6], and it would in fact be worthwhile to connect Carlip's analysis with our result. Nonetheless, we observe that the thermodynamic structure of field equations are crucially linked to the near-horizon symmetries and select out a particular class of Lagrangians, the Lanczos-Lovelock Lagrangians. This seems reasonable since it is only for this class of Lagrangians that one obtains second order equations of motion and the initial value problem is well defined. Since we do not have any criterion other than symmetry principles to analyze notions such as gravitational entropy in arbitrary theories of gravity, such a restriction to Lagrangians is important since it gives us a handle on the sort of low-energy effective actions we may expect from a full theory of quantum gravity. It must be remembered that two apparently different types of symmetries [definitely connected in some yet unknown manner] are evident in our result: the appearance of Wald entropy which arises as a Noether charge of diffeomorphism invariance (restricted to diffeomorphisms generating isometries) and the near-horizon symmetry of the field tensor. Our result above already clearly provides a direct connection of entropy so obtained with gravitational dynamics.

4. Acknowledgments

I would like to thank the organizers of Spanish Relativity Meeting ERE-2009, for giving me an opportunity to present this work at the stimulating conference in Bilbao. The research work is supported by a Fellowship from the Council of Scientific & Industrial Research (CSIR), India.

References

[1] Padmanabhan T 2002 Classical and Quantum Thermodynamics of horizons in spherically symmetric spacetimes *Class. Quantum Grav.* **19** 5387
Padmanabhan T 2005 Gravity and the Thermodynamics of Horizons *Phys. Rept.* **406** 49
[2] Kothawala D and Padmanabhan T 2009 Thermodynamic structure of Lanczos-Lovelock field equations from near-horizon symmetries *Phys. Rev.* D **79** 104020
[3] Medved A J M, Martin D and Visser M 2004 Dirty black holes: Spacetime geometry and near-horizon symmetries *Class. Quantum Grav.* **21** 3111
Medved A J M, Martin D and Visser M 2004 Dirty black holes: Symmetries at stationary non-static horizons *Phys. Rev.* D **70** 024009
[4] Wald R M 1993 Black Hole Entropy is Noether Charge *Phys.Rev.* D **48** 3427
Iyer V and Wald R M 1994 Some Properties of Noether Charge and a Proposal for Dynamical Black Hole Entropy *Phys.Rev.* D **50** 846
[5] For a recent review of these ideas, see: Padmanabhan T 2009 Thermodynamical Aspects of Gravity: New insights *Preprint* arXiv:0911.5004v1
[6] Carlip S 1999 Entropy from Conformal Field Theory at Killing Horizons *Class. Quantum Grav.* **16** 3327
Carlip S 2000 Black Hole Entropy from Horizon Conformal Field Theory *Nucl. Phys. B (Proc. Suppl.)* **88** 10

Flat deformation of a spacetime with two Killing fields

Josep Llosa[1*] **and Jaume Carot**[2]

[1] Departament de Física Fonamental, Universitat de Barcelona, Spain
[2] Departament de Física, Universitat de les Illes Balears, Spain

E-mail: [1]pitu.llosa@ub.edu, [2]jcarot@uib.cat

Abstract. It is shown that, given an analytic Lorentzian metric on a 4-manifold, g_{ab}, which admits two Killing vector fields, it exists a local deformation law $\eta_{ab} = a\, g_{ab} + b\, H_{ab}$, where H_{ab} is a 2-dimensional projector, such that η_{ab} is flat and admits the same Killing vectors.

1. Introduction

It has been recently proved [1] that given a semi-Riemannian analytic metric g_{ab} on a manifold $\mathcal{M}$, locally there exist two scalars a and b and a 2-dimensional projector H_{ab} —i.e. $H_{ab}g^{bc}H_{cd} = H_{ad}$ and $(H_{ab}g^{ab}) = 2$— such that the deformed metric, $\eta_{ab} := ag_{ab} + bH_{ab}$, is flat. We call this formula the *deformation law* associated with $(a,\, b,\, H_{ab})$. We have also proved [1] that in case that g_{ab} admits a Killing field X^a, then a deformation law can be found such that η_{ab} also admits X^a as a Killing field.

H_{ab} defines an almost-product structure [2], $g_{ab} = H_{ab} + K_{ab}$ and the deformation law differently scales the complementary planes H_{ab} and K_{ab}.

Assume now that g_{ab} admits a wider Lie algebra $\mathcal{G}$ of Killing fields, i.e. $\dim\mathcal{G} > 1$. Is it possible to find a deformation law such that η_{ab} admits any $X \in \mathcal{G}$ as a Killing field? Our conjecture is that the answer is affirmative if, and only if, $\mathcal{G}$ is a subalgebra of Poincaré algebra.

The problem we tackle here is simpler: we consider an analytic metric g_{ab} admitting two commuting Killing fields and we see that a deformation law exists such that η_{ab} admits the same Killing fields and is flat.

2. Spacetimes admiting two commuting Killing vectors

Let $\mathcal{M}$ be a spacetime with a metric η_{ab} (non necessarily flat) admitting two commuting Killing vectors X_A^a, $\mathcal{L}_{X_A}\eta_{ab} = 0\,,\quad A = 1,2$. Through any point $x \in \mathcal{M}$ there is an integral 2-submanifold $\mathcal{V}_x$, *the orbit* trough x, which is tangent to both X_1^a and X_2^a.

We denote by λ_{AB} the metric products $\lambda_{AB} := X_A^a \xi_{B\,a}$, with $\xi_{Aa} := \eta_{ab}X_A^b$ and λ^{AB} is the inverse matrix. We also have $\xi_a^A := \lambda^{AB}\xi_{Ba}$ and $X_B^a\xi_a^A = \delta_B^A$, that is $\{\xi_a^A\}_{A=1,2}$ is the dual base for $\{X_A^a\}_{A=1,2}$ on $T\mathcal{V}_x$. We assume this matrix to be hyperbolic, i. e. $\det(\lambda_{AB}) =: -\frac{\tau^2}{2} < 0$.

We shall also assume that the quotient space $\mathcal{S} := \mathcal{M}/\mathcal{V}$ is a 2-manifold. The tensor $h_b^a := \delta_b^a - X_A^a\xi_b^A$ projects tensors on $\mathcal{M}$ onto tensors on $\mathcal{S}$. Furthermore, we have the decomposition in horizontal and vertical parts: $T\mathcal{M} = T\mathcal{S} \oplus T\mathcal{V}$.

The metric $\eta_{ab} = 0$ splits also in a horizontal and a vertical part

$$\eta_{ab} = h_{ab} + \lambda_{AB}\xi_a^A\xi_b^B \tag{1}$$

and both are preserved by X_A, $A = 1, 2$.

The Killing equations, namely $\nabla_{(a}\xi_{A\,b)} = 0$, lead to

$$\nabla_c\xi_{A\,d} = \lambda_{AB\,||[c}\xi_{d]}^B + \frac{1}{2}\,\theta_A\epsilon_{cd} \qquad \text{and} \qquad \nabla_c\xi_d^A = -\lambda^{AD}\lambda_{DB|(d}\xi_{c)}^B + \frac{1}{2}\,\theta^A\epsilon_{cd} \tag{2}$$

where $\theta_A := \lambda_{AB}\theta^B$ are called *the twists* and ϵ is the volume tensor in $\mathcal{S}$. On its turn, $\Omega_{ab} := \frac{\tau}{2}\,\xi_a^1 \wedge \xi_b^2$ is the volume tensor on the orbits $\mathcal{V}_x$.

A horizontal tensor $T^{a\ldots}_{b\ldots}$ is characterized by $X^b_B T^{a\ldots}_{b\ldots} = \xi_a^A T^{a\ldots}_{b\ldots} = 0$ and $\mathcal{L}_{X_A} T^{a\ldots}_{b\ldots} = 0$, and we can define

$$D_c T^{a\ldots}_{b\ldots} := h_c^d\, h_e^a\, h_b^f\, \nabla_d T^{e\ldots}_{f\ldots}\,.$$

which turns out to be the Levi-Civita connection for h_{ab} in $T\mathcal{S}$.

The Riemann tensor for η_{ab} can be splitted into vertical, horizontal and mixed parts:

$$\begin{aligned} R_{dcab} = {} & R_1\epsilon_{dc}\epsilon_{ab} + R_2\Omega_{dc}\Omega_{ab} + 2P_{A[b}\xi^A_{a]}\epsilon_{dc} + 2P_{A[c}\xi^A_{d]}\epsilon_{ab} + 2Q_{A[b}\xi^A_{a]}\Omega_{dc} + 2Q_{A[c}\xi^A_{d]}\Omega_{ab} \\ & + 4\xi^D_{[d}P_{Dc]A[b}\xi^A_{a]} + \frac{1}{2}\,R_3\left(2\epsilon_{dc}\Omega_{ab} - \epsilon_{da}\Omega_{bc} - \epsilon_{db}\Omega_{ca} + 2\Omega_{dc}\epsilon_{ab} - \Omega_{da}\epsilon_{bc} - \Omega_{db}\epsilon_{ca}\right) \end{aligned} \tag{3}$$

with

$$R_1 = \frac{1}{2}\left(\mathcal{R} - \frac{3}{2}\,\theta^A\theta_A\right) \qquad R_2 = -\frac{1}{\tau^2}\left(\lambda_{11|c}\lambda_{22|b} - \lambda_{12|c}\lambda_{12|b}\right) h^{bc} \tag{4}$$

$$R_3 = \frac{1}{2\tau}\,\epsilon^{bc}\lambda^{TE}\lambda_{1E|c}\lambda_{2T|b} \tag{5}$$

$$P_{Ab} = \frac{1}{2}\left(D_b\theta_A + \frac{1}{2}\,\theta^T\lambda_{TA|b}\right) \qquad Q_{Ab} = -\frac{1}{\tau}\,\theta_{[1}\lambda_{2]A|c}\epsilon^c{}_b \tag{6}$$

$$P_{DcAb} = -\frac{1}{2}\left(D_b\lambda_{AD|c} - \frac{1}{2}\,\lambda^{TE}\lambda_{AE|(c}\lambda_{TD|b)} - \frac{1}{2}\,\theta_A\theta_D h_{bc}\right) \tag{7}$$

which relate R_{dcab} with the Ricci scalar $\mathcal{R}$ for h_{ab}, the twists θ_A and the vertical metric λ_{AB}.

These relations are to be taken into account in the so called 'reconstruction problem', namely to obtain η_{ab} from a given h_{ab} provided that: (a) admits X_A as Killing vectors and (b) η_{ab} is flat. Two 1-forms ξ_{Aa} must be given so that $\mathcal{L}_{X_A}\xi_{Ba} = 0$ and that (4-7) are fulfilled.

It is found [3] that h_{ab} must be (a) either flat or (b) its Ricci scalar must satisfy

$$D_{bc}\mathcal{R}^{-1/4} + \frac{1}{6}\,\mathcal{R}^{3/4}\,h_{bc} = 0$$

In the case (a), necessarily $\theta_B = 0$ and, either

$$\lambda_{AB} = \frac{1}{\sqrt{2}}\left(-\Phi_-^2\hat{m}_A\hat{m}_B + \Phi_+^2\hat{n}_A\hat{n}_B\right), \qquad \mathcal{F}^{\pm}_{ab} := D_{ab}\Phi_\pm = 0\,, \qquad h^{ab}D_a\Phi_+ D_b\Phi_- = 0 \tag{8}$$

or

$$\lambda_{AB} = \frac{\tau_0}{\sqrt{2}}\left(\hat{\lambda}_{AB} + f\hat{q}_{AB}\right), \qquad D_{ab}f = 0$$

where $\hat{m}_A := (\cos\beta, \sin\beta)$, $\hat{n}_A := (-\sin\beta, \cos\beta)$, β =constant, $\hat{\lambda}_{AB} = -\hat{m}_A\hat{m}_B + \hat{n}_A\hat{n}_B$ and $\hat{q}_{AB} = (\hat{m}_A + \hat{n}_A)\,(\hat{m}_B + \hat{n}_B)$.

3. Flat deformation

The above results are useful to prove the following

Theorem 1 *Let g_{ab} be a Lorentzian metric admitting two commuting Killing vector fields, X^a_A, $A = 1, 2$. Then, there exist two functions a, b and an elliptic 2-dimensional projector H^a_b such that the deformed metric $\eta_{ab} := a\, g_{ab} + b\, H_{ab}$, where $H_{ab} := g_{ac} H^c_b$, is flat and admits X^a_A, $A = 1, 2$, as Killing vector fields with vanishing twists.*

The proof [3] is based on the existence of solutions of the partial differential system (PDS) which follows from the flatness of η_{ab}

$$R_1 = R_2 = R_3\,, \qquad P_{Ab} = Q_{Ab} = 0\,, \qquad P_{DcAb} = 0\,, \tag{9}$$

Since X_A are KV's for both g_{ab} and η_{ab}, we easily have that $\mathcal{L}_{X_A} a = \mathcal{L}_{X_A} b = \mathcal{L}_{X_A} H_{ab} = 0$.

The unknowns a, b and H_{ab} are related to the new horizontal metric h_{ab}, vertical metric λ_{AB} and 1-forms $\kappa^A_a := \xi^A_a - \overline{\xi}^A_a \in \Lambda^1\mathcal{S}$, which determine η_{ab}. Hence, (9) is a PDS on $\mathcal{S}$ and the number of dimensions of the original problem is reduced by two.

As the twists vanish, $\theta^A = 0$, which guarantees that $P_{Ab} = Q_{Ab} = 0$ and therefore

$$d\kappa^A_a + \overline{\theta}^A \overline{\epsilon} = 0 \tag{10}$$

We also take λ_{AB} given by (8) and the 2-projector is:

$$H_{ab} = H_{AB} \overline{\xi}^A_a \overline{\xi}^B_b + \mu_a \mu_b + \nu_a \nu_b + 2\overline{\xi}^A_{(a} \kappa^B_{b)} \lambda_{AB} \tag{11}$$

where $\mu_a := (1-x)^{1/2} \hat{\mu}_a$, $\nu_a := (1-y)^{1/2} \hat{\nu}_a$, $z_1 := (a+bx)^{-1}$ and $z_2 := (a+by)^{-1}$ are the roots of $z^2 - \overline{\lambda}_{AB} \lambda^{AB} z + \overline{\tau}^2/\tau^2 = 0$, and

$$\hat{\mu}_a := \cos\psi \hat{\alpha}^3_a - \sin\psi \hat{\alpha}^4_a \qquad \text{nd} \qquad \hat{\nu}_a := \sin\psi \hat{\alpha}^3_a + \cos\psi \hat{\alpha}^4_a \tag{12}$$

where $\{\hat{\alpha}^3_a, \alpha^4_a\}$ is a $\overline{h}$-orthonormal base.

Finally the horizontal metric,

$$h_{ab} = a(a+b)\left[z_1 \hat{\mu}_a \hat{\mu}_b + z_2 \hat{\nu}_a \hat{\nu}_b\right]\,, \tag{13}$$

must be flat as a consequence of $R_1 = 0$ and $\theta^A = 0$.

The PDS resulting from (9), (10) and the substitutions (8), (11) and (13) is:

$$\begin{aligned} &R_2 = 0\,, \qquad \mathcal{F}^{\pm}_{ab} = 0\,, \qquad \mathcal{R} = 0 \\ &\mathcal{E}_a := \hat{e}_a \psi + \omega_a - \gamma_a{}^4{}_3 = 0\,, \qquad \mathcal{W} := 2\overline{D}_{[a}\omega_{b]} + \frac{1}{2}\overline{\mathcal{R}}\overline{\epsilon}_{ab} = 0 \end{aligned} \tag{14}$$

The number equations exceeds the number of unknowns. Let $\Sigma \subset \mathcal{S}$ be a Cauchy hypersurface and $\hat{\alpha}^4_a$ be the normal 1-form. We separate equations (14) into:

$$\begin{aligned} &\text{a reduced system} \qquad \mathcal{F}^{\pm}_{44} = 0\,, \quad \mathcal{E}_4 = 0\,, \quad \mathcal{W} = 0\,, \quad \mathcal{R} = 0 &(15)\\ &\text{and the constraints} \qquad \mathcal{F}^{\pm}_{3a} = 0\,, \quad \mathcal{R}_2 = 0\,, \quad \mathcal{E}_3 = 0 &(16) \end{aligned}$$

As a consequence of the second Bianchi identity for R_{dcab}, any solution of (15) fulfilling (16) on Σ also fulfils the constraints in an open neighbourhood.

A heavy-going calculation yields the characteristic determinant for the reduced system

$$\chi = -\frac{\sin^2(2\psi)\cos(2\psi)}{8ba^2(a+b)^2} \frac{z_1 - z_2}{1 - \overline{\lambda}_{AB}\lambda^{AB}(a+b) + \overline{\tau}^2(a+b)^2/\tau^2} \tag{17}$$

Hence, in order that Σ is a non-characteristic hypersurface, Cauchy data must be chosen so that

$$\psi \neq n\pi/4\,, \quad n \in \mathbb{Z}\,, \qquad \text{and} \qquad \overline{\lambda}_{AB}\lambda^{AB} \neq \frac{4\overline{\tau}^2}{\tau^2} \tag{18}$$

Cauchy data, namely $\Phi_\pm$, $\dot{\Phi}_\pm := \hat{e}_4\Phi_\pm$, a, b, $\dot{a}$, $\dot{b}$ and ψ on Σ, are also constrained by the conditions (16). Now, as Σ is a curve, these constraints are actually an ordinary differential system which always admit a solution for given initial data.

To summarize, the constructive process goes as follows:

(1) Take an analytic curve $\Sigma_0 \subset \mathcal{S}$ and pick a point $x_0 \in \Sigma_0$,
(2) **choose** a, $\dot{a}$, b and $\dot{b}$, analytic functions on Σ_0, then
(3) **give**[1] $\psi(x_0)$, $\Phi_\pm(x_0)$ and $\hat{e}_c\Phi_\pm(x_0)$, $(c = 3, 4)$.
(4) Then solve the ordinary differential system (16) to obtain ψ, $\Phi_\pm$ and $\dot{\Phi}_\pm$ on Σ, a neighbourhood of x_0 in $\mathcal{S}$.
(5) With these a, b, $\dot{a}$, $\dot{b}$, $\Phi_\pm$, $\dot{\Phi}_\pm$ and ψ as Cauchy data on Σ, the *reduced system* has an analytic solution, a, b, $\Phi_\pm$, ψ in a neighbourhood of Σ.
(6) With ψ and equation (12) construct the $\overline{h}$-orthonormal basis $\hat{\mu}_c$, $\hat{\nu}_d$.
(7) **Give** constant $\hat{m}_A$ and $\hat{n}_A$, $A = 1, 2$, such that $(\hat{m}_1\hat{n}_2 - \hat{m}_2\hat{n}_1)^2 = 1$ and substitute them in formula (8) to obtain

$$\lambda_{AB} = \frac{1}{\sqrt{2}}\left(-\Phi_-^2\,\hat{m}_A\hat{m}_A + \Phi_+^2\,\hat{n}_A\hat{n}_A\right)$$

(8) Take the covectors $\xi_a^A = \overline{\xi}_a^A + b\left(m^A\hat{\mu}_a\sqrt{1-x} + n_A\hat{\nu}_a\sqrt{1-y}\right)$, where

$$\begin{aligned} m_A\sqrt{-\left(\frac{a}{x}+b\right)} &= 2^{-1/4}\left(\Phi_-\cosh\varphi\,\hat{m}_A + \Phi_+\sinh\varphi\,\hat{n}_A\right) \\ n_A\sqrt{\frac{a}{y}+b} &= 2^{-1/4}\left(\Phi_-\sinh\varphi\,\hat{m}_A + \Phi_+\cosh\varphi\,\hat{n}_A\right) \end{aligned}$$

(9) then, using (13), derive the horizontal metric h_{ab} and
(10) finally obtain the deformed metric $\eta_{ab} = h_{ab} + \lambda_{AB}\xi_a^A\xi_b^B$.

Acknowledgement
Financial support from the Spanish Ministerio de Educación y Ciencia through grants FPA-2007-60220 and FIS2007-63034 is acknowledged, and also from the Generalitat de Catalunya (DURSI) 2009SGR-417 (J. Ll.) and from the Govern de les Illes Balears (J.C.).

References

[1] Llosa J and Carot J 2009 *Class. Quantum Grav.* **26** 055013
[2] Kobayashi S and Nomizu K 1969 *Foundations of Differential Geometry* (New York, Interscience)
[3] Llosa J and Carot J 2009 Flat deformation of a spacetime admitting two Killing fields *Preprint* gr-qc/0907.3608

[1] These data must be chosen so that the conditions (18) are fulfilled

Harmonic maps and black holes

João Lopes Costa
Lisbon University Institute – ISCTE and Mathematical Institute – University of Oxford
E-mail: jlca@iscte.pt

Abstract. We address two applications of existence and uniqueness properties of harmonic maps to the theory of stationary and axisymmetric electro-vacuum black holes. More specifically, we will consider: (1) The classification of such black hole space-times and (2) the proof of a Dain inequality with charge.

A map

$$\Phi : M \to N$$

between two Riemannian manifolds (M, g) and (N, h) is said to be *harmonic* if, given local coordinates, $\Phi = (\phi^1, \ldots, \phi^n)$ is a critical point of

$$\int_\Omega h_{ij} g^{\mu\nu} \partial_\mu \phi^i \partial_\nu \phi^j d\mu_g \ ,$$

for every conditionally compact coordinate domain Ω of M. In these local coordinates the Euler-Lagrange equations read

$$\Delta_g \phi^i + \Gamma^i_{jk} g^{\mu\nu} \partial_\mu \phi^j \partial_\nu \phi^k = 0 \ .$$

Finite energy harmonic maps were a focus of attention in the 1970s; here we will be using the results of Hildebrandt et al [8]. Later, motivated by the many black hole equilibrium problem in general relativity, Weinstein extensively studied the Dirichlet problem for harmonic maps with prescribed singularities; here we will consider the results in [10].

1. The classification of stationary space-times

CONJECTURE 1.1 *Let $(\mathcal{M}, \mathfrak{g}, F)$ be a stationary, asymptotically flat, electro-vacuum, four-dimensional regular space-time. Then the domain of outer communications $\langle\langle \mathcal{M}_{\rm ext} \rangle\rangle$ is either isometric to the domain of outer communications of a Kerr-Newman space-time or to the domain of outer communications of a (standard) Majumdar-Papapetrou space-time.*

When addressing the axisymmetric case it is possible to construct a global representation for the domain of outer communications, away from the axis of rotation, in terms of Weyl coordinates [2, 5]. In such privileged coordinates the metric takes the block diagonal form

$$\mathfrak{g} = -\rho^2 e^{2u} dt^2 + e^{-2u}(d\varphi - w dt)^2 + e^{2\lambda}(d\rho^2 + dz^2) \ , \tag{1}$$

where $\eta = \partial_\varphi$ is the axial Killing vector, and is completely determined by an axisymmetric harmonic map

$$\Phi = (u, v, \chi, \psi) : \mathbb{R}^3 \setminus \{z = 0\} \to \mathbb{H}^2_{\mathbb{C}} \ ,$$

with energy

$$H = \int_{\mathbb{R}^3} (Du)^2 + e^{4u}\left(Dv + \chi D\psi - \psi D\chi\right)^2 + e^{2u}\left((D\chi)^2 + (D\psi)^2\right) ; \tag{2}$$

the relevant electromagnetic potentials are defined by

$$\partial_\alpha \chi = F_{\mu\alpha}\eta^\mu \quad , \quad \partial_\alpha \psi = *F_{\mu\alpha}\eta^\mu \quad , \quad 2dv = \chi d\psi - \psi d\chi - *(d\eta^\flat \wedge \eta^\flat) . \tag{3}$$

Given any set of *axis data*, for instance prescribed Poincaré and Maxwell charges for each horizon component, we construct a reference axisymmetric (not necessarily harmonic) map Φ_0 modelled on the local behavior of the relevant Kerr-Newman maps near each component of the horizon and at a neighborhood of infinity. Then, there exists a unique harmonic map Φ whose hyperbolic distance satisfies [10]

$$d(\Phi, \Phi_0) \in L^\infty(\mathbb{R}^3 \setminus \{z = 0\}) \quad , \quad d(\Phi, \Phi_0) \to_{r\to+\infty} 0 . \tag{4}$$

From such harmonic maps it is always possible to construct a stationary and axisymmetric solution of the source free Einstein-Maxwell field equations [10]. Such (not necessarily I^+-regular) space-times will be referred to as *Weinstein solutions*. We then have the following partial realization of Conjecture 1.1 [2, 5]:

THEOREM 1.2 *Let $(\mathcal{M}, \mathfrak{g}, F)$ be a stationary, asymptotically flat, I^+-regular, electro-vacuum, four-dimensional analytic space-time. If each component of the event horizon is mean non-degenerate, then $\langle\langle \mathcal{M}_{\rm ext} \rangle\rangle$ is isometric to the domain of outer communications of a* Weinstein solution. *In particular, if the event horizon is connected and mean non-degenerate, then $\langle\langle \mathcal{M}_{\rm ext} \rangle\rangle$ is isometric to the domain of outer communications of a Kerr-Newman space-time.*

2. A Dain inequality with charge

The formation of event horizons is expected to occur *generically* in large families of space-times, with the exterior region approaching a Kerr-Newman metric asymptotically with time. This is the essence of the "standard picture of gravitational collapse" [9]. Since mass m is non-increasing, Maxwell charges Q_E and Q_B are conserved quantities, and angular momentum $\vec{J}$ is also conserved if one further assumes axisymmetry, the inequality

$$m \geq \sqrt{\frac{|\vec{J}|^2}{m^2} + Q_E^2 + Q_B^2} , \tag{5}$$

follows for initial data for such a collapse. Besides its intrinsic value, such inequalities provide some support for this "standard picture", and in particular for *weak cosmic censorship*.

Dain [7] proved an upper bound for angular-momentum in terms of the mass for a class of maximal, vacuum, axisymmetric initial data sets. Recently a generalized Dain inequality including electric and magnetic charges was obtained in [3, 6]. More precisely it was proven that:

THEOREM 2.1 *Let (M, g, K, E, B) be an axisymmetric electro-vacuum three-dimensional initial data set, which is* maximal *($tr_g K = 0$) and such that M is the union of a compact set and two asymptotically flat ends M_i. Let m, $\vec{J}$, Q_E and Q_B denote respectively the ADM mass, the ADM angular momentum and the total electric and magnetic charges of M_1. Then, if $M/U(1)$ is simply connected inequality 5 holds.*

From the purely geometric and topological assumptions we obtain [1] [1]

$$g = e^{-2U+2\alpha}\left(d\rho^2 + dz^2\right) + \rho^2 e^{-2U}\left(d\varphi + \rho\, W_\rho d\rho + W_z dz\right)^2 , \tag{6}$$

where $\eta = \partial_\varphi$ is again the generator of axial symmetry whose axis $\{\eta = 0\} \supset \{\rho = 0, z \neq 0\}$. This is a global representation for one of the asymptotic flat ends, say M_1, with the remaining end M_2 represented by the "puncture" $\{\rho = z = 0\}$. Moreover, the analysis of [1] provides detailed information concerning the asymptotic behavior of the fields involved. For instance, the existence of multiple ends manifests itself via

$$U = 2\log r + O(1) \ \ , \text{ for small } \ r = \sqrt{\rho^2 + z^2}\,. \tag{7}$$

Note that for extreme Kerr-Newman, for which it is impossible to construct initial data with multiple ends, one has instead

$$\tilde{U} = \log r + O(1) \ \ , \text{ for small } \ r = \sqrt{\rho^2 + z^2}\,. \tag{8}$$

Using the Einstein-Maxwell constraints we are then able to show that [3, 6]

$$m \geq \frac{1}{8\pi} I(U, v, \chi, \psi) \ ,$$

where

$$I := \int_{\mathbb{R}^3} (DU)^2 + \frac{e^{4U}}{\rho^4}\left(Dv + \chi D\psi - \psi D\chi\right)^2 + \frac{e^{2U}}{\rho^2}\left((D\chi)^2 + (D\psi)^2\right) \ ; \tag{9}$$

the electromagnetic potentials are the same as in (3). It is important to note that for stationary-axisymmetric data: $m = \frac{1}{8\pi} I(U, v, \chi, \psi)$.

There exist striking similarities between this *mass functional* and the harmonic map energy defined in (2). In fact, using the rescaling $U = u + \ln\rho$ we see that

$$I(U, v, \chi, \psi) = H(u, v, \chi, \psi) + B(U)\,. \tag{10}$$

Since I and H differ by a boundary term B they satisfy the same variational equations. It is also well known that *stationary-axisymmetric solutions of the Einstein-Maxwell equations provide critical points of* H and consequently of I as well.

So, let $(\tilde{U}, \tilde{v}, \tilde{\chi}, \tilde{\psi})$ be the harmonic map associated to the extreme Kerr-Newman satisfying

$$J_z = (v_2 - v_1)/8 \quad \text{and} \quad Q_E = (\psi_2 - \psi_1)/2\,.$$

The desired result will follow immediately if one is able to prove that:

$$I = I(U, v, \chi, \psi) \geq I(\tilde{U}, \tilde{v}, \tilde{\chi}, \tilde{\psi}) = \tilde{I}\,. \tag{11}$$

Now, using the results in [1] and in view of this new goal we see that: asymptotic flatness provides the necessary control to estimate both I and $\tilde{I}$ when the defining integrals are taken near infinity; near the axis, and away from both infinity and the origin, axisymmetry suffices; also, the necessary control near the puncture $\{\rho = z = 0\}$ follows from the referred controlled behavior of the fields near the origin and the know behavior of extreme Kerr-Newman near its horizon, here exemplified by (7) and (8), respectively. Consequently we are left with the task of establishing (11) for "large" compact sets not intersecting the singular set $\{\rho = 0\}$.

[1] It should be noted that the this ρ-coordinate does not, in general, agree with the Weyl ρ-coordinate of the previous section.

With this in mind let $\{A_{\sigma,\epsilon}\}$ be a family of such compact sets saturating $\mathbb{R}^3$ and construct associated auxiliary fields

$$(U_{\sigma,\epsilon}, v_{\sigma,\epsilon}, \chi_{\sigma,\epsilon}, \psi_{\sigma,\epsilon}) ,$$

converging pointwise to (U, v, χ, ψ), as $\sigma, \epsilon \to 0$, and satisfying the same boundary conditions as the fixed extreme Kerr-Newman map, i.e.

$$(\tilde{U}, \tilde{v}, \tilde{\chi}, \tilde{\psi}) = (U_{\sigma,\epsilon}, v_{\sigma,\epsilon}, \chi_{\sigma,\epsilon}, \psi_{\sigma,\epsilon}) \text{ on } \partial A_{\sigma,\epsilon} . \tag{12}$$

Since $\mathbb{H}^2_{\mathbb{C}}$ has negative sectional curvature, on the compact regions $A_{\sigma,\epsilon}$ we see from [8] that: Minimizers of H with Dirichlet boundary conditions exist and are smooth solutions of the variational equations. It is also well known, see for instance [4, 10], that solutions of the boundary value problem are, in the case of negatively curved target manifolds, unique. Since the map associated to extreme Kerr-Newman is necessarily a solution we see that (recall (12))

$$H_{A_{\sigma,\epsilon}}(u_{\sigma,\epsilon}, v_{\sigma,\epsilon}, \chi_{\sigma,\epsilon}, \psi_{\sigma,\epsilon}) > H_{A_{\sigma,\epsilon}}(\tilde{u}, \tilde{v}, \tilde{\chi}, \tilde{\psi}) ,$$

where by $H_{A_{\sigma,\epsilon}}$ we mean the integral (2) over the set $A_{\sigma,\epsilon}$. From (10) and relying once more on (12) we finally obtain

$$I_{A_{\sigma,\epsilon}}(U_{\sigma,\epsilon}, v_{\sigma,\epsilon}, \chi_{\sigma,\epsilon}, \psi_{\sigma,\epsilon}) > I_{A_{\sigma,\epsilon}}(\tilde{U}, \tilde{v}, \tilde{\chi}, \tilde{\psi}) ,$$

which, together with a careful asymptotic analysis in the lines previously described, allows us to establish (11).

There seem to be relevant connections between the two addressed problems which go beyond the purely technical level. In fact: this charged Dain inequality provides strong evidence that extreme Kerr-Newman initial data gives rise to the unique minimum of *mass* for fixed *angular momentum* and *charges*, within a class of axisymmetric and asymptotically flat initial data for space-times containing at most one black-hole or two asymptotic flat ends. Since the Majumdar-Papapetrou metrics provide the existence of regular and extreme multiple black hole solutions we see that the last restriction concerning the number of black holes is not merely technical. On the other hand for *pure vacuum* it is expect to be superfluous [4]. Such improvements of the known results could help us remove the undesired degeneracy assumption from Theorem 2.1 or could provide new insights concerning the stationary-axisymmetric N-body problem in relativity.

References

[1] Chruściel P T 2008 Mass and angular-momentum inequalities for axi-symmetric initial data sets. I. Positivity of mass *Annals Phys.* **323** 2566-2590 (*Preprint* arXiv:0710.3680)

[2] Chruściel P T and Costa J L 2008 On Uniqueness of Stationary Vacuum Black Holes *Gómétrie Différentielle, Physique Mathématique, Mathématique et Société, Volume en l'honneur de Jean Pierre Bourguignon* O Hjazi Ed. (Astrisque 321) p. 195-265 (*Preprint* arXiv:0806.0016)

[3] Chruściel P T and Costa J L 2009 Mass, angular-momentum, and charge inequalities for axisymmetric initial data *Class. Quantum Grav.* **26** 235013

[4] Chruściel P T, Li Y and Weinstein G 2008 Mass and angular-momentum inequalities for axi-symmetric initial data sets. II. Angular momentum, *Annals Phys.* **323** 2591-2613 (*Preprint* arXiv:0712.4064v2)

[5] Lopes Costa J 2009 On the classification of stationary electro-vacuum black holes *Class. Quantum Grav. to appear, Preprint* arXiv:0912.0834v1

[6] Lopes Costa J 2009 A Dain inequality with charge *Preprint* arXiv:0912.0838v1

[7] Dain S 2006 Proof of the angular momentum-mass inequality for axisymmetric black holes *Jour. Diff. Geom.* **79** 33-67 (*Preprint* arXiv:gr-qc/0606105)

[8] Hildebrandt S, Kaul H and Widman K 1977 An existence theorem for harmonic mappings of Riemannian manifolds *Acta Math.* **138** no. 1-2, 1–16

[9] Penrose R 2002 Gravitational collapse: the role of general relativity, Reprinted from Rivista del Nuovo Cimento 1969, Numero Speziale I, 252–276. *Gen. Rel. Grav.* **34** 1141-1165

[10] Weinstein G 1996 N-black hole stationary and axially symmetric solutions of the Einstein/Maxwell equations *Commun. Part. Diff. Eqs.* **21** 1389-1430.

Generating black strings

L A López

Dpto. de Física, Centro de Investigación y de Estudios Avanzados del I. P. N., Apdo. 14-740, D.F., México.

E-mail: `lalopez@fis.cinvestav.mx`

Abstract. It is generated a five dimensional solution of vacuum Einstein equations, with three free parameters, by means of equipping the Zipoy-Voorhees solution with a scalar field then lifting it to a fifth dimension. This solution presents as particular cases the black string with arbitrary tension and the Gregory-Laflamme black string.

The trapped surfaces, mass, tension and the condition of spherical symmetry of the solution are studied.

1. Introduction

There has been interest in study higher dimensional solutions to Einstein field equations associated whit string theory and the proposed more than dimensional universes. One example are the static vacuum solutions of black holes in five dimensions [1]. Other authors have studied the construction of higher dimensional vacuum solutions, considering the scalar fields that appear naturally in the dimensional reduction in string theory [2].

2. 5D Solution Black String

In this section we will use the Zipoy-Voorhees solution and introducing a scalar field, we will generate a 5D solution, with the black string as special case.

Let us consider the Zipoy-Voorhees (ZV) solution [3] in Weyl coordinates:

$$ds^2 = -e^{2U}dt^2 + e^{-2U+2\sigma}(d\rho^2 + dz^2) + \rho^2 e^{-2U} d\phi^2$$

$$U = \frac{1}{2}\ln\left[\frac{r_+ + r_- - 2m}{r_+ + r_- + 2m}\right]^{\frac{q}{2}} \qquad \sigma = \frac{1}{2}\ln\left[\frac{(r_+ + r_-)^2 - 4m^2}{4r_+ r_-}\right]^{\frac{q^2}{4}} \tag{1}$$

where $r_\pm^2 = \rho^2 + (z \pm m)^2$. For $q = 2$ we have Schwarzschild solution that describes a black hole, while if $q = 4$ it represents, the Darmois solution.

Adding the scalar field

$$\varphi = \frac{A}{2}\ln\left[\frac{r_+ + r_- - 2m}{r_+ + r_- + 2m}\right], \tag{2}$$

that is a solution of Laplace's equation $\varphi_{,\rho\rho} + \frac{1}{\rho}\varphi_{,\rho} + \varphi_{,zz} = 0$, it is obtained the spacetime with a scalar field φ, that is expressed as;

$$ds_{sc}^2 = -e^{2U} dt^2 + e^{-2U+2(\sigma+\sigma_{sc})}(d\rho^2 + dz^2) + \rho^2 e^{-2U} d\phi^2 \tag{3}$$

with

$$\sigma_{sc} = \frac{A^2}{2} \ln \left[\frac{(r_+ + r_-)^2 - 4m^2}{4r_+ r_-} \right] \tag{4}$$

Using the method described in [4], and considering the metric (3) as seed metric, we lift the solution to five dimensions:

$$dS_5^2 = e^{-\frac{2\varphi}{\sqrt{3}}}(ds_{sc}^2) + e^{\frac{4\varphi}{\sqrt{3}}} d\omega^2 \tag{5}$$

or explicitly

$$\begin{aligned} dS_5^2 \;=\;& -H^{\frac{q}{2}-\frac{A}{\sqrt{3}}} dt^2 + H^{-\frac{q}{2}-\frac{A}{\sqrt{3}}} \rho^2 d\phi^2 + H^{-\frac{q}{2}-\frac{A}{\sqrt{3}}} + H^{2A/\sqrt{3}} d\omega^2 \\ & + H^{-\frac{q}{2}-\frac{A}{\sqrt{3}}} \left[\frac{(r_+ + r_-)^2 - 4m^2}{4r_+ r_-} \right]^{\frac{q^2}{4}+A^2} (d\rho^2 + dz^2). \end{aligned} \tag{6}$$

where $H = \frac{r_+ + r_- - 2m}{r_+ + r_- + 2m}$.

Transforming the metric (6)to the Schwarzschild-like coordinates using;

$$\rho^2 = \left(1 - \frac{2m}{r}\right) r^2 \sin^2\theta, \qquad z = r\left(1 - \frac{m}{r}\right)\cos\theta, \tag{7}$$

we get:

$$\begin{aligned} dS_5^2 \;=\;& -G^{\frac{q}{2}-\frac{A}{\sqrt{3}}} dt^2 + G^{-\frac{q}{2}-\frac{A}{\sqrt{3}}+1} r^2 \sin^2\theta d\phi^2 + G^{\frac{2A}{\sqrt{3}}} d\omega^2 \\ & + \frac{(r-2m)^P r^Q}{(r^2 - 2mr + m^2 \sin^2\theta)^{A^2+\frac{q^2}{4}-1}} \left(\frac{dr^2}{r^2 - 2mr} + d\theta^2 \right) \end{aligned} \tag{8}$$

Where $G = (1 - \frac{2m}{r})$, $P = A^2 + \frac{q^2}{4} - \frac{A}{\sqrt{3}} - \frac{q}{2}$ and $Q = A^2 + \frac{q^2}{4} + \frac{A}{\sqrt{3}} + \frac{q}{2}$.

We note that if we want to impose spherical symmetry, we must choose $A^2 + \frac{q^2}{4} = 1$ in (8). Doing so the term with $\sin^2\theta$ does not appear in the denominator of g_{rr} and $g_{\theta\theta}$, obtaining an element proportional to $r^2 d\Omega$.

For the moment keeping free A and q, (8) is a 5D solution with three free parameters A, q and m.

3. Trapped Surfaces and Horizons

One of the most interesting properties to study in the spacetimes is the trapping surfaces and located horizons. In this section we will study the trapped surfaces and horizons of the generated 5D solution (8).

If we want to study the trapped surfaces S_{X^a} and located horizons corresponding to the spacetime (8), we need to calculate the scalar κ (see [5]).

Fixing the coordinates $x^a = \{r, t\}$ and denoting the local coordinates $x^A = \{\theta, \phi, \omega\}$ on the surface S_{X^a} the value $G \equiv \sqrt{\det g_{AB}} \equiv e^{\check{U}}$ gives the canonical volume element of the surfaces S_{X^a}.

Introducing $\mathbf{g}_a \equiv g_{aA}dx^a$ and $H_\mu = \delta^a_\mu(\check{U}_{,a} - div\overrightarrow{g_a})$

$$\kappa_{x^a} = -g^{ab}H_bH_c|_{S_{x^a}}, \tag{9}$$

the invariant κ amounts to

$$\kappa = -g^{rr}\check{U}^2_{,r}, \tag{10}$$

where

$$\check{U} = \frac{1}{2}\ln\left[\frac{r^{2(A^2+\frac{q^2}{4}+1)}\sin^2\theta(1-\frac{2m}{r})^{1-q+A^2+\frac{q^2}{4}}}{(r^2-2mr+m^2\sin^2\theta)^{A^2+\frac{q^2}{4}-1}}\right] \tag{11}$$

and therefore, κ is obtained as

$$\kappa = -\frac{\{[r-m(1+\cos\theta)][r-m(1-\cos\theta)]\}^{A^2+\frac{q^2}{4}-3}[\mathrm{Num}\check{U}_{,r}]^2}{r^{A^2+\frac{q^2}{4}+\frac{A}{\sqrt{3}}+\frac{q}{2}+1}(r-2m)^{A^2+\frac{q^2}{4}-\frac{A}{\sqrt{3}}-\frac{q}{2}+1}} \tag{12}$$

with

$$\mathrm{Num}\check{U}_{,r} = (r-m)[2r(r-2m)+m^2\sin^2\theta(A^2+\frac{q^2}{4}+1)] - mq[r^2-2mr+m^2\sin^2\theta] \tag{13}$$

Let us analyze the trapped surfaces using κ (κ=0) in (12) and the singularities. If we assume that $A > 0$ and $q > 0$, then there exists a singularity at $r = 0$ since $A^2 + \frac{q^2}{4} + \frac{A}{\sqrt{3}} + \frac{q}{2} + 1 > 0$.

Also there are two trapped surfaces in $r_\pm = m(1 \pm \cos\theta)$ nevertheless they are wrapped by the naked singularity located at $r = 2m$. Depending of the values that A and q have, $\mathrm{Num}\check{U}_{,r}$ in (12) can give a horizon that coincide with the naked singularity $r = 2m$. In other cases $\mathrm{Num}\check{U}_{,r}$ gives surfaces that do not cover completely the naked singularity.

4. Mass and Tension

The expansion of the metrical functions g_{tt} and $g_{\omega\omega}$ allow us to determinate the mass and tension of the 5D black string.

We restrict our attention to the metric (8) with $r \to \infty$. Asymptotically to order $O(1/r)$ the elements of the metric (8) are compared to the asymptotic form of metric around a stationary matter source given in [6]. Then the mass and tension of the 5D black string (8) are given by.

$$M = \frac{2qm}{4G_5}, \qquad Ma = \tau = \frac{m}{4G_5}\left(\frac{6A}{\sqrt{3}} + q\right) \tag{14}$$

Therefore when lifted to 5D the solution (1) by means of scalar field $\varphi(A,\rho,m)$, the mass and tension are function of q and A, but the tension augments in $\frac{6A}{\sqrt{3}}$.

5. Solutions with spherical symmetry

In this section we consider the condition of spherical symmetry in the 5D solution (8). Obtaining two 5D solutions of the vacuum Einstein equations that are contained in the 5D generalized black string.

Let us impose the spherical symmetry condition $A^2 + \frac{q^2}{4} = 1$ and perform in (8) the coordinate transformation

$$r = \varrho\left(1+\frac{k}{\varrho}\right)^2, \qquad \text{with} \quad \mathrm{m} = 2\kappa \tag{15}$$

in such a form (8) is transformed into:

$$\begin{aligned} dS_5^2 &= -D^{-s}dt^2 + (1-k/\varrho)^{2-\frac{1+a}{2-a}s}(1+k/\varrho)^{2+\frac{1+a}{2-a}s}(d\varrho^2+\varrho^2 d\theta^2+\varrho^2 sen^2\theta d\phi^2) \\ &\quad +D^{\frac{1-2a}{2-a}s}d\omega^2 \end{aligned} \tag{16}$$

Where $D = \frac{1+k/\varrho}{1-k/\varrho}$ and A, q, a and s are related by

$$A = -\frac{1-2a}{2\sqrt{1-a+a^2}}, \qquad q = \frac{3}{\sqrt{3(1-a+a^2)}}, \qquad \frac{2}{\sqrt{3}}A + q = s \tag{17}$$

this represents black string solution with arbitrary tension, obtained by Chul H. Lee [7].

While the invariant κ (12) with the condition of spherical symmetry becomes,

$$\kappa = \frac{1}{r^4}\left(1-\frac{2m}{r}\right)^{\frac{A}{\sqrt{3}}+\frac{q}{2}-2}(2r-2m-mq)^2 \tag{18}$$

We obtain the horizon $r = 2m$ or $r = 4k$ with $2 > a$, since in this case the highest possible value of q is 2. (16) has two free parameters k and s. Let us consider the values $A = 0$ and $q = 2$, then we recover the black string solution [8]. In this case the solution has a horizon located at $r = 2m$, the mass and tension take the values $\frac{m}{G_5}$ and $\frac{m}{2G_5}$ respectively.

6. Conclusions

In this paper we lifted the solution ZV to five dimensions, using an scalar field. So we have generated a 5D generalized black string (8). Then we studied the case of spherical symmetry (black string and Chul H. Lee's solution) for this solution, obtaining that the horizons coincide with the naked singularity $r = 2m$.

7. References

[1] Emparan R and Reall H S 2002 A rotating black ring solution in five dimensions *Phys. Rev. Lett.* **88** 101101
[2] Feinstein A and Vazquez-Mozo M A 2000 M theory resolution of four-dimensional cosmological singularities *Nucl. Phys.* B **568** 405
[3] Stephany H Kramer D MacCallum M Hoenselaers C and Herlt E 2003 *Exact solutions to Einsteins field equations* (Cambridge University Press) Second Edition
[4] Bretón N Feinstein A and López L A 2008 Generating generalized G_{D-2} solutions *Phys. Rev.* D **77** 124021
[5] Senovilla J M M 2002 Trapped surfaces, horizons and exact solution in higher dimensions *Class. Quantum Grav.* **19** L113
[6] Kim H C and Lee J 2008 Extraordinary vacuum black string solution *Phys. Rev.* D **77** 024012
[7] Lee C H 2006 Black string Solution with Arbitrary tension *Phys. Rev.* D **74** 104016
[8] Gregory R and Laflamme R 1988 Hypercylindrical black holes *Phys. Rev.* D **37** 305

Journal of Physics: Conference Series **229** (2010) 012048 doi:10.1088/1742-6596/229/1/012048

On Minkowski-like and de Sitter-like space-times

Christian Lübbe

School of Mathematical Sciences, Queen Mary University of London, E1 4NS, UK

E-mail: c.luebbe@qmul.ac.uk

Abstract. This article summarises joint work with Juan A. Valiente-Kroon on de Minkowski-like and de Sitter-like space-times and was presented by the author at ERE 2009 in Bilbao. The existence and stability problem is re-examined using extended conformal field equations. In particular we make use of a gauge based on conformal geodesics to obtain *a priori* the location of the conformal boundary from the initial data.

1. The problem

The existence and stability of Minkowski-like and de Sitter-like space-times were originally investigated and proven by Friedrich in [1]. One of the central ideas is to exploit the conformal structure of these space-times. The analysis is carried out in the unphysical space-time $(M, g_{\mu\nu})$, which is conformally related to $(\tilde{M}, \tilde{g}_{\mu\nu})$ with $g_{\mu\nu} = \Theta^2 \tilde{g}_{\mu\nu}$ and $\tilde{M} \equiv \{p \in M : \Theta(p) > 0\} \subset M$. The Einstein conformal field equations developed in [2, 3] were used formulate the Cauchy problems, which are semi-global respectively global in time, as problems local in time. The conformal gauge chosen by Friedrich was based on the unphysical Ricci scalar.

Given the importance of the conformal geometry for the analysis in [1] we re-examined the problem [4] in a gauge based on two further conformal notions: general Weyl connections and conformal geodesics. The main motivation is the resulting simplification of the analysis of the space-times and their conformal boundaries both conceptually and computationally. In particular the location of the conformal boundary $\mathscr{I} = \{p \in M : \Theta(p) = 0\}$ is obtain directly from the initial data for the congruence and can thus in parts be fixed suitably before the evolution of the space-time.

In this article we focus mainly on the analysis for Minkowski-like space-times. However our setup will be more general so that the physical space-times $(\tilde{M}, \tilde{g}_{\mu\nu})$ satisfy the Einstein field equations with cosmological constant, i.e. $\tilde{R}_{\mu\nu} = \lambda \tilde{g}_{\mu\nu}$. The discussion of the initial data and the conformal boundary will focus first on the Minkowski-like case and the de Sitter-like scenario will be investigated later on.

The initial hypersurface $\mathcal{S}$ will be space-like and diffeomorphic to $\mathbb{S}^3$. Thus the resulting space-times have the form $M = I \times \mathcal{S}$ where I is an interval on $\mathbb{R}$. For Minkowski-like space-times we use hyperboloidal initial data. This type of data is obtained by solving the conformal constraints on a simply connected set $\bar{\mathcal{S}} \subset \mathcal{S}$ with boundary $\mathcal{Z} = \partial\bar{\mathcal{S}}$ diffeomorphic to $\mathbb{S}^2$. The conformal factor Ω on $\bar{\mathcal{S}}$ vanishes on $\mathcal{Z}$ and is positive on the interior. The data and the conformal factor are then extended smoothly to the rest of $\mathcal{S}$. Note that the resulting space-time will only be conformally vacuum on the domain of dependence of $\bar{\mathcal{S}}$.

2. Conformal geometry and the conformal boundary

The evolution problem for (M, g) resp. $(\tilde{M}, \tilde{g}_{\mu\nu})$ is formulated using Weyl connections and conformal geodesics. Given a 1-form b a Weyl connection is defined by

$$\hat{\Gamma}_{\mu}{}^{\nu}{}_{\rho} = \tilde{\Gamma}_{\mu}{}^{\nu}{}_{\rho} + \left(\delta^{\nu}{}_{\mu} b_{\rho} + \delta^{\nu}{}_{\rho} b_{\mu} - g_{\mu\rho} g^{\nu\lambda} b_{\lambda}\right).$$

Conformal geodesics are conformally invariant curves $x(\tau)$ with velocity $v(\tau)$ and an associated one form $b(\tau)$ satisfying

$$\tilde{\nabla}_v v^{\mu} + 2\langle b, v\rangle v^{\mu} - \tilde{g}(v, v) b^{\mu} = 0, \tag{1}$$

$$\tilde{\nabla}_v b_{\mu} - \langle b, v\rangle b_{\mu} + \frac{1}{2} g_{\mu\nu} v^{\nu} \tilde{g}^{\sharp}(b, b) = \tilde{L}_{\lambda\mu} v^{\lambda}. \tag{2}$$

where $\tilde{L}_{\mu\nu} \equiv \frac{1}{2}\left(\tilde{R}_{\mu\nu} - \frac{1}{6}\tilde{R}\tilde{g}_{\mu\nu}\right)$ is the Schouten tensor. The solutions (1) and (2) are preserved as point sets under conformal rescaling or reparametrisation of the conformal parameter τ by a fractional linear transformation. Let $\{e_k\}$ with $e_0 = v$ be a conformally orthonormal frame that is Weyl propagated along the curves by $\hat{\nabla}_v e_k^{\mu} = 0$, where $\hat{\nabla} = \tilde{\nabla} + b$. The frame is used to define the unphysical metric by $g^{\mu\nu} = \eta^{ij} e_i^{\mu} e_j^{\nu}$. The canonical conformal factor that gives g so that $\{e_k\}$ is orthonormal is obtained by solving $\tilde{\nabla}_v \Theta = \langle b, v\rangle \Theta$ along each curve. In particular, since $\tilde{g}$ is a vacuum metric $\Theta(\tau)$ automatically takes the form

$$\Theta = \Theta_* + \dot{\Theta}_* \tau + \frac{1}{2}\ddot{\Theta}_* \tau^2 \tag{3}$$

Our calculations will use the 1-form $d = \Theta b$, as its space-like components d_a in the frame $\{e_k\}$ will be constant along each curve, while $d_0 = \dot{\Theta} = \dot{\Theta}_* + \ddot{\Theta}_* \tau$.

Initial data for v, d and Θ on $\mathcal{S}$ satisfying the constraint

$$2\Theta\ddot{\Theta} = \frac{1}{3}\lambda + g^{\sharp}(d, d) \tag{4}$$

is chosen such that

$$\Theta_* = \Omega, \quad \langle d_*, v_*\rangle = \dot{\Theta}_*, \quad d_{*a} = \Omega b_{*a} = D_a \Omega.$$

The congruence Γ of time-like conformal geodesics developed in a neighbourhood of $\mathcal{S}$ is used to drag coordinates x^A from $\mathcal{S}$ along Γ. In order for the coordinates (τ, x^A) to remain well defined throughout the evolution we need to guarantee that the congruence does not develop any caustics. This is done by studying the conformal Jacobi fields η along the congruence — see [5] for a definition. There are no conjugate points as long as the components of these fields orthogonal to the congruence do not vanish for .

The initial data for $\Theta_*, \dot{\Theta}_*, \ddot{\Theta}_*$ and d_* has to satisfy (4) and we set $v = n$, the normal of $\mathcal{S}$. On $\mathcal{Z}$ we choose $\dot{\Theta} < 0$ so that the conformal boundary represents part of $\mathscr{I}^+$. The value for $\ddot{\Theta}_*$ on $\mathcal{Z}$ is obtained by smooth extension from the interior. From this the location of $\mathscr{I}$ can be computed explicitly prior to the evolution by solving for the roots $\tau_{\pm}$ of (3). As long as $\ddot{\Theta} \neq 0$ there are always two roots and we can use the fractional linear parameter freedom of τ to make sure that all curves starting on $\bar{\mathcal{S}}$ reach $\mathscr{I}^+$ in finite time. For Minkowski-like space-times the existence and location of time-like infinity is of particular interest .

For hyperboloidal initial data sufficiently close to Minkowski data with respect to a suitable Sobolev norm, we can show that there exists a unique point $p \in \mathcal{S} \cap \{\Omega > 0\}$ such that $d_a = 0$ at p for $a = 1, 2, 3$. The conformal geodesic passing through p satisfies $\Theta = 0$ and $d\Theta = 0$ at the point i where $\tau = \tau_+ = -\dot{\Theta}_*/\ddot{\Theta}_* = -2\Omega/\dot{\Theta}_*$. It can be shown that the Hessian of Θ is non-degenerate at i and thus this is the unique point on the future conformal boundary that satisfies all the criteria of future time-like infinity of the Minkowski-like space-time.

3. The reference space-time

The Einstein cosmos is given by the manifold $M' = \mathbb{R} \times \mathbb{S}^3$ with metric

$$g_E = \mathrm{d}^2 t - \mathrm{d}^2\chi - \sin^2\chi(\mathrm{d}\theta^2 + \sin^2\theta \mathrm{d}\varphi^2). \tag{5}$$

The standard conformal embedding for Minkowski space is given by the conformal factor $\Omega_M = \cos(t + \frac{\pi}{2}) + \cos\chi$. By setting $\tau = 2\tan\frac{t}{2}$, the t-curves of the Einstein cosmos can be reparametrised as conformal geodesics for which $\Theta_E = 1 + \tau^2/4$ and $b_E = \tau/2\mathrm{d}t$. For this congruence none of the conformal Jacobi fields vanish and the coordinate system $(\tau, \chi, \theta, \varphi)$ is well defined. We denote $g' = \Theta_E^2 g_E$ and set (M', g') as the reference space-time for our stability analysis. For Minkowski hyperboloidal data on $t = 0 = \tau$ we set $\Omega_{M*} = \cos\chi$ and $d_{M*} = -(\mathrm{d}t + \sin\chi)$. Then (3) gives

$$\Theta_M = \Omega_M \Theta_E = \cos\chi\left(1 - \frac{\tau}{\cos\chi} + \frac{\tau^2}{4}\right)$$

which vanishes at $\tau = 2\frac{1\pm\sin\chi}{\cos\chi}$. The point i^+ lies on the central curve at $\tau = 2$.

From the conformal field equations we derive evolution equations along the conformal geodesic congruence. The space spinor formalism induced by the congruence can be used to show that we have a symmetric hyperbolic system for a variable u consisting of the frame, the connection, the curvature components and the conformal Jacobi fields. The difference between the solution u and the reference space-time (M', g'), represented by u', is denoted by $\breve{u}$ and the initial perturbation is accordingly denoted by $\breve{u}_0$. By considering $\breve{u}$ as a variable over the Sobolev space $(H^m(\mathbb{S}^3, \mathbb{R}^N), \|\cdot\|_m)$ for $m \geq 4$ and some N we can apply a version of Kato's theorem given in [1] to prove our existence and uniqueness results.

We can now formulate the following existence and stability result

Theorem 1. *Let $u_0 = u_0' + \breve{u}_0$ be Minkowski-like initial data. Given $T_0 > 2$ there exists $\varepsilon > 0$ such that*

(i) for $\| \breve{u}_0 \|_m < \varepsilon$ then there is a solution $u' + \breve{u}$ with minimal existence interval $\tau \in [0, T_0]$ and $u \in C^{m-2}([0, T_0] \times \mathcal{S})$;

(ii) the associated congruence of conformal geodesics contains no conjugate points in $[0, T_0]$;

(iii) for every $\breve{u}_0 \in B_\varepsilon(0)$ there is a unique point in $\overline{\mathcal{S}} \setminus \mathcal{Z}$ such that $D_k\Omega = 0$;

(iv) for all $\breve{u}_0 \in B_\varepsilon(0)$ we have $\tau_{i^+} \in [0, T_0]$.

The solution $u' + \breve{u}$ is unique on $\mathcal{D}^+(\overline{\mathcal{S}})$, the domain of dependence of $\overline{\mathcal{S}}$, and implies a C^{m-2} solution to the vacuum Einstein field equations with vanishing cosmological constant for which the set $\mathscr{I}^+$ represents future null infinity.

The point i^+ given by the conditions $\tau = -\frac{2}{\langle b_*, v_* \rangle}$ and $D_k\Omega = 0$ represents timelike infinity.

For de Sitter-like space-times we can consider two different scenarios for the initial data on $\mathcal{S}$. In the standard Cauchy problem $\mathcal{S}$ lies entirely inside the physical space-time $(\tilde{M}, \tilde{g})$ and the fractional linear freedom of the conformal parameter τ can be used to specify the initial data for Θ such that $\tau_\pm = \pm 2$ on each curve. The other scenario that we investigate is the asymptotic Cauchy problem where $\mathcal{S}$ forms one half of the conformal boundary, say $\mathscr{I}^-$, so that $\Theta_* = \Omega = 0$. In the case of asymptotic Cauchy data we set $v_* = n$ so that (4) implies

$$\dot{\Theta}_* = \langle d, v \rangle_* = \sqrt{-\frac{1}{3}\lambda} \quad \text{and} \quad d_* = \langle d, v \rangle_* v_*.$$

Then the conformal boundary is given by $\tau_- = 0$ and $\tau_+ = -2\dot{\Theta}_*/\ddot{\Theta}_*$. The initial data $\ddot{\Theta}_*$ is freely specifiable and used to fix the location of the conformal boundary by setting $\ddot{\Theta}_* = -\frac{1}{2}\dot{\Theta}_*$ so that $\tau_+ = 4$. Thus a suitable choice of data fixes the location of $\mathscr{I}^\pm$ in both cases.

We can derive an analogue version of Theorem 1 for de Sitter-like space-times.

Theorem 2. *Let $I = [a, b]$ denote an interval in $\mathbb{R}$ and let $u_0 = u_0' + \breve{u}_0$ represent either standard Cauchy data (I=[-2,2]) or asymptotic Cauchy data (I=[0,4]). There exists $\varepsilon > 0$ such that if $\| \breve{u}_0 \|_m < \varepsilon$ then*

(i) there is a unique solution $u' + \breve{u}$ with minimal existence interval $\tau \in I$ with $u \in C^{m-2}(I \times \mathcal{S})$

(ii) the associated congruence of conformal geodesics contains no conjugate points in I.

The fields $u' + \breve{u}$ imply a C^{m-2} solution to the vacuum Einstein field equations with negative cosmological constant. The sets $\mathscr{I}^- = \{a\} \times \mathcal{S}$ and $\mathscr{I}^+ = \{b\} \times \mathcal{S}$ represent, respectively, past and future null infinity.

The above approach of evolving vacuum space-times using conformal geodesics was further developed in [6] to prove semi-global existence and stability results for radiative space-times. Once more a priori knowledge of the location of the conformal boundary allowed proving that it is stable. In particular a new choice of initial data for the congruence in the perturbed space-time has further simplified the analysis of i^+ and the question of existence and uniqueness. In this new setup we write $d = \nabla\Theta + \Theta f$, where f is another 1-form. The initial data for d_{*a} is now set to be the same as for the reference space-time, i.e. as for $d_{M*} = -(\mathrm{d}t + \sin\chi)$ above, and f is used to absorb the difference between d_a and $D_a\Omega$ on $\mathcal{S}$. As a result i^+ is then clearly unique and lies on the same curve as in Minkowski.

It is currently being investigated whether similar approaches that use conformal geodesics or related concepts can help analysing stability problems for other matter models in this form.

References

[1] Friedrich H 1986 On the existence of n-geodesically complete or future complete solutions of Einstein's field equations with smooth asymptotic structure *Comm. Math. Phys.* **107** 587
[2] Friedrich H 1981 On the existence of analytic null asymptotically flat solutions of Einstein's vacuum field equations *Proc. Roy. Soc. Lond.* A **381** 361
[3] Friedrich H 1981 The asymptotic characteristic initial value problem for Einstein's vacuum field equations as an initial value problem for a first-order quasilinear symmetric hyperbolic system *Proc. Roy. Soc. Lond.* A **378** 401
[4] Lübbe C and Valiente Kroon J A 2009 On de Sitter-like and Minkowski-like spacetimes *Class. Quantum Grav.* **26** 145012
[5] Friedrich H 2003 Conformal geodesics on vacuum spacetimes *Comm. Math. Phys.* **235** 513
[6] Lübbe C and Valiente Kroon J A 2009 A stability result for purely radiative spacetimes *Preprint* arXiv:0910.4073

Invariant classification of metrics using invariant formalism

M P Machado Ramos[1]*, S B Edgar[2] and M Bradley[3]

[1] Departamento de Matemática e Aplicações, Universidade do Minho, Guimarães, Portugal
[2] Departament of Mathematics, Linköping Universitet, 581 83 Linköping, Sweden
[3] Departament of Physics, UmeåUniversitet, 901 87 Umeå, Sweden

E-mail: mpr@mct.uminho.pt[1], bredg@mai.liu.se[2], michael.bradley@physics.umu.se[3]

Abstract. Metrics obtained by integrating within the generalised invariant formalism are structured around their intrinsic coordinates and this considerably simplifies their invariant classification and symmetry analysis. We illustrate this by presenting a simple and transparent complete invariant classification of the conformally flat pure radiation metrics (except plane waves) in such intrinsic coordinates. By performing this classification we have corrected and completed statements and results by Edgar and Vickers, and by Skea, about the orders of Cartan invariants at which particular information becomes available.

1. Introduction

The integration procedure developed in the GHP formalism [1] has recently been generalised to the GIF (generalised invariant formalism [2]) [3], [4], [5], [6]. Compared to the familiar integration procedures in NP formalism [7] this procedure is much more efficient and avoids detailed complicated gauge calculations which have the potential for errors. It also supplies the metric in a natural form, with coordinates chosen as far as possible, in an intrinsic and invariant manner, permitting a comparatively simple invariant classification procedure, equivalence problem and symmetry investigations.

The equivalence problem is the problem of determining whether the metrics of two spacetimes are locally equivalent, and the original contribution of Cartan [8] directed attention to the Riemann tensor and its covariant derivatives up to $(q+1)$th order, $\mathcal{R}^{q+1}$, calculated in a particular frame. In going from $\mathcal{R}^q$ to $\mathcal{R}^{q+1}$ for a particular spacetime, if there is no new functionally independent Cartan scalar invariant and $\mathcal{R}^q$ and $\mathcal{R}^{q+1}$ have equal isotropy group, then all the local information that can be obtained about the spacetime is contained in the set $\mathcal{R}^{q+1}$. The set $\mathcal{R}^{q+1}$ is called the Cartan scalar invariants and provide the information for an invariant classification of the spacetime.

Two metrics are equivalent and represent the same spacetime if all their respective Cartan scalar invariants in $\mathcal{R}^{q+1}$ can be equated consistently. It is important to note that although there will be no new information about essential coordinates in the step from $\mathcal{R}^q$ to $\mathcal{R}^{q+1}$, there *may* be other new information, in particular about inconsistencies and also the nature of apparently non-redundant functions (including constants).

A practical method for invariant classification was developed by Karlhede [9], using fixed frames. In this algorithm the number of functionally independent quantities is kept as small as possible

at each step by putting successively the curvature and its covariant derivatives into canonical form, and only permitting those frame changes which preserve the canonical form. Although the Karlhede algorithm is more efficient than the original procedure proposed by Cartan, it may need to go as far as to $\mathcal{R}^7$ [10], and as a consequence, for some spacetimes, long complicated calculations are required, which usually need computer support, e.g., using the programme CLASSI [11], or the Maple-based GRTensor programme.
The Karlhede algorithm can be exploited to determine the structure of the isometry group of the spacetime, as well as subclasses within the spacetime which have additional isometries [12]; more recently the scheme has been the basis for an algorithm which determines whether a spacetime admits a homothetic Killing vector [13].

2. Invariant classification of Edgar-Ludwig spacetime in intrinsic coordinates from GIF

Edgar and Vickers [3] have rederived all CFPR spacetimes, which are not plane waves, using GIF, obtaining in coordinates $t,\ n,\ a,\ b$ the metric

$$\begin{aligned} \mathrm{d}s^2 \;=\;& \Big(a(2s(t)-2am(t)-a^2-b^2)-e(t)^2-n^2\Big)\mathrm{d}t^2-2a\mathrm{d}t\mathrm{d}n+ \\ & 2n\mathrm{d}t\mathrm{d}a+2e(t)\mathrm{d}t\mathrm{d}b-\mathrm{d}a^2-\mathrm{d}b^2 \end{aligned} \tag{1}$$

where $m(t),\ e(t),\ s(t)$ are non-redundant functions of the coordinate t; this form includes the possibility of any of $m(t)$ or $e(t)$ or $s(t)$ being constant. This form represents the most general metrics for CFPR spacetimes (with zero cosmological constant).
We begin by repeating the zeroth, first and second order invariants quoted in [3].
At zeroth order, there is only the one Cartan spinor invariant

$$\Phi=\frac{q^2}{a} \tag{2}$$

At first order, there are four Cartan spinor invariants

$$\text{þ}\Phi=0,\qquad \eth\Phi=\frac{pq^2}{a^2},\qquad \eth'\Phi=\frac{\bar{p}q^2}{a^2},\qquad \text{þ}'\Phi=-\frac{q^2}{a^2}\Big(\frac{qn}{a}+3p\boldsymbol{\iota}+3\bar{p}\bar{\boldsymbol{\iota}}\Big) \tag{3}$$

$\boldsymbol{\iota}$ is a second spinor which is generated in the GIF analysis, p and q are weighted scalar invariants which represent the spin and boost freedom; q is real while complex p satisfies $p\bar{p}=1/2$.
It is easy to see that we may invert (2) and (3) for a, p and q in terms of Cartan spinor invariants, and also for $(p\boldsymbol{\iota}+\bar{p}\bar{\boldsymbol{\iota}})$; therefore $\boldsymbol{\iota}$ is not uniquely determined, (and so neither is n) and at this level there clearly would remain the gauge freedom of a one parameter subgroup of null rotations. Since new information about the essential coordinates has arisen, we must go to the next order.
At second order, a complete set of independent Cartan spinor invariants is

$$\begin{aligned} \text{þ}\eth\Phi \;&=\; 0,\qquad \eth\eth\Phi=\frac{2p^2q^2}{a^3},\qquad \eth'\eth\Phi=\frac{q^2}{a^3},\qquad \text{þ}\text{þ}'\Phi=0 \\ \eth\text{þ}'\Phi \;&=\; -\frac{pq^2}{a^3}\Big(\frac{3qn}{a}+6p\boldsymbol{\iota}+8\bar{p}\bar{\boldsymbol{\iota}}\Big) \\ \text{þ}'\text{þ}'\Phi \;&=\; \frac{q^4}{a^4}\Big(s(t)-2am(t)-\frac{5}{2}a^2+\frac{1}{2}b^2+\frac{3n^2}{a}\Big) \\ &\quad+\frac{3q^2}{a^3}\Big(4p^2\boldsymbol{\iota}^2+5p\bar{p}\boldsymbol{\iota}\bar{\boldsymbol{\iota}}+\bar{p}^2\bar{\boldsymbol{\iota}}^2\Big)+12\frac{q^3n}{a^4}(p\boldsymbol{\iota}+\bar{p}\bar{\boldsymbol{\iota}}) \end{aligned} \tag{4}$$

together with complex conjugates. The GIF commutator equations enable us to concentrate on this reduced list of independent invariants.
We can now invert these equations and obtain explicit expressions for the spinor ι, as well as for n and the scalar combination $\left(s(t) - 2am(t) + \frac{1}{2}b^2\right)$, in terms of Cartan invariants. Thus, at second order, if we make this choice of ι as our second spinor, we will have fixed the frame completely (there is no isotropy freedom remaining), and we also have determined three essential base coordinates. Moreover, making this choice of ι as the second dyad spinor enables us to transfer to the simpler GHP formalism (see [3] for a fuller discussion on when this is possible) since we require only the scalar parts of the remaining non-trivial GHP Cartan invariants,

$$\begin{aligned} \eth\eth\Phi &= \frac{2p^2q^2}{a^3}, \qquad \eth'\eth\Phi = \frac{q^2}{a^3}, \qquad \eth\text{Þ}'\Phi = -\frac{pq^2}{a^3}\left(\frac{3qn}{a}\right) \\ \text{Þ}'\text{Þ}'\Phi &= \frac{q^4}{a^4}\left(s(t) - 2am(t) - \frac{5}{2}a^2 + \frac{1}{2}b^2 + \frac{3n^2}{a}\right) \end{aligned} \tag{5}$$

Since we have obtained new information about essential coordinates we need to go one step further.
At third order, a comparison of the second order expressions with the zeroth and first order ones shows that the only possibly independent new information will come from the following GHP Cartan invariants:

$$\eth\eth\text{Þ}'\Phi = -12\frac{p^2q^2}{a^4}\frac{qn}{a} \tag{6}$$

$$\eth\text{Þ}'\text{Þ}'\Phi = \frac{pq^4}{a^5}\left(4s(t) - 6am(t) - 5a^2 + 2b^2 + 15\frac{n^2}{a}\right) - \mathrm{i}\frac{pq^4}{a^4}b \tag{7}$$

$$\begin{aligned} \text{Þ}'\text{Þ}'\text{Þ}'\Phi &= \frac{q^5}{a^6}\Big(as'(t) - 2a^2m'(t) + 12anm(t) + 2ns(t) + abe(t) \\ &\quad -6a^2n - 5nb^2 - 15\frac{n^3}{a}\Big) \end{aligned} \tag{8}$$

and complex conjugates; $m'(t)$ and $s'(t)$ denotes differentiation with respect to t. The GHP commutator equations reduce the number of independent invariants.
At fourth order, a comparison of the third order expressions with the zeroth, first and second order ones shows that the only possibly new independent information will come from the operator $\text{Þ}'$ acting on the scalar X defined by

$$X = \text{Þ}'\text{Þ}'\text{Þ}'\Phi + \frac{q^5}{a^5}\left(6an + \frac{5nb^2}{a} + \frac{15n^3}{a^2} - 12m(t)n - \frac{2ns(t)}{a}\right). \tag{9}$$

This gives

$$\text{Þ}'X = \frac{q^6}{a^6}\left(s''(t) - 2am''(t) - 5\frac{ns'(t)}{a} + 9nm'(t) + be'(t) - 5\frac{nb}{a}e(t) + e(t)^2\right) \tag{10}$$

where $m''(t)$ and $s''(t)$ denote differentiation twice with respect to t, and $e'(t)$ denotes differentiation with respect to t.
Solving for a fourth coordinate is a little more complicated, since we have to go to third order, and the details will depend on the nature of the functions $s(t), m(t), e(t)$.
In summary we get the two following cases:

• if at least one of the functions $m(t), s(t), e(t)$, is not constant then all four essential base coordinates are obtained from GHP Cartan invariants at third order, and the procedure will therefore formally terminate at fourth order.

• when all of the functions $m(t), s(t), e(t)$ are constants, then only three essential base coordinates can be obtained from Cartan GHP invariants; these are obtained at second order, and the procedure will therefore formally terminate at third order.

Because of the very simple structure of this metric and the close relationship of its coordinates with its GHP Cartan invariants it has been easy to draw conclusions on its classification from a direct examination of its GHP Cartan invariants. Continuing in this manner, the three apparently non-redundant functions $m(t), s(t), e(t)$ can each be directly identified with a linear combination of the three GHP Cartan invariants $\text{þ}'\text{þ}'\Phi$, $\Re(\eth\text{þ}'\text{þ}'\Phi)$, $\text{þ}'\text{þ}'\text{þ}'\Phi$, and hence are genuinely non-redundant, and cannot be transformed away.

3. Summery

The simplicity and transparency of this GIF version (1), combined with the fact that we are able to carry it out by hand, gives us a clear unambiguous overview of the invariant classification of this class of metrics. The results obtained in Section 2 have some minor, but subtle and interesting, disagreements with the conclusions by Edgar and Vickers [3], and by Skea [14]. The traditional CLASSI analysis of the spacetime in the intrinsic coordinates was carried out in [15] confirming the results in Section 2.

In addition, using this version of the spacetime, we were able to obtain trivially the Killing vector properties, as well as the homothetic Killing vector properties by a straightforward application of the Koutras-Skea algorithm [15].

Acknowledgments

MPMR wishes to thank the Fundação Calouste Gulbenkian and CMAT (University of Minho) for funding her participation in the Spanish Relativity Meeting 2009, in Bilbao, Spain.

References

[1] Geroch R, Held A and Penrose R 1973 *J. Math. Phys.* **14** 874
[2] Machado Ramos M P and Vickers J A G 1996 *Class. Quantum Grav.* **13** 1579
[3] Edgar S B and Vickers J A G 1999 *Class. Quantum Grav.* **16** 589
[4] Edgar S B and Machado Ramos M P 2005 *Class. Quantum Grav.* **22** 791
[5] Edgar S B and Machado Ramos M P 2007 *Gen. Rel. Grav.* **39** 539
[6] Edgar S B and Machado Ramos M P 2007 *Gen. Rel. Grav.* **39** 1749
[7] Newman E and Penrose R 1962 *J. Math. Phys.* **3** 566
[8] E 1946 *Leçons sur la geometrie des espaces de Riemann* (Paris: Gauthier-Villars)
[9] A 1980 *Gen. Rel. Grav.* **12** 693
[10] Stephani H, Kramer D, MacCallum M, Hoenselaers C and Herlt E 2003 *Exact solutions of Einstein's field equations, Second edition* (Cambridge: Cambridge University Press)
[11] Aman J 1987 *Manual for CLASSI* (University of Stockholm: Institute of Theoretical Physics Technical Reports)
[12] Karlhede A and MacCallum M 1982 *Gen. Rel. Grav.* **14** 673
[13] Koutras A and Skea J 1998 *Comput. Phys. Commun.* **115** 350
[14] Skea J 1997 *Class. Quantum Grav.* **14** 2393
[15] Bradley M, Edgar S B and Machado Ramos M P 2010 *Gen. Rel. Grav.* **42** 155

Journal of Physics: Conference Series **229** (2010) 012050 doi:10.1088/1742-6596/229/1/012050

On interrelations between Sibgatullin's and Alekseev's approaches to the construction of exact solutions of the Einstein-Maxwell equations

F J Ernst[1], **V S Manko**[2*] **and E Ruiz**[3]
[1] FJE Enterprises, 511 CR 59, Potsdam, NY 13676, USA
[2] Departamento de Física, Centro de Investigación y de Estudios Avanzados del IPN, A.P. 14-740, 07000 México D.F., Mexico
[3] Instituto Universitario de Física Fundamental y Matemáticas, Universidad de Salamanca, 37008 Salamanca, Spain
E-mail: scientia@localnet.com, vsmanko@fis.cinvestav.mx, eruiz@usal.es

Abstract. The integral equations involved in Alekseev's "monodromy transform" technique are shown to be simple combinations of Sibgatullin's integral equations and normalizing conditions. An additional complex conjugation introduced by Alekseev in the integrands makes his scheme mathematically inconsistent; besides, in the electrovac case all Alekseev's principal value integrals contain an intrinsic error which has never been identified before. We also explain how operates a non-trivial double-step algorithm devised by Alekseev for rewriting, by purely algebraic manipulations and in a different (more complicated) parameter set, any particular specialization of the known analytically extended N-soliton electrovac solution obtained in 1995 with the aid of Sibgatullin's method.

1. Introduction

The integral equations and normalizing conditions involved in Sibgatullin's method have the form [1]

$$\fint_{-1}^{+1} \frac{e(\xi)+\tilde{e}(\eta)+2f(\xi)\tilde{f}(\eta)}{(\xi-\eta)\sqrt{1-\sigma^2}}\mu_a(\sigma)d\sigma = -2\pi i(\delta_a^2 - 2\tilde{f}(\eta)\delta_a^3), \tag{1}$$

$$\int_{-1}^{+1} \frac{\mu_a(\sigma)d\sigma}{\sqrt{1-\sigma^2}} = \pi\delta_a^1, \quad a=1,2,3, \tag{2}$$

and they permit one to calculate the unknown functions $\mu_a(\sigma)$ corresponding to the arbitrarily prescribed data $e(\xi)$ and $f(\xi)$, $\xi = z + i\rho\sigma$, representing local holomorphic continuations of the axis expressions $e(z) \equiv \mathcal{E}(\rho=0,z)$, $f(z) \equiv \Phi(\rho=0,z)$ of the complex Ernst potentials [2] into the complex plane. The knowledge of $\mu_a(\sigma)$ is essential for being able to subsequently obtain $\mathcal{E}(\rho,z)$, $\Phi(\rho,z)$ and metric coefficients by virtue of integral formulae (see [3] for details).

Whereas Sibgatullin's method is based on the Hauser–Ernst formulation of Kinnersley's transformation theory [4, 5] and all of its details are well known, the so-called "monodromy transform" technique first presented by Alekseev under a different name and in an axiomatic manner [6] has been always advertised by its author as an independent mathematical result and

Journal of Physics: Conference Series **229** (2010) 012050 doi:10.1088/1742-6596/229/1/012050

an alternative to the Hauser, Ernst and Sibgatullin's use of the homogeneous Riemann–Hilbert problem in their solution generating methods. In the stationary axially symmetric electrovac case, Alekseev's integral equations for the unknown functions $\varphi^a(\zeta)$ have the form

$$\frac{1}{\pi i}\int_L \frac{\mathcal{K}(\tau,\zeta)}{\zeta-\tau}\varphi^a(\zeta)d\zeta = k^a(\tau), \quad a=1,2,3, \tag{3}$$

$$\mathcal{K}(\tau,\zeta) = -[\lambda]_\zeta\left(\vec{k}(\tau)\cdot\vec{l}(\zeta)\right) = -[\lambda]_\zeta\left(1+i\zeta(u^\dagger(\zeta)-u(\tau))+4\zeta^2 v^\dagger(\zeta)v(\tau)\right), \tag{4}$$

$$\vec{k}(\tau) = \{1,u(\tau),v(\tau)\}, \quad \vec{l}(\zeta) = \begin{pmatrix} 1+i\zeta u^\dagger(\zeta) \\ -i\zeta \\ 4\zeta^2 v^\dagger(\zeta)\end{pmatrix}. \tag{5}$$

In our short communication we will show that the above equations (3)–(5) are simple combinations of Sibgatullin's equations (1) and (2). We will also describe the algebraic trick invented by Alekseev for presenting the already known solutions obtained via Sibgatullin's method as an independent product of the "monodromy transform" procedure.

2. Derivation of Alekseev's equations

We first observe that Alekseev's ζ, τ and the '$\dagger$' operation are identical, respectively, to Sibgatullin's ξ, η and the '$\sim$' operation, $\tilde{g}(\xi) \equiv \overline{g(\bar{\xi})}$. The integration in (3) along the cut L joining the points $z \pm i\rho$ reduces to the integration in (1) by parametrizing the points of the cut as $\zeta = z+i\rho\sigma$, $\sigma\in[-1,1]$, and taking into account that $d\zeta = i\rho d\sigma$.

Since in the electrovac case Alekseev's integral equations (3) and (4) are erroneous, in what follows we will first consider the pure vacuum case and then briefly comment on the mistakes in Alekseev's formulae involving the function $v(\zeta)$. In the absence of the electromagnetic field ($v(\zeta)=0$ and $f(\xi)=0$) we choose $e(\xi)$ in the form $e(\xi)=1-2i\xi u(\xi)$ and have

$$\tilde{e}(\xi)+e(\eta) = 2\{1+i\xi(\tilde{u}(\xi)-u(\eta))+i(\xi-\eta)u(\eta)\}. \tag{6}$$

Substituting (6) into equations (1) in which the additional 'tilde' operation is carried out, and getting rid of the term $i(\xi-\eta)u(\eta)$ with the aid of the normalizing conditions (2), we finally get

$$\frac{1}{\pi i}\int_{-1}^{+1}\frac{1+i\xi(\tilde{u}(\xi)-u(\eta))}{(\xi-\eta)\sqrt{1-\sigma^2}}\mu_1(\sigma)d\sigma = -u(\eta), \quad \frac{1}{\pi i}\int_{-1}^{+1}\frac{1+i\xi(\tilde{u}(\xi)-u(\eta))}{(\xi-\eta)\sqrt{1-\sigma^2}}\mu_2(\sigma)d\sigma = 1, \tag{7}$$

which are precisely Alekseev's equations (3)–(5) in the vacuum case, up to obvious redefinitions of the functions φ^1 and φ^2, and the change of the integration variable.

It should also be emphasized that due to the additional 'tilde' operation performed by Alekseev in the integral formulae his scheme turns out *mathematically inconsistent* because the final expressions for the Ernst potentials obtained in this way reduce to the initial axis data in the static case only.

In the electrovac case ($f(\xi)\neq 0$, $f(\xi)=2i\xi v(\xi)$) we have

$$\tilde{e}(\xi)+e(\eta)+2\tilde{f}(\xi)f(\eta) = 2\{1+i\xi(\tilde{u}(\xi)-u(\eta))+4\xi\eta\tilde{v}(\xi)v(\eta)+i(\xi-\eta)u(\eta)\}, \tag{8}$$

and Sibgatullin's integral equation for the function $\mu_1(\sigma)$, in which the additional 'tilde' conjugation is performed and the respective normalizing condition is used in analogy with the vacuum case, then rewrites as

$$\frac{1}{\pi i}\int_{-1}^{+1}\frac{1+i\xi(\tilde{u}(\xi)-u(\eta))+4\xi\eta\tilde{v}(\xi)v(\eta)}{(\xi-\eta)\sqrt{1-\sigma^2}}\mu_1(\sigma)d\sigma = -u(\eta), \tag{9}$$

whence it follows that the kernel $\mathcal{K}(\tau,\zeta)$ of all three Alekseev's equations (3) contains an intrinsic error – the factor ζ^2 – whose changing to the correct factor $\zeta\tau$ also implies modification of $\vec{k}(\tau)$ and $\vec{l}(\zeta)$ in (5).

3. The algebraic trick

The use of Sibgatullin's method is particularly simple in the case of the rational axis data. In 1995 the extended N–soliton electrovac solution was constructed with its help in a concise analytical form [7], the solution arising from the axis data

$$e(z) = 1 + \sum_{l=1}^{N} \frac{e_l}{z-\beta_l}, \quad f(z) = \sum_{l=1}^{N} \frac{f_l}{z-\beta_l}, \tag{10}$$

where e_l, β_l and f_l are 3N arbitrary complex constants. Two particular ($N=1$ and $N=2$) electrovac specializations of this solution were considered in a recent paper [8] within the framework of the "monodromy transform" technique. A careful analysis of the paper [8] reveals, however, that the solution generating procedure employed in [8] reduces exclusively to rewriting in different parameters the already known results using the following simple algebraic trick devised by Alekseev for avoiding the use of his problematic integral equations. Alekseev starts with the unphysical, asymptotically non–flat axis data

$$e(z) = 1 - 2i\sum_{l=1}^{N} u_l - 2i\sum_{l=1}^{N} \frac{u_l h_l}{z-h_l}, \quad f(z) = 2i\sum_{l=1}^{N} v_l + 2i\sum_{l=1}^{N} \frac{v_l h_l}{z-h_l} \tag{11}$$

(u_l, h_l and v_l are complex parameters) and formally declares that for those data he has carried out the solution construction procedure with the aid of the 'monodromy transform' technique (still having nothing in the hand) and that for writing down the final result he needs to apply to the 'obtained solution' an additional 'gauge transformation' of the form

$$\begin{aligned} \mathcal{E}' &= 1 + \frac{\mathcal{E} - \varepsilon_0 + 2\bar{\phi}_0(\Phi - \phi_0)}{p_0}, \quad \Phi' = \frac{\Phi - \phi_0}{\sqrt{p_0}}, \\ \varepsilon_0 &\equiv 1 - 2i\sum_{l=1}^{N} u_l, \quad \phi_0 \equiv 2i\sum_{l=1}^{N} v_l, \quad \bar{\phi}_0 \equiv -2i\sum_{l=1}^{N} \bar{v}_l, \quad p_0 \equiv 1 - i\sum_{l=1}^{N}(u_l - \bar{u}_l) + \phi_0\bar{\phi}_0. \end{aligned} \tag{12}$$

The only purpose of the above transformation is to cast the unphysical axis data (11) to the new data $e'(z)$ and $f'(z)$ identical to (10) but written in a different parameter set:

$$e'(z) = 1 + \sum_{l=1}^{N} \frac{2ih_l(-u_l + 2\bar{\phi}_0 v_l)}{p_0(z-h_l)}, \quad f'(z) = \sum_{l=1}^{N} \frac{2iv_l h_l}{\sqrt{p_0}(z-h_l)}, \tag{13}$$

and these expressions provide Alekseev with the desired relations between his parameters u_l, h_l, v_l and the parameters e_l, β_l, f_l entering the axis data (10) of the known analytically extended N–soliton solution:

$$e_l = \frac{2ih_l}{p_0}(-u_l + 2\bar{\phi}_0 v_l), \quad \beta_l = h_l, \quad f_l = \frac{2iv_l h_l}{\sqrt{p_0}}. \tag{14}$$

The inverted formulae are

$$u_l = \frac{ip_0}{2\beta_l}\left(e_l - 2f_l\sum_{k=1}^{N} \frac{\bar{f}_k}{\bar{\beta}_k}\right), \quad h_l = \beta_l, \quad v_l = -\frac{if_l\sqrt{p_0}}{2\beta_l}, \quad p_0 \equiv \frac{\prod_{l=1}^{N} \beta_l\bar{\beta}_l}{\prod_{n=1}^{2N} \alpha_n}, \tag{15}$$

where α_n are $2N$ formal roots of the algebraic equation $e(\xi) + \tilde{e}(\xi) + 2f(\xi)\tilde{f}(\xi) = 0$.

Since the N–soliton electrovac solution in terms of the parameters e_l, β_l, f_l is known, the substitution of (14) into the formulae of the paper [7] immediately supplies Alekseev with the explicit form of the multisoliton solution written in terms of the parameters u_l, h_l, v_l. Apparently, the algebraic trick described above, if not known, causes an illusion of a true usage of the integral equations in obtaining Alekseev's results and novelty of the latter. We point out that writing the soliton solution in terms of the parameter set $\{u_l, h_l, v_l\}$ instead of the canonical set $\{e_l, \beta_l, f_l\}$ considerably complicates the solution's form and hence does not have any scientific justification.

In section 2 of [8] the above formulae (14) and (15) are used for reproducing the results of the paper [9] devoted to the derivation of the physical representation of the electrovac Demiański–Newman solution [10]. Just to have an idea about the advantages of the original work over its remake, it suffices to say that the axis data employed in [9] have the form

$$e(z) = \frac{z - m - i(a+\nu)}{z + m - i(a-\nu)}, \qquad f(z) = \frac{q + ib}{z + m - i(a-\nu)}, \tag{16}$$

while the initial expression for $e(z)$ in terms of the same parameters considered in [8] is the following complicated function:

$$e(z) = -1 - \frac{2ia[m - i(a-\nu)]}{m^2 + \nu^2 - a^2 - q^2 - b^2} + \frac{2[m - i(a-\nu)]}{z + m - i(a-\nu)}\left(1 + \frac{ia[m - i(a-\nu)]}{m^2 + \nu^2 - a^2 - q^2 - b^2}\right). \tag{17}$$

Turning now to the two–pole electrostatic solution considered in section 3 of [8], it should be first of all pointed out that the solution defined by formulae (34)–(38) of that paper, contrary to the authors' claim, *does not correspond* to the axis data (20), (21) of [8]. The algebraic trick is used this time for rewriting the Bretón *et al* double–Reissner–Nordström solution [11]. Formulae (20)–(29) of [8] utilize the relations

$$u_l = \frac{ip_0}{2\beta_l}\left(e_l - 2f_l \sum_{k=1}^{2} \frac{f_k}{\beta_k}\right), \quad h_l = \beta_l, \quad v_l = -\frac{if_l\sqrt{p_0}}{2\beta_l}, \quad p_0 \equiv \frac{\beta_l^2 \beta_2^2}{\alpha_1\alpha_2\alpha_3\alpha_4}, \tag{18}$$

to get the form of u_l and v_l in terms of h_l, w_l, $\tilde{w}_l$ (β_l, α_n in the notions of [11]) by substituting into (18) the explicit expressions for e_l and f_l derived in [11]. We mention in conclusion that formulae (34)–(38) of [8] describe a 5–parameter representation of the BMA solution in physical parameters first obtained in [12] and then just rewritten in [8] in different coordinates.

References

[1] Sibgatullin N R 1984 *Oscillations and Waves in Strong Gravitational and Electromagnetic Fields* (Moscow: Nauka)
[2] Ernst F J 1968 *Phys. Rev.* **168** 1415
[3] Manko V S and Sibgatullin N R 1993 *Class. Quantum Grav.* **10** 1383
[4] Hauser I and Ernst F J 1979 *Phys. Rev.* D **20** 1783
[5] Kinnersley W 1977 *J. Math. Phys.* **18** 1529
[6] Alekseev G A 1985 *Sov. Phys. Dokl.* **30** 565
[7] Ruiz E, Manko V S and Martín J 1995 *Phys. Rev.* D **51** 4192
[8] Alekseev A G and Belinski V A 2008 *Proceedings of the 11th Marcel Grossmann Meeting on General Gelativity*, ed H Kleinert, R Jantzen and R Ruffini (Singapore, World Scientific) p 543
[9] Aguilar–Sánchez J A, García A A and Manko V S 2001 *Gravitation and Cosmology* **7** 149 (*Preprint* gr-qc/0106011)
[10] Demiański M and Newman E T 1966 *Bull. Acad. Polon. Sci. Ser. Math. Astron. Phys.* **14** 653
[11] Bretón N, Manko V S and Aguilar–Sánchez J A 1998 *Class. Quantum Grav.* **15** 3071
[12] Manko V S 2007 *Phys. Rev.* D **76** 124032

Journal of Physics: Conference Series **229** (2010) 012051 doi:10.1088/1742-6596/229/1/012051

Unifying exact scalar field cosmologies

José P Mimoso[1]* and Tiago Charters[2]

Centro de Astronomia e Astrofísica da Universidade de Lisboa, Avenida Professor Gama Pinto 2, P-1649-003 Lisbon, Portugal

E-mail: [1] jpmimoso@cii.fc.ul.pt, [2] tca@cii.fc.ul.pt

Abstract. We present a mechanism that generates the exact solutions of scalar field cosmologies in an unified way. This applies both to standard and phantom scalar fields in the framework of Einsteins General Relativity, and also to extended scenarios based on generalized gravity theories such as Scalar-Tensor theories. We provide a classification of the solutions in terms a single generating function. The existence of form-invariant dualities between the solutions is also discussed.

1. Introduction

During approximately the past three decades there has been considerable interest in finding exact solutions of the well known scalar field equations in a spatially flat Friedman-Robertson-Walker (FRW) space-time. Inflation has provided the underlying motivation [1]. Since it is very difficult to obtain exact solutions for specific potentials emerging from particle physics, the focus was directed to the solutions of particular cases which had remarkable properties, and which still shared a general particle physics motivation, e.g., exponential potentials [2]. Some methods to obtain exact solutions were devised such as the use of the scalar field, ϕ, as the independent variable, and the study of Lie symmetries of the equations [3]. A skillful combination of various, separate methods has since lead some authors to derive some classes of new solutions.

Here we are concerned with the question of unifying all the previous solutions in a single, and simple framework, which moreover permits the derivation of novel solutions[4]. Our procedure applies to models with self-interacting scalar fields in the framework of General Relativity (GR), encompassing both standard models of scalar field cosmologies, i.e., those in which the scalar field has a positive kinetic energy, phantom models where the latter is negative, and scalar-tensor models as well.

2. Equations and procedure

Although our procedure applies to spacetimes of $(N+1)$ dimensions, here we restrict ourselves to GR, and consider a $(3+1)$-dimensional spatially flat Friedmann spacetime and we assume that the matter content is a minimally coupled scalar field. The Einstein equations are (we adopt units in which $8\pi G = c^2 = 1$)

$$\ddot{\phi} + 3H\dot{\phi} + \varepsilon V_{,\phi} = 0, \tag{1}$$

$$3H^2 = \varepsilon \frac{1}{2}\dot{\phi}^2 + V(\phi), \tag{2}$$

where the overdot represents the time derivative, $a(t)$ is the scale factor, $H = \dot{a}/a$ is the Hubble parameter, and ε parameter is a discriminator that takes the values $\varepsilon = \pm 1$ distinguishing standard scalar field models from phantom field models. The value $\varepsilon = +1$ corresponds to the former models and $\varepsilon = -1$ to the latter.

Introducing the new time variable $d\tau = Hdt$ and the dimensionless variable $x = \frac{\dot{\phi}}{H}$, we obtain

$$\frac{dx}{d\phi} = \frac{x'}{\phi'} = \frac{\left(3 - \varepsilon\frac{1}{2}x^2\right)\left(x + \varepsilon\frac{V_{,\phi}}{V}\right)}{x}. \tag{3}$$

The latter Eq. (3) can be approached in two alternative ways. On the one hand, given the potential $V(\phi)$ we can solve it to obtain $x(\phi)$. On the other hand, we can instead arbitrarily choose $x = x(\phi)$, and derive the corresponding scalar field potential

$$V(\phi) = A\left(3 - \varepsilon\frac{1}{2}x^2(\phi)\right)e^{-\frac{2\varepsilon}{2}\int x(\phi)d\phi}. \tag{4}$$

where A is an arbitrary integration constant, and also

$$H(\phi) = \pm\sqrt{A}e^{-\frac{\varepsilon}{2}\int x(\phi)d\phi}, \tag{5}$$

$$\int dt = \int \frac{d\phi}{\pm\sqrt{A}x(\phi)\exp\left(-\frac{\varepsilon}{N-1}\int x(\phi)d\phi\right)}, \tag{6}$$

which is the final quadrature.

Therefore, selecting a suitable $x = x(\phi)$, one easily integrates the equations for the scalar field model and derives an exact solution in parametric form, or in closed form whenever (6) is invertible. From the definition (2) we see that $x(\phi)$ is related to the usual barotropic index $\gamma = (\rho_\phi + p_\phi)/\rho_\phi$ in the following way

$$x(\phi) = \pm\sqrt{3\gamma} \tag{7}$$

and thus choosing $x(\phi)$ amounts to a specification of the equation of state. On the other hand $x(\phi)$ relates to the slow roll parameters [5] in the following, simple way

$$\epsilon(\phi) = \frac{x^2(\phi)}{2} = 2\frac{H'(\phi)}{H(\phi)} \tag{8}$$

$$\bar{\eta}(\phi) = x'(\phi) + \frac{x^2(\phi)}{2} = 2\frac{H''(\phi)}{H(\phi)}. \tag{9}$$

3. Exact GR Solutions

We illustrate our procedure by giving, for $\varepsilon = 1$ (i.e., standard models) a brief list of cases yielding well-known solutions, and also by deriving a novel exact solution. Known solutions arise from the following correspondences:

- $x(\phi) = \lambda$ with $\lambda =$ constant, which gives the power-law solutions associated with the exponential potential $V(\phi) = V_0\, e^{-\lambda\phi}$[2];
- $x(\phi) = \lambda\phi$ yields the Easther solution [6]for which $V(\phi) = A\,(3 - \phi^2/2)\, e^{-\phi^2/2}$, and which gives a constant scalar spectral index equal to 3;

- $x(\phi) = \beta/\phi$ yields Barrow's intermediate inflation solutions [7] associated with $V(\phi) = \left(3 - \frac{\beta^2}{2\phi^2}\right)\phi^{-\beta}$;
- $x(\phi) = (N-1)\frac{G(\phi)}{\int G(\phi)}$ with $G(\phi) = AH(\phi) + L(\phi)$, where H are linear generating functions and $L(\phi) = c_1\phi + c_2\phi^2$, where c_1, c_2 are real constants, corresponds to Chimento's solutions[8];
- More cases can be found in Ref. [4].

A new solution can be derived from the prescription $x(\phi) = -2W'(\phi)$, where $W(\phi)$ is the Lambert function [9] defined as

$$W(\phi)e^{W(\phi)} = \phi. \tag{10}$$

When ϕ is real, for $-1/e \leq \phi < 0$ there are two possible real values of $W(\phi)$. We denote just by $W(\phi)$ the branch satisfying $-1 \leq W(\phi)$. In this case one gets for the self-interacting potential $V(\phi)$ the expression

$$V(\phi) = A\left(3 - \frac{1}{2}\frac{W(\phi)^2}{\phi^2(1+W(\phi))^2}\right)\frac{\phi^2}{W^2(\phi)}. \tag{11}$$

The Hubble parameter and the velocity of the ϕ field are

$$H(\phi) = \pm\sqrt{A}e^{W(\phi)}, \tag{12}$$

$$\dot{\phi} = \pm\sqrt{A}\frac{\phi W'(\phi)}{W(\phi)} = \pm\sqrt{A}\frac{1}{1+W(\phi)}, \tag{13}$$

and thus one gets

$$\pm\sqrt{A}(t - t_0) = \phi\left(W(\phi) + \frac{1}{W(\phi)}\right). \tag{14}$$

The examples given here show not only that diverse solutions correspond to different choices for $x(\phi)$, but also that this provides the basis for a classification of the solutions. To conclude this section, we refer that for $\varepsilon = -1$ it is possible to obtain phantom exact solutions as counterparts for all choices of $x(\phi)$ yielding standard solutions. To the best of our knowledge this gives a plethora of new phantom exact solutions.

4. Form-invariance dualities

Recently Chimento and Lazkoz have discussed [10] invertible maps that preserve the form of the equations of motion of flat FRW models. We generalise their results for the case of general scalar field models.

In our present case the transformations arise from $(x(\phi), \phi) \to (\bar{x}(\bar{\phi}), \bar{\phi})$, satisfying the two conditions

$$\frac{\bar{x}^2}{x^2} = \pm\left(\frac{H}{\bar{H}}\right)^2 \frac{\mathrm{d}\bar{H}}{\mathrm{d}H} \quad , \quad \left(\frac{d\bar{\phi}}{d\phi}\right)^2 = \frac{d\bar{H}}{dH} \tag{15}$$

which leave invariant the equations of motion. In parallel, we also have

$$\frac{d\bar{\phi}}{\bar{x}\bar{H}} = \frac{d\phi}{xH} \tag{16}$$

which translates the fact that the form invariance transformation preserves the time variable (6).

According to our procedure, the choices of $x(\phi)$ and $\bar{x}(\bar{\phi})$ determine the corresponding potentials. If we choose for instance $x(\phi) = \lambda$ and $\bar{x} = \bar{\lambda}$ we obtain

$$\bar{\phi} = \frac{\lambda}{\bar{\lambda}}\,\phi + \bar{\phi}_0 \qquad \text{and} \quad \bar{H} = \left(\frac{\lambda}{\bar{\lambda}}\right)^2 H\,, \tag{17}$$

and thus we recover the results of Refs. [10]. If $x = \lambda$, and $\bar{x} = \bar{\lambda}/\bar{\phi}$ we derive

$$\bar{\phi}^{\beta} = A\left(\exp(\frac{\lambda}{N-1}\phi) + \nu\right) \tag{18}$$

where

$$\beta = 2 + \frac{\bar{\lambda}}{N-1} \quad , \qquad A = \frac{(N-1)\beta}{\lambda}\left(\frac{\bar{H}_0\bar{\lambda}}{H_0\lambda}\right) \tag{19}$$

and ν is once again an integration constant. When $\nu = 0$, $\bar{H}(H)$ is simply

$$\bar{H} = \frac{\bar{H}_0}{H_0}\,A^{\beta-2}H^{1-\beta}\,. \tag{20}$$

The multiplying factor on the right-hand side can be negative in which case the transformation can map standard scalar field cosmology into a generalised phantom one.

5. Discussion

The procedure investigated here permits to recover all known exact scalar field solutions, and allows the derivation of new exact solutions, such as exact solutions of phantom cosmologies. Everything depends on the choice of a single generating function $x(\phi)$ that hence sets the basis for a classification of the solutions. It also permits us to establish form-invariance dualities by means of a simple mapping between different $x(\phi)$.

The details of the work reported here, and of a follow up devoted to scalar-tensor theories can be found in[4, 11].

Acknowledgments

The authors are grateful to Luis Chimento for helpful comments. JPM also acknowledges the ERE LOC for a most enjoyable conference.

References

[1] Olive K A 1990 *Phys. Rept.* **190** 307
Liddle A R and Lyth D H 2000 *Cosmological Inflation and Large-Scale-Structure* (Cambridge University Press)
[2] Lucchin F and Matarrese S 1985 *Phys. Rev.* D **32** 1316
[3] Muslimov A G 1990 *Class. Quantum Grav.* **7** 231
Lidsey J E 1991 *Phys. Lett.* B **273** 42
Capozziello S, De Ritis R and Scudellaro P 1994 *Nuovo Cim.* B **109** 159
[4] Charters T and Mimoso J P Self-interacting scalar field cosmologies: Unified exact solutions and symmetries *Prepint* arXiv:0909.2282 [hep-ph]
[5] Liddle A R, Parsons P and Barrow J D 1994 *Phys. Rev.* D **50** 7222
[6] Easther R 1996 *Class. Quantum Grav.* **13** 1775
[7] Barrow J D 1990 *Phys. Lett.* B **235** 40
Barrow J D and Saich P 1990 *Phys. Lett.* B **249** 406
[8] Chimento L P, Zuccala N and Mendez V 1999 *Class. Quantum Grav.* **16** 3749
[9] Corless R M, Gonnet G H, Hare D E G, Jeffrey D J and Knuth D E 1996 *Adv. Comput. Math.* **5** 329-359
[10] Chimento L P 2002 *Phys. Rev.* D **65** 063517
Chimento L P and Lazkoz R 2003 *Phys. Rev. Lett.* **91** 211301
[11] Mimoso J P and Charters T, Scalar-tensor Exact Solutions: A Unified Approach, in preparation.

From causality to time and back

Ettore Minguzzi

Dipartimento di Matematica Applicata, Università degli Studi di Firenze, Via S. Marta 3, I-50139 Firenze, Italy

E-mail: ettore.minguzzi@unifi.it

Abstract. In this work the problem of the existence of a (semi-)time function on spacetime is investigated together with the problem of recovering the causal structure from the set of time functions allowed by the spacetime. These problems are solved thanks also to a mathematical correspondence with utility theory.

1. Introduction

When reasoning about time in general relativity we can distinguish between two different notions which intuitively represent what in everyday life we may term the *metric* and the *order* sides of time. The former notion is given by *proper time* τ which allows us to give a rigorous meaning to the 'time interval' between two events on an observer's worldline. The latter notion is given by the *time function*, $t : M \to \mathbb{R}$, namely a function defined all over the spacetime which is continuous and increases over causal curves, that is, $x < y \Rightarrow t(x) < t(y)$.

In this work we shall deal with the latter aspect of time. In particular we investigate under which causality conditions a time function exists. If a time function exists it needs not to be unique, a fact which will bring us to the question as to whether the causal structure of a spacetime can be recovered from the set of time functions. We shall point out that mathematically these problems have analogs in the field of microeconomics known as *Utility Theory*. The drawn connection will allow us to obtain interesting new results on the spacetime side. For a more detailed study with full proofs the reader is referred to [1].

2. Order relations

Let us recall some definitions from the theory of relations. Let X be a set and let $\Delta = \{(x, x) : x \in X\}$. A *preorder* $\le$ on a set X is a reflexive ($x \le x$) and transitive relation. A *strict partial order* is an irreflexive ($x \nleq x$) and transitive relation. A *partial order* is a preorder which is antisymmetric ($x \le y$ and $y \le x \Rightarrow x = y$). A total preorder is a preorder in which every two elements $x, y \in X$ are comparable, that is (totality) $x \le y$ or $y \le x$. A total order is a partial order which is total, or equivalently, a total preorder which is antisymmetric.

Szpilrajn [2, 3] proved[1]

Theorem 2.1 *(Szpilrajn) Every partial order can be extended to a total order. Moreover, every partial order is the intersection of all the total orders that extend it.*

[1] Actually he gave the proof for strict partial orders but this version can be proved similarly.

In a sense in this paper we are going to find an analogous result in which the partial order is replaced by the Seifert relation (see below) and the total orders are replaced by preorderings induced by time functions.

On a spacetime (time oriented Lorentzian manifold) (M, g) let us write $x < y$ if there is a future directed causal curve from x to y and $x \le y$ if $x < y$ or $x = y$. Let us write $x \ll y$ if there is a future directed timelike curve from x to y. On a spacetime manifold it is natural to define the following relations

$$I^+ = \{(x, y) : x \ll y\},$$
$$J^+ = \{(x, y) : x \le y\}.$$

called respectively the chronological relation and the causal relation of the spacetime. In a chronological spacetime I^+ is a strict partial order. The relation J^+ is a preorder which becomes a partial order in causal spacetimes. With respect to the product topology on $M \times M$, I^+ is open [4, 5] while J^+ is not necessarily closed (e.g. remove a point from Minkowski spacetime).

Given two Lorentzian metrics on the manifold M we write $g < g'$ if at any point $x \in M$, the future timelike cone of g' contains the future causal cone of g. A spacetime is *stably causal* if causality is preserved under sufficiently small perturbations of the metric. More precisely, a spacetime is stably causal if there is $g' > g$ such that (M, g') is causal [6]. Stable causality is stronger than causality because it not only demands the absence of closed causal curves but also states that the light cones can be widened all over the spacetime without introducing closed causal curves.

Perhaps one of the most well behaved relations that can be defined on a spacetime is the Seifert relation [7]

$$J_S^+ = \bigcap_{g'>g} J_{g'}^+$$

which turns out to be reflexive, transitive and closed[8]. Moreover, it is a partial order if and only if the spacetime is stably causal [8, 9].

Another well behaved causality relation is that defined by Sorkin and Woolgar [10] and denoted K^+. By definition K^+ is the smallest relation which is closed, transitive and contains J^+. Thus the relation K^+ is defined exactly through its good properties. They also defined the spacetime to be K-causal if K^+ is antisymmetric. It turns out [11] that stable causality is equivalent to K-causality and that in this case $K^+ = J_S^+$. For this reason in most of what follows K^+ can replace J_S^+ and the other way around.

Let us now introduce the concept of time function. A *time function* is a continuous map $t : M \to \mathbb{R}$ such that $x < y \Rightarrow t(x) < t(y)$. The definition of *semi-time function* is similar with $x < y$ replaced by $x \ll y$. In [6, 4] Hawking proved that a stably causal spacetime admits a time function. The proof was based on the following argument. The spacetime M can be endowed with a complete Riemannian metric to which corresponds a volume form μ. Moreover, the Riemannian metric can be chosen in such a way that, $\mu(M) = 1$. Let g_λ, $\lambda \in [1, 2]$, be a one parameter family of causal metrics such that $g_\lambda > g$, $\lambda' > \lambda \Rightarrow g_{\lambda'} > g_\lambda$. The function $t_\lambda^-(x) = \mu(I_{g_\lambda}^-(x))$ satisfies $x < y \Rightarrow t_\lambda^-(x) < t_\lambda^-(y)$ but is only lower semi-continuous. Nevertheless the average $t = \int_1^2 t_\lambda^- \mathrm{d}\lambda$ can be proved to be continuous [4].

I shall give here a different and more general argument for the existence of a time function under stable causality. The idea is that this existence follows from the *closure* of the relation whose antisymmetry implies stable causality, that is J_S^+ or K^+. According to this approach the existence of a time function is in a sense independent of the Lorentzian geometric nature of the manifold. Before we come to this argument let me review some developments in utility theory.

3. Utility theory and its use in Lorentzian geometry

Utility theory begun with the solution given by Gabriel Cramer and Daniel Bernoulli to a problem conceived by Daniel's cousin, Nicolas Bernoulli, in 1713. The problem is today known as St. Petersburg's paradox. Nicolas considered two people A and B playing the following game. Player A pays a fixed fee F to enter the game, then the players toss repeatedly a coin till they get an head. If they get the first head at the n-th tossing then A wins 2^n dollars. What would be a fair price F for entering the game? Without the fee player A is expected to win

$$\frac{1}{2}\cdot 2+\frac{1}{4}\cdot 4+\cdot+\frac{1}{2^n}\cdot 2^n+\cdots=+\infty,$$

thus when offered to play A should accept whatever is the finite fee. This fact goes contrary to experiments in which it seems that people would hardly pay a mere 25$ fee to enter the game. According to D. Bernoulli the reason is that, roughly speaking, "happiness is not proportional to money", so it is not the expected value that should be averaged but rather a new function, called *utility* that quantifies how much an option is preferred over another. Thus in St. Petersburg game the average utility would actually converge if u is bounded as a function of money

$$\frac{1}{2}\cdot u(2)+\frac{1}{4}\cdot u(4)+\cdot+\frac{1}{2^n}\cdot u(2^n)+\cdots<+\infty$$

From the discussion around this problem started the utilitarian approach of economics, according to which an individual behaves so as to maximize the expected utility. In mathematical terms the utility is a function $u : A \to \mathbb{R}$ over a totally preordered space A which satisfies $x \le y$ iff $u(x) \le u(y)$. The space A is the space of alternatives, and the preorder expresses the preferences of the individual for those alternatives. The preorder is total because it is believed that the individual can always tell, when comparing two options, which one is preferred.

Actually the totality property was later challenged by economists [12] who dropped it in order to accommodate some evidence of indecisiveness in the individual behavior. With a preorder the definition of utility had to be changed as follows

$$x \le y \Rightarrow u(x) \le u(y), \qquad x<y \Rightarrow u(x)<u(y)$$

which reduces to the above definition for total preorders.

The alternative or prospects space A can be endowed with a topology which in this way introduces a notion of 'closeness' between alternatives. As a consequence in mathematical economics it became important to establish whether a given prospect space admitted a continuous utility function. In particular it was important to investigate the conditions to be imposed on the preorder so as to obtain the existence of a continuous utility function.

The most important theorems in this direction were Peleg's [13] and Levin's [14] theorems. We comment on Levin's theorem because it can be shown that, at least for our purposes, Peleg's theorem gives results which are weaker than Levin's. Levin's theorem in short states that the closure of the order relation is a sufficient requirement. More precisely we have the following

Theorem 3.1 *(Levin) Let X be a second countable locally compact Hausdorff space, and R a closed preorder on X, then there exists a continuous utility function. Moreover, denoting with $\mathcal{U}$ the set of continuous utilities we have that the preorder R can be recovered from the continuous utility functions, namely there is a multi-utility representation*

$$(x,y)\in R \Leftrightarrow \forall u \in \mathcal{U},\ u(x)\le u(y). \qquad (1)$$

Actually, the last statement was not included in Levin's article although its proof is tacitly given in that paper. This utility representation has been pointed out only recently in a preprint by Evren and Ok [15].

In [1] I have investigated the concept of utility for the relations I^+ and K^+ proving the following

Theorem 3.2

(i) In a chronological spacetime the continuous I^+-utilities are the semi-time functions.

(ii) In a stably causal spacetime the continuous K^+-utilities are the time functions.

Using point (i) I obtained what is possibly the only result available on the the existence of semi-time functions.

Theorem 3.3 *A chronological spacetime in which $\bar{J}^+$ is transitive admits a semi-time function.*

In order to apply this result it can be useful to recall that in reflective ($x \in \bar{I}^-(y)$ iff $y \in \bar{I}^+(x)$) spacetimes $\bar{J}^+$ is transitive [11] and that spacetimes admitting a complete timelike vector field are reflective [16].

I also proved, without using smoothing techniques, that the existence of a time function implies K-causality [11]. As K^+ is closed, joining this result with Levin's theorem we get

Theorem 3.4 *A spacetime is K-causal if and only if it admits a time function (as a consequence time functions are always K^+-utilities). In this case, denoting with $\mathcal{A}$ the set of time functions we have that the partial order K^+ can be recovered from the time functions, that is*

$$(x, y) \in K^+ \Leftrightarrow \forall t \in \mathcal{A},\ t(x) \le t(y). \tag{2}$$

I also showed that in this theorem *time function* can be replaced with *temporal function*, where a temporal function is a time function which is smooth with a timelike gradient. This theorem tells us under which conditions it is possible to recover the causality or the chronology relation from the set of time (or temporal) functions. In a causally simple spacetime it is possible to recover the causal relation because in these spacetimes $K^+ = J^+$, as J^+ is closed. In a class of spacetime which I called *causally easy* [11] which are characterized by strong causality and the transitivity of $\bar{J}^+$, and which stay between stable causality and causal continuity, it is possible to recover the chronology relation because $K^+ = \bar{J}^+$ and hence $I^+ = \text{Int}\bar{J}^+ = \text{Int}K^+$. Nevertheless, this study suggests that neither I^+ not J^+ should be regarded as fundamental and that instead at a more fundamental level the spacetime should be regarded as a topological ordered space where the order is provided by the K^+ relation.

Acknowledgments

This work has been partially supported by GNFM of INDAM and by FQXi.

References

[1] Minguzzi E 2009 Time functions as utilities *Preprint* arXiv:0909.0890
[2] Szpilrajn E 1930 *Fund. Math.* **16** 386–389
[3] Dushnik B and Miller E 1941 *Amer. J. Math.* **63** 600–610
[4] Hawking S W and Ellis G F R 1973 *The Large Scale Structure of Space-Time* (Cambridge: Cambridge University Press)
[5] Minguzzi E and Sánchez M 2008 *The causal hierarchy of spacetimes* (*ESI Lect. Math. Phys.* vol H. Baum, D. Alekseevsky (eds.), Recent developments in pseudo-Riemannian geometry) (Zurich: Eur. Math. Soc. Publ. House) pp 299–358, *Preprint* gr-qc/0609119
[6] Hawking S W 1968 *Proc. Roy. Soc. London, series A* **308** 433–435
[7] Seifert H 1971 *Gen. Relat. Gravit.* **1** 247–259
[8] Minguzzi E 2008 *Class. Quantum Grav.* **25** 015010
[9] Hawking S W and Sachs R K 1974 *Commun. Math. Phys.* **35** 287–296
[10] Sorkin R D and Woolgar E 1996 *Class. Quantum Grav.* **13** 1971–1993
[11] Minguzzi E 2009 *Commun. Math. Phys.* **290** 239–248
[12] Aumann R J 1962 *Econometrica* **30** 445–462
[13] Peleg B 1970 *Econometrica* **38** 93–96
[14] Levin V L 1983 *Soviet Math. Dokl.* **28** 715–718
[15] Evren O and Ok E A On the multi-utility representation of preference relations, preprint 2009
[16] Clarke C J S and Joshi P S 1988 *Class. Quantum Grav.* **5** 19–25

Creatable universes: a new approach

Ramon Lapiedra and Juan Antonio Morales-Lladosa*

Departament d'Astronomia i Astrofísica, Universitat de València, C/ Dr. Moliner 50, E-46100 Burjassot, València, Spain

E-mail: `ramon.lapiedra@uv.es`, `antonio.morales@uv.es`

Abstract. We are interested in the non asymptotically flat space-times for which all the momenta (energy, 3-momentum and angular 4-momenta) are conserved in time. We call *universes* such space-times. Starting from the Weinberg definition of the momenta associated to a spacelike 3-surface, we give a coordinate prescription to properly define the energy of a universe. The prescription includes the vanishing of linear and angular 3-momenta. This result allows us to consider the case of universes with vanishing 4-momenta (*creatable universes*) in a consistent way.

1. Introduction

People have speculated about the possibility that our Universe came out from a vacuum quantum fluctuation [1, 2]. In this case, one could expect that the resulting universe had vanishing energy, P^0, and ever better, vanishing energy-momentum, P^α, and vanishing angular 4-momentum $J^{\alpha\beta}$.

But the problem is the dramatic dependence of these momenta on the coordinate system used, letting aside the particular case of asymptotically flat space-times where one knows how to choose the good coordinate systems, as it is well known [3, 4]. Nevertheless, if we want to deal with a space-time, V_4, which could represent our Universe, we cannot always impose this asymptotic flatness.

To circumvent this problem of lack of uniqueness in the definition of these momenta, in [5] we explain why we think that we must deal with Gauss coordinates in V_4, based on some space-like 3-surface Σ_3. Even more, we discussed why we should take coordinates which, at the same time, allow to write the corresponding instantaneous 3-space metric in a conformally flat way on the boundary, Σ_2, of Σ_3.

But, this choice of coordinates let still a lot of freedom. Then, in the present paper, we impose further that our coordinates be such that the corresponding 3-momenta, P^i and J^{ij}, vanish. To do this, we previously prove that in any *universe* (i. e. a space-time where, once the convenient coordinate system has been choose, the defined 4-momenta remain constant in time) we always can change Σ_3, without changing Σ_2, such that $P^i = J^{ij} = 0$.

Finally, in a general and consistent way, we can define a *creatable universe* as a *universe* where the 4-momenta, defined in the above coordinate systems, vanish. As we have just said, these are coordinate systems which are Gauss coordinate systems, where $P^i = J^{ij} = 0$, and where the instantaneous space metric can be written in a conformally flat way on Σ_2. We will call these coordinate systems *intrinsic* coordinate systems.

Journal of Physics: Conference Series **229** (2010) 012053 doi:10.1088/1742-6596/229/1/012053

2. The good coordinate systems

As it has been stated in the Introduction, the big problem of the Universe energy and momenta definition is the lack of uniqueness: they are coordinate dependent quantities (whatever energy-momentum complex we take). In asymptotically flat space-times, there is no problem: P^α and $J^{\alpha\beta}$, are uniquely defined (in asymptotic Minkowskian coordinate systems). But, since we are interested in the energy and momenta of the Universe, we need to go beyond asymptotically flat space-times.

So, our first goal will be to reduce as much as possible the number of acceptable coordinate systems to properly define the Universe energy and momenta.

2.1. Gauss coordinates in V_4 from an space-like 3-surface, Σ_3.

We expect any possible universe to be a *universe* in the sense defined in the Introduction: a space-time whose well defined energy and momenta, i.e., P^α and $J^{\alpha\beta}$, are finite and conserved in time. Even better, we expect them to be zero for the actual Universe, since we expect it to be *creatable*.

Which coordinate time, then, to begin with? A physical universal time. That is, in order to properly define P^α and $J^{\alpha\beta}$, we need Gauss coordinates $\{x^\alpha\} = \{t, x^i\}$:

$$ds^2 = -dt^2 + g_{ij}dx^i dx^j \qquad (i, j = 1, 2, 3). \tag{1}$$

Since we treat all momenta, linear and angular, at the same level, we need a symmetric energy-momentum complex. We will take the Weinberg one [3]. Then, one obtains in Gauss coordinates:

- Linear 4-momentum:

$$P^0 = \frac{1}{\kappa}\int(\partial_j g_{ij} - \partial_i g)d\Sigma_{2i}\,, \quad P^i = \frac{1}{\kappa}\int(\partial_0 g\delta_{ij} - \partial_0 g_{ij})d\Sigma_{2j} \tag{2}$$

- Angular 4-momentum:

$$J^{jk} = \frac{1}{\kappa}\int(x_k\partial_0 g_{ij} - x_j\partial_0 g_{ki})d\Sigma_{2i}\,, \tag{3}$$

$$J^{0i} = P^i t - \frac{1}{\kappa}\int[(\partial_k g_{kj} - \partial_j g)x_i + g\delta_{ij} - g_{ij}]d\Sigma_{2j} \tag{4}$$

Notation: $g \equiv \delta^{ij}g_{ij}$; the 2-surface Σ_2 is the boundary of Σ_3; $d\Sigma_{2i}$ is the corresponding surface element. Indices are raised and lowered with the Minkowski tensor, and $\kappa \equiv 16\pi G$.

2.2. Conformal coordinates for the 3-space metric on the boundary Σ_2 of Σ_3.

Since P^α and $J^{\alpha\beta}$ are supposed to be conserved, we only need the metric in the neighborhood of Σ_3 $(t = 0)$. Furthermore, since P^α and $J^{\alpha\beta}$ are written as surface integrals on Σ_2, or as a limit of a sequence of surface integrals, all we need is a coordinate system on Σ_2 and its immediate neighborhood. In [5], the following result was obtained:

▷ Let Σ_2 be a 2-surface of a spacelike hypersurface Σ_3. Then, there exists a Gauss coordinate system $\{t, x^i\}$, based on Σ_3 $(t = 0)$, such that Σ_2 is given by $t = x^3 = 0$, and the space-time metric is written as

$$ds^2 = -dt^2 + g_{ij}(t, x^k)dx^i dx^j, \quad \text{with} \quad g_{ij}(t = x^3 = 0) = f(x^1, x^2)\delta_{ij} \tag{5}$$

That is, on Σ_2, the metric of Σ_3 may be expressed in conformally flat form.

In this new Gauss chart, at $t = 0$, the above expressions for the momenta take this simple form:

- Linear 4-momentum:

$$P^0 = -\frac{1}{\kappa}\int \partial_3 g_{aa}\, dx^1 dx^2\,, \quad P^3 = \frac{1}{\kappa}\int \partial_0 g_{aa}\, dx^1 dx^2 \tag{6}$$

$$P^a = -\frac{1}{\kappa}\int \partial_0 g_{3a}\, dx^1 dx^2 \quad (a = 1, 2) \tag{7}$$

where $g_{aa} = g_{11} + g_{22}$.

- Angular 4-momentum:

$$J^{ij} = \frac{1}{\kappa}\int (x_j \partial_0 g_{3i} - x_i \partial_0 g_{3j})\, dx^1 dx^2\,, \tag{8}$$

$$J^{0a} = \frac{1}{\kappa}\int x^a \partial_3 g_{bb}\, dx^1 dx^2\,, \tag{9}$$

$$J^{03} = -\frac{1}{\kappa}\int g_{aa}\, dx^1 dx^2\,. \tag{10}$$

3. Universes. Intrinsic coordinates

Let us consider a *universe*: a space-time where both 4-momenta, P^α and $J^{\alpha\beta}$, have been well defined and are conserved quantities.

Then, the question is: for any *universe*, there exist Gauss coordinate systems such that both 3-momenta, P^i and J^{ij}, irrespective of the momentum origin vanish? The answer is YES, and consequently, the energy P^0 and momenta J^{0i} have an "intrinsic" meaning in these coordinate systems. More precisely, the following statement occurs:

▷ For any *universe*, a Gauss coordinate system $\{t, x^i\}$ exists such that the metric is written in the form (5) and, in addition, the lineal and angular 3-momenta of the hypersurface $t = 0$ vanish, that is

$$P^i = 0 \quad \text{and} \quad J^{ij} = 0. \tag{11}$$

We call here such a Gauss coordinate system a universe *intrinsic* coordinate system. Intrinsic coordinate systems are the appropriated ones in order to calculate the energy, P^0, and mixed momenta, J^{0i}, of a *universe*.

4. The precise definition of creatable universes

Let it be a given *universe*. Let us choose some intrinsic coordinate system in the precise sense we have just defined. Imagine that the universe is such that P^0 and J^{0i} vanish. Then, we will have in all $P^\alpha = J^{\alpha\beta} = 0$ and we will call this universe a creatable universe.

This definition drive us to do the following comment: should not this name of creatable universes be reserved for universes such that their linear and angular 4-momenta vanish irrespective of the intrinsic coordinates chosen? The answer is no, since it can be seen that even the Minkowski space, which has $P^\alpha = J^{\alpha\beta} = 0$ for some intrinsic coordinates (for Lorentzian coordinates), cannot preserve this vanishing property when we move to another arbitrary intrinsic coordinate system. The reason of this is that, in doing so, we can leave the coordinates "adapted" to the symmetries of the metric: the ones tied to the ten parameters of the Poincaré group.

Thus, given a universe which has $P^\alpha = J^{\alpha\beta} = 0$ for some intrinsic coordinate system, if there are other intrinsic coordinates where this vanishing is not preserved, we should consider that this non preservation expresses the fact that the new intrinsic coordinates are not well adapted to some basic metric symmetries. Adapted to which symmetries? To the ones which allow us to have just vanishing 4-momenta for some intrinsic coordinate system.

5. Final considerations

One of the main interests of the present paper could be to give a criterion to reject from the very beginning a very large family of space-times as candidates to represent our actual Universe. According to this criterion it could only be represented by a creatable universe. Thus, in [6], it was proved that, within the inflationary perturbed Friedmann-Lemaître-Robertson-Walker universes, only the closed case corresponds to a creatable universe.

Acknowledgments

This work has been also supported by the MEC-FEDER project No. FIS2006-06062 and the MICINN-FEDER project No. FIS2009-07705.

References

[1] Albrow M G 1973 *Nature* **241** 56
[2] Tryon E P 1973 *Nature* **246**, 396
[3] Weinberg S 1972 *Gravitation and Cosmology* (New York: John Wiley & Sons)
[4] Murchadha N Ó 1986 *J. Math. Phys.* **27** 2111
[5] Ferrando J J, Lapiedra R and Morales J A 2007 *Phys. Rev.* D **75** 124003
[6] Lapiedra R and Sáez D 2008 *Phys. Rev.* D **77** 104011

Journal of Physics: Conference Series **229** (2010) 012054 doi:10.1088/1742-6596/229/1/012054

Secondarily-induced gravitational wave background in modified gravity

Masaaki Morita* **and Hirotaka Takahashi**

Nagaoka University of Technology, 1603-1 Kamitomioka, Nagaoka, Niigata 940-2188, Japan

E-mail: `morita@mst.nagaokaut.ac.jp`, `hirotaka@oberon.nagaokaut.ac.jp`

Abstract. We explore the effect of structure formation on gravitational wave background in dark energy cosmology. Second-order tensor perturbations induced by first-order scalar mode are considered in a modified gravity dark energy model. We present explicitly the procedure for an approximate estimation of the second-order tensor perturbations caused by primary structure formation.

1. Introduction

It is more than ten years since accelerated expansion in the late-time universe was first discovered by observations of type Ia supernovae [1, 2, 3]. In order to account for the cosmic acceleration, theoretical possibilities have been explored with a large number of dark energy models [4]. A 'modified gravity' dark energy model, in which some corrections are added to the Einstein-Hilbert action, is one of theoretically possible models. Although it is of interest to consider the link between gravitational theories and dark energy, modified gravity models have generally severe constraints that come from local gravity tests and cosmological observations [5]. For such constraints, matter density perturbations have been studied extensively [6], but in future measurements of relic gravitational waves will also be useful [7].

In this article, we consider cosmological gravitational wave background generated by primary structure formation. We present how to compute second-order tensor perturbations induced by first-order scalar perturbations in the framework of a modified gravity dark energy model. The appearance of the effect of gravity modification is presumed with the help of the results of the second-order perturbations in a standard, spatially flat FLRW universe model.

2. Basic equations in modified gravity

Let us consider modified gravity models with the action

$$\mathcal{S} = \frac{1}{16\pi G}\int f(R)\sqrt{-g}\,\mathrm{d}^4x \;+\mathcal{S}_{\mathrm{m}}\,, \tag{1}$$

where G is Newton's gravitational constant, R is the four-dimensional scalar curvature, g is the determinant of the space-time metric $g_{\mu\nu}$, and $\mathcal{S}_{\mathrm{m}}$ denotes the action of pressureless matter. The action principle yields the following field equation:

$$F(R)R_{\mu\nu} - \frac{1}{2}f(R)g_{\mu\nu} - \mathcal{D}_\mu\mathcal{D}_\nu F(R) + g_{\mu\nu}\Box F(R) = 8\pi G T^{(\mathrm{m})}_{\mu\nu}\,, \tag{2}$$

where $F := \partial f/\partial R$, $\mathcal{D}$ represents the four-dimensional covariant derivative, $\Box := g^{\mu\nu}\mathcal{D}_\mu\mathcal{D}_\nu$, and $T^{(\mathrm{m})}_{\mu\nu}$ is the energy-momentum tensor of pressureless matter. Hereafter we write the function $f(R)$ as $R - \xi(R)$, where $\xi(R)$ plays the role of an effective dark energy component, and then $F(R) = 1 - \xi'(R)$ with a prime $(\,')$ $:= \partial/\partial R$.

In a spatially flat, homogeneous and isotropic universe model with the scale factor $a(t)$ and the energy density ρ_{m} of pressureless matter, the field equation (2) reads

$$3H^2 = 8\pi G\rho_{\mathrm{m}} + \frac{1}{2}\xi(R_{\mathrm{b}}) + 3H\left[\xi'(R_{\mathrm{b}})\right]^{\cdot} - 3\xi'(R_{\mathrm{b}})\left(\dot{H} + H^2\right), \tag{3}$$

$$2\dot{H} = -8\pi G\rho_{\mathrm{m}} + 2\xi'(R_{\mathrm{b}})\dot{H} + \left[\xi'(R_{\mathrm{b}})\right]^{\cdot\cdot} - H\left[\xi'(R_{\mathrm{b}})\right]^{\cdot}, \tag{4}$$

where an overdot $(\,\dot{}\,)$ denotes the time derivative $\partial/\partial t$, $H := \dot{a}/a$ is the Hubble parameter, and $R_{\mathrm{b}} := 6\dot{H} + 12H^2$. Note that these equations are reduced to the Friedmann equations in the standard FLRW universe model if we set $\xi(R) = 2\Lambda$ with a positive cosmological constant Λ.

3. Linear tensor perturbations in a modified gravity model

Let us proceed to linear perturbations in a modified gravity dark energy model. We choose a comoving and synchronous coordinate condition with the line element

$$\mathrm{d}s^2 = -\mathrm{d}t^2 + {}^{(3)}g_{ij}\,\mathrm{d}x^i\mathrm{d}x^j, \tag{5}$$

and the four-velocity $u^\mu = (1, \mathbf{0})$ of pressureless matter. Writing the spatial metric ${}^{(3)}g_{ij}$ as ${}^{(3)}g_{ij} = a(t)^2(\gamma_{ij} + h_{ij})$, where γ_{ij} and h_{ij} are metric components of a background universe and a linearly perturbed universe, respectively, the four-dimensional scalar curvature is written as

$$R = R_{\mathrm{b}} + R_1 + \mathcal{O}(h^2)\,; \qquad R_1 := \ddot{h} + 4H\dot{h} + \frac{1}{a^2}\left(h^k{}_{\ell|k}{}^{|\ell} - \nabla^2 h\right), \tag{6}$$

where $h := \gamma^{ij}h_{ij}$ is the trace of the metric perturbation h_{ij}, and $\nabla^2(\cdot) := \gamma^{ij}(\cdot)_{|ij}$ (a vertical bar $|$ represents the covariant derivative with respect to γ_{ij}). By applying the Taylor expansion, we also have

$$F(R) = F_{\mathrm{b}} + F_1 + \mathcal{O}(h^2)\,; \quad F_{\mathrm{b}} := F(R_{\mathrm{b}}) = 1 - \xi'(R_{\mathrm{b}}), \quad F_1 := F'(R_{\mathrm{b}})R_1 = -\xi''(R_{\mathrm{b}})R_1\,. \tag{7}$$

Taking the part of order h of the field equation (2) with the help of equation (7), we obtain the evolution equation for h_{ij}:

$$\ddot{h}_{ij} + \left(3H + \frac{\dot{F}_{\mathrm{b}}}{F_{\mathrm{b}}}\right)\dot{h}_{ij} + \frac{2}{a^2}\left(\mathcal{R}_{ij} - \frac{1}{4}\mathcal{R}\gamma_{ij}\right) + \frac{1}{F_{\mathrm{b}}}\left[\ddot{F}_1 + 2H\dot{F}_1 - \left(\dot{H} + 3H^2\right)F_1\right]\gamma_{ij} - \frac{2}{a^2F_{\mathrm{b}}}F_{1\,|ij} = 0\,, \tag{8}$$

where $\mathcal{R}_{ij} := \frac{1}{2}(h^k{}_{i|jk} + h^k{}_{j|ik} - \nabla^2 h_{ij} - h_{|ij})$, and $\mathcal{R} := \gamma^{ij}\mathcal{R}_{ij}$. The transverse and traceless part of equation (8) gives

$$\ddot{h}^{\mathrm{TT}}_{ij} + \left(3H + \frac{\dot{F}_{\mathrm{b}}}{F_{\mathrm{b}}}\right)\dot{h}^{\mathrm{TT}}_{ij} - \frac{1}{a^2}\nabla^2 h^{\mathrm{TT}}_{ij} = 0\,, \tag{9}$$

where h^{TT}_{ij} is a transverse and traceless part of h_{ij}. Equation (9) tells us that the effect of gravity modification appears only in the friction term in the wave equation for the linear tensor perturbations. Solutions for equation (9) cannot be expressed in terms of analytic functions in general. However, in matter-dominated era, where the effective dark energy component can be neglected, solutions for the linear tensor perturbations in a Fourier space are approximately given as [8]

$$h_{ij}^{\mathrm{TT}}(k,a) \propto a^{-3/4}\mathcal{J}_{\pm 3/2}\left(\frac{2ka^{1/2}}{H_0\Omega_{\mathrm{m0}}^{1/2}}\right), \tag{10}$$

and in dark-energy-dominated era, they should become asymptotically

$$h_{ij}^{\mathrm{TT}}(k,a) \propto a^{-3/2}\mathcal{J}_{\pm 3/2}\left(\frac{ka^{-1}}{H_0\Omega_{\Lambda 0}^{1/2}}\right), \tag{11}$$

where $\mathcal{J}_\nu$ is Bessel's function of order ν, k is a wavenumber, and H_0, Ω_{m0} and $\Omega_{\Lambda 0}$ denote the present values of the Hubble parameter, the matter density parameter and the amount of the effective dark energy component, respectively. These approximate solutions imply that the presence of the effective dark energy component leads to the changes of decaying rate and oscillation periods of the linear tensor perturbations.

4. Second-order tensor perturbations in a modified gravity model

Next let us consider second-order perturbations in a modified gravity dark energy model, and write the spatial metric ${}^{(3)}g_{ij}$ as ${}^{(3)}g_{ij} = a(t)^2(\gamma_{ij} + h_{ij}^{(1)} + h_{ij}^{(2)})$, where $h_{ij}^{(1)}$ and $h_{ij}^{(2)}$ represent the first-order and second-order metric perturbations, respectively. Here we focus on second-order tensor perturbations generated by primary structure formation so that $h_{ij}^{(1)}$ contains only a scalar mode. Following the procedure described in the case of the linear tensor perturbations, we obtain the evolution equation for $h_{ij}^{(2)}$:

$$\ddot{h}_{ij}^{(2)} + \left(3H + \frac{\dot{F}_{\mathrm{b}}}{F_{\mathrm{b}}}\right)\dot{h}_{ij}^{(2)} - \frac{1}{a^2}\nabla^2 h_{ij}^{(2)} = \mathcal{Q}_{ij}, \tag{12}$$

where $\mathcal{Q}_{ij}$ is a source term that consists of pieces of quadratic terms with respect to $h_{ij}^{(1)}$. In a spatially flat FLRW universe model under Einstein's gravity without a cosmological constant, a scalar-mode solution for linear metric perturbations is written as

$$h_{ij}^{(1)} = \frac{20}{9}\varphi\gamma_{ij} + 2a(t)\varphi_{|ij}, \tag{13}$$

where $\varphi = \varphi(\mathbf{x})$ is an initial gravitational potential. We can adopt the above scalar-mode solution as $h_{ij}^{(1)}$ if primary structure formation proceeds mainly in matter-dominated era, where the effect of dark energy component is not important. Substituting equation (13) into the source term $\mathcal{Q}_{ij}$, we find [9, 10, 11]

$$\mathcal{Q}_{ij} = \frac{3}{7}\left(\nabla^2\mu\gamma_{ij} + \mu_{|ij} - 4\nabla^2\mu_{ij}\right), \tag{14}$$

where

$$\mu_{ij} := \frac{1}{2}\left(\nabla^2\varphi\varphi_{|ij} - \varphi_{|ik}\varphi^{|k}{}_{|j}\right), \quad \mu := \gamma^{ij}\mu_{ij}. \tag{15}$$

In the case of Einstein's gravity, equation (12) is solved in a Fourier space using the method of a Green function as [11]

$$h_{ij}^{(2)}(k,t) = a(t)^2\frac{\mathcal{Q}_{ij}(k)}{k^2} + a(t)\frac{28\mathcal{Q}_{ij}(k)}{9k^4} + \left[1 - a^{-3/4}\mathcal{J}_{\pm 3/2}\left(\frac{2ka^{1/2}}{H_0\Omega_{\mathrm{m0}}^{1/2}}\right)\right]\frac{280\mathcal{Q}_{ij}(k)}{81k^6}. \tag{16}$$

Note that only the last term in equation (16) is observable as a gravitational wave because of its oscillatory behaviour, and its temporal dependence is the same as that of the solution (10) in the linear case. Hence we presume that the effect of gravity modification will also appear in the same way as the linear case.

5. Summary and outlook

In this article, cosmological gravitational wave background has been considered within a modified gravity dark energy model up to the second order in perturbations. We have shown how to compute the second-order tensor perturbations induced by the first-order scalar perturbations. In the level of linear perturbations, the effect of gravity modification appears only in the friction term in the wave equation, leading to the changes of decaying rate and oscillation periods. In the second-order perturbations, the appearance of the effect of gravity modification is approximately the same as in the linear level, if primary structure formation proceeds mainly in matter-dominated era. In order to make the results more rigorous, the effect of gravity modification should be taken into account in the calculation of the second-order source term $\mathcal{Q}_{ij}$. Also, the effect of scalar-type gravitational waves should be considered, which will arise typically in modified gravity models [7]. These are now under investigation.

Acknowledgments

The authors would like to thank the organizers of ERE2009 for providing them an opportunity to present this work in the Meeting, and Fernando Atrio-Barandela for valuable remarks on tensor perturbations. The work was partially supported by JSPS Grant-in-Aid for Scientific Research, No. 21740199 (MM) and No. 20540260 (HT). HT was also supported in part by the Uchida Energy Science Promotion Foundation and Sasagawa Scientific Research Grant.

References

[1] Perlmutter S *et al* 1999 *Astrophys. J.* **517** 565
[2] Riess A G *et al* 1998 *Astron. J.* **116** 1009
[3] Riess A G *et al* 1999 *Astron. J.* **117** 707
[4] Copeland E J, Sami M and Tsujikawa S 2006 *Int. J. Mod. Phys.* D **15** 1753
[5] Tsujikawa S 2008 *Phys. Rev.* D **77** 023507
[6] Tsujikawa S, Gannouji R, Moraes B and Polarski D 2009 *Phys. Rev.* D **80** 084044
[7] Capozziello S, De Laurentis M, Nojiri S and Odintsov S D 2009 *Gen. Rel. Grav.* **41** 2313
[8] Weinberg S 1972 *Gravitation and Cosmology* (New York: Wiley) pp 584–586
[9] Tomita K 1967 *Prog. Theor. Phys.* **37** 831
[10] Russ H, Morita M, Kasai M and Börner G 1996 *Phys. Rev.* D **53** 6881
[11] Matarrese S, Mollerach S and Bruni M 1998 *Phys. Rev.* D **58** 043504

Journal of Physics: Conference Series **229** (2010) 012055 doi:10.1088/1742-6596/229/1/012055

On the use of curvature invariants in numerical relativity

A Nerozzi[1,2]

[1] Centro Multidisciplinar de Astrofísica — CENTRA, Departamento de Física, Instituto Superior Técnico — IST, Av. Rovisco Pais 1, 1049-001 Lisboa, Portugal

[2] Institut für Angewandte Mathematik, Friedrich-Schiller-Universität, Ernst-Abbe-Platz 2, 07743 Jena, Germany

E-mail: andrea.nerozzi@ist.utl.pt

Abstract. The recent breakthroughs in numerical relativity have made it possible to generate gravitational wave signals coming from binary systems. The waveforms are extracted using the Newman-Penrose formalism and in particular the Weyl scalar Ψ_4. In order for Ψ_4 to contain the physical information on the gravitational wave signal, the tetrad chosen for its calculation should converge to the Kinnersley tetrad in the limit of single black hole space-time.

In this work we present a compact expression for Ψ_4 that automatically enforces this condition; the result is a simple function of curvature invariants that can be easily calculated in a numerical code.

1. Introduction

Given the recent successes of numerical relativity and the ability of current codes [1, 2, 3] to simulate a binary black hole merger, the accuracy of generated waveforms has become of primary importance. This is the so called problem of wave extraction which consists of calculating specific quantities from the numerically evolved variables that are directly related to the gravitational wave degrees of freedom. To achieve this, most groups use the Newman-Penrose (NP) formalism and in particular the Weyl scalar Ψ_4.

The relevant quantities in the NP formalism are the mentioned Weyl scalars, given by the contraction of the Weyl tensor over a specific combination of the four null tetrad vectors, and the connection coefficients (spin coefficients) related to the covariant derivatives of the same tetrad vectors. All of these quantities are scalars, making the choice of the coordinate system irrelevant for their calculation; they are however dependent on the tetrad choice which constitutes the gauge freedom in this formalism.

It is well known that these variable, under certain assumptions, acquire a precise physical meaning. For example Ψ_0 and Ψ_4 are related to the ingoing and outgoing gravitational wave contribution, while Ψ_2 is related to the background contribution to the curvature. This explains the use of Ψ_4 for numerical simulations of Einstein's equations. However, a suitable tetrad choice is fundamental for the calculation of these quantities. The importance of a robust wave extraction technique has been underlined by some recent articles [4, 5, 6, 7]. Recent works [8, 9, 10] have identified in *transverse tetrads*, i.e. those tetrads satisfying the condition $\Psi_1 = \Psi_3 = 0$, a convenient candidate for wave extraction and in general for a better understanding of the equations governing the Newman-Penrose formalism.

The advantages of the choice $\Psi_1 = \Psi_3 = 0$ have already been shown in [8, 9, 10], namely one of these three frames naturally converges to the frame where $\Psi_0 = \Psi_4 = 0$ when the spacetime approaches Petrov type D. This property ensures that the values for the two scalars Ψ_0 and Ψ_4, in the linear regime, are at first order tetrad invariant, and directly associated with the gravitational wave signal [11].

To determine the tetrad completely, however, one also needs to fix the spin-boost degree of freedom. The Kinnersley tetrad [12] identifies the spin-boost parameter by imposing the condition on one spin coefficient $\epsilon = 0$.

In this paper we show that, choosing the tetrads where $\Psi_1 = \Psi_3 = 0$, it is possible to find a connection between the spin-boost parameter and the expressions for the spin coefficients. This will allow us to impose the condition $\epsilon = 0$ and find the correspondent value for the spin-boost parameter. Applying this spin boost parameter to the expression of Ψ_4 will give us a compact result which is function of curvature invariants.

2. General definitions

Weyl scalars are given by contraction of the Weyl tensor over a certain combination of four null vectors, two real (ℓ^μ and n^μ) and two complex conjugates (m^μ and $\bar{m}^\mu$), according to

$$\Psi_0 = -C_{abcd}\ell^a m^b \ell^c m^d, \quad (1a)$$
$$\Psi_1 = -C_{abcd}\ell^a n^b \ell^c m^d, \quad (1b)$$
$$\Psi_2 = -C_{abcd}\ell^a m^b \bar{m}^c n^d, \quad (1c)$$
$$\Psi_3 = -C_{abcd}\ell^a n^b \bar{m}^c n^d, \quad (1d)$$
$$\Psi_4 = -C_{abcd}n^a \bar{m}^b n^c \bar{m}^d. \quad (1e)$$

These expressions are obviously tetrad dependent, however it is possible to construct two quantities which are no longer dependent on the tetrad choice. Such quantities are the curvature invariants I and J, and their expressions as functions of the Weyl scalars are given by $I = \Psi_4\Psi_0 - 4\Psi_1\Psi_3 + 3\Psi_2^2$ and $J = \Psi_4\left(\Psi_0\Psi_2 - \Psi_1^2\right) - \Psi_3\left(\Psi_3\Psi_0 - \Psi_1\Psi_2\right) + \Psi_2\left(\Psi_1\Psi_3 - \Psi_2^2\right)$.

In transverse frames (i.e. when $\Psi_1 = \Psi_3 = 0$) the non-vanishing Weyl scalars can be written directly as functions of curvature invariants, yielding

$$\Psi_0 = -\frac{i\mathcal{B}^{-2}}{2} \cdot \Psi_-, \quad (2a)$$
$$\Psi_2 = -\frac{1}{2\sqrt{3}} \cdot \Psi_+, \quad (2b)$$
$$\Psi_4 = -\frac{i\mathcal{B}^{2}}{2} \cdot \Psi_-, \quad (2c)$$

where

$$\Psi_\pm = I^{\frac{1}{2}}\left(e^{\frac{2\pi i k}{3}}\Theta \pm e^{-\frac{2\pi i k}{3}}\Theta^{-1}\right). \quad (3)$$

In these expressions $\Theta = \sqrt{3}PI^{-\frac{1}{2}}$, $P = \left[J + \sqrt{J^2 - (I/3)^3}\right]^{\frac{1}{3}}$ and k is an integer number assuming the values $\{0,1,2\}$ corresponding to the three different transverse frames. The spin-boost parameter is given by $\mathcal{B} = \left(\frac{\Psi_4}{\Psi_0}\right)^{\frac{1}{4}}$. The limit of type D corresponds to $\Theta \to 1$.

Journal of Physics: Conference Series **229** (2010) 012055 doi:10.1088/1742-6596/229/1/012055

3. The connection between spin coefficients and spin-boost parameter

In this section, we will use the Ricci identities to understand how the spin coefficients ϵ, γ, α and β relate to the spin-boost parameter $\mathcal{B}$.

We assume to be in the Petrov type D limit, where $\Psi_0 = \Psi_1 = \Psi_3 = \Psi_4 = 0$ and also, as a consequence of the Goldberg-Sachs theorem, the four spin coefficients λ, σ, ν and κ vanish. We begin with the following Ricci identity

$$D\tilde{\beta} - \delta\tilde{\epsilon} = \tilde{\epsilon}\,(\pi^* - \alpha^* - \beta) + \tilde{\beta}\,(\rho^* + \epsilon - \epsilon^*)\,. \tag{4}$$

which is expressed in terms of the reduced spin coefficients $\tilde{\epsilon} = \epsilon + \frac{1}{2}D\ln\mathcal{B}$ and $\tilde{\beta} = \beta + \frac{1}{2}\delta\ln\mathcal{B}$.

Comparing Eq. (4) with the expression of the commutator $[D,\delta]$ (again assuming $\sigma = \kappa = 0$)

$$[D,\delta] = (\pi^* - \alpha^* - \beta)\,D + (\rho^* + \epsilon - \epsilon^*)\,\delta, \tag{5}$$

it is possible to see that the Ricci identity is consistent with having $\tilde{\epsilon} = D\mathcal{H}_1$ and $\tilde{\beta} = \delta\mathcal{H}_1$, where $\mathcal{H}_1$ is a function to be determined. Using the equivalent Ricci identity obtained after exchanging the tetrad vectors $\ell \leftrightarrow n$ and $m \leftrightarrow \bar{m}$ we obtain an equivalent result for the reduced spin coefficients $\tilde{\gamma} = \gamma + \frac{1}{2}\Delta\ln\mathcal{B}$ and $\tilde{\alpha} = \alpha + \frac{1}{2}\delta^*\ln\mathcal{B}$ and conclude that they also can be expressed as $\tilde{\gamma} = \Delta\mathcal{H}_2$ and $\tilde{\alpha} = \delta^*\mathcal{H}_2$, and from the properties of transformation of spin coefficients under exchange operation one must have $\mathcal{H}_1 = -\mathcal{H}_2 = \mathcal{H}$. The expression for $\mathcal{H}$ can be found using the following two other Ricci identities

$$\begin{aligned} D\tilde{\gamma} - \Delta\tilde{\epsilon} &= \tilde{\alpha}\,(\tau + \pi^*) + \tilde{\beta}\,(\tau^* + \pi) + \tau\pi \\ &- \tilde{\gamma}\,(\epsilon + \epsilon^*) - \tilde{\epsilon}\,(\gamma + \gamma^*) + \Psi_2, \end{aligned} \tag{6a}$$

$$\begin{aligned} \delta\tilde{\alpha} - \delta^*\tilde{\beta} &= \mu\rho + \tilde{\alpha}\alpha^* + \tilde{\beta}\beta^* - 2\tilde{\alpha}\tilde{\beta} \\ &+ \tilde{\gamma}\,(\rho - \rho^*) + \tilde{\epsilon}\,(\mu - \mu^*) - \Psi_2, \end{aligned} \tag{6b}$$

which, together with the just found properties of these spin coefficients as directional derivatives of the function $\mathcal{H}$, lead to the following equation for $\mathcal{H}$:

$$\nabla^\mu\nabla_\mu\mathcal{H} + \nabla^\mu\ln\left(I^{\frac{1}{6}}\right)\nabla_\mu\left(2\mathcal{H} + \ln I^{\frac{1}{12}}\right) = -2\Psi_2, \tag{7}$$

where we have also used the fact that in the Petrov type D limit $\rho = D\ln I^{\frac{1}{6}}$, $\mu = -\Delta\ln I^{\frac{1}{6}}$, $\tau = \delta\ln I^{\frac{1}{6}}$ and $\pi = -\delta^*\ln I^{\frac{1}{6}}$. The solution for Eq. (7) in the case of Kerr space-time using Boyer-Lindquist coordinates, is given by

$$\mathcal{H} = \frac{1}{2}\ln\left(\Gamma^{\frac{1}{2}}I^{\frac{1}{6}}\sin\theta\right), \tag{8}$$

where $\Gamma = r^2 - 2Mr + a^2$, M and a being mass and spin parameter of the black hole. The spin coefficient ϵ is therefore given by

$$\epsilon = \frac{1}{2}D\ln\left(\Gamma^{\frac{1}{2}}I^{\frac{1}{6}}\mathcal{B}^{-1}\sin\theta\right). \tag{9}$$

Setting $\epsilon = 0$ leads to the following expression for the spin-boost parameter

$$\mathcal{B} = \mathcal{B}_0 f\,(\theta)\,I^{\frac{1}{6}}\Gamma^{\frac{1}{2}}\sin\theta, \tag{10}$$

where $\mathcal{B}_0$ is an integration constant and the function $f(\theta)$ can be found from the expression of the spin coefficient β and is given by $f\,(\theta) = \sin^{-1}\theta$. The final result for $\mathcal{B}$ is therefore $\mathcal{B} = \mathcal{B}_0 I^{\frac{1}{6}}\Gamma^{\frac{1}{2}}$.

When applied to the expression for the Weyl scalars given in Eq. (2), Eq. (10) yields

$$\Psi_0 = \mathcal{B}_0^{-2} \cdot \Gamma^{-1} I^{\frac{1}{6}} \left(\Theta - \Theta^{-1}\right), \tag{11a}$$

$$\Psi_4 = \mathcal{B}_0^{2} \cdot \Gamma I^{\frac{5}{6}} \left(\Theta - \Theta^{-1}\right), \tag{11b}$$

for ingoing waves, and

$$\Psi_0 = \mathcal{B}_0^{2} \cdot \Gamma I^{\frac{5}{6}} \left(\Theta - \Theta^{-1}\right), \tag{12a}$$

$$\Psi_4 = \mathcal{B}_0^{-2} \cdot \Gamma^{-1} I^{\frac{1}{6}} \left(\Theta - \Theta^{-1}\right), \tag{12b}$$

for outgoing waves.

These expressions for the scalars immediately give the correct radial fall-offs at future null infinity once the peeling behavior of the Weyl tensor is assumed, corresponding to $I \propto r^{-6}$ and $J \propto r^{-9}$. The function Γ is defined in the limit of Petrov type D and gives no radial contribution at future null infinity; the variable Θ gives no contribution either as it is the ratio of quantities that have the same radial behavior at future null infinity. In conclusion, Eq. (12) gives $\Psi_0 \propto r^{-5}$ and $\Psi_4 \propto r^{-1}$ for outgoing waves, as is expected from perturbation theory.

Future work will address the determination of the integration constant $\mathcal{B}_0$ as well as relating the function Γ to other invariant properties of the space-time under consideration.

Acknowledgments

The author is funded by FCT through grant SFRH/BPD/47955/2008 and by the DFG grant SFB/Transregio 7 "Gravitational Wave Astronomy". This work was partially supported by FCT - Portugal through projects PTDC/FIS/64175/2006, PTDC/FIS/098025/2008, PTDC/FIS/098032/2008, PTDC/CTE-AST/098034/2008 and CERN/FP/109290/2009. Computations were performed on the TeraGrid clusters ranger and kraken and at Magerit in Madrid. The authors thankfully acknowledge the computer resources, technical expertise and assistance provided by the Barcelona Supercomputing Centre—Centro Nacional de Supercomputación.

References

[1] Pretorius F 2005 *Phys. Rev. Lett.* **95** 121101
[2] Baker J G, Centrella J, Choi D I, Koppitz M and van Meter J 2006 *Phys. Rev.* D **73** 104002
[3] Campanelli M, Lousto C O, Marronetti P and Zlochower Y 2006 *Phys. Rev. Lett.* **96** 111101 (*Preprint* gr-qc/0511048)
[4] Lehner L and Moreschi O M 2007 *Phys. Rev.* D **76** 124040 (*Preprint* 0706.1319)
[5] Gallo E, Lehner L and Moreschi O M 2008 *Preprint* 0810.0666
[6] Gallo E, Lehner L and Moreschi O 2008 *Phys. Rev.* D **78** 084027
[7] Pazos E *et al.* 2007 *Class. Quantum Grav.* **24** S341-S368 (*Preprint* gr-qc/0612149)
[8] Beetle C, Bruni M, Burko L M and Nerozzi A 2005 *Phys. Rev.* D **72** 024013
[9] Nerozzi A, Beetle C, Bruni M, Burko L M and Pollney D 2005 *Phys. Rev.* D **72** 024014
[10] Nerozzi A, Bruni M, Re V and Burko L M 2006 *Phys. Rev.* D **73** 044020
[11] Teukolsky S A 1973 *Astrophys. J.* **185** 635–647
[12] Kinnersley W 1969 *J. Math. Phys* **10** 1195

Journal of Physics: Conference Series **229** (2010) 012056 doi:10.1088/1742-6596/229/1/012056

Stability of black holes from hydrodynamics

Masato Nozawa

Department of Physics, Waseda University, Tokyo 169-8555, Japan

E-mail: `nozawa@gravity.phys.waseda.ac.jp`

Abstract. It has been well known from old days as a "membrane paradigm" that black holes exhibit laws analogous to laws of fluids. Recent development of AdS/CFT correspondence enhances this similarity into an exact duality, according to which the fluid lumps solving Navier-Stokes equations corresponds holographically to black holes in Scherk-Schwarz (SS) AdS spacetime solving Einstein's equations in the bulk. Black rings in SS-AdS$_5$ are considered to be gravitational duals of stationary and axisymmetric fluid annuli found by Lahiri and Minwalla. We will present a study on the stability of higher dimensional black holes via this dual configurations and further discuss new multiple plasma configurations.

1. Preliminaries

Higher dimensional black holes are of much interest in various points of view. Among other things, stability against gravitational perturbations is a very fundamental but difficult issue to be answered [1]. The AdS/CFT correspondence may provide a useful insight to this problem in a new approach. In the AdS/CFT viewpoint, certain conformal field theory of finite temperature is dual to a black hole immersed in AdS spacetime. At high energies, field theories have an effective hydrodynamic description: an approximation at long-wavelength expansion. The similarity of physics between black holes and hydrodynamics has been known as a "membrane paradigm," which is now enhanced to exact duality within the context of AdS/CFT duality. For example, it was demonstrated in [2] that large black holes in AdS space is mapped to the conformal fluid (filling the boundary) which solves the Navier-Stokes equation on the boundary. Furthermore, Aharony *et al.* have shown that the *lumps* of plasma are holographic duals of black holes in the SS-compactified AdS$_{d+2}$ space [3]. Topology of these black holes is given by the S^1-fibre over the corresponding plasma lump. To be specific, the plasmaballs and plasmarings in d-dimensions are dual to the black holes and black rings in SS-AdS$_{d+2}$ [4]. These bounded objects appear in field theories that admit deconfinement phase transition of first-order. Lumps of fluids are used as a testbed to gain insight into black holes in AdS bulk (see [5, 6]). The analysis of stability is a primary concern in this short article.

2. Stability of plasmaballs

Let us consider equilibrium fluid lumps in $d = n + 2$ dimensional flat space, $\mathrm{d}s^2 = -\mathrm{d}t^2 + \mathrm{d}r^2 + r^2\mathrm{d}\Omega_n^2$, where $\mathrm{d}\Omega_n^2$ denotes the line element of a unit n-sphere. Dynamical variables in hydrodynamics are the number flux N^a and the stress-energy tensor T_{ab} of fluid, which obey divergence-type equation of motion $\nabla_a N^a = \nabla_a T^{ab} = 0$ and the constitutive equations $\nabla_a X^{abc} = I^{bc}$, where X^{abc} nad I^{ab} are functional of N^a and T^{ab} with $X^{ab}{}_b = X^{a[bc]} = I^{[ab]} = 0$.

The dissipative effects (entropy generation) are encoded in the constitutive equations. The positive divergence of an entropy current, S^a, specifies the direction of time.

We mean by 'equilibrium' as a state fulfilling the following 4-conditions: (i) vanishing entropy generation $\nabla_a S^a = 0$, (ii) there exists a unique hydrodynamic velocity u^a for which no spatially preferable directions, (iii) $P := T^i{}_i/(d-1)$ satisfies Euler's relation (no bulk viscous pressure), (iv) entropy density takes a maximum value. Condition (i) implies that all vector quantities are proportional to u^a: $N^a = nu^a$, $S^a = Su^a$, etc. Condition (ii) means that the stress-tensor is diagonalizable and condition (iv) leads to $\nabla_a(\delta S^a) = 0$, implying $\nabla_a \alpha = 0 = \nabla_{(a}\beta_{b)}$, where $\alpha := (P+\rho)/nT - (S/n)$ is the chemical potential per temperature and $\beta_a := T^{-1}u_a$ is the inverse temperature flow. Constant α is indicative of Tolmann's effect. It follows that the equilibrium condition is necessarily stationary ($\mathcal{L}_\beta g_{ab} = 0$ and $\beta^a\beta_a < 0$), which forbids dissipative terms (expansion and shear). Hence, we shall confine attention to the stationary perfect fluid.

Away from the boundary of the fluid, the fluid lumps satisfy the continuity equation and the relativistic Euler's equations

$$u^a\nabla_a\rho + (P+\rho)\nabla_a u^a = 0\,, \qquad (P+\rho)u^b\nabla_b u^a = -(g^{ab}+u^a u^b)\nabla_b P\,. \tag{1}$$

The plasma lumps are sustained by surface tension σ approximated by a delta-function, which contributes to the Young-Laplace equation,

$$P_< - P_> = \sigma K\,, \qquad K = h^{ab}\nabla_a n_b\,. \tag{2}$$

Here n^a is a unit outward normal to the boundary. The SS-equation of state reads $P = \frac{1}{d}\rho - \frac{d+1}{d}\rho_0$, where ρ_0 (>0) is a constant [3].

Let us discuss the static plasmaballs with an SO(n)-symmetry specified by $r = R_0$ surface, for which the field equations are solved by

$$u^a = (\partial/\partial t)^a\,, \qquad K = \frac{n}{R_0}\,, \qquad P = n\frac{\sigma}{R_0}. \tag{3}$$

Specifically, the boundary surface is described by a minimal surface. The static plasmaball is characterized by conserved charges $(\tilde{E}, \tilde{S})$ ("tilde" is used for the dimensionless quantities) given by

$$\tilde{E} = \frac{\rho_0^n}{V_n\sigma^{n+1}}E = (n+3)\tilde{R}_0^{n+1} + (n^2+3n+1)\tilde{R}_0^n \tag{4}$$

$$\tilde{S} = \frac{\rho_0^{(n^2+3n+1)/(n+3)}S}{V_n\alpha^{1/(n+3)}\sigma^{n+1}} = (n+3)\tilde{R}_0^{n+1}\left(1+\frac{n}{\tilde{R}_0}\right)^{(n+2)/(n+3)}, \tag{5}$$

satisfying the first law,

$$\mathrm{d}\tilde{E} = \tilde{T}\mathrm{d}\tilde{S}, \qquad \tilde{T} = \left(\frac{\alpha}{\rho_0}\right)^{n+3} T = \left(1+\frac{n}{\tilde{R}_0}\right)^{1/(n+3)}, \tag{6}$$

where α is a constant [for $d=3$, $\alpha = \pi^2 N^2/(8T_c)$]. It then immediately follows that Smarr's formula is reproduced in the limit of large spacetime dimensions, $n \to \infty$.

We now turn to the analysis of linearized stability of these plasmaballs. We shall consider the perturbations $(\delta\rho, \delta P, \delta u^a)$ of an equilibrium plasmaball (hereafter 0th-order quantities are marked by naught), where $\delta\rho$ and δP are related to the linearized SS-equation of state. We shall decompose the fluctuation modes in terms of a spherical harmonics, $Y_{\mathbf{k}}$, of a unit n-sphere satisfying $(D^iD_i + \mathbf{k}^2)Y_{\mathbf{k}}(\Omega_n) = 0$, where D_i is a covariant derivative of $\mathrm{d}\Omega_n^2$ and $\mathbf{k}^2 = \ell(\ell+n-1)$

is an eigenvalue. The "vector" and "tensor" modes turn out to be trivial. Then, the perturbed quantities $\delta Q = \{\delta u^a, \delta\rho, \delta P\}$ of "scalar-type" are expressed as

$$\delta\bar{Q} = \delta\bar{Q}_{\mathbf{k}}(r)e^{\omega t}Y_{\mathbf{k}}(\Omega_n); \quad \delta\bar{Q} = \{\delta u^t, \delta u^r, \delta\rho, \delta P\}, \quad \delta u^i = \delta u_{\mathbf{k}}(r)e^{\omega t}r^{-2}D^iY(\Omega_n). \tag{7}$$

In what follows we omit the index $\mathbf{k}$ specifying the each mode. Under the above first-order fluctuations, the boundary of the plasma lump changes also at first order. We specify it by

$$r = R(t, r, \Omega_n) = R_0 + \epsilon e^{\omega t}Y_{\mathbf{k}}(\Omega_n). \tag{8}$$

These perturbed quantities δQ evolve according to the linearized energy conservation equation and the linearized Euler's equations, subjected to the linearized boundary conditions,

$$\delta P|_{\text{boundary}} = \sigma\left(K|_{R(t,r,\Omega_n)} - K|_{R_0}\right) - \left[\left(P_<^{(0)} - P_>^{(0)}\right)_{R(t,r,\Omega)} - \left(P_<^{(0)} - P_>^{(0)}\right)_{R_0}\right], \tag{9}$$

and $\delta n_a u^a_{(0)} + n_a^{(0)}\delta u^a = 0$. Express δu^t and δu^r in terms of δP by solving linearized Euler's equations, and insert the result into the linearized continuity equation, we obtain

$$\frac{\mathrm{d}^2}{\mathrm{d}r^2}\delta P + \frac{n}{r}\frac{\mathrm{d}}{\mathrm{d}r}\delta P - \left(\frac{\mathbf{k}^2}{r^2} + (n+2)\omega^2\right)\delta P = 0. \tag{10}$$

In terms of a modified Bessel function of the first kind, $I_\nu(x)$, this equation solves as

$$\delta P(r) = A\frac{1}{r^{(n-1)/2}}I_{\ell+(n-1)/2}(pr), \qquad p = \sqrt{n+2}\omega. \tag{11}$$

Here, A is a constant to be determined by the boundary conditions. Substitute (11) into the linearized boundary conditions (9), we obtain the dispersion relation,

$$-\omega^2 = C\left(\frac{pI_{\ell+(n+1)/2}(pR_0)}{I_{\ell+(n-1)/2}(pR_0)} + \ell\right)\left(\omega^2 + \frac{(\ell-1)(\ell+n)}{R_0^2}\right), \tag{12}$$

where $C = \sigma/[(P_{(0)} + \rho_{(0)})R_0]$. This equation does not allow solutions of real ω. Hence, one concludes that the static fluid lump is stable under linear perturbations in arbitrary dimensions. For the static mode ($\omega = 0$), equation (10) is solved as,

$$\delta P(r) = C_1 r^{\alpha_+} + C_2 r^{\alpha_-}, \qquad \alpha_\pm = \frac{1}{2}\left(-n + 1 \pm \sqrt{(n-1)^2 + 4\mathbf{k}^2}\right) \gtrless 0. \tag{13}$$

Hence, either of two linearly independent solutions is divergent at origin or at infinity unless $\ell = 0$. That is, the static mode with $\ell \geq 1$ is not normalizable. Whereas the pressure perturbation for the static s-wave is constant, i.e., it is not a dynamical perturbation.

In the nonrelativistic limit $\omega R_0 \ll 1$ and $P_{(0)}/\rho_{(0)} \ll 1$, the above equation (12) reduces to

$$-\omega^2 = \frac{\sigma\ell(\ell-1)(\ell+n)}{\rho_{(0)}R_0^3}. \tag{14}$$

It then follows that the $\ell = 0, 1$ modes are not excited in this limit since they amount to the trivial addition of mass and change of the center of mass coordinate, respectively.

Since the static plasmaballs are linearly stable against perturbations, the corresponding (yet unknown) static black holes in SS-AdS$_{d+2}$ are expected to be stable. Moreover, the lack of normalizable static modes implies that the static SS-AdS$_{d+2}$ black holes are perturbatively unique [7].

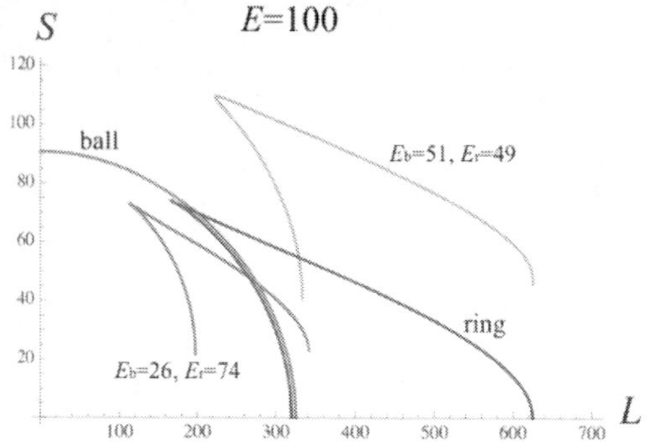

Figure 1. Phase diagram of plasma lumps in 2+1 dimensions. The red and blue curves denote a plasmaball and plasmaring, respectively. We write two saturns with distinct mass distribution between ball and ring. The magenta curve is $\tilde{R}_b = 2$ corresponding to $(E_b = 26, E_r = 74)$, while the green curve for $\tilde{R}_b = 3$ corresponding to $(E_b = 51, E_r = 49)$ has larger entropy.

3. Lumps of multiple plasma

In higher dimensions, black holes are able to remain equilibrium without any external forces supplied by electromagnetic force or conical singularity, just by the finely tuned balance between gravity and angular momentum. These multiple black objects reflect the solitonic character of circular system for vacuum Einstein's equations.

We can construct corresponding duals of multiple plasma lumps. To illustrate, let us consider a simplest example of a *plasmasaturn* (coexistence of a plasmaball and a plasmaring enclosing it) with inner plasmaball being non-rotating. As for the black saturn, plasmasaturns exhibit a two-fold non-uniqueness corresponding to the distribution of total mass and total angular momentum. Figure 1 depicts an entropy vs angular momentum diagram with fixed energy $\tilde{E} = 100$. As the ball becomes large, the entropy tends to increase. As argued in [8], the saturn configuration with thin ring may be entropically favoured among balls and rings. Also, these saturn configurations necessarily have distinct temperature on the ball and on the ring.

4. Concluding remarks

We have explored the stability of plasmaballs in arbitrary dimensions and showed that they are indeed stable. The AdS/CFT dictionary tells us that the corresponding higher-dimensional black holes are stable. We have also constructed a stationary multi-plasma objects. We believe that further analysis in this direction provides intriguing aspects of higher-dimensional gravity and updates a new entry in the holographic dictionary.

Acknowledgments

The author is grateful to Umpei Miyamoto for discussions.

References

[1] Kodama H 2009 *Lect. Notes Phys.* **769** 427
[2] Bhattacharyya S, Lahiri S, Loganayagam R and Minwalla S 2008 *J. High Energy Phys.* JHEP09(2008)054
[3] Aharony O, Minwalla S and Wiseman T 2006 *Class. Quantum Grav.* **23** 2171
[4] Lahiri S and Minwalla S 2008 *J. High Energy Phys.* JHEP05(2008)001
[5] Caldarelli M M, Dias O J C, Emparan R and Klemm D 2009 *J. High Energy Phys.* JHEP04(2009)024
[6] Cardoso V and Dias O J C 2009 *J. High Energy Phys.* JHEP04(2009)125
[7] Kodama h 2004 *Prog. Theor. Phys.* **112** 249
[8] Elvang H, Emparan R and Figueras P 2007 *J. High Energy Phys.* JHEP05(2007) 056

Polarized Electrogowdy spacetimes censored

Ernesto Nungesser
Max-Planck-Institut für Gravitationsphysik, Albert-Einstein-Institut, Am Mühlenberg 1, 14476 Potsdam, Germany
E-mail: `ernesto.nungesser@aei.mpg.de`

Abstract. A sketch of the proof of strong cosmic censorship is presented for a class of solutions of the Einstein-Maxwell equations, those with polarized Gowdy symmetry. A key element of the argument is the observation that by means of a suitable choice of variables the central equations in this problem can be written in a form where they are identical to the central equations for general (i.e. non-polarized) vacuum Gowdy spacetimes. Using this it is seen that the results of Ringström on strong cosmic censorship in the vacuum case have implications for the Einstein-Maxwell case. Working out the geometrical meaning of these analytical results leads to the main conclusion.

1. Strong Cosmic Censorship (SCC)

From the singularity theorems of Hawking and Penrose we know that singularities appear under general physically reasonable assumptions, but we do not know what happens in the singularities without a theory of quantum gravity. So if the singularities can influence what is outside, there would be a breakdown of predictability. A conjecture to solve this problem is the strong cosmic censorship hypothesis of Penrose [1]. We will use the following definition of SCC taken from [2]:

Definition 1 *(SCC) The maximal Cauchy development of all initial data for the Einstein-matter system belonging to an open dense subset is inextendible provided the matter model is well-behaved.*

This ensures that solutions which are non-generic or "unstable" and with a "badly-behaved" matter model are not taken into account (see chapter 9.4 of [2] for a discussion of that). For the reader interested in non-generic solutions within the Gowdy-class we recommend [3]. Since the maximal Cauchy development is the largest region of spacetime which is uniquely determined by initial data, if this region is inextendible, general relativity would remain a deterministic theory. In practice one has to look at the structure of the singularities (initial singularity) or to show that there do not exist (in the future for example). Strictly speaking not related to SCC in general, but assuming symmetries one can look what can be shown and often the techniques used can be generalized at least in principle.

2. Gowdy symmetry

Gowdy spacetimes are vacuum spacetimes which model closed universes filled with gravitational waves of two polarizations. These spacetimes are a good toy model for the understanding of inhomogeneous and anisotropic cosmological models. Since we will assume some matter content, in particular a Maxwell field, we will use the terminology of Gowdy-symmetric spacetimes instead of Gowdy spacetimes where the Gowdy symmetry is defined as follows:

Definition 2 *(Gowdy symmetry)*

(i) It is a T^2 $(U(1) \times U(1))$ symmetry with the group action generated by two commuting spacelike Killing vectors ($\frac{\partial}{\partial x}$ and $\frac{\partial}{\partial y}$)

(ii) The transformation which simultaneously maps x to $-x$ and y to $-y$ is an isometry

With this group action the possible spatial topologies are basically S^3, $S^2 \times S^1$ and T^3 and in the following the spatial topology is assumed to be the three-dimensional torus. A metric with Gowdy symmetry is said to be *polarized* if the *individual* transformations mapping x to $-x$ and y to $-y$ are symmetries, which has the physical interpretation that the gravitational waves have only one polarization. We will assume that our spacetime has a polarized Gowdy-symmetry.

2.1. Polarized Gowdy metric

It has been shown that the area spanned by the two Killing fields can be used as a time coordinate and moreover that there exists a Cauchy hypersurface of constant t. Using this *areal time* coordinate the polarized Gowdy metric can be put in the following manner:

$$t^{-\frac{1}{2}} e^{\frac{\lambda(t,\theta)}{2}} (-dt^2 + d\theta^2) + t(e^{P(t,\theta)} dx^2 + e^{-P(t,\theta)} dy^2) \tag{1}$$

Here θ, x and y are periodic coordinates on T^3 and λ and P are smooth functions.

3. Einstein-Maxwell equations

3.1. Maxwell field and new variables

The matter content will be described by a Maxwell field which is defined by a four-potential A_α. We choose a field which is consistent with the symmetries and denote the remaining components of the potential by

$$A_2 = \omega \tag{2}$$
$$A_3 = \chi. \tag{3}$$

Before coming to the equations we introduce new variables

$$\bar{P} = \frac{1}{2}(P - \log t) \tag{4}$$
$$\bar{\lambda} = \frac{1}{4}(\lambda - \log t) - \bar{P}. \tag{5}$$

What we obtain is the same PDE-system as in the non-polarized Gowdy case! That this choice of variables linking the two cases exists was already known long time ago (see for instance [6]). This enabled us to use the work of Ringström and others obtained for the Gowdy spacetimes which has culminated in [4] and [5].

3.2. Basic equations for Einstein-Maxwell assuming polarized Gowdy symmetry

The basic equations are the following:

$$-\bar{P}_{tt} - t^{-1}\bar{P}_t + \bar{P}_{\theta\theta} = e^{2\bar{P}}(-\chi_t^2 + \chi_\theta^2) \tag{6}$$
$$-\chi_{tt} - t^{-1}\chi_t + \chi_{\theta\theta} = -2(-\bar{P}_t\chi_t + \bar{P}_\theta\chi_\theta) \tag{7}$$
$$\bar{\lambda}_t = t[\bar{P}_t^2 + \bar{P}_\theta^2 + e^{2\bar{P}}(\chi_t^2 + \chi_\theta^2)] \tag{8}$$
$$\bar{\lambda}_\theta = 2t[\bar{P}_t\bar{P}_\theta + e^{2\bar{P}}\chi_t\chi_\theta] \tag{9}$$

Journal of Physics: Conference Series **229** (2010) 012057 doi:10.1088/1742-6596/229/1/012057

where the first two are the evolution and the last two the constraint equations. There is also a consistency condition which lead us to assume that $\omega = 0$ and other equations involving higher derivatives of λ which are not important here. Also an integral constraint coming from the fact that λ is periodic has to be satisfied. In terms of our new variables the metric is

$$e^{2(\bar{\lambda}+\bar{P})}(-dt^2 + d\theta^2) + t^2 e^{2\bar{P}} dx^2 + e^{-2\bar{P}} dy^2. \tag{10}$$

4. Central results

4.1. Key steps

There is a large class of Gowdy solutions called the low velocity solutions which, when translated to the case of interest here, admit asymptotic expansions of the following form

$$\bar{P}(t,\theta) = -v_\infty(\theta)\log t + \dots \tag{11}$$

$$\chi(t,\theta) = q(\theta) + \psi(\theta)t^{2v_\infty} + \dots \tag{12}$$

where the terms omitted are lower order as $t \to 0$. q and ψ are smooth functions and $v_\infty(\theta)$ is called the asymptotic velocity. There are also more complicated solutions with so called true and false spikes which had to be treated separately. In order to show that the spacetime is not extendible we looked at invariants and hoped that they become unbounded. We looked at $F^{\alpha\beta}F_{\alpha\beta}$, but this did not work in general with the available information. $F^{\alpha\beta*}F_{\alpha\beta}$ vanishes in our case so this invariant is not helpful. Finally it worked with the Kretschmann scalar $R_{\alpha\beta\gamma\delta}R^{\alpha\beta\gamma\delta}$.

The argument used to prove geodesic completeness in the vacuum case basically also applies to the Einstein-Maxwell case, proving geodesic completeness in that case too. Moreover the solutions exhibit oscillatory behaviour at late times analogous to that found in [4] for Gowdy spacetimes.

4.2. Central results

In the following theorems the topology on the set of initial data used is the C^∞ topology. $\mathcal{G}_c$ is a generic set of initial data satisfying an integral constraint. Data belonging to $\mathcal{G}_c$ are called generic. This subset is open and dense in the set of initial data in the C^∞ topology.

Theorem 1 *There is an open dense subset $\mathcal{G}_c$ of the set of smooth initial data for the Einstein-Maxwell equations with polarized Gowdy symmetry and constant areal time such that the Kretschmann scalar tends to infinity along any inextendible past-directed causal geodesic.*

Another result is that strong cosmic censorship holds for polarized Electrogowdy, more precisely

Theorem 2 *For data belonging to the open dense subset $\mathcal{G}_c$ of the set of smooth initial data for the Einstein-Maxwell equations with polarized Gowdy symmetry and constant area radius the corresponding maximal Cauchy development is inextendible.*

The proofs and details can be found in [7] and [8].

5. Conclusions and Outlook

The analysis of the structure of the sigularities in our case, the case of gravitational waves with only one polarization and an electromagnetic field has revealed to be neither simpler nor more difficult than the vacuum case of gravitational waves with two polarizations (once the PDE-system is solved of course). Generalizations of the results to polarized Gowdy-symmetric spacetimes with a Maxwell field which does not come from a potential seem quite difficult because in this case the evolution and constraint equations are coupled. Something similar happens in the case of a positive cosmological constant and in the case of a negative cosmological constant

Journal of Physics: Conference Series **229** (2010) 012057 doi:10.1088/1742-6596/229/1/012057

even the existence of a global time coordinate is problematic. However there are different results for the case of a positive cosmological constant in the context of T^2-symmetric spacetimes (see for instance [9], [10]). The most natural generalization of our results would be the case of general solutions of the Einstein-Maxwell equations with Gowdy symmetry, i.e. without the restriction of polarization. In this direction already some results have been obtained. An energy decay has been shown by Ringström [11] which could be an important first step. The difficulties in the general case come from the fact that in the corresponding wave map formulation the target space is the complex hyperbolic space instead of the hyperbolic space. A discussion of that and clues concerning the asymptotic behaviour near the singularity can be found in [12].

Acknowledgments
The author would like to thank Alan D. Rendall for a lot of helpful discussions, especially those which take place after work in the train.

References

[1] Penrose R 1979 *In General Relativity, an Einstein Centenary Survey, ed. S. W. Hawking and W. Israel.* (Cambridge University Press)
[2] Rendall A D 2008 *Partial Differential Equations in General Relativity* (Oxford University Press)
[3] Quevedo H 2006 *Gen. Rel. Grav.* **38** 599-606
[4] Ringström H 2004 *Commun. Pure Appl. Math.* **57** 657–703
[5] Ringström H Strong cosmic censorship in T^3-Gowdy spacetimes *Preprint. To appear in Ann. Math.*
[6] Carmeli M, Charach Ch and Malin S 1981 *Phys. Rep.* **76** 79–156
[7] Nungesser E 2008 Strong cosmic censorship in polarized T^3-Gowdy symmetric spacetimes with a Maxwell field. Diploma thesis, Free University, Berlin
[8] Nungesser E and Rendall A D 2009 Strong cosmic censorship for solutions of the Einstein-Maxwell field equations with polarized Gowdy symmetry *Class. Quantum Grav.* **26** 105019
[9] Clausen A and Isenberg J 2007 *J. Math. Phys.* **48** 082501
[10] Smulevici J 2008 *Ann. H. Poincaré* **9** 1425-1453.
[11] Ringström H 2006 *Ann. H. Poincaré* **7** 1-20.
[12] Mansfield P A 1989 PhD thesis Yale University

Inflation, renormalization, and CMB anisotropies

I Agulló[1], J Navarro-Salas[2], G J Olmo[3]* and L Parker[1]

[1] Physics Department, University of Wisconsin-Milwaukee, P.O.Box 413, Milwaukee, WI 53201 USA

[2] Departamento de Física Teórica and IFIC, Centro Mixto Universidad de Valencia-CSIC. Facultad de Física, Universidad de Valencia, Burjassot-46100, Valencia, Spain

[3] Instituto de Estructura de la Materia, CSIC, Serrano 121, 28006 Madrid, Spain

E-mail: ivan.agullo@uv.es, jnavarro@ific.uv.es, olmo@iem.cfmac.csic.es, leonard@uwm.edu

Abstract. In single-field, slow-roll inflationary models, scalar and tensorial (Gaussian) perturbations are both characterized by a zero mean and a non-zero variance. In position space, the corresponding variance of those fields diverges in the ultraviolet. The requirement of a finite variance in position space forces its regularization via quantum field renormalization in an expanding universe. This has an important impact on the predicted scalar and tensorial power spectra for wavelengths that today are at observable scales. In particular, we find a non-trivial change in the consistency condition that relates the tensor-to-scalar ratio r to the spectral indices. For instance, an exact scale-invariant tensorial power spectrum, $n_t = 0$, is now compatible with a non-zero ratio $r \approx 0.12 \pm 0.06$, which is forbidden by the standard prediction ($r = -8n_t$). Forthcoming observations of the influence of relic gravitational waves on the CMB will offer a non-trivial test of the new predictions.

1. Introduction

Inflation has become a fundamental piece of the standard cosmological model [1]. Not only it addresses and solves the classical problems of the big bang model but it also provides an elegant quantum mechanical mechanism for the origin of primordial perturbations, essential for explaining the structures that we see today [2]. In the simplest models of inflation, a scalar field slowly rolling down its potential causes an exponentially large expansion. In this background, quantum vacuum fluctuations of the inflaton field itself and of purely tensorial modes (gravitational waves) acquire classical properties due to their interaction with the expanding geometry. As a result, a classical field of scalar (and tensorial) inhomogeneities arises. Cosmological perturbation theory tells us that these primordial perturbations are responsible for the existence and richness of the structure that we observe in the temperature anisotropies of the cosmic microwave background (CMB) and in the large scale distribution of galaxies.

A disturbing aspect of the spectra of scalar and tensorial perturbations generated during inflation is that they have a divergent variance. In the case of classical perturbations, divergences of the variance are usually removed by means of window functions that filter out the wavelengths that cause the problem. However, the primordial spectra have a quantum origin and their divergences should be removed, on grounds of theoretical consistency, by means of the well-established methods of renormalization in curved spaces[3]. In this work we show that the regularization of the variance of scalar and tensorial perturbations during inflation has a non-trivial impact on the primordial power spectra of those fields [4, 5]. Though the resulting spectra are almost scale free within the slow-roll approximation, the relation between spectral indices and the slow-roll parameters is significantly changed as compared to the standard derivation. The ratio of tensor to scalar amplitudes is also modified and gives rise to a number of possibilities which are explicitly forbidden according to the standard predictions. In particular, the tensorial spectral index is allowed to take a zero value without implying a zero tensor to scalar ratio, or may even take positive values.

2. Spectrum of fluctuations from inflation

Let us first focus on the production of relic gravitational waves by considering fluctuating tensorial modes $h_{ij}(\vec{x},t)$ in a spatially flat FRW universe $ds^2 = -dt^2 + a^2(t)(\delta_{ij} + h_{ij})dx^i dx^j$. The wave equation obeyed by these modes is given by

$$-a^2\ddot{h}_{ij} - 3a\dot{a}\dot{h}_{ij} + \nabla^2 h_{ij} = 0\ . \tag{1}$$

The fluctuating fields h_{ij} can be decomposed in two independent polarization modes $h^{(p)}(\vec{x},t)e_{ij}^{(p)}$, where the index p represents the two polarizations $+$ and $\times$, and $e_{ij}^{(p)}$ are two independent constant matrices that obey the conditions $e_{ij} = e_{ji}$, $\sum_i e_{ii} = 0$, and $\sum_i k_i e_{ij} = 0$. Expanding the fields $h^{(p)}(\vec{x},t)$ in modes and omitting the polarization label p, eq. (1) yields

$$\ddot{h}_k + 3H\dot{h}_k + \frac{k^2}{a^2}h_k = 0\ , \tag{2}$$

with $k \equiv |\vec{k}|$. This equation indicates that the relevant part of the tensorial modes is represented by two independent massless scalar fields $h^{(+,\times)}(\vec{x},t)$. The solutions that satisfy the asymptotic adiabatic condition [6, 3] for very high k in a slow-roll inflationary background are $h_k(t) = (-16\pi G\tau\pi/4(2\pi)^3a^2)^{1/2}H_\nu^{(1)}(-k\tau)$, where $\tau = \int dt/a(t)$ is the conformal time, the index of the Bessel function is $\nu = 3/2+\epsilon$, and $\epsilon = \frac{M_P^2}{2}\left(\frac{V'}{V}\right)^2$ is one of the slow-roll parameters. With these solutions it is easy to evaluate the variance of the gravitational wave fields $h^{(+,\times)}$

$$\langle h^2\rangle = \int_0^\infty k^2 dk \int d\Omega |h_k|^2 = \int_0^\infty \frac{dk}{k}\Delta_h^2(k,t)\ , \tag{3}$$

where we have defined the power spectrum of the field h as $\Delta_h^2(k,t) \equiv 4\pi k^3|h_k|^2$. As we advanced in the introduction, the large k behavior of the modes makes this integral divergent

$$\langle h^2\rangle = \int_0^\infty \frac{dk}{k}\frac{16\pi G k^3}{4\pi^2 a^3}\left[\frac{a}{k}[1 + \frac{(2+3\epsilon)}{2k^2\tau^2}] + ...\right]\ . \tag{4}$$

Since the power spectrum is expressed in momentum space, the natural renormalization scheme to apply is the so-called adiabatic subtraction [4], as it renormalizes the theory in momentum space. Adiabatic renormalization [3, 7] removes the divergences present in the formal expression (3) by subtracting counterterms mode by mode in the integrand of (3)

$$\langle h^2\rangle_{ren} = \int_0^\infty \frac{dk}{k}\tilde{\Delta}_h^2(k,t) = \int_0^\infty \frac{dk}{k}\left[4\pi k^3|h_{\vec{k}}|^2 - \frac{16\pi G k^3}{4\pi^2 a^3}(w_k^{-1} + (W_k^{-1})^{(2)})\right]\ , \tag{5}$$

with $w_k = k/a(t)$. The subtraction of the first term $(16\pi G k^3/4\pi^2 a^3 w_k)$ cancels the typical flat space vacuum fluctuations, which are responsible for the quadratic divergence in the integral (4). The additional term, proportional to $(W_k^{-1})^{(2)} = -\frac{1}{w_k^2}\left[\frac{1}{2}w_k^{-1/2}\frac{d^2}{dt^2}w_k^{-1/2} - \frac{1}{2}w_k^{-1}\left(\frac{3}{4}\frac{\dot{a}^2}{a^2} + \frac{3}{2}\frac{\ddot{a}}{a}\right)\right]$

and which involves $\dot{a}^2$ and $\ddot{a}$, is necessary to properly perform the renormalization in an expanding universe, since it cancels the logarithmic divergence in (4).
According to expression (3), the unrenormalized power spectrum of tensor perturbations evaluated a few e-folds after the Hubble horizon exit time t_k (defined by $k/a(t_k) = H(t_k)$) is given by

$$P_t(k) = 4\Delta_h^2(k) = 16\pi k^3 |h_k|^2 = \frac{8}{M_P^2}\left(\frac{H(t_k)}{2\pi}\right)^2 \tag{6}$$

The k dependence of $H^2(t_k)$, $d\ln H(t_k)/d\ln k = -\epsilon$, leads to $P_t(k,t_k) = P_t(k_0)\left(\frac{k}{k_0}\right)^{-2\epsilon}$, where k_0 is a pivot scale. The tensorial spectral index n_t is defined as the exponent in that expression. So $n_t = -2\epsilon(t_k)$. Using the renormalized variance (5) to define the renormalized power spectrum and taking into account the time dependence of the counterterms, we find

$$P_t^{ren}(k,n) = 4\tilde{\Delta}_h^2(k,n) \approx \frac{8}{M_P^2}\left(\frac{H(t_k)}{2\pi}\right)^2 \epsilon(t_k)(2n - 3/2)\ , \tag{7}$$

where n represents the number of e-folds after the Hubble exit at which the spectrum is evaluated (we assume $n > 1$ but $n\epsilon \ll 1$). We note that if one evaluates the power spectra at the end of the slow-roll era (where $n\epsilon \sim 1$) the contribution of the counterterms is still significant. However, we find it more natural to evaluate the spectra soon after t_k, when the modes have already acquired classical properties. The corresponding tensorial spectral index is (here ϵ and $\eta = M_P^2(V''/V)$ are evaluated at t_k)

$$n_t^{ren} \equiv \frac{d\ln P_t^{ren}}{d\ln k} = 2(\epsilon - \eta)\ . \tag{8}$$

The discussion just presented regarding the tensorial modes can be parallelled to the case of scalar perturbations. In this case one deals with the gauge invariant scalar $\mathcal{R} = \Psi + \frac{H}{\dot{\phi}_0}\delta\phi$, where Ψ is the curvature perturbation ($R^{(3)} = 4\nabla^2\Psi/a^2$) of the spatial metric $g_{ij} = a^2[(1-2\Psi)\delta_{ij} + 2\partial_{ij}E]$. In momentum space, $\mathcal{R}$ obeys the equation [1]

$$\frac{d^2\mathcal{R}_k}{d\tau^2} + \frac{2}{z}\frac{dz}{d\tau}\frac{d\mathcal{R}_k}{d\tau} + k^2\mathcal{R}_k = 0\ , \tag{9}$$

where $z \equiv a\dot{\phi}_0/H$. In the slow roll approximation $z^{-1}dz/d\tau = aH(1 + 2\epsilon - \eta)$, and the solutions obeying the adiabatic condition (and the de Sitter symmetry for H constant) are $\mathcal{R}_k(t) = (-\pi\tau/4(2\pi)^3 z^2)^{1/2} H_\mu^{(1)}(-\tau k)$, where $\mu = 3/2 + 3\epsilon - \eta$. The resulting unrenormalized and renormalized scalar power spectra and spectral indices are ($P_{\mathcal{R}}(k) = P_{\mathcal{R}}(k_0)\left(\frac{k}{k_0}\right)^{n_s - 1}$)

$$P_{\mathcal{R}}(k) = \frac{1}{2M_P^2\epsilon}\left(\frac{H(t_k)}{2\pi}\right)^2 \quad , \quad n_s - 1 = -6\epsilon + 2\eta \tag{10}$$

$$P_{\mathcal{R}}^{ren}(k,n) \approx \frac{(3\epsilon - \eta)}{2M_P^2\epsilon}\left(\frac{H(t_k)}{2\pi}\right)^2 (2n - 3/2) \quad , \quad n_s^{ren} - 1 = -6\epsilon + 2\eta + \frac{(12\epsilon^2 - 8\epsilon\eta + \xi)}{3\epsilon - \eta} \tag{11}$$

where ξ is another slow roll parameter, $\xi \equiv M_P^4(V'V'''/V^2)$, which can be re-expressed in terms of ϵ, η and the running of the renormalized tensorial index $n_t' \equiv dn_t/d\ln k$ as $n_t' = 8\epsilon(\epsilon - \eta) + 2\xi$.

3. Conclusions

The expressions derived above for the power spectra and spectral indices show that renormalization has a significant impact on the predictions of slow-roll inflation. Unlike in the standard derivation, we see that the renormalized tensorial power spectrum is no longer directly related to the Hubble scale during inflation. The amplitude of this spectrum is now modulated by the slow-roll parameter ϵ, which is zero in exact de Sitter inflation. On the other hand, the appearance of the combination $3\epsilon - \eta$ in the numerator of the renormalized scalar spectrum suggests that $P_{\mathcal{R}}^{ren}(k,n) \sim \frac{1}{2M_P^2}\left(\frac{H(t_k)}{2\pi}\right)^2$ (up to the numerical factor $2n - 3/2$), which indicates that $H(t_k) \sim 10^{15}$GeV. For the exponential potential [1], we find that $H \sim 1.5 \times 10^{15}/\sqrt{4n-3}$ GeV, roughly up to an order of magnitude larger than the unrenormalized prediction.

The most dramatic difference between the standard predictions and the predictions after renormalization presented here affects the so-called consistency condition. The tensor to scalar ratio in the standard approach gives $r = P_t/P_{\mathcal{R}} = 16\epsilon = -8n_t$. This implies that if a background of gravitational waves is observed, then its spectral index must be constrained by the ratio of the tensor to scalar amplitudes. If this constraint is not satisfied, then single field slow-roll inflation would be ruled out. However, the consistency condition after renormalization becomes

$$r = 4(1 - n_s - n_t) + \frac{4n_t'}{n_t^2 - 2n_t'}\left(1 - n_s - \sqrt{2n_t' + (1-n_s)^2 - n_t^2}\right) , \tag{12}$$

where we have omitted the indices "ren". This expression is much more involved than the standard one and its implications are far reaching. It allows for a zero n_t while having a non-zero ratio r. Also, the values of n_t are not constrained to be negative. In addition, according to the standard derivation, the running of n_t is fully determined by the values of n_s and n_t, since then one finds $n_t' = -n_t(1 - n_s + n_t)$. On the contrary, the manipulations that lead to (12) indicate that n_t' is now an independent quantity that needs to be measured in order to check the new consistency relation (12). This aspect could make more challenging the experimental verification of the consistency condition of (single-field) slow roll inflation.

To conclude, if single field slow-roll inflation is correct, then observations will (hopefully) tell us if nature have chosen the renormalized spectra as the seeds of perturbations or not. Quantum field theory in curved spacetimes will thus face a crucial experimental test.

Acknowledgements. This work has been partially supported by the spanish grant FIS2008-06078-C03-02. I.A. and L.P. have been partly supported by NSF grants PHY-0071044 and PHY-0503366 and by a UWM RGI grant. G.O. thanks MICINN for a JdC contract and the "José Castillejo" program for funding a stay at the University of Wisconsin-Milwaukee.

References

[1] Lyth D H and Liddle A R 2009 *The primordial density perturbation*, (Cambridge University Press)
Weinberg S 2008 *Cosmology* (Oxford University Press)
Dodelson S 2003 *Modern Cosmology* (Academic Press)
[2] Mukhanov V F and Chibisov G V 1981 *JETP Letters* **33** 532
Hawking S W 1982 *Phys. Lett.* B**115** 295
Guth A and Pi S -Y 1982 *Phys. Rev. Lett.* **49** 1110
Starobinsky A A 1982 *Phys. Lett.* B**117** 175
Bardeen J M, Steinhardt P J and Turner M S 1983 *Phys. Rev.* D**28** 679
[3] Parker L and Toms D J 2009 *Quantum field theory in curved spacetime: quantized fields and gravity* (Cambridge University Press)
[4] Parker L Amplitude of perturbations from inflation *Preprint* hep-th/0702216
[5] Agulló I, Navarro-Salas J, Olmo G J and Parker L 2009 *Phys. Rev. Lett.* **103** 061301
Agulló I, Navarro-Salas J, Olmo G J and Parker L 2009 *Gen. Rel. Grav.* **41** 2301
Agulló I, Navarro-Salas J, Olmo G J and Parker L 2008 *Phys. Rev. Lett.* **101** 171301
see also *Preprint* hep-th/0911.0961
[6] Parker L 1968 *Phys.Rev.Lett.* **21** 562
Parker L 1969 *Phys. Rev.* **183** 1057
Parker L 1966 *The creation of particles in an expanding universe*, Ph.D. thesis, Harvard University
[7] Birrel N D and Davies P C W 1982 *Quantum fields in curved space* (Cambridge University Press)

Numerical treatment of the light propagation problem in the post-Newtonian formalism

A San Miguel, F Vicente and J-F Pascual-Sánchez*

Dept. de Matemática Aplicada, Facultad de Ciencias, Universidad de Valladolid, 47005 Valladolid, Spain

E-mail: asmiguel@maf.uva.es, fvicente@maf.uva.es, jfpascua@maf.uva.es

Abstract. The geometry of a light wavefront, evolving from a initial flat wavefront in the 3-space associated with a post-Newtonian relativistic spacetime, is studied numerically by means of the ray tracing method. For a discretization of the bidimensional light wavefront, a surface fitting technique is used to determine the curvature of this surface. The relationship between the intrinsic curvature of the wavefront and the change of the arrival time at different points on the Earth is also numerically discussed.

1. Introduction

According to Einstein's theory of general relativity, a generic gravitational field has, at a time, three main effects on light: the Shapiro time delay, the bending of rays and the curvature of wavefronts. The gravitational field acts respectively as a retarder delaying a light wavefront, as a prism tilting a light wavefront, and as a lens curving a light wavefront. In special relativity, time delays can be also induced by inertial motion of the observer or the source as in the Doppler effect and uniform motion of a observer can result in aberration, i.e., an apparent change in the position of the source.

Hence, the curvature of a initially plane light wavefront by a gravity field is a purely general relativistic effect that has not any analogue in special relativity. In order to obtain an experimental measurement of the curvature of a wave front, Samuel [1] proposed a method based on the relation between the differences of arrival time recorded at four points on the Earth and the volume of a parallelepiped determined by four points in the curved wavefront surface. We will establish a discretized model of the wavefront surface by means of a regular triangulation for the study of the curvature(s) of this surface. The main methods and results have been recently published in [2], in this work we will only expose a summary of them.

2. Light propagation in a gravitational field

Let us consider a spacetime $(\mathcal{M}, g)$ corresponding to a weak gravitational field determined by a metric tensor given in a global coordinate system $\{(\boldsymbol{z}, ct)\}$ by

$$g_{\alpha\beta} = \eta_{\alpha\beta} + h_{\alpha\beta}, \qquad \text{with} \quad \eta_{\alpha\beta} = \text{diag}\,(1, 1, 1, -1). \tag{1}$$

where the coordinate components of the metric deviation $h_{\alpha\beta}$ are:

$$h_{ab} = 2c^{-2}\kappa\|\boldsymbol{z}\|^{-1}\delta_{ab}, \quad h_{a4} = -4c^{-3}\kappa\|\boldsymbol{z}\|^{-1}\dot{Z}_a, \quad h_{44} = 2c^{-2}\kappa\|\boldsymbol{z}\|^{-1}, \tag{2}$$

here $\kappa := GM$ represents the gravitational constant of the Sun, located at $Z^a(t)$, and c represents the vacuum light speed.

The null geodesics $z(t) = (\boldsymbol{z}(t), t)$, satisfy the following equations (see [3]):

$$\begin{aligned} \ddot{\boldsymbol{z}}^a &= \tfrac{1}{2}c^2 h_{44,a} - [\tfrac{1}{2}h_{44,t}\delta^a_k + h_{ak,t} + c(h_{4a,k} - h_{4k,a})]\dot{\boldsymbol{z}}^k \\ &\quad -(h_{44,k}\delta^a_l + h_{ak,l} - \tfrac{1}{2}h_{kl,a})\dot{\boldsymbol{z}}^k\dot{\boldsymbol{z}}^l \\ &\quad -(c^{-1}h_{4k,j} - \tfrac{1}{2}c^{-2}h_{jk,t})\dot{\boldsymbol{z}}^j\dot{\boldsymbol{z}}^k\dot{\boldsymbol{z}}^a, \\ 0 &= g_{\alpha\beta}\dot{z}^\alpha\dot{z}^\beta. \end{aligned}$$

where, in the first equation, the first and third terms in the right are of order $O(1)$, while the remaining terms are $O(c^{-1})$. The second equation is the isotropy constraint satisfied by the null geodesics.

3. Numerical description of a spacelike bidimensional light wavefront

3.1. Discretization of the initial wavefront

We consider a flat initial surface $\mathcal{S}_0$ far from the Sun formed by points $(z_1, z_2, -\zeta)$ (with $\zeta > 0$) in an asymptotically Cartesian coordinate system $\{z\}$. For the discretization of $\mathcal{S}_0$ a triangulation is constructed in such a form that each vertex is represented by a complex number of the set:

$$\bar{\mathcal{V}} := \{z = a_1 + a_2\omega + a_3\omega^2 \mid a_1, a_2, a_3 \in \mathcal{A}, \ \omega := \exp(2\pi \mathrm{i}/3)\}, \tag{3}$$

The complex plane and the plane $\mathcal{S}_0$ may be identified by means of the mapping $\iota : z \mapsto (\Re(z), \Im(z), -\zeta)$. Thus a regular triangulation $\mathcal{V}$ of the initial wavefront $\mathcal{S}_0$ is determined. At each vertex in $\mathcal{V}$ a photon $\boldsymbol{z}_0 := \boldsymbol{z}(0) \in \mathcal{V}$ with velocity $\dot{\boldsymbol{z}}_0 := (0, 0, c)$ is located. The null geodesics equation may be written as a first order differential system $\dot{\boldsymbol{u}} = \boldsymbol{F}(\boldsymbol{u}, t)$, in phase space $\boldsymbol{u} = (\boldsymbol{z}, \dot{\boldsymbol{z}})$, which determines in $\mathcal{E}$ (the quotient space of $\mathcal{M}$ by the global timelike vector field ∂_t associated to the global coordinate system used in the post-Newtonian formalism) a flow: $\boldsymbol{z}(t) = \varphi_t(\boldsymbol{z}_0, \dot{\boldsymbol{z}}_0)$. For each time t, the image of $\mathcal{S}_0$ under the flow φ_t determines a curved wavefront $\mathcal{S}_t$. The initial triangulation by $\mathcal{V}$ induces a triangulation on the final wavefront $\mathcal{S}_t$, whose vertices we enumerate using the same labels used for the corresponding vertices in $\mathcal{S}_0$.

Let $\{y^j\}_{j=1}^3$ be a *normal coordinate system* with pole at the point $P \in \mathcal{S}_t$ and associated normal reference frame $\{\boldsymbol{e}_i\}_{i=1}^3$. Under the coordinate transformation $z^i \mapsto y^i$, from post-Newtonian to normal coordinates, the metric tensor γ on the space $\mathcal{E}$ is determined (up to terms of first order in $\tilde{\Gamma}$) from $\tilde{\gamma}_{ab} := g_{ab} - \frac{g_{a4}g_{b4}}{g_{44}}$ by

$$\gamma_{ij} = \frac{\partial z^a}{\partial y^i}\frac{\partial z^b}{\partial y^j}\tilde{\gamma}_{ab} \quad \text{where} \quad z^a = a_0^a + \Lambda_i^a(y^i - \tfrac{1}{2}(\tilde{\Gamma}^i_{jk})_0 y^j y^k). \tag{4}$$

3.2. Local approximation of the wavefront

To compute differential magnitudes of the wavefront surface corresponding to the mesh $\mathcal{V}$ at each inner vertex $z_j^0 \in \bar{\mathcal{V}}, j = 1, \ldots, J$ we consider the 1–ring $[z_j^0; z_{j_k}^0]_{j=1,\ldots,J,k=0,\ldots,5}$ formed by the six vertices $z_{j_k}^0$ closest to z_j^0: For each 1–ring $[z_j^*; z_{j_k}^*]$ one obtains on the mesh $\mathcal{V}$ the image 1–ring $[z_j; z_{j_k}]$ under the flow φ_t.

In a neighbourhood of the image point z_j the wavefront can be approximated by a least-squares fitting of the data $[z_j; z_{j_k}]$ as the quadric $\bar{\mathcal{S}}_t$:

$$y^3 = f(y^1, y^2) := \tfrac{1}{2}a_1(y^1)^2 + a_2 y^1 y^2 + \tfrac{1}{2}a_3(y^2)^2. \tag{5}$$

using adapted normal coordinates $\{y^i\}$.

3.3. Curvature of the wavefront

Using coordinates x^1, x^2 adapted to the quadric $\tilde{\mathcal{S}}_t$: the metric $\boldsymbol{\gamma}$ on $\mathcal{E}$ induces a metric on $\tilde{\mathcal{S}}_j$ of the form

$$g_{AB} := \gamma_{ij} \frac{\partial y^i}{\partial x^A} \frac{\partial y^j}{\partial x^B}. \tag{6}$$

By means of a generalized Gauss formula, the difference of sectional curvatures K and $\bar{K}$ associated with the plane generated by the tangent vectors $\{\boldsymbol{v}_1, \boldsymbol{v}_2\}$, in $\mathcal{S}$ and $\mathcal{E}$ respectively, is the relative sectional curvature $K_{\rm rel}$ (see [4]):

$$K_{\rm rel} := K(\boldsymbol{v}_1, \boldsymbol{v}_2) - \bar{K}(\boldsymbol{v}_1, \boldsymbol{v}_2) = \lambda_1 \lambda_2, \tag{7}$$

whereas the mean curvature H is determined by half the trace of the second fundamental form II:

$$H = \tfrac{1}{2}(\lambda_1 + \lambda_2) \tag{8}$$

4. Application of a numerical integrator to the case of a static Sun

The ray tracing has been carried out using an integrator based on the classic Taylor series method for ordinary differential equations (see [5]). This integrator presents the following advantages: allows the control of both the order and the step size employed in the method. In this integrator one may use extended arithmetic precision for the highly accurate computation required in this problem. (Precision of 120 binary digits and a tolerance `Tol` $= 1.\mathrm{E}-20$ are used to solve this problem.) At each step the null constraint equation is preserved.

Hereafter, we consider the simplest gravitational model generated by a static Sun, considered as a point. However, it can be applied to more complex and realistic gravitational fields of the solar system, as considered in [6]. We now apply the ray tracing method to a light wavefront region propagating along a tubular neighborhood around the Oz^3-axis and where the inner and outer hexagons have radii of lengths $R_\odot/25$ and $2R_\odot$, respectively. The initial flat wave surface $\mathcal{S}_0$, perpendicular to the Oz^3-axis, is positioned at -100 A.U.. The wavefront $\mathcal{S}_T$ is then determined after a trip of 101 A.U., i.e., at the position of the Earth. Remember that the focal length for a light ray grazing an opaque Sun is located at 548.30 A.U. from the Sun.

In Figure 1, the surface $\mathcal{S}_T$ at the time when the wavefront arrives at the Earth is shown using a gray-scale to represent the relative sectional curvature (note we have used a different scale on the Oz^3-axis). One sees in this figure that the absolute value of the relative sectional curvature function defined on $\mathcal{S}_T$ increases as the distance between the photon and the Oz^3–axis decreases.

5. Variation of the time of arrival and curvature of the light wavefront

Suppose four receiving stations located at four points on an Earth hemisphere. The arrival time differences between these stations depend on the curvature of the wavefront. Assumed known the measurements of arrival times corresponding to four points $\{Q_a\}_{a=1}^4$ on the wavefront it is possible to determine an approximation of the wavefront curvature in a region far enough from the Sun (say the Earth), without resorting to the ray tracing method.

An estimation of the Gaussian curvature of the wavefront surface can be obtained using the notion of the Wald curvature of a metric space established in Distance Geometry (see [7]) that in the case of 2-dimensional manifolds agrees with the Gaussian curvature. The Wald curvature is determined as the limit of the embedding curvatures of metric quadruples isometrically embedded in surfaces of constant curvature (the Euclidean plane $\mathbb{R}^2$, the 2–sphere $\mathbb{S}^2_{\sqrt{\kappa}}$ or the

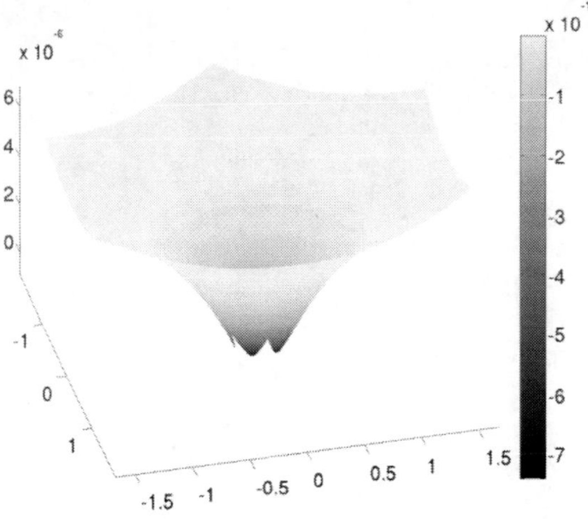

Figure 1. Wavefront surface and relative sectional curvature (gray scale) deformed by a spherical gravitational field (a different scale is used for the vertical axis).

hyperbolic space $\mathbb{H}^2_{\sqrt{-\kappa}}$).

In the hyperbolic plane $\mathbb{H}^2_r$ of curvature $-1/r^2$, represented by the Blumenthal model ([7]) we consider the metric quadruple associated with the points $\{Q_a\}$ assuming that the geometry of the 3–space in the vicinity of the Earth is Euclidean.

The Wald curvature associated to the chosen quadruple $\{Q_a\}$ prove to be

$$\kappa := -\frac{1}{r^2} = -0.16\,\mathrm{E}{-10}. \tag{9}$$

This result gives an approximation of the total curvature of the wavefront surface under the assumption that locally this surface may be identified with a hyperbolic plane in which the quadruple considered is isometrically embedded.

Acknowledgments

This research was partially supported by the Spanish Ministerio de Educación y Ciencia, MEC-FEDER grant ESP2006-01263.

References

[1] Samuel J 2004 *Class. Quantum Grav.* **21** L83-L88
[2] San Miguel A, Vicente F and Pascual-Sánchez J F 2009 *Class. Quantum Grav.* **26** 235004
[3] Brumberg V A 1991 *Essential Relativistic Celestial Mechanics* (Bristol: Adam Hilger)
[4] do Carmo M P 1992 *Riemannian Geometry* (Boston: Birkhäuser)
[5] Jorba À and Zou M 2005 *Experimental Mathematics* **14** 99-117
[6] Klioner S A and Peip M 2003 *Astron. Astrophys.* **410** 1063-1074
[7] Blumenthal L M 1970 *Theory and Applications of Distance Geometry* (New York: Chelsea Publishing Company, 2nd edition)

Spanish Relativity Meeting (ERE 2009) IOP Publishing
Journal of Physics: Conference Series **229** (2010) 012060 doi:10.1088/1742-6596/229/1/012060

Critical phenomena and information geometry in black hole physics

Jan E Åman and Narit Pidokrajt*

Department of Physics, Stockholm University, 106 91 Stockholm (Sweden)

E-mail: `ja@fysik.su.se`, `narit@fysik.su.se`

Abstract. We discuss the use of information geometry in black hole physics and present the outcomes. The type of information geometry we utilize in this approach is the thermodynamic (Ruppeiner) geometry defined on the state space of a given thermodynamic system in equilibrium. The Ruppeiner geometry can be used to analyze stability and critical phenomena in black hole physics with results consistent with those from the Poincaré stability analysis for black holes and black rings. Furthermore other physical phenomena are well encoded in the Ruppeiner metric such as the sign of specific heat and the extremality of the solutions. The black hole families we discuss in particular in this manuscript are the Myers-Perry black holes.

1. Introduction

In this talk we discuss the use of information geometry, in particular the *thermodynamic geometry* [1], also known as Ruppeiner geometry, of various black hole (BH) families. This has been studied over the past few years in *e.g.* [2], [3], [3], [5], [6], [7] and recently in [8] among several dozens of papers devoted to the use of this method to study BHs. Our results so far have been physically suggestive [9], particularly in the Myers-Perry (MP) Kerr BH case where the curvature singularities signal the initial onset of thermodynamic instability of such BH. The geometrical patterns are given by the curvature of the Ruppeiner metric[1] defined as the Hessian of the entropy on the state space of the thermodynamic system

$$g^R_{ij} = -\partial_i \partial_j S(M, N^a), \tag{1}$$

where M denotes mass (internal energy) and N^a are other parameters such as charge and spin. The minus sign in the definition is due to concavity of the entropy function. Interpretations of the geometries associated with the metric are discussed in [10] plus references therein. It has been argued that the curvature scalar of the Ruppeiner metric measures the complexity of the underlying interactions of the system, *i.e.* the metric is flat for the ideal gas whereas it has curvature singularities for the Van der Waals gas. We take note that most interesting Ruppeiner metrics that we encounter have curvature singularities which are physically suggestive, but there are some known flat Ruppeiner metrics that can be understood from a mathematical point of view *i.e.* we proved in Åman *et al.* [11] a flatness theorem which states that Riemann curvature

[1] This metric is conformal to the so-called Weinhold metric via $g^W_{ij} = Tg^R_{ij}$ where T is thermodynamic temperature of the system of interest.

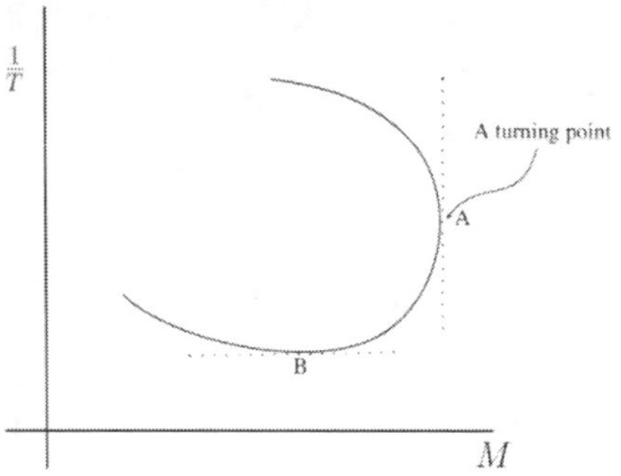

Figure 1. *In the Poincaré method one plots the conjugacy diagram i.e. the inverse temperature versus mass in our case. The turning point according to the Poincaré method is where there is a vertical tangent to the curve, i.e. point A. This is where a change of stability can take place. No change of stability happens at point B.*

tensor constructed out of the negative of the Hessian of the entropy of the form

$$S = M^k f(Q/M) \tag{2}$$

will vanish, where f is an arbitrary analytic function and $k \neq 1$. The latter condition is necessary in order for the metric to be nondegenerate. This theorem has proven useful in our work on the dilaton BHs [12] as it allows us to see the local geometry already by glancing at the entropy function. We also note that the signature of the Ruppeiner metric corresponds to the sign of the system's specific heat, *i.e.* the signature is Lorentzian for system with negative specific heat and Euclidean when all specific heats are positive.

We have observed that the Ruppeiner curvature scalar diverges in the extremal limits[2] of the Kerr, MP and Reissner-Nordström (RN) AdS BH whilst it is vanishing for RN, BTZ and dilaton BHs [8]. Despite the flat thermodynamic geometry one can extract useful information on the BH solutions by plotting the state space of such the flat geometry. Normally it is done by first transforming the metric into a manifestly flat form and then bringing it into a recognizably Minskowskian form.

Incidentally there is a so-called *Poincaré*'s linear series method for analyzing stability in non-extensive systems. The simplicity of this method is owing to the fact that it utilizes only a few thermodynamic functions such as the fundamental relations in order to study/analyze (in)stabilities. This method can thus be applied to BHs although they are non-extensive systems. Non-extensitivity in BHs is due to self-gravitation and furthermore the BHs cannot be subdivided and the BH entropy scales with area instead of the volume[3]. For the analysis we use the recipes given in [13] without further elaboration. In the Poincaré method[4] one plots a conjugacy diagram *e.g.* an inverse temperature versus mass and we can infer some information about the existence of instability if the turning point is present (see Fig. 1).

2. Higher-dimensional black holes

In higher dimensions BHs are more interesting objects due to richer rotation dynamics, and the appearance of extended black objects such as black strings. For more detailed and complete

[2] *i.e.* where the BH temperatures vanish.

[3] This makes the issue of dynamical and thermodynamic (in)stabilities subtler. In extensive systems thermodynamic stability normally implies dynamical stability.

[4] The proof of this is given in [14] and [15].

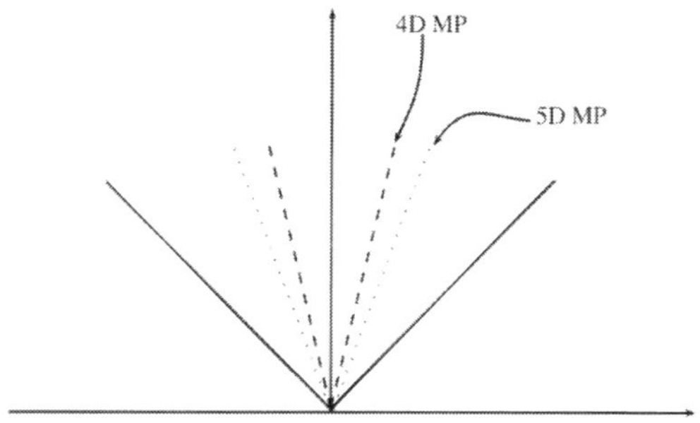

Figure 2. *The state space of the 4D and 5D MP BHs. Note that as the number of dimensions increases the wider the opening angle becomes. The extremal limits lie on the edge of the wedge, and on the thermodynamic cone entropy is vanishing.*

information see *e.g.* [16]. The higher-dimensional generalization of the 4D BHs is the Myers-Perry BHs[5]. In 5D there is a black ring solution [17] which is a BH with a horizon topology $S^2 \times S^1$ in asymptotically flat spacetime.

Critical phenomena in higher-dimensional black holes

Black strings and black branes suffer from instabilities known as the Gregory-Laflamme instability [18]. The MP black holes with only one angular momentum turned on when spinning ultrafast also suffer from instabilities. The Ruppeiner curvature scalar [9] for the MP BH with one spin is singular at

$$4\frac{J^2}{S^2} = \frac{d-3}{d-5}. \tag{3}$$

which coincides with that found by Emparan and Myers in [19] but in different coordinates[6] and the BH temperature reaches its minimum at this point. This is where the MP BH starts behaving like black membrane (which is an unstable object) but the instability is believed to set in somewhat later when the mass of the system decreases plus that there are unstable dynamical modes due to metric perturbations which the Poincaré method does not see such as those recently investigated (in various dimensions) by Dias *et al.* in [20]. Attempts to analyze such the instabilities by means of thermodynamic geometry are being made in [21].

The Weinhold metric for the MP BH is flat and can be brought into a manifestly flat form by coordinate transformations. The state space of the MP BHs appears as a wedge embedded in a Minkowski space and we call this diagram a *thermodynamic cone*. As mentioned above one can obtain some information from the thermodynamic cone *e.g.* the BH entropy is vanishing on the light cone whereas the edge of the wedge is where $T = 0$ which is the BH's extremal limit. The opening angle is unique for each BH and it increases as the number of dimensions increases. As we increase the number of dimensions the opening angle tends to the right angle (see Fig. 2).

Black rings

In [13] the authors show that the method of thermodynamic geometry is consistent with the Poincaré stability analysis, and they are able to prove that one of the black ring[7] branches is always locally unstable, showing that there is a change of stability at the point where the two

[5] The MP BHs can be regarded as the higher-dimensional versions of the Kerr solution.

[6] Note that this is for the 5D MP BH with only one spin turned on. In principle there can be two spins for the MP BH in 5D.

[7] They consider large and small black rings categorized by the angular momentum the black ring possesses.

black ring branches meet. Their results using two different methods (Ruppeiner and Poincaré) are consistent with each other.

3. Summary and outlook

Information geometry is a new approach for studying BH thermodynamics and possibly BH instabilities. This approach opens up new perspectives and sheds light on critical phenomena in BH systems. The signature of the thermodynamic metric and the curvature singularities correspond to the sign of the specific heat and the extremal limit of the BHs respectively. The curvature singularity is physically suggestive in that it signals the initial onset of instability in higher dimensional BHs. This method has so far been consistent with the Poincaré analysis[8]. Geometrical patterns of BH thermodynamics uncovered may play an important role in the context of quantum gravity.

Acknowledgments

NP would like to acknowledge the organizers of the ERE2009 in Bilbao for their kind support, and he would also like to thank Roberto Emparan for useful discussions. NP and JÅ thank Ingemar Bengtsson on various comments. Finally NP warmly thanks the KoF group and Department of Physics, Stockholm University for the kind hospitality.

References

[1] Ruppeiner G 1979 *Phys. Rev.* A **20** 1608
[2] Aman J E, Bengtsson I and Pidokrajt N 2003 *Gen. Rel. Grav.* **35** 1733
[3] Shen J y, Cai R G, Wang B and Su R K 2007 *Int. J. Mod. Phys.* A **22** 11
[4] Quevedo H and Vazquez A 2008 *AIP Conf. Proc.* **977** 165
[5] Medved A J M 2008 *Mod. Phys. Lett.* A **23** 2149
[6] Chakraborty S and Bandyopadhyay T 2008 *Class. Quantum Grav.* **25** 245015
[7] Sarkar T, Sengupta G and Tiwari B N 2008 *J. High Energy Phys.* JHEP10(2008)076
[8] Pidokrajt N 2009 "Information geometries in black hole physics", PhD Thesis, Department of Physics, Stockholm University, Sweden
[9] Aman J E and Pidokrajt N 2006 *Phys. Rev.* D **73** 024017
[10] Ruppeiner G 1995 *Rev. Mod. Phys.* **67** 605; Erratum 1996 **68** 313
[11] Åman J E, Bengtsson I and Pidokrajt N 2006 *Gen. Rel. Grav.* **38** 1305
[12] Åman J E, Pidokrajt N and Ward J 2008 *EAS Publ. Ser.* **30** 279
[13] Arcioni G and Lozano-Tellechea E 2005 *Phys. Rev.* D **72** 104021
[14] Katz J 1978 *Mon. Not. R. Astr. Soc.* **183** 765
Katz J 1979 *Mon. Not. R. Astr. Soc.* **189** 817
[15] Sorkin R 1981 *Astrophys. J.* **249** 254
[16] Emparan R and Reall H S 2008 *Living Rev. Rel.* **11** 6
[17] Emparan R and Reall H S 2002 *Phys. Rev. Lett.* **88** 101101
[18] Gregory R and Laflamme R 1993 *Phys. Rev. Lett.* **70** 2837
[19] Emparan R and Myers R C 2003 *J. High Energy Phys.* JHEP09(2003)025
[20] Dias O J C, Figueras P, Monteiro R, Santos J E and Emparan R 2009 *Phys. Rev.* D **80** 111701
[21] Åman J E and Pidokrajt N in preparation.

[8] We would like to point out that a divergence in Ruppeiner curvature scalar does not necessarily imply a turning point in the conjugacy diagram and *vice versa*. We note also that one cannot in general anticipate to obtain all the information on dynamic stability from the Poincaré method but should the turning point appear in a system in question it would be worth investigating in depth what it implies. One can take the Poincaré and the Ruppeiner method on the same footing that it helps detect instabilities.

Study on internal structure of Maxwell-Gauss-Bonnet black hole

K A Rannu[1*], S O Alexeyev[1] and A Barrau[2,3]

[1] Sternberg Astronomical Institute, Lomonosov Moscow State University, Universitetsy Prospect, 13, Moscow 119991, RUSSIA
[2] Laboratoire de Physique Subatomique et de Cosmologie, UJF-INPG-CNRS, 53, avenue des Martyrs, 38026 Grenoble cedex, France
[3] Institut de Hautes Etudes Scientifiques, 35, route de Chartres, 91440, Bures-sur-Yvette, France

E-mail: rannu@xray.sai.msu.ru

Abstract. The influence of the Maxwell field on a static, asymptotically flat and spherically-symmetric Gauss-Bonnet black hole is considered. Numerical computations suggest that if the charge increases beyond a critical value, the inner determinant singularity is replaced by an inner singular horizon.

1. Introduction

The internal structure of black holes described by the action

$$S = \frac{1}{16\pi} \int d^4x\sqrt{-g} \left[m_{pl}^2(-R + 2\partial_\mu\phi\partial^\mu\phi) - e^{-2\phi}F_{\mu\nu}F^{\mu\nu} + \lambda e^{-2\phi}S_{GB} \right], \quad (1)$$

where m_{pl} is the Plank mass, ϕ is the dilaton field, R is the scalar curvature, $S_{GB} = R_{ijkl}R^{ijkl} - 4R_{ij}R^{ij} + R^2$ is the Gauss-Bonnet term, $F_{\mu\nu}F^{\mu\nu}$ is the Maxwell field and λ is the string coupling constant. The influence of the magnetic charge of the black hole on the behavior of the metric functions was considered and it was shown that there exists a "critical value" of the charge beyond which the influence of the Maxwell term becomes more important than the Gauss-Bonnet one. The inner determinant singularity at $r = r_s$ is then replaced by a smooth local minimum. The focus was made on the behavior of the curvature invariant $R_{ijkl}R^{ijkl}$ near this critical point and in the vicinity of the main singularity at $r = r_x$.

2. Field equations, curvature invariant and calculation aspects

Considering a static, asymptotically flat and spherically symmetric black hole solution, we focus on the following metric:

$$ds^2 = \Delta dt^2 - \frac{\sigma^2}{\Delta}dr^2 - f^2\left(d\theta^2 + \sin^2\theta d\varphi^2\right), \quad (2)$$

where Δ, σ and f are functions that depend on the radial coordinate r only. To simplify the problem, only the magnetic charge was taken into account. Therefore, for the Maxwell tensor

$F_{\mu\nu}$, one can use the ansatz $F = q\sin\theta\ d\theta \wedge d\varphi$. The corresponding field equations in the GHS gauge ($\sigma(r) = 1$) are as follows:

$$m_{Pl}^2[ff'' + f^2(\phi')^2] + 4e^{-2\phi}\lambda[\phi'' - 2(\phi')^2]\Delta(f')^2 - 1 + 4e^{-2\phi}\lambda\phi' 2\Delta f' f'' = 0, \tag{3}$$

$$m_{Pl}^2[1 + \Delta f^2(\phi')^2 - \Delta' ff' - \Delta(f')^2] + 4e^{-2\phi}\lambda\Delta'\phi'[1 - 3\Delta(f')^2] - e^{-2\phi}q^2 f^{-2} = 0, \tag{4}$$

$$\begin{aligned} m_{Pl}^2[\Delta'' f + 2\Delta' f' + 2\Delta f'' + 2\Delta f(\phi')^2] + 4e^{-2\phi}\lambda[\phi'' - 2(\phi')^2]2\Delta\Delta' f' + \\ + 4e^{-2\phi}\lambda\phi' 2[(\Delta')^2 f' + \Delta\Delta'' f' + \Delta\Delta' f''] - 2e^{-2\phi}q^2 f^{-3} = 0, \end{aligned} \tag{5}$$

$$\begin{aligned} -2m_{Pl}^2[\Delta' f^2\phi' + 2\Delta ff'\phi' + \Delta f^2\phi''] + 4e^{-2\phi}\lambda[(\Delta')^2(f')^2 + \Delta\Delta''(f')^2 + 2\Delta\Delta' f' f'' - \Delta''] - \\ - 2e^{-2\phi}q^2 f^{-2} = 0. \end{aligned} \tag{6}$$

The behavior of the metric functions and of the dilatonic field near the horizon are described by a simple Taylor expansion:

$$\begin{aligned} \Delta &= d_1 x + d_2 x^2 + O(x^2), \\ f &= f_0 + f_1 x + f_2 x^2 + O(x^2), \\ e^{-2\phi} &= e^{-2\phi_0} + \phi_1 x + \phi_2 x^2 + O(x^2), \end{aligned} \tag{7}$$

where ($x = r - r_h, \ll 1$).

Without the Gauss-Bonnet term, the Gibbons-Maeda-Garfinkle-Horowitz-Strominger solution (GM-GHS) should be recovered as the basic solution of the Einstein equations with the dilaton and Maxwell terms. This solution is given by:

$$\begin{aligned} ds^2 &= \left(1 - \frac{2M}{r}\right)dt^2 - \left(1 - \frac{2M}{r}\right)^{-1} dr^2 \\ &- r\left(r - \frac{q^2\exp(2\phi_0)}{M}\right) d\Omega, \\ \exp(-2\phi) &= \exp(-2\phi_0) - \frac{q^2}{Mr}, \end{aligned} \tag{8}$$

where M stands for the black hole mass. In the limit $\lambda \to 0$, the solution of equations (3)–(6) at infinity should coincide with Eq. (8).

In order to determine the two metric functions and the dilatonic field, three equations are required. Among the four equations (3)-(6), only equations (3), (5) and (6), which contain the second derivative of the metric functions and the dilaton, are used. In contrast, Eq. (4), which contains the first derivative only, is considered as a constraint to check the solution.

To solve the system (3)-(5)-(6), the equations are rewritten using $E = e^{-2\phi}$ instead of the dilaton itself. Furthermore, the case $\lambda = 1$ is considered. In the chosen metric gauge, the squared Riemann tensor is given by:

$$\begin{aligned} R_{ijkl}R^{ijkl} &= \Delta''^2 + 4\Delta'^2\frac{f'^2}{f^2} + 8\Delta^2\frac{f''^2}{f^2} + 8\Delta\Delta'\frac{f'f''}{f^2} \\ &+ \frac{4}{f^4} - 8\Delta\frac{f'^2}{f^4} + 4\Delta^2\frac{f'^4}{f^4}. \end{aligned} \tag{9}$$

The computation process was divided into two parts. First, the GM-GHS solution (8) was taken as the initial condition at infinity. Solutions for the metric functions and the dilaton outside the event horizon were found. Then, the results near the horizon were taken as new initial conditions.

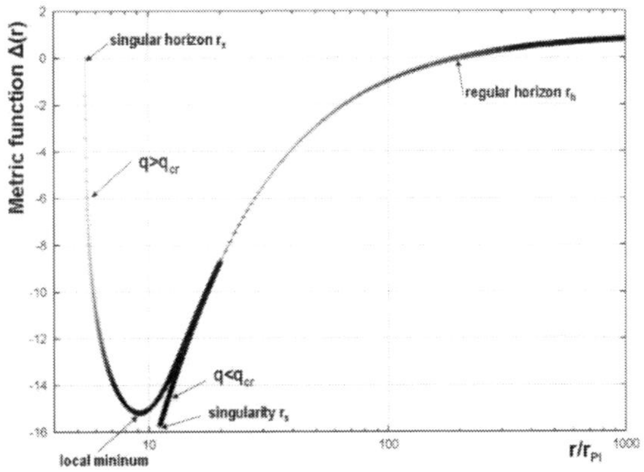

Figure 1. Metric function Δ as a function of the radial coordinate r for $q = 21.50 < q_{cr}$ (left curve) and $q = 24.81 > q_{cr}$ (right curve) when $r_h = 200.0$ Planck units.

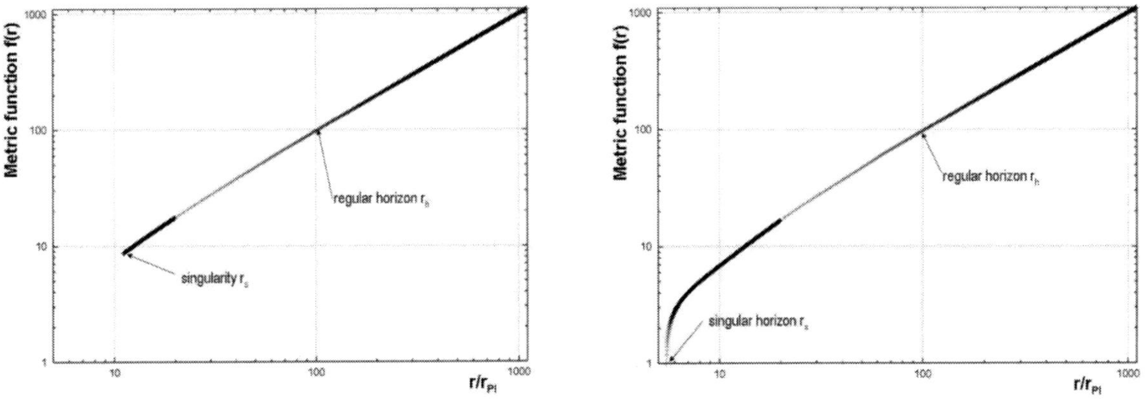

Figure 2. Metric function f as a function of the radial coordinate r for $q = 21.50 < q_{cr}$ (left plot) and $q = 24.81 > q_{cr}$ (right plot) when $r_h = 200.0$ Planck units.

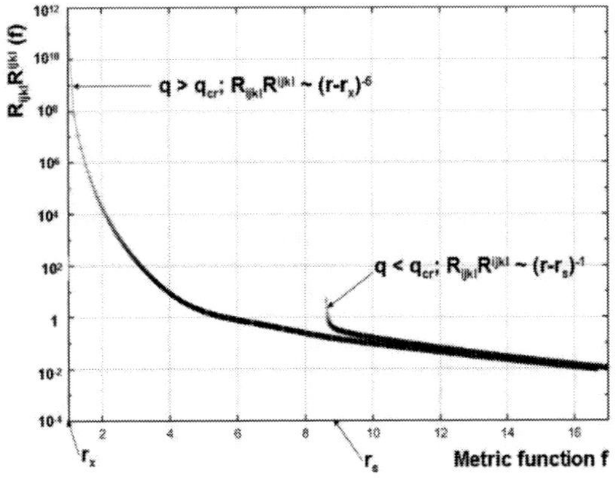

Figure 3. Curvature invariant $R_{ijkl}R^{ijkl}$ as a function of the metric function f for $q = 21.50 < q_{cr}$ (left curve) and $q = 24.81 > q_{cr}$ (right curve), with $r_h = 200.0$ Planck units.

3. Results

Metric function Δ has a particular point that causes the inner singularity r_s when the charge is small. However if the charge is larger than some critical value q_{cr} the inner singularity disappears and function Δ displays the local minimum in a point not far from the function's f zero, as it can be seen on Fig. 1. This is the difference between the given soluton and the GM-GHS.

Functions' f and $e^{-2\phi}$ behavior is analogous to the GM-GHS case. If $q < q_{cr}$ they decrease monotonously till $r = r_s$. Metric function f can be approximately described as $f = r$, but it reaches it's zero before r does even if $q < q_{cr}$. Function $e^{-2\phi}$ also approaches it's zero near the function's f zero. This leads to a fact that near the function's f zero influence of the Maxwell and Gauss-Bonnet terms become negligible and ordinary Einstein gravitation is realised.

It was confirmed that the behavior of the curvature invariant $R_{ijkl}R^{ijkl}$ under the black hole's event horizon is about zero almost everywhere and near the r_s $R_{ijkl}R^{ijkl} \to \infty$. When black hole charge q reaches it's critical value and metric function's Δ local minimum replaces singularity in r_s. In this case the value of curvature invariant $R_{ijkl}R^{ijkl}$ does not increase (Fig. 3). So it is obvious that metric function's Δ local minimum is not singular.

Function f plays the role of the radial coordinate in our solution. When r_s vanishes, the new point r_x in which f reaches it's zero, appears. In this case curvature invariant increases near the r_x. So we can consider r_x to be the singular horizon. When $q < q_{cr}$ this horizon belongs to a second branch of the system's (3)–(6) solution. This branch is nonphysical.

Near the singular horizon r_x curvature invariant increases much more rapidly than near the singularity r_s.

$$\begin{aligned}
f(r \to r_s) &= f_s + f_{s2}(\sqrt{r - r_s})^2 + f_{s3}(\sqrt{r - r_s})^3 + \dots \\
f(r \to r_x) &= f_x + f_{x1}\sqrt{r - r_x} + f_{x2}(\sqrt{r - r_s})^2 + \dots \\
\text{for} \quad f \to f_s \quad & R_{ijkl}R^{ijkl} \sim \text{const}_1 \times (f - f_s)^{-1} \\
\text{for} \quad f \to f_x \quad & R_{ijkl}R^{ijkl} \sim \text{const}_2 \times (f - f_x)^{-5}
\end{aligned}$$

So we can conclude that if singularity r_s is replaced by the function's Δ local minimum the singularity in r_x is much stronger than the one in r_s.

4. Conclusions

When the black hole charge becomes larger than the critical value the singularity r_s is replaced by a local minimum of the function $\Delta(r)$ and the solution exists till the singular horizon r_x.

Function $f(r)$ is the radius of S^2, so it plays the role of the radial coordinate. If $q < q_{cr}$ it decreases monotonously till $r = r_s$ like in GHS. When r_s disappears the function $f(r)$ reaches its zero in the new point r_x.

Curvature invariant increases much more rapidly (as $(r - r_x)^{-5}$) near the singular horizon r_x than near the singularity r_s (as $(r - r_s)^{-1}$), so the singularity in r_x is much stronger than the one in r_s.

New kind of singularity inside black hole was found. Unfortunately Maxwell-Gauss-Bonnet black hole cannot help wormholes' or multiverse theories because this singularity is very strong.

References

[1] Alexeyev S O, Barrau A and Rannu K A 2009 *Phys. Rev.* D **79** 067503
[2] Alexeyev S O and Pomazanov M V 1997 *Phys. Rev.* D **55** 2110
[3] Alexeyev S O and Sazhin M V 1998 *Gen. Relativ. Grav.* **30** 1187
[4] Alexeyev S O, Sazhin M V and Pomazanov M V 2001 *J. Mod. Phys.* D **10** 225
[5] Gibbons G W and Maeda K 1988 *Nucl. Phys.* B **298** 741
[6] Garfincle D, Horowitz G and Strominger A 1991 *Phys. Rev.* D **43** 3140
[7] Garfincle D, Horowitz G and Strominger A 1992 *Phys. Rev.* D **45** 3888

Physical properties of the double-Kerr solution

Miguel S Costa, Carlos A R Herdeiro and Carmen Rebelo*
Departamento de Física e Astronomia e Centro de Física do Porto, Faculdade de Ciências da Universidade do Porto, Rua do Campo Alegre, 687, 4169-007 Porto, Portugal

E-mail: miguelc@fc.up.pt, crherdei@fc.up.pt, mrebelo@fc.up.pt

Abstract. We analysed two special cases of the double-Kerr solution describing a system of two stationary co-axial Kerr black holes, with equal mass and either the same or opposite angular momentum. These cases are asymptotically flat and obey the axis condition, but there is always a strut (i.e. a conical singularity) between the black holes, preventing the collapse of the system. Studying the force associated to the strut, we show that the force between the two black holes is more attractive in the counter-rotating than in the co-rotating case. We also show that, for both the counter-rotating and co-rotating cases, 1) the angular velocity of the two black holes decreases as they approach one another, for fixed mass and angular momentum; 2) the extremal limit J/M^2 varies with the distance between the black holes and may even be larger than one. These results are interpreted in terms of rotational dragging effects.

1. Introduction

The main purpose of this communication is to summarise some curious features of the double-Kerr solution. This solution, which describes the nonlinear superposition of two Kerr black holes, was originally derived by Kramer and Neugebauer [1]. Subsequent analyses [2, 3, 4, 5] revealed that, for two under-extreme objects, the solution cannot be made free of singularities. The metric has always a conical singularity along the axis between the black holes. This conical singularity, that can be seen as a massless strut, balances the attraction force, avoiding the collapse of the system. This may be interpreted as follows: while in the the known Majumdar-Papetrou solution, describing a system of extremal Reissner-Nordström black-holes in equilibrium, the repulsive Coulomb force can balance the attractive gravitational force [6], in the double-Kerr solution the spin-spin repulsion is not strong enough to overcome gravitational attraction and hence two aligned Kerr black holes cannot be in stationary equilibrium.

Probably due to the presence of the strut, which implies that the solution cannot be seen as a realistic physical situation, little more was known about this particular solution. Recent studies [7, 8], focused on two special limits of the general double-Kerr solution where black holes have equal mass and either the same or opposite angular momentum, revealed that:

- the force between the two black holes is more attractive in the counter-rotating than in the co-rotating case;
- whenever the black holes are co- or counter-rotating, the angular velocity always decreases as the black holes approach one another and its maximum is always smaller than the angular velocity of an extremal isolated black hole with the same mass.

This peculiar behaviour of the angular velocity can be interpreted in terms of rotational dragging effects and allows the black holes to have a $J > M^2$.

2. Angular velocity

The co- and counter-rotating limits of the double-Kerr solution are characterised by three parameters: the mass M, the angular momentum J and the physical (i.e. proper) distance d between the black holes. As can be seen in figure 1, in both the counter-rotating and the co-rotating cases, the angular velocity Ω of the horizon, for each black hole, decreases when the two black holes get closer, keeping fixed their mass and the angular momentum, computed as Komar integrals.

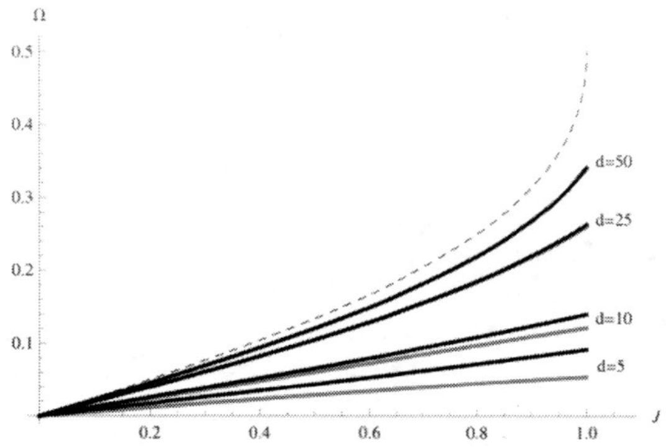

Figure 1. Horizon angular velocity as function of the angular momentum, for fixed mass $M = 1$, in the double Kerr system. The dashed line corresponds to the infinite distance limit (i.e. the usual Kerr solution). The black (grey) lines correspond to the co-rotating (counter-rotating) case for different values of the physical distance d. Note that in the co-rotating (counter-rotating) case, $\Omega = \Omega_1 = \Omega_2$ ($\Omega = \Omega_1 = -\Omega_2$). This figure was first shown in [9].

Moreover, it shows that the angular velocity is smaller, at the same physical distance, in the counter-rotating than in the co-rotating case. In the counter-rotating case, since black holes are rotating in opposite directions they are being counter-dragged and therefore they slow down. Using the same line of reasoning, one expects that the mutual rotational dragging, in the co-rotating case, speeds up the horizon angular velocity. Indeed one verifies that it is larger than in the counter-rotating case. However there must be a dominant effect responsible for the general slow down of the angular velocity. In [8, 9] it was suggested that each black hole is simultaneously dragging and being dragged; and dragging the other black hole always slows down the black hole which is dragging. This novel effect can only be seen in exact multi-black hole solutions and it is not an artifact of the strut. In fact, we can also observe this general slow down of the angular velocity in the balanced black Saturn [9], which is a regular (on and outside an event horizon) solution of the vacuum Einstein's equations in 5 dimensions. Considering a fat ring [1], with fixed mass and angular momentum, when the mass of the central Myers-Perry black hole is increased, the angular velocity of the ring horizon decreases. This happens because the black ring has to drag a heavy black hole.

3. Extremal Limit

The decrease in the angular velocity, due to frame dragging effects, allows the black holes in binary solutions to carry a higher angular momentum than isolated ones. This can be easily seen analysing Smarr's relation $M = TA_H/2 + 2\Omega J$, where T and A_H denote respectively the temperature and horizon area of each black hole. For the counter-rotating case it is possible to find the explicit expression to the extremal limit in terms of M and of the coordinate distance ζ [2] (in Weyl canonical coordinates), more specifically

$$\frac{|J_{ext}|}{M^2} = \sqrt{\frac{\zeta + 2M}{\zeta - 2M}},$$

[1] Just like Kerr black holes, fat rings have a positive moment of inertia, i.e. a higher J leads to a higher Ω.

[2] Note that ζ is a good measure of the physical distance, in the sense that d grows monotonically with ζ. In Weyl canonical coordinates $d = \int_{-\zeta/2}^{\zeta/2} dz \sqrt{g_{zz}}|_{\rho=0}$.

which gets larger as the counter-rotating black holes are approached.

The ratio J_{ext}/M^2 in the co-rotating case is displayed in figure 2, where it varies from one when the black holes are far away to two at the merging limit, corresponding to an extremal Kerr black hole with mass $2M$ and angular momentum $2J$, that is formed after merging.

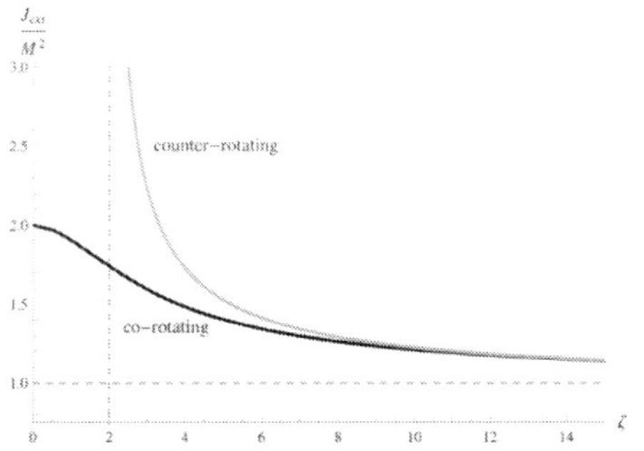

Figure 2. J_{ext}/M^2 in terms of the coordinate distance, for the co-rotating case and the counter-rotating case. We fixed $M = 1$ which determines the merging limit in the counter-rotating case to be at $\zeta = 2$. This figure was first shown in [8].

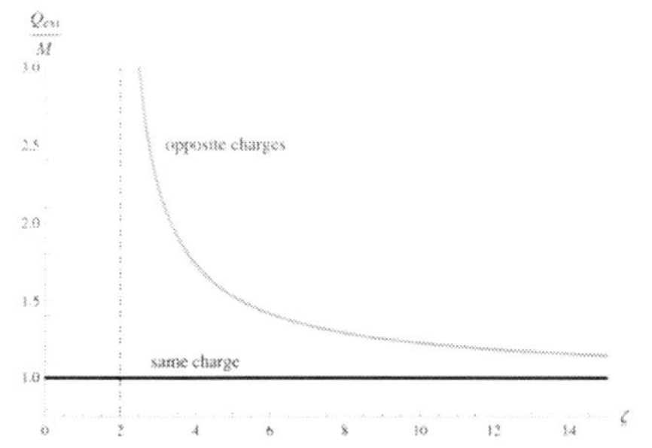

Figure 3. Q_{ext}/M in terms of the coordinate distance, for the cases where balck holes have equal mass and either the same or the opposite charge. We fixed $M = 1$ which determines the merging limit in the dihole case to be at $\zeta = 2$.

It should be point out that this property concerning the upper bound for the angular momentum has a counterpart for charged solutions. Considering a system of two charged black holes with equal mass M and either the same [6] or the opposite [10] charge Q, one can see that in the case of opposite charge case one finds

$$\frac{|Q_{ext}|}{M} = \sqrt{\frac{\zeta + 2M}{\zeta - 2M}},$$

just like the ratio J_{ext}/M^2 in the counter-rotating case. Moreover, just like it happens with the horizon angular velocity in the double Kerr system, in the charged dihole solution [10] the electrostatic potential at the horizon decreases as black holes approach one another and therefore each black hole supports more charge without reaching the extremal limit. When the black holes have the same charge [6] the electrostatic potential at the horizons is independent of the distance between the black holes and, as can be seen in figure 3, the ratio Q_{ext}/M is always equal to one; this is the expect Q_{ext}/M value to the Reissner-Nordström black hole with mass $2M$ and charge $2Q$, that is formed after the merging and, as previously pointed out, whenever $Q = M$ the black holes are in equilibrium.

4. Interaction Force

The force associated to the strut reflects the interaction force between the black holes. With the aim of studying the spin-spin interaction, we computed the force between the black holes in terms of J, for fixed mass and physical distance. Figure 4 shows that in both co- and counter-rotating cases, the force increases, in modulus, as J increases, meaning that we are in the presence of an attractive force.

However, it must be taken into account that when we vary J, for fixed d and M, we are also varying the horizon shape and therefore higher gravitational multiple moments should contribute to the force at the same order as the spin-spin interaction. This may explain why even in co-rotating case, where the spin-spin force is expected to be repulsive, the total force becomes more attractive when we increase J.

It is, still, reassuring to note that the force is greater, in modulus, in the counter-rotating case than in the co-rotating one. This should be expected from the properties of the spin-spin force which also explains why more energy can be extracted in a head on collision of Kerr black holes, if the spins are anti-aligned (counter-rotation), rather than aligned (co-rotation), with the direction of separation [11].

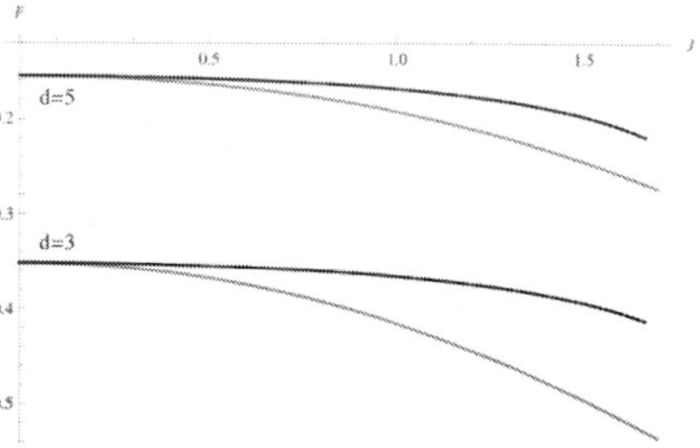

Figure 4. Force between the two black holes for two physical distances d = 3 and d = 5 as a function of J for M = 1, for the counter-rotating (grey lines) and co-rotating case (black lines). This figure was first shown in [8].

5. Conclusion

We have shown that due to frame dragging effects the angular velocity of two aligned black holes always decreases when the black holes are approached, even if they are co-rotating. Also due to this slow down in the angular velocity, black holes in binaries can carry more angular momentum than isolated black holes with the same mass. Thus, extremal black holes have $J > M^2$ in the double-Kerr system. Regarding the interaction force between the black holes we have observed that it is more attractive in the counter-rotating than in the co-rotating case. This makes clear why, in particular, it is always impossible to reach equilibrium.

Acknowledgments

C.R. is funded by FCT through grant SFRH/BD/18502/2004. C.R. would like to thank the organization of the "Spanish Relativity Meeting 2009" the grant to attend this meeting.

References

[1] Kramer D and Neugebauer G 1980 *Phys. Lett.* **75A** 259
[2] Dietz W and Hoenselaers C 1985 *Ann. Phys.* **165** 319
[3] Manko V S, Ruiz E and Sanabria-Gómez J D 2000 *Class. Quantum Grav.* **17** 3881
[4] Manko V S and Ruiz E 2001 *Class. Quantum Grav.* **18** L11
[5] Neugebauer G and Hennig J 2009 *Gen. Rel. Grav.* **41** 2113
[6] Azuma T and Koikawa T 1994 *Prog. Theor. Phys.* **92** 1095
[7] Herdeiro C A R and Rebelo C 2008 *J. High Energy Phys.* JHEP10(2008)017
[8] Costa M S, Herdeiro C A R and Rebelo C 2009 *Phys. Rev.* D **79** 123508
[9] Herdeiro C A R, Rebelo C and Warnick C M 2009 *Phys. Rev.* D **80** 084037
[10] Emparan R and Teo E 2001 *Nucl. Phys.* B **610** 190
[11] Hawking S W 1971 *Phys. Rev. Lett.* **26** 1344

Journal of Physics: Conference Series **229** (2010) 012063 doi:10.1088/1742-6596/229/1/012063

Power spectrum of large-scale structure cosmological models in the framework of scalar-tensor theories

M A Rodríguez-Meza

(Instituto Avanzado de Cosmología, IAC)
Instituto Nacional de Investigaciones Nucleares, Col. Escandón, Apdo. Postal 18-1027, 11801 México D.F., México

E-mail: `marioalberto.rodriguez@inin.gob.mx`; `http://astro.inin.mx/mar`

Abstract. We study the large-scale structure formation in the Universe in the frame of scalar-tensor theories as an alternative to general relativity. We review briefly the Newtonian limit of non-minimally coupled scalar-tensor theories and the evolution equations of the N-body system that is appropriate to study large-scale structure formation in the Universe. We compute the power-spectrum of the universe at present epoch and show how the large-scale structure depends on the scalar field contribution.

1. Introduction

During the last two decades high precision observations have established the existence of two component of the universe. These are known as dark energy and dark matter. What are their fundamental nature? this one of the questions we need to answer[1]. One of the most successful models to explain observations is the ΛCDM but it still has some problems.

In the past we have proposed a dark matter model in the framework of the Newtonian limit of a scalar-tensor theory[2]. We have done N-body simulations using a numerical code developed to take into account the scalar field (SF) force contribution which is of Yukawa type in its Newtonian limit and we have use it to study several situations, like collision of protogalaxies[3], dynamics and collision of galaxies[4, 5], and to study cosmological simulations[6, 7, 8]. Recently we have used it to address a possible solution to two of the main problems the ΛCDM has, i.e., the tendency to over populate the formed halos of galaxies with satellites and the flatness problem[10, 11].

Several authors have also been using similar dark matter models to study large-scale structure formation, see for instance[12, 13, 14].

One of the tools to characterize the clustering properties of a universe model is the power spectrum. Constraints on cosmological models and on cosmological parameters have been derived from measuring power spectrum. Therefore, in this work we consider a dark matter model build from the Newtonian limit of a scalar-tensor theory[2] and use it to study the large-scale structure formation of the universe by analyzing the power-spectrum versus the parameters of the model.

2. Evolution equations for a ΛCMD universe within the framework of a scalar-tensor theory

Giving that we are interested in the study of the large-scale structure formation of the universe we will use the Newtonian limit of the scalar-tensor theory. This limit applies in a region small compared to the Hubble length cH^{-1}, with c the speed of light and H the Hubble factor, and large compared to the Schwarzschild radii of any collapsed objects.

In this case the Vlasov-Poisson equation in an expanding universe describes the evolution of the six-dimensional, one-particle distribution function, $f(\mathbf{x}, \mathbf{p})$[15]. The Vlasov equation is,

$$\frac{\partial f}{\partial t} + \frac{\mathbf{p}}{ma^2} \cdot \frac{\partial f}{\partial \mathbf{x}} - m\nabla\Phi_N(\mathbf{x}) \cdot \frac{\partial f}{\partial \mathbf{p}} = 0 \tag{1}$$

where $\mathbf{x}$ is the comoving coordinate, $\mathbf{p} = ma^2\dot{\mathbf{x}}$, m is the particle mass, and Φ_N is the self-consistent gravitational potential given by the Poisson equation,

$$\nabla^2\Phi_N(\mathbf{x}) = 4\pi G_N\, a^2\, [\rho(\mathbf{x}) - \rho_b(t)]] \tag{2}$$

where ρ_b is the background mass density. Eqs. (1) and (2) form the Vlasov-Poisson equation, constitutes a collisionless, mean-field approximation to the evolution of the full N-body distribution. An N-body code attempts to solve Eqs. (1) and (2) by representing the one-particle distribution function as

$$f(\mathbf{x}, \mathbf{p}) = \sum_{i=1}^{N} \delta(\mathbf{x} - \mathbf{x}_i)\, \delta(\mathbf{p} - \mathbf{p}_i) \tag{3}$$

Substitution of (3) in the Vlasov-Poisson system of equations yields the exact Newton's equations for a system of N gravitating particles.

In the Newtonian limit of STT of gravity, the Newtonian motion equation for a particle i is written as[7]

$$\ddot{\mathbf{x}}_i + 2\, H\, \mathbf{x}_i = -\frac{1}{a^3}\frac{G_N}{1+\alpha} \sum_{j\neq i} \frac{m_j(\mathbf{x}_i - \mathbf{x}_j)}{|\mathbf{x}_i - \mathbf{x}_j|^3}\, F_{SF}(|\mathbf{x}_i - \mathbf{x}_j|, \alpha, \lambda) \tag{4}$$

where the sum includes all periodic images of particle j, and $F_{SF}(r, \alpha, \lambda)$ is

$$F_{SF}(r, \alpha, \lambda) = 1 + \alpha\left(1 + \frac{r}{\lambda}\right) e^{-r/\lambda} \tag{5}$$

which, for small distances compared to λ, is $F_{SF}(r \ll \lambda, \alpha, \lambda) \approx 1 + \alpha\,(1 + \frac{r}{\lambda})$ and, for long distances, is $F_{SF}(r \gg \lambda, \alpha, \lambda) \approx 1$, as in Newtonian physics.

To simulate cosmological systems, the expansion of the universe has to be taken into account. Also, to determine the nature of the cosmological model we need to determine the composition of the universe, i. e., we need to give the values of Ω_i for each component i, taking into account in this way all forms of energy densities that exist at present.

In this work we will consider a model with only two energy density contribution. One which is a pressureless and nonbaryonic dark matter with $\Omega_{DM} \approx 0.3$ that does not couple with radiation. Other, that will be a cosmological constant contribution $\Omega_\Lambda \approx 0.7$. Also, the model must be consistent with a static SF according to the Newtonian limit we are using in this work. Thus, the scale factor, $a(t)$, is given by the following Friedman model[8],

$$a^3H^2 = H_0^2\left[\frac{\Omega_{DM0} + \Omega_{\Lambda 0}\, a^3}{1+\alpha} + \left(1 - \frac{\Omega_{DM0} + \Omega_{\Lambda 0}}{1+\alpha}\right) a\right] \tag{6}$$

where $H = \dot{a}/a$, Ω_{DM0} and $\Omega_{\Lambda 0}$ are the matter and energy density evaluated at present, respectively. We have also demanded that the universe be a flat one. To be consistent with the CMB spectrum and structure formation numerical experiments, cosmological constraints must be applied on α in order for it to be within the range $(-1, 1)$ [16, 17, 18, 19].

Journal of Physics: Conference Series **229** (2010) 012063 doi:10.1088/1742-6596/229/1/012063

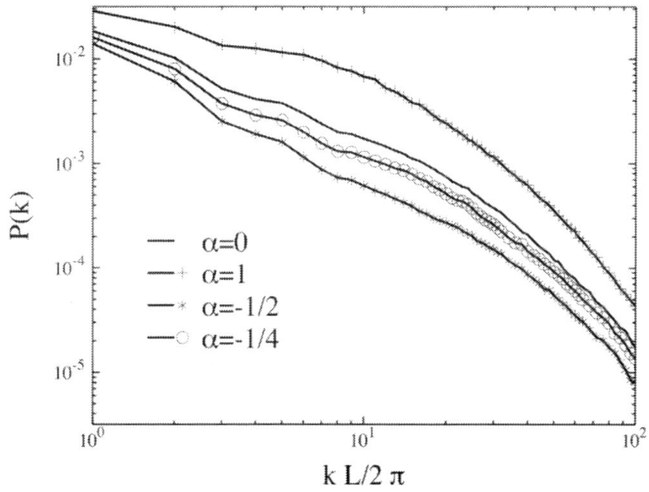

Figure 1. Power spectrum of our scalar field universe model without and with SF. See text for details.

3. Results

In this section, we present results of cosmological simulations of a ΛCDM universe with and without SF contribution. We start our simulations with an initial distribution of $N = 2 \times 32^3$ particles in a box with sides of $L = 50h^{-1}$ Mpc at $z = 10$. This case is similar to the one that comes with Gadget 1 [20]. At present epoch, $\Omega_{m0} = 0.3$, $\Omega_{\Lambda 0} = 0.7$, $H_0 = 100h$ km/s/Mpc, $h = 0.7$. We restrict the values of α to the interval $(-1, 1)$ [16, 17, 18, 19] and use $\lambda = 5$ Mpc, since this scale turns out to be an intermediate scale between the size of the clump groups and the separation of the formed groups. However, we have analyzed elsewhere several values of λ[8]. Other universe models have been analyzed in [6, 7, 9].

Because the visible component is the smaller one and given our interest to test the consequences of including a SF contribution to the evolution equations, our model excludes gas particles, but all its mass has been added to the dark matter. In Fig. 1 we show the power spectrum at redshift $z = 0$ of our ΛCDM model for several values of α. The line without symbols shows the case of a ΛCDM model without the contribution of the scalar field. The line with $(+)$ is the case of a ΛCDM model with scalar field with $\alpha = 1$, the line with $(*)$ uses $\alpha = -1/2$, and the line with (o) uses $\alpha = -1/4$. One notes clearly how the SF modifies the clustering of matter of the system. The most dramatic cases are the models with $\alpha = -1/2$ and $\alpha = 1$.

We now analyze the general effect the parameter α has on the clustering. For a given λ the role of α in our approach can be seen in equations (4)-(5) and is contained in the factor $F_{SF}(r, \alpha, \lambda)$. The factor F_{SF} augments (diminishes) for positive (negative) values of α for small distances compared to λ, resulting in more (less) structure formation for positive (negative) values of α compared to the ΛCDM model. In the case of the line with $(+)$, for $r \ll \lambda$, the effective gravitational pull has been augmented by a factor of 2, in contrast to case shown with line and $(*)$ where it has diminished by a factor of $1/2$; in the model with the line and (o) the pull diminishes only by a factor of $3/4$. That is why one observes for $r < \lambda$ more structure formation in case with line and $(+)$, less in case with line and $(*)$, and lesser in model with line and (o). The effect is then, for a growing positive α, to speed up the growth of perturbations, then of halos and then of clusters, whereas negative α values $(\alpha \to -1)$ tend to slow down the growth.

4. Conclusions

The general gravitational effect of the scalar-tensor dark matter model is that the interaction changes by a factor $F_{SF}(r, \alpha, \lambda) \approx 1 + \alpha(1 + r/\lambda)$ for $r < \lambda$ in comparison with the Newtonian case. For negative values of α the effect is to diminish the formation of structures and the opposite occurs for positive values of α. For $r > \lambda$ the dynamics is essentially Newtonian. By computing the power spectrum we were able to determine the effects of the scalar field contribution to the galaxy clustering. Of course we will need to do a systematic and with greater resolution study and make detailed comparisons with observations to support or to rule out a dark model based on scalar-tensor theories.

Acknowledgments

This work was supported by CONACYT, grant numbers CB-2007-84133-F, I0101/131/07 C-234/07, IAC collaboration. The simulations were performed in the UNAM HP cluster *Kan-Balam*.

References

[1] Bretón N, Cervantes-Cota J L and Salgado M (Eds.) 2004 *The Early Universe and Observational Cosmology* (Springer-Verlag, Berlin Heidelberg)
[2] Rodríguez-Meza M A and Cervantes-Cota J L 2004 *Mon. Not. Roy. Astron. Soc.* **350** 671
[3] Rodríguez-Meza M A, Klapp J, Cervantes-Cota J L and Dehnen H in *Exact solutions and scalar fields in gravity: Recent developments*, Edited by A. Macias, J.L. Cervantes-Cota, and C. Lämmerzahl (Kluwer Academic/Plenum Publishers, NY, 2001) p. 213. (*Preprint* arXiv:0907.2466)
[4] Gabbasov R F, Rodríguez-Meza M A, Klapp J and Cervantes-Cota J L 2006 *Astron. & Astrophys.* **449** 1043
[5] Cervantes-Cota J L, Rodríguez-Meza M A, Gabbasov R F and Klapp J 2007 *Rev. Mex. Fis. S* **53** 22 (*Preprint* arXiv: 0704.2579)
[6] Rodríguez-Meza M A, González-Morales A X, Gabbasov R F and Cervantes-Cota J L 2007 *J. Phys.: Conf. Series* **91** 012012 (*Preprint* arXiv:0708.3605)
[7] Rodríguez-Meza M A 2008 *AIP Conference Proceedings* **977**, 302 (*Preprint* arXiv:0802.1170)
[8] Rodríguez-Meza M A 2009 *AIP Conference Proceedings* **1116** 171 (*Preprint* astro-ph/0907.2898)
[9] Rodríguez-Meza M A 2009 *AIP Conference Proceedings* **1083** 190 (*Preprint* astro-ph/0810.0491)
[10] Cervantes-Cota J L, Rodríguez-Meza M A and Nuñez D 2007 *J. Phys.: Conf. Series* **91** 012007
[11] Cervantes-Cota J L, Rodríguez-Meza M A and Nuñez D 2009 *Phys. Rev.* D **79** 064011
[12] Hellwing W A and Juszkiewicz R 2009 *Phys. Rev.* D **80** 083522
[13] Brandao C S S and de Araujo J C N *Preprint* arXiv: 0910.1550
[14] Keselman J A, Nusser A and Peebles P J E *Preprint* arXiv: 0912.4177
[15] Bertschinger E 1998 *Annual Review of Astronomy and Astrophysics* **36** 599
[16] Nagata R, Chiba T and Sugiyama N 2002 *Phys. Rev.* D **66** 103510
[17] Nagata R, Chiba T and Sugiyama N 2004 *Phys. Rev.* D **69** 083512
[18] Shirata A, Shiromizu T, Yoshida N and Suto Y 2005 *Phys. Rev.* D **71** 064030
[19] Umezu K, Ichiki K and Yahiro M 2005 *Phys. Rev.* D **72** 044010
[20] Springel V, Yoshida N and White S D M 2001 *New Astronomy* **6** 79

Journal of Physics: Conference Series **229** (2010) 012064 doi:10.1088/1742-6596/229/1/012064

Late-time tails of self-gravitating massless fields

Andrzej Rostworowski

M. Smoluchowski Institute of Physics, Jagiellonian University, Kraków, Poland

E-mail: `arostwor@th.if.uj.edu.pl`

Abstract. In this talk I summarize briefly recent results of joint work with P. Bizoń, T. Chmaj and S. Zając, on the nonlinear origin of the power-law tails in the long-time evolution of self-gravitating massless fields. We focus on a spherically symmetric massless scalar field and wave map matter coupled to gravity. Using third-order perturbation method we derive explicit expressions for the tail (the decay rate and the amplitude) for solutions starting from small initial data and we verify this prediction via numerical integration of the full system of Einstein field equations. Our results show that the nonlinear effects can dominate the late time asymptotics.

1. Introduction.

Self-gravitating massless fields have been serving as a toy-model of gravitational collapse for a few decades, leading to valuable insights about the validity of the weak cosmic censorship and no-hair conjectures. In particular, for the case of spherically symmetric self-gravitating massless scalar field, Christodoulou proved that there are two generic endstates of evolution: Minkowski spacetime for small initial data [1] and Schwarzschild black hole for large initial data [2]. In both cases the upper bound for the rate of relaxation to the endstate inside the light cone is t^{-3} (this was proved in [1] for the dispersive solutions and by Dafermos and Rodnianski [3] for the collapsing solutions). In view of these rigorous results one might wonder what is the point of studying this problem again.

In general our motivation is to understand the mechanism of relaxation to stable static solutions of nonlinear wave equations (for example Minkowski or Schwarzschild spacetime in the case of Einstein's equations). To illustrate the problem let us consider a spherically symmetric wave equation with a potential and the cubic power nonlinearity in 3 spatial dimensions:

$$\partial_t^2 F(t,r) - \Delta F(t,r) + V(r)F(t,r) + F^3(t,r) = 0. \qquad (1)$$

Assume the potential to be spherically symmetric, nonnegative and decaying as $1/r^3$ for r going to infinity. We prepare smooth, compactly supported and spherically symmetric Cauchy data and watch how the solution behaves at some fixed distance from the center at large time t. The potential is repulsive and the nonlinearity is defocussing as well, so at any fixed point r the solution $F(t,r)$ tends to zero as times goes to infinity. In fact, at intermediate times the solution decays exponentially (this stage is usually called quasinormal ringing) but at late times polynomial decay is revealed (this stage is usually called a tail). Quasinormal ringing is a well understood linear phenomenon (related to resonances of the potential V) and since at later times the solution gets even smaller, it is tempting to describe the tail by the linearised theory as well.

It turns out that this is wrong and the cubic term cannot be neglected. The heuristic picture which explains the tail phenomena in the linearised approximation is the following. In the first approximation the signal which arrives at late time t at some small distance r_{obs}, comes from a single scattering off the potential. The amplitude of the signal is proportional to the amplitude of the potential at the scattering point. The signal spends roughly one half of its time travelling forth and second half travelling back thus it gets scattered at the distance $r_{scatt} \approx t/2$. As the potential falls off as $1/r^3$ the amplitude of the potential at the scattering point is proportional to $1/t^3$. Now we should compare two terms: a linear term $V\,F$ and a cubic term $F^3 = F^2\,F$, hence V versus F^2. The general solution of a free wave equation decays as $1/r$ along the light-cone. Therefore on the light-cone we have $V \sim 1/r^3$ versus $F^2 \sim 1/r^2$ and it is F^2 which decays slower and sets the decay rate of the tail to be $1/t^2$, while a linear tail decays as $1/t^3$!

Despite the above simple heuristics, the nonlinear effects in the late-time asymptotics have been frequently overlooked in the physical literature. Let us take a self-gravitating massless scalar field as an example. In 1972 Price [4] argued that the ℓ multipole of the massless scalar field evolving on a fixed Schwarzschild background in $3+1$ spacetime dimensions decays as $1/t^{3+2\ell}$. Then in 1994 Gundlach, Price and Pullin published two influential papers [5, 6]. In [5] they checked Price's prediction numerically and indeed obtained $1/t^{3+2\ell}$ decay for an ℓ-multipole of the massless scalar field evolving on a fixed Schwarzschild background. In [6] they evolved the full system of Einstein equations for spherically symmetric self-gravitating massless scalar field and again found $1/t^3$ decay regardless of whether the scalar field collapses and forms a black hole or disperses to infinity. As this decay rate coincides with a decay rate of the tail on the fixed Schwarzschild background they suggested that the linearised theory might apply also in the nonlinear regime: "*We found that the predictions for power-law tails of perturbations of Schwarzschild spacetime hold to reasonable approximation, even quantitatively, in a variety of situations to which the predictions might seem initially not to apply*". As the time passed this rather cautious statement was repeated with increasing sureness in numerous citations of [6]. In fact, there are hints already in [6] that the decay rates of linear tails on static background are not generic and the true decay rates may be slower. In sec.V of [6] two scalar fields are considered: a self-gravitating field and a test field evolving on a background provided by the self-gravitating field. For the test field the problem is linear (and it does make sense to expand the test field into ℓ multipoles), however the background provided by self-gravitating field is not static and $1/t^{2\ell+3}$ prediction does not apply (see Fig. 12 in [6]). Another hint is that the tails exist for arbitrary small initial data and for such data one should not expect a tail in linearised approximation at all.

2. Self-gravitating massless scalar field.

As the tails exist for arbitrary small initial data we approach the problem from Minkowski side . We generalise the problem slightly and work in an odd number of spatial dimensions $d = 3+2\ell$, parametrised with a nonnegative integer ℓ. We use the letter ℓ deliberately to stress the link between the solutions of the spherically symmetric wave equation in $d = 3+2\ell$ spatial dimensions and the solutions of the wave equation in 3 spatial dimensions for an ℓ multipole:

$$\Box_{(d)} \frac{F(t,r)}{r^\ell} = \frac{1}{r^\ell} \Box_{(\ell)} F(t,r), \tag{2}$$

where

$$\Box_{(d)} = \left(\partial_t^2 - \partial_r^2 - \frac{d-1}{r}\partial_r\right) \text{ and } \Box_{(\ell)} = \left(\partial_t^2 - \partial_r^2 - \frac{2}{r}\partial_r + \frac{\ell(\ell+1)}{r^2}\right). \tag{3}$$

We assume spherical symmetry and parametrize the metric using two metric functions $m(t,r)$ and $\beta(t,r)$:

$$ds^2 = \left(1 - \frac{m}{r^{d-2}}\right)^{-1} \left(-e^{2\beta} dt^2 + dr^2\right) + r^2 d\Omega_{d-1}^2 \tag{4}$$

Einstein equations set a link between metric functions m and β and massless scalar field ϕ

$$G_{\alpha\beta} = 8\pi T_{\alpha\beta}, \quad T_{\alpha\beta} = \nabla_\alpha\phi\nabla_\beta\phi - \frac{1}{2}g_{\alpha\beta}\left(\nabla_\mu\phi\nabla^\mu\phi\right), \tag{5}$$

which together with equation of motion for the massless field $\nabla_\mu\nabla^\mu\phi = 0$ constitutes a closed system of equations:

$$\begin{aligned}
m' &= \kappa\, r^{d-1}\left(1-\frac{m}{r^{d-2}}\right)\left(\phi'^2 + e^{-2\beta}\dot\phi^2\right), && \text{(hamiltonian constraint)} \quad (6)\\
\dot m &= 2\kappa\, r^{d-1}\left(1-\frac{m}{r^{d-2}}\right)\dot\phi\,\phi', && \text{(momentum constraint)} \quad (7)\\
\beta' &= (d-2)\frac{m}{r^{d-1}}\left(1-\frac{m}{r^{d-2}}\right)^{-1}, && \text{(slicing condition)} \quad (8)\\
\left(e^{-\beta}\dot\phi\right)^{\cdot} &= \frac{1}{r^{d-1}}\left(r^{d-1}e^{\beta}\phi'\right)', && \text{(wave equation)} \quad (9)
\end{aligned}$$

where $\kappa = 8\pi/(d-1)$, and primes and dots denote partial derivatives with respect to r and t, respectively. We treat this system perturbatively, taking as perturbative parameter the amplitude of Cauchy data. Consider small and compactly supported initial data $(\phi, \dot\phi)_{t=0} = (\varepsilon f(r), \varepsilon g(r))$. Then, up to the order $\mathcal{O}(\varepsilon^3)$, we have

$$\phi = \varepsilon\phi_1 + \varepsilon^3\phi_3, \qquad m = \varepsilon^2 m_2, \qquad \beta = \varepsilon^2\beta_2, \tag{10}$$

where ϕ_1 satisfies the flat-space radial wave equation

$$\Box_{(d)}\phi_1 = 0, \qquad (\phi_1, \dot\phi_1)_{t=0} = (f(r), g(r)), \tag{11}$$

the second-order perturbations of the metric functions satisfy

$$m_2' = \kappa\, r^{d-1}\left(\dot\phi_1^2 + {\phi_1'}^2\right), \qquad \dot m_2 = 2\kappa\, r^{d-1}\dot\phi_1\phi_1', \qquad \beta_2' = \frac{(d-2)m_2}{r^{d-1}} \tag{12}$$

(these equations can be easily integrated once ϕ_1 is known) and finally ϕ_3 satisfies the inhomogeneous wave equation (with zero initial data)

$$\Box_{(d)}\phi_3 = 2\beta_2\ddot\phi_1 + \dot\beta_2\dot\phi_1 + \beta_2'\phi_1' =: S(t,r)\,. \tag{13}$$

Thus the tools to deal with the system (11-13) are elementary. They consist of the formula for the general solution of the equation (11)

$$\phi_1(t,r) = \frac{1}{r^{\ell+1}}\sum_{k=0}^{\ell}\frac{(2\ell-k)!}{k!(\ell-k)!}\,\frac{\left(a^{(k)}(t-r) - (-1)^k a^{(k)}(t+r)\right)}{(2r)^{\ell-k}} \tag{14}$$

(where a compactly supported function $a(x)$ is set by initial data in (11)), and the Duhamel formula for the equation (13)

$$\phi_3(t,r) = \frac{1}{2r^{\ell+1}}\int_0^t d\tau \int_{|t-r-\tau|}^{t+r-\tau} \rho^{\ell+1}P_\ell\left(\frac{r^2+\rho^2-(t-\tau)^2}{2r\rho}\right) S(\tau,\rho)\,d\rho \tag{15}$$

(here $P_\ell(x)$ is a Legendre polynomial of degree ℓ). In this way we can get not only the decay rate but also the amplitude of the tail both at timelike and null infinity. The asymptotic behavior at timlike infinity reads

$$\phi_3(t,r) = \frac{1}{t^{3\ell+3}}\left[A_\ell + \mathcal{O}\left(\frac{1}{t}\right)\right], \tag{16}$$

where

$$A_0 = -2^5\pi \int_{-\infty}^{+\infty} a(u) \int_{u}^{\infty} \left(a'(s)\right)^2 \, ds \, du\,, \tag{17}$$

$$A_\ell = (-1)^{\ell+1} 2^{3\ell+5}\pi \int_{-\infty}^{+\infty} a^{(\ell-1)}(u) \left(a^{(\ell+1)}(u)\right)^2 du, \quad \text{for } \ell > 0 \tag{18}$$

are the only trace of initial data. We refer the reader to [7] for more details about this calculation and numerical evidence.

Now, it is interesting to compare our predictions with the predictions for the tails on a fixed Schwarzschild − Tangherlini black hole in $2\ell+3$ spatial dimensions. For dispersing solutions the tail has genuinely nonlinear origin, while for collapsing solutions it has two contributions: a nonlinear one and a linear one coming from the scattering on the fixed background curvature. In 3 spatial dimensions both tails decay as $1/t^3$, but for $\ell > 0$ the nonlinear tail decays as $1/t^{3\ell+3}$, while the linear tail decays as $1/t^{6\ell+4}$ [8]. Thus in higher dimensions it is the nonlinear tail which dominates.

3. Wave map matter.

As a second toy-model to study late time asymptotics in gravitating system we considered wave map matter coupled to gravity [9]. This model is interesting as the wave map index ℓ plays a role similar to angular momentum, but can be built consistently into fully nonlinear Einstein equations. Thus it may serve as a toy-model to study non spherically symmetric gravitational collapse. For $\ell > 0$ the asymptotics reads $r^\ell / \left(t^2 - r^2\right)^{\ell+1}$ (see [9] for details).

This result suggest that in $3+1$ spacetime dimensions the linear tail of an $\ell > 0$ multipole of a massless scalar field evolving on a fixed but time dependent background decays generically as $1/t^{2\ell+2}$ (as numerically reported in [6] Sec.V), that is by one power slower then Price's law prediction $1/t^{2\ell+3}$ for a fixed static background [4, 5]. This issue is discussed in [10].

4. Summary.

We discussed the non-linear origin of the power-law tails in the long-time evolution of a spherically symmetric self-gravitating massless fields in even-dimensional spacetime. Our results show that the agreement between decay rates of linear (on static background) and nonlinear tails of self-gravitating massless scalar field observed in $3+1$ spacetime dimensions [6] is accidental and holds neither in higher dimensions nor in other models (e.g. wave maps).

Acknowledgments

We acknowledge support by the MNII grants: NN202079235 and 189/6.PRUE/2007/7.

References

[1] Christodoulou D 1986 *Commun. Math. Phys.* **105** 337
[2] Christodoulou D 1991 *Comm. Pure Appl. Math.* **44** 339
[3] Dafermos M and Rodnianski I 2005 *Invent. Math.* **162** 381
[4] Price R H 1972 *Phys. Rev.* D **5** 2419
[5] Gundlach C, Price R H and Pullin J 1994 *Phys. Rev.* D **49** 883
[6] Gundlach C, Price R H and Pullin J 1994 *Phys. Rev.* D **49** 890
[7] Bizoń P, Chmaj T and Rostworowski A 2009 *Class. Quantum Grav.* **26** 175006
[8] Bizoń P, Chmaj T and Rostworowski A 2007 *Phys. Rev.* D **76** 124035
[9] Bizoń P, Chmaj T, Rostworowski A and Zając S 2009 *Class. Quant. Grav.* **26** 225015
[10] Bizoń P and Rostworowski A 2009 *Preprint* arXiv:0912.3474

Holographic interacting tachyon model

Alberto Rozas-Fernández[1]*, David Brizuela[2] and Norman Cruz[3]

[1] Instituto de Física Fundamental, CSIC, Serrano 121, 28006 Madrid, Spain
[2] Instituto de Estructura de la Materia, CSIC, Serrano 121, 28006 Madrid, Spain
[3] Departamento de Física, Facultad de Ciencia, Universidad de Santiago de Chile, Casilla 307, Santiago, Chile

E-mail: a.rozas@cfmac.csic.es

Abstract. We propose an interacting holographic tachyon model of dark energy. A correspondence between the tachyon energy density and the interacting holographic dark energy is established. As a result, we reconstruct the potential of the interacting holographic tachyon field and the dynamics of the tachyon field, in a flat Friedmann-Robertson-Walker background. We show that the evolution of the universe for $-1 < w < 0$ can be completely described by the resulting interacting tachyon model.

1. Introduction

There is a mounting observational evidence in favour of a present accelerating universe. Within the framework of the standard Friedmann-Robertson-Walker (FRW) cosmology, this present acceleration requires the existence of a negative pressure fluid, dubbed dark energy (DE), whose pressure p_Λ and density ρ_Λ satisfy $\omega_\Lambda = p_\Lambda/\rho_\Lambda < -1/3$. Nevertheless, the underlying physical mechanism behind this phenomenon remains unknown.

On the other hand, based on the validity of effective local quantum field theory in a box of size L, Cohen et al [1] suggested a relationship between the ultraviolet (UV) and the infrared (IR) cutoffs due to the limit set by the formation of a black hole. The $UV - IR$ relationship gives an upper bound on the zero point energy density

$$\rho_\Lambda \leq L^{-2} M_p^2, \tag{1}$$

where L acts as an IR cutoff and M_p is the reduced Planck mass in natural units. The largest L is chosen by saturating the bound in Eq.(1) so that we obtain the holographic dark energy density

$$\rho_\Lambda = 3c^2 M_p^2 L^{-2}, \tag{2}$$

where c is a dimensionless $O(1)$ parameter. Interestingly, this ρ_Λ is comparable to the observed dark energy density $\sim 10^{-10} eV^4$ for the Hubble parameter at the present epoch $H = H_0 \sim 10^{-33} eV$.

However, L should be taken as the size of the future event horizon of the universe in order to have an accelerated universe [2]

$$R_{\rm eh}(a) \equiv a \int_t^\infty \frac{dt'}{a(t')} = a \int_a^\infty \frac{da'}{Ha'^2}\ , \tag{3}$$

where a is the scale factor of the universe. This allows to construct a satisfactory holographic dark energy (HDE) model.

A further development was to consider a possible interaction between dark matter (DM) and the HDE [3].

2. Interacting holographic tachyon dark energy model

The tachyon can act as a source of dark energy and be described by the following effective action [4]

$$S = \int d^4x\sqrt{-g}\left[\frac{R}{16\pi G} - V(\phi)\sqrt{1+g^{\mu\nu}\partial_\mu\phi\partial_\nu\phi}\right], \tag{4}$$

where $V(\phi)$ is the tachyon potential. In the flat FRW background the energy density ρ_t and the pressure p_t are given by

$$\rho_t = \frac{V(\phi)}{\sqrt{1-\dot{\phi}^2}}, \tag{5}$$

$$p_t = -V(\phi)\sqrt{1-\dot{\phi}^2}. \tag{6}$$

From Eqs. (5) and (6) we obtain the tachyon equation of state parameter

$$w_t = \frac{p_t}{\rho_t} = \dot{\phi}^2 - 1. \tag{7}$$

In order to have a real energy density for the tachyon we require $0 < \dot{\phi}^2 < 1$ which implies, from Eq. (7), that the equation of state parameter is constrained to $-1 < w_t < 0$.

In order to impose the holographic nature to the tachyon, we should identify ρ_t with ρ_Λ. We consider a spatially flat FRW universe filled with DM and HDE. The Friedmann equation reads

$$3M_p^2H^2 = \rho_{\rm m} + \rho_{\rm t}. \tag{8}$$

In the case of an interaction between HDE and DM, their energy densities obey

$$\dot{\rho}_m + 3H\rho_m = Q\,, \tag{9}$$

$$\dot{\rho}_t + 3H(1+\omega_t)\rho_t = -Q, \tag{10}$$

where Q is an interaction term. Here we consider $Q = 3b^2H(\rho_{\rm m} + \rho_{\rm t})$, where b^2 is a coupling constant which also measures to what extent the different evolution of the DM due to its interaction with the DE gives rise to a different expansion history of the Universe. A positive b^2 corresponds to a decay of DE into DM.

Combining the definition of HDE (3) and that of the future event horizon (4) we take the derivative with respect to $x = \ln a$ and obtain

$$\rho_t' \equiv \frac{d\rho_t}{dx} = -6M_p^2H^2\Omega_t(1 - \frac{\sqrt{\Omega_t}}{c}), \tag{11}$$

where $\Omega_t = \rho_t/(3M_p^2H^2)$. Given that $\dot{\rho}_t \equiv d\rho_t/dt = \rho_t'H$ and making use of the Friedmann equation (8), Eq. (10) can be written as

$$\rho_t' + 3(1+w_t)\rho_t = -9M_p^2b^2H^2. \tag{12}$$

Combining the last two equations, we are led to the equation of state of this IHDE model,

$$w_t = -\frac{1}{3} - \frac{2}{3}\frac{\sqrt{\Omega_t}}{c} - \frac{b^2}{\Omega_t}. \tag{13}$$

Inserting Eq. (13) into Eq. (12) and using the definition of Ω_t, we arrive at

$$\frac{H'}{H} = -\frac{\Omega_t'}{2\Omega_t} + \frac{\sqrt{\Omega_t}}{c} - 1. \tag{14}$$

On the other hand, replacing $\dot{H} = H'H$ and $p_t = w_t\rho_t$ into the derivative of the Friedmann equation with respect to cosmic time $\dot{H} = -\frac{1}{2M_p^2}(\rho + p)$ (where ρ and p are the total energy density and pressure respectively), we have

$$\frac{H'}{H} = \frac{1}{2}\Omega_t + \frac{\Omega_t^{3/2}}{c} + \frac{3}{2}b^2 - \frac{3}{2}. \tag{15}$$

If we combine now the last two equations, we find the evolution equation for Ω_t

$$\frac{d\Omega_t}{dx} = \Omega_t(1-\Omega_t)\left(1 + \frac{2\sqrt{\Omega_t}}{c} - \frac{3b^2}{1-\Omega_t}\right). \tag{16}$$

Since $\frac{d}{dt} = H\frac{d}{dx} = -H(1+z)\frac{d}{dz}$ we can rewrite the above equation with respect to z as

$$\frac{d\Omega_t}{dz} = -(1+z)^{-1}\Omega_t(1-\Omega_t)\left(1 + \frac{2\sqrt{\Omega_t}}{c} - \frac{3b^2}{1-\Omega_t}\right). \tag{17}$$

Therefore, the differential equation for the Hubble parameter $H(z)$ can be expressed as

$$\frac{dH}{dz} = -(1+z)^{-1}H\left(\frac{1}{2}\Omega_t + \frac{\Omega_t^{3/2}}{c} + \frac{3}{2}b^2 - \frac{3}{2}\right). \tag{18}$$

The above equations can be solved numerically to obtain the evolution of Ω_t and H as a function of the redshift.

Using Eqs. (5), (7) and (18), we derive the interacting holographic tachyon potential

$$\frac{V(\phi)}{\rho_{cr,0}} = H^2\Omega_t\sqrt{-w_t}, \tag{19}$$

where Ω_t and w_t are respectively given by Eqs. (17) and (13), being $\rho_{cr,0} = 3M_p^2H_0^2$ the critical energy density of the universe at the present epoch. Besides, using Eqs. (7) and (18), the derivative of the interacting holographic tachyon scalar field ϕ with respect to the redshift z can be expressed as

$$\frac{\phi'}{H_0^{-1}} = \pm\frac{\sqrt{1+w_t}}{H(1+z)}. \tag{20}$$

The sign is in fact arbitrary as it can be changed by a redefinition of the field $\phi \to -\phi$.

The evolutionary form of the interacting holographic tachyon field can be easily obtained integrating it numerically from $z = 0$ to a given value z. The field amplitude at the present epoch ($z = 0$) is taken to vanish, $\phi(0) = 0$.

As already discussed in [5] the interaction Q is very weak and positive and the parameters b^2 and c are not totally free; they need to satisfy some constraints. Following the latest observational results for the IHDE models [6, 7], we take $0 \leq b^2 \leq 0.03$ and $\sqrt{\Omega_t} < c < 1.255$.

The reconstructed $V(\phi)$ is plotted in Fig. 2. The tachyon scalar field $\phi(z)$ is also shown in Fig. 1. Figs. 1 and 2 display the dynamics of the interacting tachyon scalar field explicitly. All the potentials are more steep in the early epoch, tending to be flat near today. Consequently, ϕ rolls down the potential more slowly as the universe expands and the equation of state parameter tends to negative values close to -1 as $\dot{\phi} \to 0$. As a result $dw_t/d\text{lna} < 0$. Note that $\phi(z)$ decreases as the universe expands.

A similar behaviour has been obtained in [9] for a holographic tachyon model without interaction.

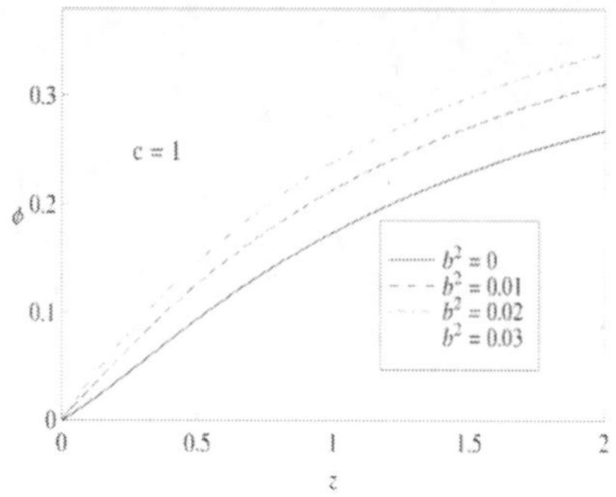

Figure 1. The evolution of $\phi(z)$, where ϕ is in units of H_0^{-1}, for $c = 1$ and different values of the coupling. We take here $\Omega_{m0} = 0.27$.

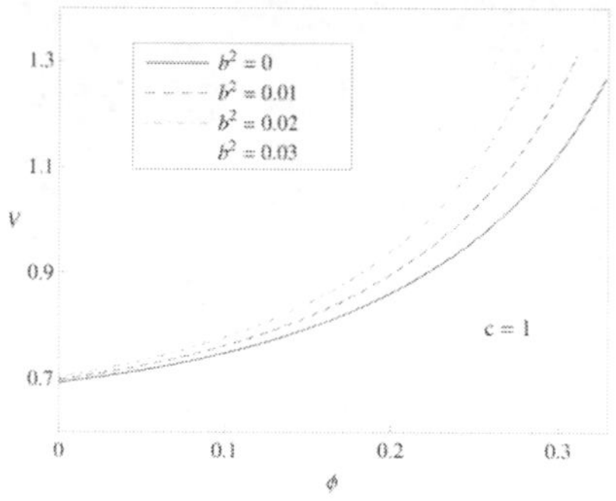

Figure 2. V vs ϕ, where ϕ is in units of H_0^{-1} and $V(\phi)$ in $\varrho_{cr,0}$, for $c = 1$ and different values of the coupling. We take here $\Omega_{m0} = 0.27$.

3. Conclusions

We have proposed an interacting holographic tachyon model of dark energy with the future event horizon as infrared cut-off. We have also carried out a throughout analysis of its evolution and deduce its cosmological consequences. We have used the tachyon scalar field model to mimic the evolving behaviour of the IHDE. As a result, we have reconstructed the interacting holographic tachyon model in the region $-1 < w < 0$.

Acknowledgments

This work was supported by the Spanish MICINN Project FIS2008-06078-C03-03 and by DICYT 040831 CM, Universidad de Santiago de Chile. DB also acknowledges financial support by the FPI Program of the Regional Government of Madrid.

References

[1] Cohen A G, Kaplan D B and Nelson A E 1999 *Phys. Rev. Lett.* **82** 4971
[2] Li M 2004 *Phys. Lett.* B **603** 1
[3] Pavon D and Zimdahl W 2005 *Phys. Lett.* B **628** 206
[4] Sen A 2002 *J. High Energy Phys.* JHEP04(2002)048
[5] Wang B, Gong Y G and SU R K 2005 *Phys. Lett.* B **605** 9
[6] Feng C, Wang B, Gong Y and Su R K 2007 *J. Cosmol. Astropart. P.* JCAP09(2007)005
[7] Wu Q, Gong Y, Wang A and Alcaniz J S 2008 *Phys. Lett.* B **659** 34
[8] Zhang J, Zhang X and Liu H 2008 *Phys. Lett.* B **659** 26
[9] Zhang J, Zhang X and Liu H 2007 *Phys. Lett.* B **651** 84

Journal of Physics: Conference Series **229** (2010) 012066 doi:10.1088/1742-6596/229/1/012066

Reconstructing cosmological solutions in F(R) gravity. Towards a unified model of the Universe evolution

Diego Sáez-Gómez[1]

[1]Institut de Ciències de l'Espai ICE/CSIC-IEEC, Campus UAB, Facultat de Ciències, Torre C5-Parell-2a pl, E-08193 Bellaterra (Barcelona) Spain

Abstract. It is accepted by the majority of scientific community that the Universe is currently in an accelerated epoch. In order to explain this *shock* of late 90's, a lot of dark energy candidates have been proposed. We study in the context of $f(R)$ gravity, how some modifications on General relativity could reproduce such behavior of the cosmic evolution. It is showed that in general, an $f(R)$ theory can be reconstructed from a non-minimal scalar-tensor theory, where a desired cosmological solution can be achieved. Some viable models are studied as well as its cosmological evolution, and the possibility that inflationary epoch is also a result of these fine modifications.

1. Reconstructing F(R) gravity from scalar-tensor theory

In this section, it is showed how in general an arbitrary cosmological solution can be reconstructed in f(R) gravity from a non-minimal coupled scalar-tensor theory (see Ref.[1] and [2]). Let us start by writing the action that describes a general non-minimal scalar-tensor theory,

$$S = \int d^4x\sqrt{-g}P(\phi)R + Q(\phi) + L_m \ , \tag{1}$$

where L_m is the Lagrangian for the matter fields. Note that the scalar field ϕ has no kinetic term for convenience in order to be able to reconstruct $f(R)$ gravity from the action (1). By varying the action on the metric tensor $g_{\mu\nu}$ and on the scalar field ϕ, the gravity and scalar field equations are obtained,

$$\frac{1}{2}g_{\mu\nu}\left(P(\phi)R + Q(\phi)\right) + P(\phi)R_{\mu\nu} + g_{\mu\nu}\nabla_\sigma\nabla^\sigma P(\phi) - \nabla_\mu\nabla_\nu P(\phi) = \frac{\kappa^2}{2}T^{(m)}_{\mu\nu} \ ,$$

$$P'(\phi)R + Q'(\phi) = 0 \ , \tag{2}$$

here the primes denote derivatives respect ϕ. The second equation may be resolved with the scalar field as a function of R, $\phi = \phi(R)$, and by replacing this result in the action (1), the action for $f(R)$ is obtained with $f(R) = P(\phi(R))\,R + Q(\phi(R))$, and it yields $S = \int d^4x\sqrt{-g}\,(f(R) + L_m)$, which is the general action for $f(R)$ gravity, and whose field equation are obtained by varying the action respect the metric $g_{\mu\nu}$,

$$R_{\mu\nu}f'(R) - \frac{1}{2}g_{\mu\nu}f(R) + g_{\mu\nu}\nabla_\sigma\nabla^\sigma f'(R) - \nabla_\mu\nabla_\nu f'(R) = \frac{\kappa^2}{2}T^{(m)}_{\mu\nu} \ , \tag{3}$$

Hence, in general, any cosmological model could be solved by the first field equation in (2) and the function $f(R)$ is reconstructed by the scalar field equation. We are interested in solutions for a flat FRW Universe, this means for a metric $ds^2 = -dt^2 + a^2(t)\sum_{i=1}^{3} dx_i^2$, in such a case the Friedmann equations for (2) read as [2],

$$3H\frac{dP(\phi)}{dt} + 3H^2P(\phi) + \frac{1}{2}Q(\phi) - \frac{\rho_m}{2} = 0\ ,$$

$$2\frac{d^2P(\phi)}{dt^2} + 4H\frac{dP(\phi)}{dt} + (4\dot{H} + 6H^2)P(\phi) + Q(\phi) + p_m = 0\ . \tag{4}$$

In order to get a general solution for the Hubble parameter , $H(t) = \dot{a}/a$, we redefine for simplicity the scalar field to be the time coordinate $\phi = t$. The perfect fluid defined by $T_{\mu\nu}^{(m)}$ has an equation of state (EoS) given by $p_m = w_m\rho_m$, and its evolution is described by the continuity equation $\dot{\rho}_m + 3H(1 + w_m)\rho_m = 0$. Then, the Hubble parameter may be calculated as a function of the scalar field ϕ, $H = g(\phi)$, and by combining equations (4), the following solution is obtained

$$g(\phi) = -\sqrt{P(\phi)}\int d\phi \frac{P''(\phi)}{2P^{3/2}(\phi)} + H_0\sqrt{P(\phi)}\ , \tag{5}$$

where H_0 is an integration constant, and we have considered no matter contribution for simplicity. Then, the solution for the Hubble parameter is calculated $H = g(\phi(t)) = g(t)$. As we have fixed the solution for the scalar field to be $\phi = t$, the equations in (4) act as a constraint on $Q(\phi)$. Let us consider a simple example to show how one could find a cosmological solution in scalar-tensor theory, and the corresponded $f(R)$ function. We consider (see Ref.[1]) $P(\phi) = \phi^2$, the solution is given by (5),

$$H(t) = H_0 t + \frac{1}{2}\frac{1}{t}\ , \tag{6}$$

which describes a Universe that passes through two phases, a first decelerated epoch identified as the radiation/matter dominated epochs, and a second accelerated, which corresponds with the current epoch. The function $Q(\phi)$ can be defined by (4), and by inverting the scalar field equation (2), it is calculated $\phi = \phi(R)$, and the function $f(R)$ yields,

$$f(R) = [R - 6(H_0(H_0 + 1) + 5/2)]\,\frac{R - 2k \pm \sqrt{R(1 - 2H_0)}}{2} + const\ . \tag{7}$$

Thus, this expression for the function $f(R)$ reproduces the current cosmic acceleration with no need to introduce any kind of dark energy. Nevertheless, the modification of the law of gravity implies other serious problems as the violation of local tests, which can be avoided for a class of $f(R)$ theories called viable $f(R)$ gravity that are studied in the next section.

2. Cosmological evolution from viable f(R) gravity

The modifications introduced by the function $f(R)$ in General Relativity may resolve the dark energy problem, although it could imply other problems as the violation of Solar System tests, where GR is very well constrained, or problems on stability around matter distributions. Some models have been proposed where these problems are avoided (see Ref.[3] and [4]), and where also the inflation can be achieved. This kind of models assumes a theory of gravity described by an action $S = \int d^4\sqrt{-g}R + F(R)$, such that $F(R)$ is negligible for small scales (and the Hilbert-Einstein action is recovered) but it starts to become important for large scales, where its effects contributes to the cosmic acceleration. Let us consider this kind of action with $F(R)$ given by

$$F(R) = \frac{R^2(\alpha R^2 - \beta)}{1 + \gamma R^2} . \tag{8}$$

This model has been analyzed in Ref.[5], where it is found that passes through at least two de Sitter points, which can be corresponded with the inflation and cosmic acceleration eras (the idea of unified inflation and cosmic acceleration under $f(R)$ theory was suggested in Ref.[6]). Let us write the first Friedmann equation for a flat FRW universe, from (3) with $f(R) = R + F(R)$, it yields for the function (8)

$$3H^2 = -\frac{R^2(\alpha R^2 - \beta)}{2(1+\gamma R^2)} + 3(H^2 + \dot{H})\frac{2R(\alpha\gamma R^4 - 2\alpha R^2 - \beta)}{(1+\gamma R^2)^2} -$$

$$-18F''(R)(H^2\dot{H} + H\ddot{H})\frac{2\alpha\gamma^2 R^6 + 20\alpha\gamma R^4 + 6(\beta\gamma - \alpha)R^2 - 2\beta}{(1+\gamma R^2)^3} , \tag{9}$$

where the matter has been neglected for simplicity. This equation can be rewritten as a dynamical system with $\dot{H} = C$, $\dot{C} = F_1(H, C)$, where the critical points are clearly those for $\dot{H} = 0$, i.e. de Sitter points (see Ref.[7]). For our model, the existence of these critical points can be achieved by studying the equation (9) for $H = H_0$ a constant, in such a case the equation reduces to

$$\frac{\gamma}{4}R_0^5 - \gamma\beta R_0^4 + \frac{\gamma}{2}R_0^3 + \frac{1}{4}R_0 = 0 , \tag{10}$$

where $R_0 = 12H_0^2$. The positive roots for R_0 correspond to de Sitter points. By a simple analysis of equation (10), and using the Descartes' rule of signs, one concludes that the equation (10) can have either two or no positive roots, where each one can be identified with the inflation and cosmic acceleration epoch. As the function (8) posses a minimum, which can be seen as an effective cosmological constant, it is clear that Universe passes through two de Sitter points. Let us now study in more detail this model during the current epoch, when matter is included $p_m = w_m \rho_m$. As it was pointed above, the function (8) has a minimum, which can be easily calculated assuming $\beta\gamma/\alpha >> 1$. It gives $R_0 \sim (\beta/\alpha\gamma)^{1/4}$ where $F'(R_0) = 0$ and $F(R_0) = -2\tilde{R}_0 \sim -\frac{\beta}{\gamma}$. Then, we can study the cosmological evolution around this point $R = R_0$, where the function (8) can be expressed as $F(R) = -2\tilde{R}_0 + f_0(R - R_0)^2 + O\left((R - R_0)^3\right)$. The solution for the Friedmann equation will have the form $H(t) = H_0(t) + \delta H$, which at zero order is the same as in the case of a cosmological constant,

$$H^2(z) = \frac{\tilde{R}_0}{3}\left[A^2(1+z)^{3(w_m+1)} + 1\right] . \tag{11}$$

where the Hubble parameter is expressed as a function of the redshift z instead of time. The perturbations δH are irrelevant around the minimum R_0. This solution is the same to ΛCDM model, although the evolution of the cosmological parameters is not exactly the same, as it is showed in Fig.1. We can fit the free parameters of the model (α, β, γ) by using the observable values $(H_0,\ \Omega_m^0)$ at the current epoch. Then, by taking the minimum R_0 at $z = 0$, we can study the cosmological evolution for small redshifts, where the derivatives of $F(R)$ are negligible compared with the value of the function and the solution (11) is valid. In order to account the effects of the geometrical terms from $F(R)$, an effective cosmological parameter is defined,

$$\Omega_{F(R)}(z) = \frac{\rho_{F(R)}}{3H^2(z)} = -\frac{F(R)}{6H^2(z)} + \left[1 + \frac{\dot{H}(z)}{H^2(z)}\right]F'(R) - 18F''(R)\left[\dot{H}(z) + \frac{\ddot{H}(z)}{H(z)}\right] , \tag{12}$$

This function is plotted in Fig.1 versus the redshift, and compared with ΛCDM model for $z = 0$ to $z = 1.8$. Note that the evolution is very similar, both models have common points at $z = 0$ and $z = 1.74$, while the behavior between such points is completely different. This difference comes form the fact that while a cosmological constant posses a constant EoS parameter, in $F(R)$ gravity, the EoS parameter defined as $w_{F(R)} = p_{F(R)}/\rho_{F(R)}$, changes with the redshift, as it goes from -1 for $z = 0$ to 0 for $z = 1.5$.

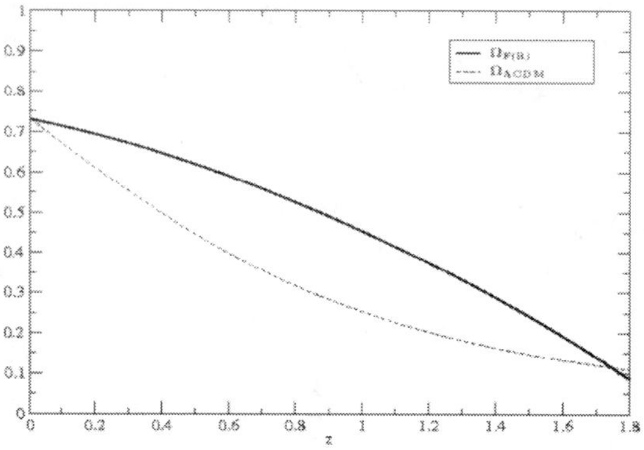

Figure 1. Evolution of the cosmological parameter from dark energy versus redshift, such for $F(R)$ theory as for ΛCDM model.

3. Conclusions

It has been showed that a general solution in a flat FRW Universe can be achieved in $f(R)$ gravity with the help of an auxiliary scalar field. Then, the acceleration epochs of the Universe could be explained just in the context of gravity with no need to introduce new forms of the so-called dark energy. As the modifications on GR produce some undesirable effects, one has to consider some complex $f(R)$ functions which could avoid them. In this context, the current cosmological evolution in $f(R)$ is compared with the popular ΛCDM model, and it is seen that a similar behavior is achieved for low redshifts. Alternative methods of reconstruction have been recently proposed in Ref.[8], which could imply a more powerful technique and give interesting properties.

References

[1] Sáez-Gómez D 2009 *Gen. Rel. Grav.* **41** 1527 (*Preprint* hep-th/0809.1311)
[2] Nojiri S and Odintsov S D 2006 *Phys.Rev.* D **74** 086005 (*Preprint* hep-th/0608008 and hep-th/0611071)
[3] Nojiri S and Odintsov S D 2008 *Phys. Rev.* D **77** 026007 (*Preprint* hep-th/0710.1738)
[4] Hu W and Sawicki I 2007 *Phys. Rev.* D **76** 064004 (*Preprint* astro-ph/0705.1158)
[5] Elizalde E and Sáez-Gómez D 2009 *Phys. Rev.* D **80** 044030 (*Preprint* hep-th/0903.2732)
[6] Nojiri S, Odintsov S D, 2003 *Phys. Rev.* D **68** 123512 (*Preprint* hep-th/0307288)
[7] Cognola G, Elizalde E, Odintsov S D, Tretyakov P and Zerbini S 2009 *Phys. Rev.* D **79** 044001 (*Preprint* gr-qc/0810.4989)
[8] Nojiri S, Odintsov S D, Sáez-Gómez D 2009 *Phys. Lett.* B **681** 74 (*Preprint* hep-th/0908.1269)

Journal of Physics: Conference Series **229** (2010) 012067 doi:10.1088/1742-6596/229/1/012067

Two static charged masses in the external field

J Sánchez-Mondragón[1*], V S Manko[1] and E Ruiz[2]

[1] Departamento de Física, Centro de Investigación y de Estudios Avanzados del IPN, A.P. 14-740, 07000 México D.F., Mexico

[2] Instituto Universitario de Física Fundamental y Matemáticas, Universidad de Salamanca, 37008 Salamanca, Spain

E-mail: jsanchez@fis.cinvestav.mx

Abstract. We consider a binary system of magnetically charged black holes in the external uniform magnetic field by applying the Harrison transformation to the magnetic analog of the double–Reissner–Nordström solution. The balance of subextreme components is only possible in the case of *asymmetric black diholes* – configurations composed of two non-extreme black holes possessing unequal masses and charges equal in magnitude but opposite in sign.

1. Introduction

In 2000 Emparan demonstrated [1] that the well–known Bonnor magnetostatic solution [2] can be interpreted as describing a system of two extreme black holes with opposite magnetic charges. He also suggested the name 'black diholes' for the two–body configurations consisting of separated equal black holes carrying opposite charges and showed that the extremal constituents in the Bonnor–type dihole can be balanced by placing the latter into the external uniform magnetic field [3, 4] via the Harrison transformation [5].

In the present communication we shall consider a system of two arbitrary magnetically charged static black holes. The external magnetic field will be shown to ensure equilibrium of the black–hole constituents only in the case of *asymmetric black diholes* – configurations composed of two black holes with unequal masses but equal and opposite charges.

2. The double–Reissner–Nordström solution in magnetostatics

The magnetic analog of the general double–Reissner–Nordström solution is defined by the Ernst potentials $\mathcal{E}$ and Φ [6] of the form [7]

$$\begin{aligned}
\mathcal{E} &= \frac{\mathcal{A}-\mathcal{B}}{\mathcal{A}+\mathcal{B}}, \quad \Phi = \frac{i\mathcal{C}}{\mathcal{A}+\mathcal{B}}, \\
\mathcal{A} &= \sigma_1\sigma_2[\nu(R_+ + R_-)(r_+ + r_-) + 4\kappa(R_+R_- + r_+r_-)] \\
&\quad -(\mu^2\nu - 2\kappa^2)(R_+ - R_-)(r_+ - r_-), \\
\mathcal{B} &= 2\sigma_1\sigma_2[(\nu m_2 + 2\kappa m_1)(R_+ + R_-) + (\nu m_1 + 2\kappa m_2)(r_+ + r_-)] \\
&\quad +2\sigma_2[\nu\mu(q_1 - \mu) - 2\kappa(Rm_1 - \mu q_2 - \mu^2)](R_+ - R_-) \\
&\quad +2\sigma_1[\nu\mu(q_2 + \mu) + 2\kappa(Rm_2 + \mu q_1 - \mu^2)](r_+ - r_-), \\
\mathcal{C} &= 2\sigma_1\sigma_2\{[\nu(q_2 + \mu) + 2\kappa(q_1 - \mu)](R_+ + R_-) + [\nu(q_1 - \mu) + 2\kappa(q_2 + \mu)](r_+ + r_-)\} \\
&\quad +2\sigma_2[\mu\nu m_1 + 2\kappa(\mu m_2 - Rq_1 + \mu R)](R_+ - R_-)
\end{aligned}$$

$$+2\sigma_1[\mu\nu m_2 + 2\kappa(\mu m_1 + Rq_2 + \mu R)](r_+ - r_-),$$

$$R_\pm = \sqrt{\rho^2 + \left(z + \frac{1}{2}R \pm \sigma_1\right)^2}, \quad r_\pm = \sqrt{\rho^2 + \left(z - \frac{1}{2}R \pm \sigma_2\right)^2}, \tag{1}$$

where R is the distance between the centers of the black holes, and the constants σ_1, σ_2, ν and κ are defined by the formulae

$$\begin{aligned} \sigma_1 &= \sqrt{m_1^2 - q_1^2 + 2\mu q_1}, \quad \sigma_2 = \sqrt{m_2^2 - q_2^2 - 2\mu q_2}, \quad \mu := \frac{m_2 q_1 - m_1 q_2}{R + m_1 + m_2}, \\ \nu &= R^2 - m_1^2 - m_2^2 + (q_1 - \mu)^2 + (q_2 + \mu)^2, \quad \kappa = m_1 m_2 - (q_1 - \mu)(q_2 + \mu). \end{aligned} \tag{2}$$

The metric coefficients f and γ from the Weyl line element

$$ds^2 = f^{-1}[e^{2\gamma}(d\rho^2 + dz^2) + \rho^2 d\varphi^2] - f dt^2 \tag{3}$$

in the case of the solution (1) have the following form involving the functions $\mathcal{A}$, $\mathcal{B}$ and $\mathcal{C}$:

$$\begin{aligned} f &= \frac{\mathcal{A}^2 - \mathcal{B}^2 + \mathcal{C}^2}{(\mathcal{A} + \mathcal{B})^2}, \quad e^{2\gamma} = \frac{\mathcal{A}^2 - \mathcal{B}^2 + \mathcal{C}^2}{K_0^2 R_+ R_- r_+ r_-}, \\ K_0 &= 4\sigma_1\sigma_2[R^2 - (m_1 - m_2)^2 + (q_1 - q_2 - 2\mu)^2], \end{aligned} \tag{4}$$

while the corresponding magnetic potential A_φ has the form [7]

$$\begin{aligned} A_\varphi &= Q + \frac{\mathcal{I} - z\mathcal{C}}{\mathcal{A} + \mathcal{B}}, \\ \mathcal{I} &= 4\sigma_1\sigma_2\{[m_2 R(q_1 - \mu) - \mu(\kappa + m_2^2) + \mu(q_1 - \mu)^2]r_+ r_- \\ &\quad -[m_1 R(q_2 + \mu) + \mu(\kappa + m_1^2) - \mu(q_2 + \mu)^2]R_+ R_-\} \\ &\quad +\mu(\nu - 2\kappa)[(\kappa + \mu^2)(R_+ - R_-)(r_+ - r_-) - \sigma_1\sigma_2(R_+ + R_-)(r_+ + r_-)] \\ &\quad -(\nu - 2\sigma_1\sigma_2 - 2\mu^2)[m_1\sigma_2(q_1 - \mu) - m_2\sigma_1(q_2 + \mu)](R_+ r_- - R_- r_+) \\ &\quad +(\nu + 2\sigma_1\sigma_2 - 2\mu^2)[m_1\sigma_2(q_1 - \mu) + m_2\sigma_1(q_2 + \mu)](R_- r_- - R_+ r_+) \\ &\quad +\sigma_1[m_2\mu\nu + 2m_1\mu\kappa + 2R\kappa(q_2 + \mu)][2\sigma_2(r_+ + r_-) - R(r_+ - r_-)] \\ &\quad +\sigma_2[m_1\mu\nu + 2m_2\mu\kappa - 2R\kappa(q_1 - \mu)][2\sigma_1(R_+ + R_-) + R(R_+ - R_-)] \\ &\quad +\sigma_1\sigma_2[\nu(q_1 - \mu) + 2\kappa(q_2 + \mu)][2\sigma_2(r_+ - r_-) - R(r_+ + r_-)] \\ &\quad +\sigma_1\sigma_2[\nu(q_2 + \mu) + 2\kappa(q_1 - \mu)][2\sigma_1(R_+ - R_-) + R(R_+ - R_-)] \\ &\quad -2\mu(\nu - 2\kappa)\{\sigma_2[m_1 R + \mu(q_1 - q_2 - 2\mu)](R_+ - R_-) \\ &\quad -\sigma_1[m_2 R + \mu(q_1 - q_2 - 2\mu)](r_+ - r_-) \\ &\quad +\sigma_1\sigma_2[(m_1 - m_2)(r_+ + r_- - R_+ - R_-) + 2(\nu + 2\kappa)]\}, \end{aligned} \tag{5}$$

where $Q \equiv q_1 + q_2$ is the total charge of the system.

3. Asymmetric black diholes

In analogy with the paper [1], the application of the Harrison transformation to the magnetostatic solution considered in the previous section yields the new functions $\tilde{f}$, $\tilde{\gamma}$ and $\tilde{A}_\varphi$ of the form

$$\begin{aligned} &\tilde{f} = \lambda^2 f, \quad e^{2\tilde{\gamma}} = \lambda^4 e^{2\gamma}, \quad \tilde{A}_\varphi = \frac{2}{B}\left[\frac{1}{\lambda}\left(1 + \frac{1}{2}BA_\varphi\right) - 1\right], \\ &\lambda = \left(1 + \frac{1}{2}BA_\varphi\right)^2 + \frac{1}{4}B^2\rho^2 f^{-1}, \end{aligned} \tag{6}$$

which describe two magnetically charged masses in the external uniform magnetic field determined by the constant B.

In order to see whether the black–hole constituents can be balanced by the external field, it is necessary to consider for them the balance conditions which reduce to the requirement of the regularity of the parts of the symmetry axis outside the location of the sources, i.e., $\tilde{\gamma}^{(I,II,III)} = 0$, where the segments I, II and III of the symmetry axis are defined as $\rho = 0$, $z > \frac{1}{2}R + \sigma_2$ (the upper part), $\rho = 0$, $-\frac{1}{2}R + \sigma_1 < z < \frac{1}{2}R - \sigma_2$ (the intermediate part), and $\rho = 0$, $z < -\frac{1}{2}R - \sigma_1$ (the lower part), respectively. From the equations (1), (4) and (6) it follows that the system of the balance equations takes the form

$$A_\varphi^{(I)} = 0, \quad \left(1 + \frac{1}{2}BA_\varphi^{(II)}\right)^8 e^{2\gamma^{(II)}} = 1, \quad A_\varphi^{(III)} = 0, \tag{7}$$

where $A_\varphi^{(I,II,III)}$ denote the values of the magnetic potential at the respective parts of the z–axis. On the other hand, from (5) we find that

$$A_\varphi^{(I)} = 0, \quad A_\varphi^{(II)} = 2q_2, \quad A_\varphi^{(III)} = 2Q, \tag{8}$$

which implies vanishing of the total charge $Q \Leftrightarrow q_1 = -q_2$. Then, solving the second equation in (7) for B, we obtain the value of the external magnetic field leading to the equilibrium:

$$B = \frac{1}{q_2}\left(\sqrt[4]{\frac{\nu + 2\kappa}{\nu - 2\kappa}} - 1\right), \quad Q = 0. \tag{9}$$

The condition $Q = 0$ naturally singles out the family of *asymmetric black diholes* as the only static subextreme two–black–hole configurations which can be balanced by the external magnetic field. In this case, by setting $q_1 = -q_2 = q$ in (2), we get

$$\begin{aligned}
\sigma_1 &= \sqrt{m_1^2 - q^2(1-2\mu)}, \quad \sigma_2 = \sqrt{m_2^2 - q^2(1-2\mu)}, \quad \mu := \frac{m_1 + m_2}{R + m_1 + m_2}, \\
\nu &= R^2 - m_1^2 - m_2^2 + 2q^2(1-\mu)^2, \quad \kappa = m_1 m_2 + q^2(1-\mu)^2,
\end{aligned} \tag{10}$$

while formulae of the previous section for the metric functions f, γ and for the magnetic potential A_φ can be rewritten as

$$\begin{aligned}
f &= \frac{\mathcal{A}^2 - 4\mathcal{B}^2 + 4q^2\mathcal{C}^2}{(\mathcal{A} + 2\mathcal{B})^2}, \quad e^{2\gamma} = \frac{\mathcal{A}^2 - 4\mathcal{B}^2 + 4q^2\mathcal{C}^2}{K_0^2 R_+ R_- r_+ r_-}, \quad A_\varphi = \frac{q(\mathcal{I} - 2z\mathcal{C})}{\mathcal{A} + 2\mathcal{B}}, \\
\mathcal{A} &= \sigma_1\sigma_2[\nu(R_+ + R_-)(r_+ + r_-) + 4\kappa(R_+R_- + r_+r_-)] \\
&\quad -(q^2\mu^2\nu - 2\kappa^2)(R_+ - R_-)(r_+ - r_-), \\
\mathcal{B} &= \sigma_1\sigma_2[(m_2\nu + 2m_1\kappa)(R_+ + R_-) + (m_1\nu + 2m_2\kappa)(r_+ + r_-)] \\
&\quad +q^2(\mu - \mu^2)(\nu - 2\kappa)[\sigma_2(R_+ - R_-) - \sigma_1(r_+ - r_-)] \\
&\quad -2R\kappa[m_1\sigma_2(R_+ - R_-) - m_2\sigma_1(r_+ - r_-)], \\
\mathcal{C} &= \sigma_1\sigma_2(1-\mu)(\nu - 2\kappa)(r_+ + r_- - R_+ - R_-) \\
&\quad +\sigma_2[m_1\mu\nu + 2\kappa(m_2\mu - R + R\mu)](R_+ - R_-) \\
&\quad +\sigma_1[m_2\mu\nu + 2\kappa(m_1\mu - R + R\mu)](r_+ - r_-), \\
R_\pm &= \sqrt{\rho^2 + \left(z + \frac{1}{2}R \pm \sigma_1\right)^2}, \quad r_\pm = \sqrt{\rho^2 + \left(z - \frac{1}{2}R \pm \sigma_2\right)^2}, \\
K_0 &= 4\sigma_1\sigma_2[R^2 - (m_1 - m_2)^2 + 4q^2(1-\mu)^2],
\end{aligned} \tag{11}$$

with

$$\begin{aligned}
\mathcal{I} \;=\; & 4\sigma_1\sigma_2[R(1-\mu)-\mu(m_1+m_2)](m_1R_+R_- + m_2r_+r_-) \\
& +\mu(\nu-2\kappa)[(\kappa+\mu^2q^2)(R_+-R_-)(r_+-r_-)-\sigma_1\sigma_2(R_++R_-)(r_++r_-)] \\
& -(1-\mu)\{[R^2-(\sigma_1+\sigma_2)^2](m_1\sigma_2+m_2\sigma_1)(R_+r_- - R_-r_+) \\
& +[R^2-(\sigma_1-\sigma_2)^2](m_1\sigma_2-m_2\sigma_1)(R_+r_+ - R_-r_-)\} \\
& +\sigma_1[m_2\mu\nu+2m_1\mu\kappa-2R\kappa(1-\mu)][2\sigma_2(r_++r_-)-R(r_+-r_-)] \\
& +\sigma_2[m_1\mu\nu+2m_2\mu\kappa-2R\kappa(1-\mu)][2\sigma_1(R_++R_-)+R(R_+-R_-)] \\
& -\sigma_1\sigma_2R(1-\mu)(\nu-2\kappa)(R_++R_-+r_++r_-) \\
& -2(\nu-2\kappa)\{\sigma_2[m_1R\mu+(\sigma_1^2+2q^2\mu^2)(1-\mu)](R_+-R_-) \\
& -\sigma_1[m_2R\mu+(\sigma_2^2+2q^2\mu^2)(1-\mu)](r_+-r_-) \\
& +\mu\sigma_1\sigma_2[(m_1-m_2)(r_++r_--R_+-R_-)+2(\nu+2\kappa)]\}. \qquad (12)
\end{aligned}$$

Lastly, the explicit expression for the parameter B at which balance occurs assumes the form

$$B = -\frac{1}{q}\left(\sqrt{\frac{R^2-(m_1-m_2)^2+4q^2(1-\mu)^2}{R^2-(m_1+m_2)^2}}-1\right). \qquad (13)$$

We remark in conclusion that in the future paper we will generalize the results of the present work by including into the consideration the additional dilatonic field.

References

[1] Emparan R 2000 *Phys. Rev.* D **61** 104009
[2] Bonnor W B 1966 *Z. Phys.* **190** 444
[3] Bonnor W B 1954 *Proc. Phys. Soc. Lond. A* **67** 225
[4] Melvin M A 1964 *Phys. Lett.* **8** 65
[5] Harrison B K 1968 *J. Math. Phys.* **9** 1744
[6] Ernst F J 1968 *Phys. Rev.* **168** 1415
[7] Manko V S, Ruiz E and Sánchez-Mondragón J 2009 *Phys. Rev.* D **79** 084024

Journal of Physics: Conference Series **229** (2010) 012068 doi:10.1088/1742-6596/229/1/012068

DBI models for the unification of dark matter and dark energy

L P Chimento, R Lazkoz and I Sendra*

Fisika Teorikoa, Zientzia eta Teknologia Fakultatea, Euskal Herriko Unibertsitatea, 644 Posta Kutxatila, 48080 Bilbao, Spain

E-mail: chimento@df.uba.ar, ruth.lazkoz@ehu.es, irene_sendra@ehu.es

Abstract. We propose a model based on a DBI action for the unification of dark matter and dark energy. This is supported by the results of the study of its background behavior at early and late times. We also perform a Bayesian analysis to set observational constraints on the parameters of the model using type Ia SN, CMB shift and BAO data. Finally, to complete the study we investigate its kinematic aspects.

1. Introduction

The current standard model of cosmology proposes that the universe is mainly filled with dark matter and dark energy. If one believes these two entities really fill our universe, it remains to discover what their nature is. Unified models of these components represent a interesting explanation for the current acceleration of the universe which at the same time would simplify the problem. Following the tradition of trying to find a connection between the world of Particle Physics and Cosmology, it is customary to try and view unified dark energy models as scalar field scenarios. In that context, we consider a scalar field setup, and fixing some of its degrees of freedom we obtain an expansionary cosmology which mimics a dark matter dominated background at early times and a dark energy dominated one at the late stages of its history.

It has been suggested that acceleration can be explained within brane theory. In this context, we propose to describe the current acceleration with a Dirac-Born-Infeld action: $\mathcal{S} = -\int d^4x\, a(t)^3 (f(\phi)^{-1}(\sqrt{1 - f(\phi)\dot{\phi}^2} - 1) + V(\phi))$, characterized by two functions, the warp factor which should be positive, $f(\phi) > 0$, and the inflation potential, $V(\phi)$. Any conclusions about the capability of our model to represent a solid alternative to other dark energy scenarios must ideally be reached from both the theoretical and the observational perspectives, as we do in this work.

2. The model

Our scenario is a four-dimensional spatially flat FRW spacetime filled with a non-canonical scalar field of DBI type. Using the customary perfect fluid interpretation we set $\rho = (\gamma - 1)/f + V(\phi)$ and $p = (\gamma - 1)/\gamma f - V(\phi)$, with $\gamma = 1/\sqrt{1 - f(\phi)\dot{\phi}^2}$, where, in principle, f and V are arbitrary functions. In this paper we explore the case in which both $f = f_0$ are $V = V_0$ constants. The goal is to obtain a purely kinetic model as in other unified dark sector models [1, 2, 3, 4, 5], so that the field ϕ depends solely on the scale factor; and as a consequence the same holds for

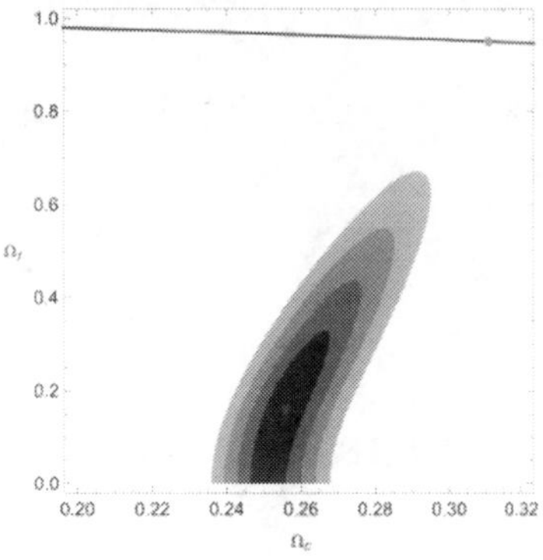

Figure 1. Credible intervals from the combination of SN(Union)+CMB+BAO observations.

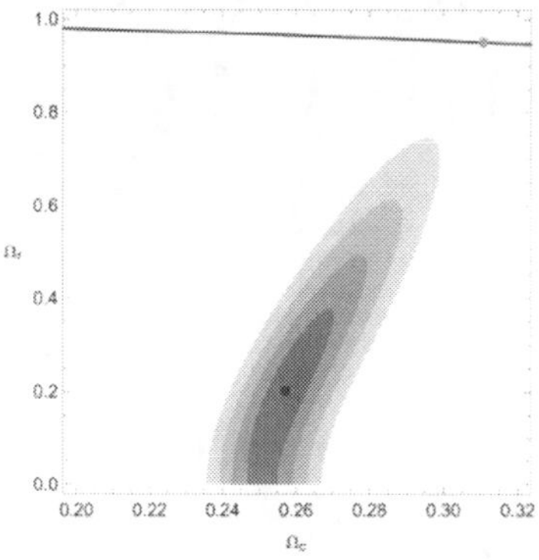

Figure 2. Credible intervals from the combination of SN(Essence)+CMB+BAO observations.

the effective pressure and energy density. From the conservation equation, $\dot{\rho} + 3H\Gamma\rho = 0$, one obtains a differential equation: $\ddot{\phi}/\dot{\phi} + 3H + \dot{\gamma}/\gamma = 0$, which has the following first integral

$$\gamma^2 = 1 + c^2 f_0 \left(a_0/a\right)^6, \tag{1}$$

with c an arbitrary integration constant and a_0 the value of the scale factor today, which we fix as $a_0 = 1$. On the one hand we have accomplished our goal of obtaining a DBI model which is purely kinetic throughout the evolution. On the other hand, the behavior of the model obtained is quite appealing. Using (1) it can be seen that at very late times $\gamma \simeq 1$, i.e. in the regime $a \gg a_0$, one has $\rho \sim V_0$, whereas at early epochs, i.e. for $a \ll a_0$, one has $\gamma \simeq a^{-3}$ instead, which in turn gives $\rho \sim 1/a^3$. Thus, the solution found interpolates between a dust and a de Sitter model, with the piece V_0 acting as a cosmological constant and no trace left of the genuinely DBI degree of freedom f_0. The background evolution of our unified model happens to mimic that induced by the joint contribution of a conventional Chaplygin gas (Cg) and a cosmological constant (cc), but the novelty is that such background behavior is realized with a single fluid.

3. Observational constraints

In this section we will set constraints on the parameters of the model from a Bayesian perspective. Our analysis will use geometrical tests: the SN type Ia luminosity test, the CMB shift test [6, 7], and the BAO test [8]. As these two last tests involve early universe quantities, one must consider a slightly more general setup and include radiation, which must to be conserved independently from the DBI fluid. This way, the Friedmann equation turns out to be $H^2/H_0^2 = \sqrt{\Omega_f^2 + \Omega_c^2(1+z)^6} + \Omega_\Lambda + \Omega_r(1+z)^4$, where $\Omega_f = 1/3H_0^2 f_0$, $\Omega_\Lambda = \left(f_0 V_0 - 1\right)/3H_0^2 f_0$ and $\Omega_r = \rho_{r0}/3H_0^2$. In addition, the CMB and BAO tests require that we identify a combination of parameters which behaves effectively as Ω_m in the high energy regime. In our case this mimicry is played by Ω_c. As this paper is a first approach to this model, we are setting constraints only on Ω_c and Ω_f. In contrast, we fix a prior for Ω_b, taking the WMAP 5-year best fit, $\Omega_b = 0.0432$. We have used two different compilations for SNIa data: ESSENCE [9, 10] and UNION [11]. In the case of the UNION sample, the best values obtained are $\Omega_c = 0.256^{+0.012}_{-0.010}$, $\Omega_f = 0.160^{+0.171}_{-0.160}$ and for the ESSENCE sample $\Omega_c = 0.257^{+0.013}_{-0.011}$ and $\Omega_f = 0.202^{+0.177}_{-0.202}$ with the corresponding 68.30% uncertainties. The lines in the upper sides of the plots in Figs.1 and 2 represent the locations on the parameter space which correspond to Chaplygin gas cases and the points in each of the lines indicate the case with the lowest χ^2 value. From visual inspection one can infer

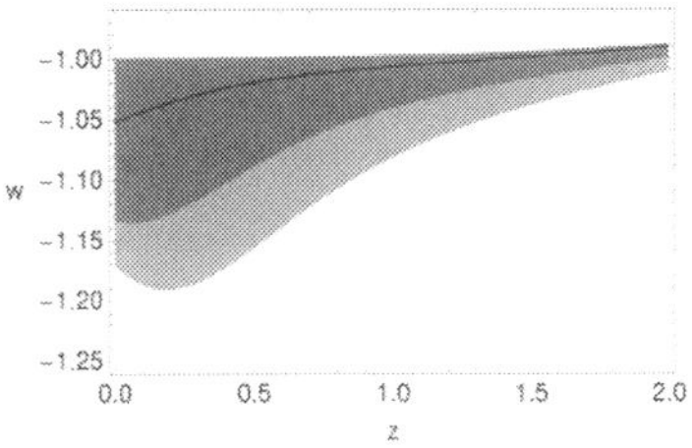

Figure 3. $w(z)$ for UNION compilation.

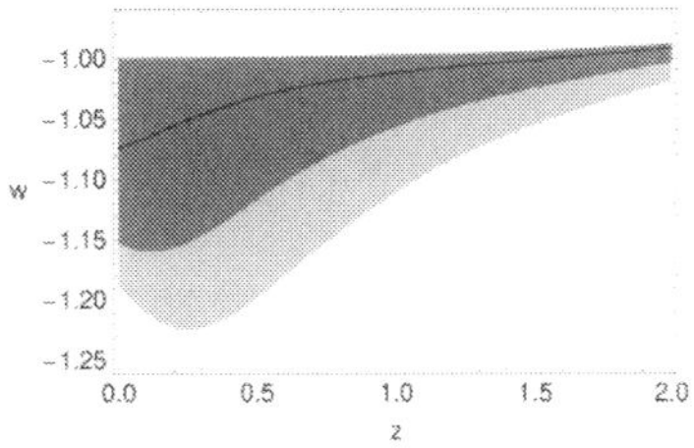

Figure 4. $w(z)$ for ESSENCE compilation.

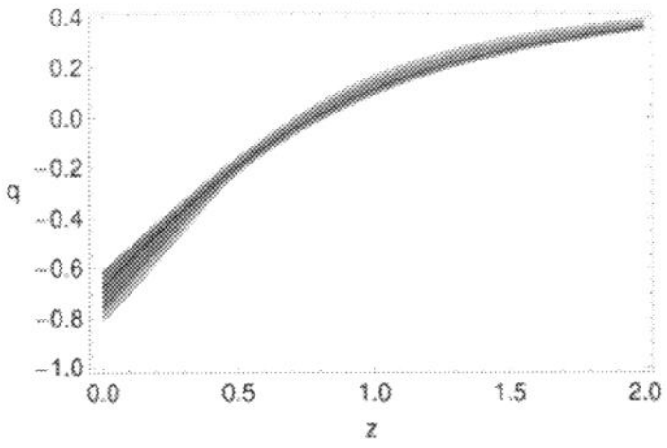

Figure 5. $q(z)$ for UNION compilation.

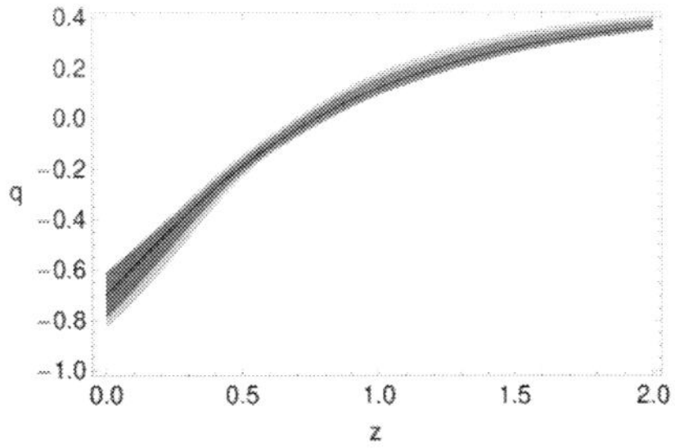

Figure 6. $q(z)$ for ESSENCE compilation.

that the Chaplygin gas is rejected by our model. In contrast, LCDM (the $\Omega_f = 0$ locations) is not significantly excluded, as for a certain range of Ω_m, LCDM cases lie in the 68.30% likelihood credible interval. All in all, our model provides better fits.

4. Model kinematics

As this is a new model it should be examined from different perspectives, the kinematic one being a specially relevant one. We investigate the redshift dependence of the effective equation of state parameter, $w(z)$, and derived quantities such as the acceleration parameter, $q(z)$, or the transition redshift, z_t. In order to obtain the behavior of $w(z)$, we use the expression that relates it with the Friedman equation [12, 13], $w(z) = (2/3(d\ln H/dz)(1+z)-1)/(1-(H_0/H)^2\,\Omega_c(1+z)^3)$. Analyzing the dependence of the equation of state parameter with the redshift we can see that the current observational data in the context of our model restricts it to be smaller than -1 with $dw/dz|_{z=0} > 0$ for the current time. More precisely, we obtain $w(z=0) = -1.052^{+0.052}_{-0.081}$, $dw/dz|_{z=0} = 0.081^{+0.081}_{-0.140}$ with the UNION sample and $w(z=0) = -1.074^{+0.074}_{-0.078}$, $dw/dz|_{z=0} = 0.074^{+0.074}_{-0.199}$ with the ESSENCE sample, with the uncertainties corresponding to the 68.30% interval of confidence.

In addition, we study the acceleration parameter, $q(z)$ which for instance can be expressed as $q(z) = (3\left(1-\Omega_c\,(1+z)^3/H^2\right)w(z)+1)/2$. Figs. 3, 4, 5 and 6 depict the shape of the equation of state and acceleration parameters with their corresponding 68.30% and 95.45% errors. From the analysis of the acceleration parameter we gather that there is a strong evidence of the transition from a deceleration to an acceleration stage. For a better insight on this matter, we have inferred the redshift at which the transition happens in different ways. Since we have an expression for $q(z)$, we can compute $q(z_t) = 0$ to obtain the transition redshift, z_t. For the ESSENCE compilation of SNIa we have $z_t = 0.766^{+0.041}_{-0.047}$, and for the UNION sample $z_t = 0.778^{+0.036}_{-0.048}$. In [14] another approach to obtain it was proposed. It involves expanding the acceleration parameter, $q(z)$, into two terms: $q(z) = q_0 + z dq/dz|_{z=0}$. Under this definition, we get with the UNION sample $z_t = 0.649^{+0.096}_{-0.079}$ and with the ESSENCE sample $z_t = 0.650^{+0.139}_{-0.078}$. These results are in good agreement with the results obtained in [15, 14, 16]. Yet another parameterization was considered in [17, 18]: $q(z) = q_0 + q_1 z/(1+z)$. With this parameterization we get that the value of the transition redshift for ESSENCE is $z_t = 0.674^{+0.092}_{-0.062}$ and for UNION $z_t = 0.697^{+0.088}_{-0.085}$. As the value of z_t obtained directly by the explicit equation of the acceleration equation is bigger that the approximate one, we infer that the approximations are not good enough for accounting accurately for the tendency of our DBI fluid to induce a phantom stage. The result obtained by our procedure tells us that the acceleration-deceleration transition happens before than the other definitions allow to estimate.

5. Conclusions

This model offers a worthy alternative to the popular models which attempt at a unification of the dark matter and dark energy components of the Universe. One of its features is that is based on a purely kinetic DBI action and extends them to late time universe. Moreover the transition from a $H^2 \sim a^{-3}$ regime to a de Sitter phase is realized with a single fluid. From the observational analysis perspective, our model is better suited to the observations that the most popular unified dark sector model: the Chaplygin gas [1, 2]. Our results also indicate a modest preference of our model as compared to the LCDM one. And perhaps the most remarkable outcome is that the best fit corresponds to a phantom behavior and it is achieved without an explicit phantom component in the model. Finally, the study of the model from different relevant angles and the results achieved, convinces us that our model represents an alternative model for the unification of the dark sector and suggests the interest of exploring generalizations of this model, probably by relaxing the assumption of a purely kinetic Lagrangian, as perhaps further degrees of freedom would allow and even better suitability to astronomical observations.

Acknowledgments

We thank A. Díez-Tejedor for comments. L.P.C. is partially supported by the University of Buenos Aires for partial support under project X224, and the Consejo Nacional de Investigaciones Científicas y Técnicas under project 5169. R.L. and I.S. are supported by the former Spanish Ministry of Science and Innovation through research grant FIS2007-61800. R.L. has also the support of the University of the Basque Country through research grant GIU06/37.

References

[1] Kamenshchik A Y, Moschella U and Pasquier V 2001 *Phys. Lett.* B **511** 265–268 (*Preprint* gr-qc/0103004)
[2] Bento M C, Bertolami O and Sen A A 2002 *Phys. Rev.* D **66** 043507 (*Preprint* gr-qc/0202064)
[3] Bilic N, Tupper G and Viollier R 2002 *Phys. Lett.* B **535** 17 – 21
[4] Scherrer R J 2004 *Phys. Rev. Lett.* **93** 011301 (*Preprint* astro-ph/0402316)
[5] Chimento L P, Forte M and Lazkoz R 2005 *Mod. Phys. Lett.* A **20** 2075 (*Preprint* astro-ph/0407288)
[6] Wang Y and Mukherjee P 2007 *Phys. Rev.* D **76**
[7] Komatsu E *et al.* (WMAP) 2009 *Astrophys. J. Suppl.* **180** 330–376 (*Preprint* 0803.0547)
[8] Percival W J *et al.* 2007 *Mon. Not. Roy. Astron. Soc* 1053–1066
[9] Krisciunas K 2008 (*Preprint* 0809.2612)
[10] Davis T M *et al.* 2007 *Astrophys. J.* **666** 716 (*Preprint* astro-ph/0701510)
[11] Kowalski M *et al.* (Supernova Cosmology Project) 2008 *Astrophys. J.* **686** 749–778 (*Preprint* 0804.4142)
[12] Saini T D, Raychaudhury S, Sahni V and Starobinsky A A 2000 *Phys. Rev. Lett.* **85** 1162–1165
[13] Nesseris S and Perivolaropoulos L 2004 *Phys. Rev.* D **70** 043531
[14] Riess A G *et al.* (Supernova Search Team) 2004 *Astrophys. J.* **607** 665–687 (*Preprint* astro-ph/0402512)
[15] Wang F Y and Dai Z G 2006 *Chin. J. Astron. Astrophys.* **6** 561–571
[16] Gong Y and Wang A 2007 *Phys. Rev.* D **75** 043520 (pages 7)
[17] Xu L I, Zhang C W, Chang B R and Liu H Y 2007 *TP-DUT-2007-2* 5 (*Preprint* astro-ph/0701519)
[18] Cunha J V and Lima J A S 2008 *Mon. Not. Roy. Astron. Soc.* **390** 210–217 (*Preprint* 0805.1261)

Journal of Physics: Conference Series **229** (2010) 012069 doi:10.1088/1742-6596/229/1/012069

Stellar oscillations in TeVeS

Hajime Sotani

Theoretical Astrophysics, University of Tübingen, 72076 Tübingen, Germany

E-mail: sotani@astro.auth.gr

Abstract. In this work we examine the oscillation spectra of neutron stars in TeVeS. As a result with Cowling approximation, we find that the frequencies of fundamental modes in TeVeS could become lager than those expected in general relativity, while the dependence of frequency of higher overtone on gravitational theory is stronger than that of lower modes. Additionally, as a results of calculations with the assumption that the vector field is unperturbed, we also find that the dependences of frequencies of spacetime oscillations (w modes) on the stellar compactness are almost independent from the adopted equation of state and the parameter in TeVeS, and that these dependences of frequencies of axial w modes in TeVeS is obviously different from those expected in the general relativity. These imprints of TeVeS make it possible to distinguish the gravitational theory in strong-field regime via the observations of gravitational waves, which can provide unique confirmation of the existence of scalar field.

1. Introduction

The tests of gravitational theories in the strong-field regime are quite important, because the gravitational theories in the strong-field regime are still largely unconstrained by observations in contrast to those in the weak-field regime. However, owing to the developments in observational technology, it is becoming possible to observe compact objects with high accuracy. Via the observations of X-rays and/or gamma rays emitted from compact objects, one can know the properties of compact objects and could use as a direct test of the gravitational theory in the strong-field regime [1]. As an alternative way to observe compact objects, the gravitational waves are also expected to obtain the raw information of the compact objects [2, 3].

The tensor-vector-scalar (TeVeS) theory has attracted considerable attention, which is proposed by Bekenstein [4] as a relativistic theory of Modified Newtonian Dynamics (MOND) [5]. In TeVeS, it is possible to explain many galactic and cosmological observations without the need for dark matter [4]. While, in the strong-field regime, the Schwarzschild solution in TeVeS and the Reissner-Nordström solution in TeVeS have been found [6, 7], while Lasky *et al.* derived the TOV equations and they showed the possibility of distinguishing TeVeS from GR by way of redshift observations [8]. In this work, we examine whether observations of gravitational waves associated with the neutron star oscillations can provide an alternative way of probing the gravitational theory in the strong-field regime. The more detailed study can be seen in [9, 10].

2. Stellar model in TeVeS

TeVeS is based on three dynamical gravitational fields, such as an Einstein metric $g_{\mu\nu}$, a timelike 4-vector field $\mathcal{U}^\mu$, and a scalar field φ, in addition to a nondynamical scalar field σ. By varying the total action with respect to $g_{\mu\nu}$, $\mathcal{U}_\mu$, and φ, one obtains the field equations for the metric, vector, and scalar fields, which include two positive dimensionless parameters, k and $\mathcal{K}$. To construct the stellar models in TeVeS, three parameters, k, $\mathcal{K}$ and φ_c, are introduced with respect to GR, where φ_c denotes the cosmological value of the scalar field. The value of k has a tightly constraint as $k \sim 0.03$ by both cosmological models and also planetary motions in the solar system, while the cosmological considerations imply that the value of φ_c is restricted to $0 \leq \varphi_c \ll 1$ [4]. With respect to the value of φ_c, it was shown that φ_c could have a minimum value of around 0.001, which is based on the causality issues inside the star [8]. Anyway, since it was also found in [8] that the neutron star models are almost independent from the values of k and φ_c, in this work we make examinations with $k = 0.03$ and $\varphi_c = 0.003$. While, so far there is no severe restriction on $\mathcal{K}$ except that $\mathcal{K}$ should be in the range of $0 < \mathcal{K} < 2$ [7, 8]. Thus we adopt various values of $\mathcal{K}$ in the above range and study the dependence of gravitational waves on $\mathcal{K}$. Furthermore, as equilibrium stellar models, we adopt the similar models in [9].

3. Fluid Oscillations

First, with the Cowling approximation we examine the fluid oscillations of neutron stars. Especially, we focus on the stellar models whose central density is in the range from $\tilde{\rho}_0 = 10^{14}$ g/cm^3 up to the value given the maximum ADM mass. In general, the oscillation spectrum is directly related to the stellar parameter, but the frequencies of fundamental oscillation modes can be connected to the stellar average density. Actually, in GR, it is well known the empirical formula for the frequency of f mode as a function of stellar average density, which is almost independent from the adopted EOS to construct the stellar models. While, the frequencies of f mode for the stellar models in TeVeS with above two EOS are shown in Fig.1. The deviation from GR is clearly cognized for typical neutron stars and depending on the value of parameter $\mathcal{K}$, the frequencies become around 20 % larger than those expected in a general relativistic neutron star. This can be an observable effect and one might distinguish the gravitational theory in strong gravitational field by using the observations of gravitational waves.

The possibility to probe the gravitational theory by using observations of gravitational waves can be also seen in Fig.2, where we plot the normalized frequencies of f and p_1 modes as functions of ADM mass. One can easily observe that the frequencies expected in TeVeS are quite different from those in GR. Since this distinction results from the difference of gravitational theory, which creates due to the presence of scalar field, observing more than one mode of gravitational wave could tell us the existence of the scalar field. Through this figure, we can find that with a help of observation of stellar mass, it is possible to probe the gravitational theory in the strong-field regime by using observations of gravitational waves.

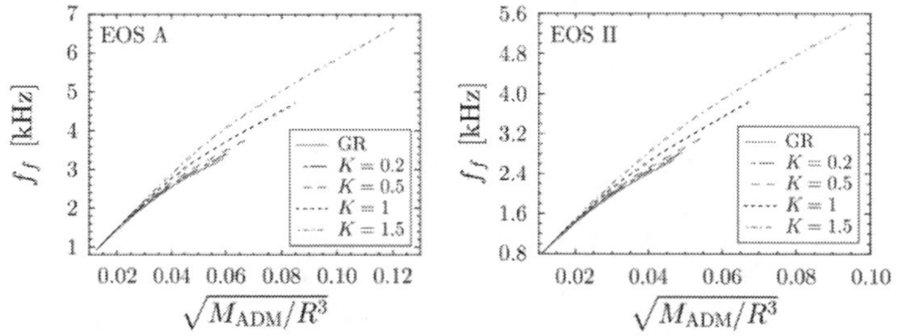

Figure 1. The frequency of f mode as a function of the stellar average density.

4. Spacetime Oscillations

Next, with the assumption that the vector field is fixed, we examine the spacetime oscillations. Similar to the fluid modes, in GR it is well known that the frequencies of w and $w_{\rm II}$ modes can

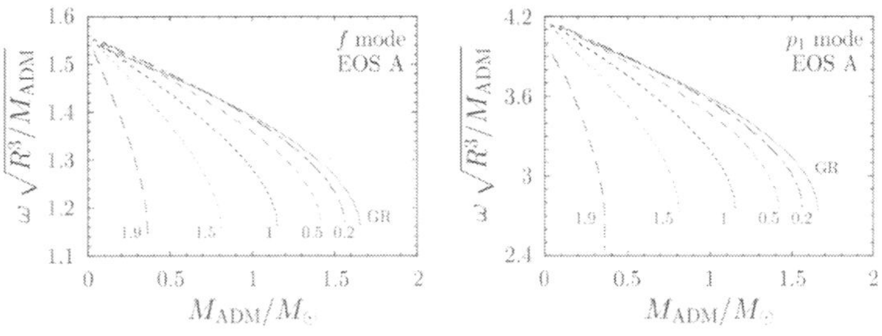

Figure 2. The frequencies of first two fluid modes are plotted as functions of the ADM mass.

be described as a function of stellar compactness M_{ADM}/R. So, we will examine the dependence of frequencies of w_{II} and w modes on the stellar compactness. Figs.3 and 4 show the frequencies of axial w_{II} and w_I as functions of the total stellar compactness, where the solid symbols are results with EOS A and the open ones are those with EOS II. From Fig.3, we can see that the dependence of frequencies of w_{II} modes on the stellar compactness are almost independent from the values of $\mathcal{K}$ and the adopted EOS. Additionally, it is found that the deviation from the frequencies expected in GR is very little. We can see little difference between the expectations in GR and in TeVeS in the damping rates (imaginary part of complex frequencies) for the stars with larger compactness and in the oscillation frequencies (real part of complex frequencies) for the stars with weaker compactness. For the oscillation frequencies in TeVeS, we can get the empirical formula, such as

$$Re\,(\omega M_{ADM}) = -0.0412 + 0.7084\left(\frac{M_{ADM}}{R}\right) + 3.305\left(\frac{M_{ADM}}{R}\right)^2, \qquad (1)$$

and the oscillation frequencies can be in very good agreement with this empirical formula.

On the other hand, in Fig.4 for the frequencies of w_I modes, we can observe same feature as the frequencies of w_{II} modes, i.e., if we see those frequencies as functions of the stellar compactness, those are almost independent from the values of $\mathcal{K}$ and the adopted EOS. However, in the case of w_I modes, there exists the crucial different feature in contrast to the w_{II} modes. That is, the dependence of the frequencies in TeVeS is obviously different from those expected in GR. In other words, with this different dependence of frequencies on the gravitational theory, one can distinguish the gravitational theory in strong-field regime by using the gravitational waves observations. In fact, the oscillation frequencies of axial w_I modes in GR and TeVeS can be expected with high accuracy via the following empirical formula

$$Re\,(\omega R) = \alpha - \beta\left(\frac{M_{ADM}}{R}\right), \qquad (2)$$

where $(\alpha, \beta) = (2.797, 4.255)$ in GR and $(\alpha, \beta) = (2.533, 3.714)$ in TeVeS.

5. Conclusion

In this work, we examine the effect of the Tensor-Vector-Scalar (TeVeS) Theory on the oscillation spectra of neutron stars. Depending on the parameter of TeVeS, the frequencies of fundamental oscillation could be off the well-known empirical formula in GR and they become lager than those expected in GR. We can also see the deviation from GR in the frequencies of higher overtones and they have stronger dependence on the parameter $\mathcal{K}$ than the lower oscillation modes. Additionally, by calculating the complex eigenfrequencies of axial w_{II} and w modes, where the real and imaginary parts of complex frequencies are corresponding to the oscillation

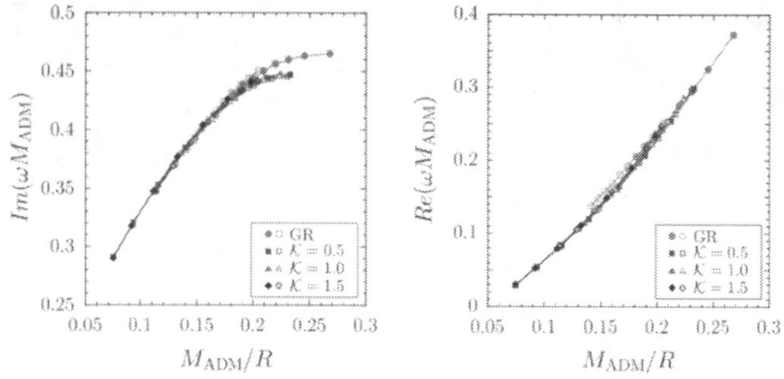

Figure 3. For $w_{\rm II}$ modes, the damping rates (left panel) and the oscillation frequencies (right panel) as functions of stellar compactness $M_{\rm ADM}/R$. The solid symbols correspond to the results for EOS A and the open ones to those for EOS II.

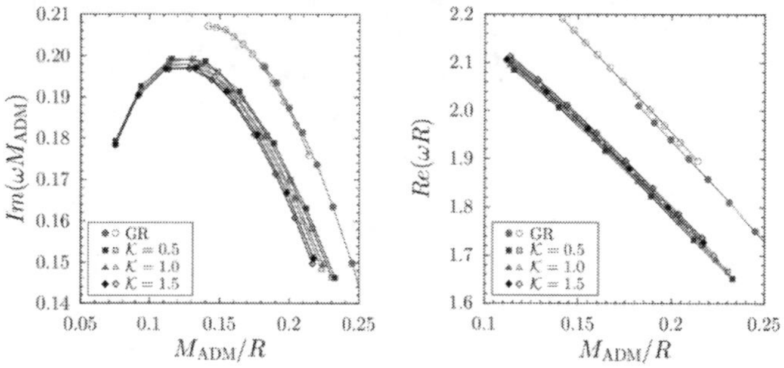

Figure 4. Similar to Fig.3, but for $w_{\rm I}$ modes.

frequencies and the damping rates of the emitted gravitational waves, respectively, we find that the dependences of the both frequencies of $w_{\rm II}$ and $w_{\rm I}$ modes in TeVeS on the stellar compactness are almost independent from the parameter $\mathcal{K}$ and the adopted EOS. The dependences of $w_{\rm I}$ modes are obviously different from that expected in GR, while the dependences of $w_{\rm II}$ modes are similar to that in GR. Owing to these differences of dependence on the gravitational theory, one can distinguish the gravitational theory in the strong-field regime by using the direct observations of gravitational waves emitted from the neutron stars. Since these imprints of TeVeS come from the presence of scalar field, by using the observations of gravitational waves associated with the stellar oscillations, it will be possible not only to distinguish the gravitational theory in the strong-field regime, but also to probe the existence of the scalar field.

References

[1] Psaltis D 2008 *Living Rev. Relativity* **9**
[2] Sotani H and Kokkotas K D 2004 *Phys. Rev.* D **70** 084026
[3] Sotani H and Kokkotas K D 2005 *Phys. Rev.* D **71** 124038
[4] Bekenstein J D 2004 *Phys. Rev.* D **70** 083509
[5] Milgrom M 1983 *Astrophys. J.* **270** 365
[6] Giannios D 2005 *Phys. Rev.* D **71** 103511
[7] Sagi E and Bekenstein D 2008 *Phys. Rev.* D **77** 024010
[8] Lasky P D, Sotani H and Giannios D 2008 *Phys. Rev.* D **78** 104019
[9] Sotani H 2009 *Phys. Rev.* D **79** 064033
[10] Sotani H 2009 *Phys. Rev.* D **80** 064035

Journal of Physics: Conference Series **229** (2010) 012070 doi:10.1088/1742-6596/229/1/012070

Features of gravitational waves in higher dimensions

Otakar Svítek

Institute of Theoretical Physics, Charles University in Prague, V Holesovickach 2, 180 00 Praha 8 - Holesovice, Czech Republic

E-mail: `ota@matfyz.cz`

Abstract. There are several fundamental differences between four-dimensional and higher-dimensional gravitational waves, namely in the so called braneworld set-up. One of them is their asymptotic behaviour within the Cauchy problem. This study is connected with the so called Hadamard problem, which aims at the question of Huygens principle validity. We investigate the effect of braneworld scenarios on the character of propagation of gravitational waves on FRW background.

1. Introduction

Gravitational waves in higher dimensions have several features that might distinguish them from the waves propagating on the four-dimensional manifold only. Therefore, many features were already studied, including quasi-normal modes, modification of late-time tails, and also the higher-dimensional exact radiative spacetimes.

Aside from these studies in general higher-dimensional spacetimes, special attention has been paid to specific braneworld models describing our universe as a hypersurface (brane) in a higher-dimensional spacetime (bulk). Apart from the quickly discovered short-distance modification of the Newtonian gravitational potential, a nontrivial correction to the quadrupole radiation formula has been derived [1]. There are also the notorious KK modes that are effectively massive on the brane.

2. Huygens principle and tails

Huygens principle (and related Hadamard problem) deals with the dependence of field values on the Cauchy data. Namely, whether it suffices to know the data only in an arbitrarily small neighbourhood of the intersection of the past null cone with the Cauchy surface. In numerous mathematical studies of the problem (namely works by Hadamard [2], Günther [3] and McLenaghan [4]) it was shown that the result depends on both the differential operator properties and underlying spacetime. This makes it a possible candidate for studying deviations that might occur when we move to higher-dimensional spacetimes.

Related to the above described qualitative description there is a problem of presence and characteristic decay of late-time tails. It quantifies the fall-of of the initially sharp signal from distant source as seen by the observer at fixed position. In the context of physically interesting situations in four dimensions the study was started mainly by Price [5], in higher dimensions it was recently studied for example in [6].

The usual explanation of these late-time tails is the effect of backscatering on the curvature (or the potential), but one should be careful to use this explanation for any conclusions.

3. Brane model

Our concern here is to analyze how the specific brane model affects the validity of Huygens principle of gravitational waves on FRW spacetime. In braneworld scenarios gravity feels the higher dimensional spacetime while other fields are confined only to four-dimensional brane. There are several different models and we will assume the Randall-Sundrum single brane set-up [7].

For the description of the model we use the Gaussian embedding technique [8] with the following form of the metric

$$ds^2 = dw^2 + i_{\mu\nu}dx^\mu dx^\nu \ . \tag{1}$$

We assume Z_2 symmetry with bulk cosmological constant. The five-dimensional Einstein equations have the form

$$G_{MN} = \kappa_5^2[-\Lambda_5 g_{MN} + \tau_{\mu\nu}\delta^\mu_M\delta^\nu_N\delta(w)] \ . \tag{2}$$

The matter content on the brane is given by the Israel junction conditions

$$K^+_{\mu\nu} = -K^-_{\mu\nu} = \kappa_5^2\left[\tau_{\mu\nu} - \frac{1}{3}i_{\mu\nu}(0)\tau^\alpha{}_\alpha\right] \ , \tag{3}$$

where the brane stress energy tensor is decomposed in the following way

$$\tau_{\mu\nu} = -\lambda i_{\mu\nu}(0) + S_{\mu\nu} \ . \tag{4}$$

Four-dimensional Einstein equations then reduce to the form

$$^4G_{\mu\nu} = -\Lambda_4 i_{\mu\nu}(0) + \kappa_4^2 S_{\mu\nu} + \kappa_5^4\pi_{\mu\nu} + \bar{E}_{\mu\nu}(0) \tag{5}$$

where $\Lambda_4 = \frac{\kappa_5^2}{12}(\kappa_5^2\lambda^2 + 6\Lambda_5)$, $\kappa_4^2 = \frac{\kappa_5^4\lambda}{6}$, $\pi \sim S^2$, and $\bar{E}_{\mu\nu}(0)$ is a continuous part of certain Weyl tensor projection at the brane.

Since we would like to study cosmological model we assume the perfect fluid on the brane

$$S_{\mu\nu} = (p+\rho)u_\mu u_\nu + p i_{\mu\nu} \ . \tag{6}$$

Then even $\pi_{\mu\nu}$ can be put to perfect fluid form with parameters quadratic in p, ρ.

4. Modified Friedmann equation

Restricting $i_{\mu\nu}(0)$ to the form of a standard FRW metric on the brane and assuming $\bar{E}_{00}(0) = \Lambda_4 = 0$ we arrive at the modified Friedmann equation corresponding to the previously derived results [9]

$$\frac{3(\dot{a}^2 + k)}{a^2} = \kappa_4^2\rho + \frac{\kappa_5^4\rho^2}{12} \ . \tag{7}$$

For example the behaviour of expansion function in the radiation era ($\rho = \rho_0 a^{-4}$) with $k = 0$ is the following

$$a(t) \sim t^{1/2}\left[1 + \frac{\ell}{2t}\right]^{1/4} \ , \tag{8}$$

where the modification due to brane model is clearly visible as the second term in the bracket.

5. Perturbations of the brane metric

We consider FRW metric in the following form

$$ds^2 = a^2(-d\eta^2 + dr^2 + \sigma^2 d\Omega^2) , \tag{9}$$

with conformal time η and radial function $\sigma = r, \sin(r), \sinh(r)$ (depending on the spatial curvature $k = 0, 1, -1$).

Axial perturbations of gravitational field in the Regge-Wheeler gauge were derived in [10] and two polarizations might be parametrized as

$$\begin{aligned} h_{0\phi} &= h_0(\eta, \rho)\sin(\theta)\ \partial_\theta Y_{lm} \\ h_{1\phi} &= h_1(\eta, \rho)\sin(\theta)\ \partial_\theta Y_{lm} \end{aligned} \tag{10}$$

If compatibility condition

$$(\sigma\sigma_{,r})_{,r} + 2k\sigma^2 - 1 = 0$$

is satisfied, then equations for both modes can be reduced to one equivalent scalar equation. Defining $F = \frac{h_1}{\sigma a}$ this equation is

$$\partial^2_\eta F - \partial^2_r F + \frac{l(l+1)}{\sigma^2} F - kF - \frac{\partial^2_\eta a}{a} F = 0 \tag{11}$$

From the modified Friedmann equation (7) with conformal time and assuming radiation era we get

$$\partial^2_\eta a = -ka + \frac{\kappa_5^4 \rho^2 a^3}{18} , \tag{12}$$

which after combining with (11) leads to

$$\partial^2_\eta F - \partial^2_r F + \frac{l(l+1)}{\sigma^2} F - \frac{\kappa_5^4 \rho^2 a^2}{18} F = 0 . \tag{13}$$

The behaviour of the solutions to this equation will now be analyzed both in the case of standard and modified Friedmann evolution equations.

6. Behaviour analysis

For standard cosmology the term $\sim \rho^2$ is missing in (13) and the solution is of the form

$$F_l \sim \sigma^l \underbrace{\partial_r \sigma^{-1} \cdots \partial_r \sigma^{-1}}_{l-1} \partial_r \left(\frac{f+g}{\sigma}\right) , \tag{14}$$

with functions f, g depending only on combinations $r - \eta$ and $r + \eta$ and so the gravitational radiation is bounded by null cones. Therefore, Huygens principle is valid in this case.

In our braneworld scenario the analysis is more difficult and the approach from [10] will be used. We start with the decomposition of solution $F = F_0 + f$, where F_0 satisfies the reduced equation (without the brane-induced modification) with initial data of compact support $\sigma(r) \leq \sigma(r_0)$ and f fulfils

$$\partial^2_\eta f - \partial^2_r f + \frac{l(l+1)}{\sigma^2} f = \frac{\kappa_5^4 \rho^2 a^2}{18}(f + F_0) \tag{15}$$

with homogeneous initial conditions.

Next, we introduce the energy functional

$$E(F) \equiv \frac{1}{2}\int \left[(\partial_\eta F)^2 - (\partial_r F)^2 + \frac{l(l+1)}{\sigma^2}F^2 \right] dr \tag{16}$$

which is conserved for F_0 but not for f because we have

$$\frac{d}{d\eta}E(f) = \int \frac{\kappa_5^4 \rho^2 a^2}{18}(F_0 + f)\partial_\eta f \; dr \; . \tag{17}$$

Since $\sigma(r_0 + \eta) \geq \sigma(r)$ (assuming $\sigma(r+\eta) \leq \frac{\pi}{2}$ for $k = 1$) and using the Schwartz inequality together with the bounds implied by $E(f), E(F_0)$ we arrive at the following inequality

$$e^{\alpha}\frac{d}{d\eta}(e^{-\alpha}\sqrt{E(f)}) \leq \frac{d\alpha}{d\eta}\sqrt{E(F_0)} \; , \tag{18}$$

where $\alpha = \int_{\eta_0}^{\eta} \frac{\kappa_5^4 \rho^2 a^2}{18\sqrt{l(l+1)}}\sigma(r_0 + \bar{\eta})d\bar{\eta}$.

We continue by estimation of parameter α with the help of the behaviour of functions σ, H and the relation $\frac{\kappa_5^4 \rho^2}{36} \lesssim H^2$ (for $k = 1$ one has to assume $\frac{k}{a^2} \ll H^2$).

Finally, we obtain

$$\alpha \leq \sigma(r_0 + \eta)a(\eta)\frac{H(\eta)}{2\sqrt{l(l+1)}} \equiv \beta \tag{19}$$

After further manipulation we get the following estimate for the energy functional of f

$$E(f) \leq E(F_0)(e^{\beta} - 1)^2 \tag{20}$$

Similar estimates can be obtained for derivatives of f. Using Sobolev estimates the field f can be controlled pointwisely provided we have initial data of compact support and $\beta \ll 1$. The last condition is satisfied only for high enough mode number l.

Conclusion

We have shown that due to the modification of Friedmann equation in brane model the Huygens principle for gravitational waves seems to be satisfied only for high enough multipoles. This is in contrast with the standard cosmology. However, one should be aware that the presented analysis is something like a zero approximation of the more difficult analysis that would involve complete spectrum of brane perturbations. Here we have analyzed just the so called massless mode and reduced the brane influence to the modification of Friedmann equation.

Acknowledgements

This work was supported by grants GACR 202/07/P284, GACR 202/09/0772 and the Czech Ministry of Education project Center of Theoretical Astrophysics LC06014.

References

[1] Kinoshita S et al 2005 *Class. Quantum Grav.* **22** 3911-3922
[2] Hadamard J 1923 *Lectures on Cauchys problem in linear partial differential equations* (Yale, New Haven: Yale University Press)
[3] Günther F 1988 *Huygens Principle and Hyperbolic Equations* (Boston, MA: Academic)
[4] McLenaghan R G 1982 *Ann. Inst. H. Poincare* **37** 211
[5] Price R H 1972 *Phys. Rev.* D **5** 2419, 2439
[6] Bizoń P, Chmaj T and Rostworowski A 2009 *Class. Quantum Grav.* **26** 175006
[7] Randall L and Sundrum R 1999 *Phys. Rev. Lett.* **83** 4690
[8] Shiromizu T, Maeda K and Sasaki M 2000 *Phys. Rev.* D **62** 024012
[9] Binetruy P, Deffayet C and Langlois D 2000 *Nucl. Phys.* B **565** 269-287
[10] Malec E and Wylężek G 2005 *Class. Quantum Grav.* **22** 3549-3553

Journal of Physics: Conference Series **229** (2010) 012071 doi:10.1088/1742-6596/229/1/012071

Integration without integration: new Killing spinor spacetimes

Norbert Van den Bergh
Department of Mathematical Analysis TW16, University of Ghent, Galglaan 2, Ghent, Belgium

E-mail: norbert.vandenbergh@ugent.be

Abstract. Non-conformally flat spacetimes admitting a non-null two-index Killing spinor are investigated by means of the Geroch-Held-Penrose formalism. Claims appearing in the literature that such spacetimes are all explicitly known are incorrect. This was shown in [5] for the family where, in the canonical frame, the spin coefficients ρ or μ vanish. Here the general case with non-vanishing ρ, μ, π and τ is re-considered. It is shown that the construction in [4] hinges on the tacit assumption that certain integrability conditions hold, implying two algebraic relations for the spin coefficients and the components of the Ricci spinor. All (conformal classes of) spacetimes, in which one of these conditions is violated, are obtained by invariant integration. The resulting classes are each other's Sachs transform and are characterised by one free function. They admit in general no Killing vectors, but still admit a conformal gauge (different from the trivial unitary gauge) in which a Killing tensor exists.

1. Introduction

The history of the subject which I present goes back to Carter's work [1] on the complete integration of charged particle orbits in the Kerr metric, making use of what he called "the unexpected separability of the Hamilton-Jacobi equation ... " and to the subsequent construction of all Abelian G_2 Hamilton-Jacobi separable spacetimes. The relationship between separability and the existence of Killing tensors led Walker and Penrose [2] to investigate vacuum type D spacetimes admitting quadratic first integrals of the null geodesic equations. They showed that the separability property was linked to the existence of an object which they called a *Killing spinor*, which is a symmetric valence 2 spinor $X_{AB} = X_{(AB)}$, satisfying the twistor equation

$$\nabla_{A'(A}X_{BC)} = 0. \tag{1}$$

The tensor analogue of a Killing spinor is a tensor $D_{ab} = D_{[ab]} = X_{AB}\epsilon_{A'B'} + \bar{X}_{A'B'}\epsilon_{AB}$, which nowadays usually is called a conformal Killing-Yano (CKY) tensor and which is a two-form satisfying the CKY equation

$$\nabla_{(a}D_{b)c} = \frac{1}{3}\nabla_d(g_{ab}D^d{}_c - D^d{}_{(a}g_{b)c}). \tag{2}$$

The Killing spinor X_{AB} was used by Walker and Penrose to construct a second rank so called conformal Killing tensor, namely a symmetric tensor P_{ab} obeying the equation

$$\nabla_{(a}P_{bc)} = g_{(ab}p_{c)} \tag{3}$$

and on which depends the construction of a full Killing tensor in all but the Kinnersley case III metrics (namely the C and C-NUT metric).

In the wake of these investigations a lot of effort was devoted to the characterisation and construction of spacetimes admitting Killing spinors, conformal Killing tensors, Killing-Yano tensors etc. (see [3] for references). For Killing spinors the situation is at first sight fairly easy: for a non-null Killing spinor $X o_{(A} \iota_{B)}$ and a non-conformally flat spacetime, the Petrov type is necessarily D, with the principal spinors of X and of the Weyl spinor being aligned, geodesic and shearfree. Adapting a Geroch-Held-Penrose tetrad to the latter, one obtains

$$\kappa = \sigma = \kappa' = \sigma' = 0 \tag{4}$$

and the equations (1) become

$$Þ X = -\rho X, \ Þ' X = -\rho' X \tag{5}$$
$$ð X = -\tau X, \ ð' X = -\tau' X \tag{6}$$

(recall that in the GHP formalism the $'$-operator interchanges the real null vectors $\mathbf{k}$ and $\mathbf{l}$, as well as $\mathbf{m}$ and $\overline{\mathbf{m}}$, hence $\rho' = -\mu$ and $\tau' = -\pi$).

Obviously, as these equations are conformally invariant, the spacetime is only defined up to a conformal factor. However one can always construct a unique conformal representant $(\mathcal{M}, \hat{g})$ in which $|\hat{X}| = 1$. In this representant one has

$$\hat{\rho} + \overline{\hat{\rho}} = \hat{\rho}' + \overline{\hat{\rho}}' = \hat{\tau} + \overline{\hat{\tau}}' = 0 \tag{7}$$

and a general representant is now related to $(\mathcal{M}, \hat{g})$ by $g = \Omega^2 \hat{g}, \ X = \Omega \hat{X}$. In $(\mathcal{M}, \hat{g})$ the trace of the conformal Killing tensor P_{ab} associated to X is constant and P_{ab} is a Killing tensor with constant eigenvalues. In the invariant construction of Killing spinor spacetimes these eigenvalues can play a significant role (turning up as possibly independent coordinates) and therefore one tries to construct representants in which the existence of a Killing tensor is preserved under conformal transformations. This requires that the conformal Killing tensor P_{ab} should be the trace-free part of a Killing tensor itself: a necessary and sufficient condition is that a solution exists of the equation

$$\nabla_a P^a{}_c + \frac{3}{4} \nabla_c K = 0, \tag{8}$$

or, in terms of the eigenvalues of the Killing tensor $K_{cd} = 2am_{(c}\overline{m}_{d)} + 2b\ell_{(c}n_{d)}$,

$$\mathrm{d}a = -(a+b)[(\rho + \overline{\rho})\boldsymbol{\omega}^4 + (\rho' + \overline{\rho}')\boldsymbol{\omega}^3] \tag{9}$$
$$\mathrm{d}b = -(a+b)[(\tau + \overline{\tau}')\boldsymbol{\omega}^1 + (\overline{\tau} + \tau')\boldsymbol{\omega}^2], \tag{10}$$

with $a = (\Omega^2 - K)/4$ and $b = (\Omega^2 + K)/4$. In order that this system should have non-constant solutions, the following integrability conditions must be satisfied:

$$\rho' Þ a - \rho Þ' a = 0 \tag{11}$$
$$\rho' \overline{\phi}' Þ a + \rho \phi Þ' a = 0 \tag{12}$$

and

$$\tau' ð b - \tau ð' b = 0 \tag{13}$$
$$\tau' \overline{\phi} ð b + \tau \phi ð' b = 0, \tag{14}$$

with ϕ and ϕ' defined by

$$\Phi_{01} = -3\rho\tau - 2\frac{\rho}{\tau'}\phi \,, \Phi_{21} = -3\rho'\tau' - 2\rho'\phi'/\tau. \tag{15}$$

It follows that when $\tau\tau' = 0$ the classification given in [4] is essentially correct. When $\tau\tau' \neq 0$ and $\rho\rho' = 0$ an extra spacetime arises [5], admitting a G_3 (but without Abelian subgroup) and in which the b-equations only have constant solutions. In the general case ($\tau\tau'\rho\rho' \neq 0$) a Killing tensor with non-constant eigenvalues can only exist when the following conditions hold, which I henceforth call KS_1 and KS_2:

$$KS_1: \ \exists \text{ non-constant } b \Longleftrightarrow \phi + \overline{\phi} = 0 = \phi' + \overline{\phi}' \tag{16}$$

$$KS_2: \ \exists \text{ non-constant } a \Longleftrightarrow \phi + \overline{\phi}' = 0. \tag{17}$$

In this contribution all spacetimes in the sets $KS_1 \setminus KS_2$ and $KS_2 \setminus KS_1$ are presented.

2. Construction of KS_1 and KS_2

In the KS_1 spacetimes one first deduces from the GHP commutator equations, Ricci equations and Bianchi equations the following integrable system:

$$\begin{aligned} Þ\rho &= 0 \\ Þ'\rho &= 2\rho\rho'/|\tau|^2(\phi + \phi') \\ ð\rho &= \rho/\tau'(|\tau|^2 - 2\phi), \end{aligned} \tag{18}$$

together with similar equations for τ, ϕ, ϕ' and $\Psi_2 = E + iH$. These systems provide one immediately with two coordinate-candidates, r and m, defined by $r^2 = Q\rho\rho'$, $m = |\tau|$ $(Q = \pm 1)$. Herewith ϕ, ϕ' become functions of m and r:

$$\phi = m^2 r^{-2}\phi_0, \phi' = m^2 r^{-2}\chi_0, \tag{19}$$

with $\chi_0 = \phi_0 - 2ic_1$ and $(r^2 - c_2)^2 - (\phi_0 - ic_1)^2 = c_1^2 + c_2^2$ ($c_1 \neq 0$ and $|r^2 - c_2| < \sqrt{c_1^2 + c_2^2}$).

This leads to the following explicit expressions for the exterior derivatives of r, m and b:

$$\mathrm{d}r = (\phi_0 + \chi_0)[\frac{m}{r}(\frac{\tau}{m}\boldsymbol{\omega}^1 - \frac{m}{\tau}\boldsymbol{\omega}^2) + Q\frac{r}{\rho}\boldsymbol{\omega}^3 + \frac{\rho}{r}\boldsymbol{\omega}^4] \tag{20}$$

$$\mathrm{d}m = \frac{i}{2}(2iQ(\phi_0 + \chi_0) - H)[\frac{\tau}{m}\boldsymbol{\omega}^1 - \frac{m}{\tau}\boldsymbol{\omega}^2] + \frac{2m}{r}[Q\frac{r}{\rho}\chi_0\boldsymbol{\omega}^3 + \frac{\rho}{r}\phi_0\boldsymbol{\omega}^4] \tag{21}$$

$$\mathrm{d}b = i(\frac{\tau}{m}\boldsymbol{\omega}^1 - \frac{m}{\tau}\boldsymbol{\omega}^2), \tag{22}$$

Calculating the triple wedge product

$$\mathrm{d}r \wedge \mathrm{d}m \wedge \mathrm{d}b = 2iQ\frac{m(\phi_0^2 - \chi_0^2)}{r}(\frac{\tau}{m}\boldsymbol{\omega}^3 \wedge \boldsymbol{\omega}^4 \wedge \boldsymbol{\omega}^1 - \frac{m}{\tau}\boldsymbol{\omega}^3 \wedge \boldsymbol{\omega}^4 \wedge \boldsymbol{\omega}^2) \neq 0. \tag{23}$$

shows that r, m and b are functionally independent scalars. Although one could in principle use the magnetic part H of Ψ_2 as a fourth coordinate (and hence obtain a complete invariant integration of the Cartan equations), it is more practical to define a fourth coordinate x by

$$\frac{1}{2}(\frac{\tau}{m}\boldsymbol{\omega}^1 + \frac{m}{\tau}\boldsymbol{\omega}^2) = h_1\mathrm{d}b + h_2\mathrm{d}r + h_3\mathrm{d}m + \mathrm{d}x, \tag{24}$$

leading one to the following expression for the metric:

$$\frac{\tau}{m}\omega^1 = \mathrm{d}x + \frac{iQmr}{4c_1(\phi_0 - ic_1)}\mathrm{d}r + (\frac{H_0 x}{2} - 4Qc_1x^2 - \frac{2ic_1 + Qm^2}{4c_1})\mathrm{d}b \tag{25}$$

$$\frac{1}{\rho}\omega^3 = \frac{iQ}{4c_1}(\frac{\mathrm{d}m}{m} - \frac{\phi_0}{\phi_0 - ic_1}\frac{\mathrm{d}r}{r}) + i(Q\frac{r^2H_0 - 4mi\phi_0}{8r^2c_1} - 2x)\mathrm{d}b \tag{26}$$

$$\rho\omega^4 = -\frac{ir^2}{4c_1}(\frac{\mathrm{d}m}{m} + \frac{\chi_0}{\chi_0 + ic_1}\frac{\mathrm{d}r}{r}) - i[\frac{r^2H_0 - 4mi\phi_0}{8c_1} - m - 2Qr^2x]\mathrm{d}b. \tag{27}$$

The corresponding curvature components are $\Phi_{00} = -\rho^2, \Phi_{22} = -\frac{r^4}{\rho^2}, \Phi_{02} = -\tau^2$ and

$$R = -12(Qr^2 + m^2 + E), \tag{28}$$

$$E = -2m^2 - \frac{8}{3}Qr^2 + 32c_1^2x^2 + \frac{1}{12}H_0^2 - \frac{1}{6}\frac{dH_0}{db} - 4Qc_1xH_0 - \frac{4Q}{3r^2}\phi_0\chi_0, \tag{29}$$

$$H = (H_0 - 16Qc_1x)m + 2iQ(\phi_0 + \chi_0), \tag{30}$$

$$\Phi_{11} = -\frac{9}{2}Qr^2 - \frac{7}{2}m^2 - \frac{3}{2}E - \frac{4Q\phi_0\chi_0}{r^2}, \tag{31}$$

$$\Phi_{01} = \frac{\rho\tau}{r^2}(2\phi_0 - 3r^2), \tag{32}$$

$$\Phi_{12} = -\frac{Q\tau}{\rho}(2\chi_0 + 3r^2). \tag{33}$$

For KS_2 one can proceed along exactly the same lines, but it is more instructive to note that the Sachs asterisk operation [9] $þ^* = ð, þ'^* = -ð', ð^* = -þ, ð'^* = þ'$ transforms the two situations into each other. As a consequence it is possible to obtain the KS_2 metric via a complex transformation of KS_1 (see [3]).

A remaining problem stems from a corollary in [7], showing that all Petrov type D spacetimes admitting a conformal Killing tensor of Segre type [(11)(11)] necessarily are conformally Killing-Yano. As a consequence the KS_1 and KS_2 spacetimes ought to belong to the classes of spacetimes considered in [8]. Although the title of [8] gives the impression that the paper deals with non-constant eigenvalues only, the authors claim that all results remain valid also when one of the eigenvalues is constant. However the latter metrics appear not to be equivalent to the KS_1 and KS_2 metrics! A further problem is to determine all possible Killing spinor spacetimes in the complement of KS_1 and KS_2 [1].

References

[1] Carter B 1968 *Comm. Math. Phys.* **10** 280
[2] Walker M and Penrose R 1970 *Comm. Math. Phys.* **18** 265
[3] Van den Bergh N 2009 *Class. Quantum Grav.* (to be published)
[4] Jeffryes BP 1984 *Proc. R. Soc. A* **392** 323
[5] McLenaghan RG and Van den Bergh N 1993 *Class. Quantum Grav.* **10** 2179
[6] Dietz W and Rüdiger W 1980 *Proc. R. Soc. A* **375** 361
[7] Coll B, Ferrando JJ and Saez JA 2006 *J. Math. Phys.* **47** 062503
[8] Hauser I and Malhiot RJ 1976 *J. Math. Phys.* **17** 1306
[9] Geroch R, Held A and Penrose R 1973 *J. Math. Phys.* **14** 874

[1] at the time of writing at least one example has been found of a Killing spinor spacetime in which *both* conditions KS_1 and KS_2 are violated

Journal of Physics: Conference Series **229** (2010) 012072 doi:10.1088/1742-6596/229/1/012072

Black holes in a box

Helvi Witek[1]*, Vitor Cardoso[1,2], Leonardo Gualtieri[3], Carlos Herdeiro[4], Andrea Nerozzi[1], Ulrich Sperhake[5] and Miguel Zilhão[4]

[1] Centro Multidisciplinar de Astrofísica - CENTRA, Dept. de Física, Instituto Superior Técnico, Av. Rovisco Pais 1, 1049-001 Lisboa, Portugal
[2] Department of Physics and Astronomy, The University of Mississippi, University, MS 38677-1848, USA
[3] Dipartimento di Fisica, Università di Roma "Sapienza" and Sezione INFN Roma1, P.A. Moro 5, 00185, Roma, Italy
[4] Centro de Física do Porto e Departamento de Física da Faculdade de Ciências da Universidade do Porto, Rua do Campo Alegre 687, 4169-007 Porto, Portugal
[5] California Institute of Technology, Pasadena, CA 91109, USA

E-mail: helvi.witek@ist.utl.pt, vitor.cardoso@ist.utl.pt, Leonardo.Gualtieri@roma1.infn.it, crherdei@fc.up.pt, andrea.nerozzi@ist.utl.pt, sperhake@tapir.caltech.edu, mzilhao@fc.up.pt

Abstract. The evolution of BHs in "confining boxes" is interesting for a number of reasons, particularly because it mimics some aspects of anti-de Sitter spacetimes. These admit no Cauchy surface and are a simple example of a non-globally hyperbolic spacetime. We are here interested in the potential role that boundary conditions play in the evolution of a BH system. For that, we imprison a binary BH in a box, at which boundary we set mirror-like boundary conditions.

1. Introduction

Black holes (BHs) have become of increasing importance in fundamental physics, including not only astrophysics but also high-energy physics and quantum gravity. In high-energy physics, the duality between gauge theory and gravity in anti-de Sitter (AdS) spacetimes has created a powerful framework for the study of strongly coupled gauge theories and found applications in connection with the experimental program on heavy ion collisions at Brookhaven's Relativistic Heavy Ion Collider [1] and at the Large Hadron Collider at CERN [2, 3], among many others. One peculiar feature of asymptotically AdS space is the "active role" played by its boundary. In AdS, null geodesics reach the boundary in finite coordinate time. One thus often refers to an asymptotically AdS space as a box [4, 5, 6]. Following the remarkable breakthroughs of Numerical Relativity (NR) in the course of the past few years, it is now time to explore numerically the dynamics of BHs in different background spacetimes, from higher dimensions [7, 8] to anti-de Sitter backgrounds. Here, we wish to begin exploring BHs in AdS by extending NR methods so as to encompass AdS-like boundary conditions. We will model AdS backgrounds by a confining box with reflecting walls in which a BH binary evolves. This will allow us to study a number of interesting phenomena. The final outcome of a generic BH binary merger is a rotating BH. Since the whole system is inside a box, we expect superradiant effects on the waves generated during the merger to play a role; if the frequency ω of the impinging wave satisfies $\omega < m\Omega$, with the azimuthal number m and the angular velocity of the horizon Ω, this radiation is amplified as it scatters off the BH. Thus, energy and angular momentum of the rotating hole are extracted by superradiant modes. The bouncing back and forth of the waves at the reflecting wall and its subsequent amplification by superradiance close to the ergoregion are expected to turn it into a BH bomb [9, 10, 11]. This mechanism makes small rotating BHs in AdS unstable [10] and may presumably play a role in these simulations.

2. Numerical Setup

The numerical simulations were carried out using Sperhake's LEAN code [12]. We make use of puncture initial data computed by the pseudo-spectral TWOPUNCTURES code [13, 14]. The evolution of the Einstein equations in the "3+1" framework is based on the χ-version of the Baumgarte-Shapiro-Shibata-Nakamura (BSSN) [15, 16] system, with the moving puncture approach [17, 18]. The punctures have initial mass parameter $m_i = 0.483$, initial position parameter $x_i = \pm 3.257$ and linear momentum parameter $p_{y_i} = 0.133$. If not denoted otherwise the simulations have been performed using the resolution $h_c = 1/48M$ near the puncture.

In order to mimic the numerical evolution of BHs in AdS spacetimes, we surround the BH binary by a "mirror-like" box at each spatial hypersurface. We impose a spherical box with reflecting boundary conditions at a certain distance r_B around the BH binary. Because of the non-physical "junk"-radiation the system is evolved using outgoing boundary condition until this pulse has left the numerical domain of interest. Afterwards we switch to reflecting boundary conditions. We consider an inspiral of initially non-spinning BHs, within a "box" with radius $r_B = 48M$.

In order to get some insight into the physical properties of the system we consider gravitational radiation as well as properties of the apparent horizon (AH) computed by Thornburg's AHFINDERDIRECT [19, 20]. Gravitational wave information is obtained by computing the Weyl scalar Ψ_4 in the Newman-Penrose formalism and its multipolar components [21].

3. Results

3.1. Wave extraction

In Fig. 1 we present the real part of the dominant $l = m = 2$ mode of Ψ_4, extracted at $r_{ex} = 40M$. We should stress that this is not an ordinary spacetime. As such, quantities such as Ψ_4 do *not* have their usual physical meaning, though for large enough distances they should. The gravitational wave emitted during the inspiral and plunge of the BH binary is reflected back at the boundary, interacts with the final BH and travels outwards again which can be seen in the second and third cycle in Fig. 1.

One of the most important issues regards the well-posedness of this problem. It is not known whether the BSSN evolution system in combination with a reflecting boundary conditions imposes a well-posed initial value boundary problem. We thus perform a convergence analysis using the 22-mode of Ψ_4. The inspiral is simulated at three different resolutions $h_c = 1/48M$, $h_m = 1/52$ and $h_f = 1/56$. The difference between the mid and fine resolution waveform has been scaled by the factor $Q = 1.47$ and $Q = 1.26$ which indicate fourth and second order convergence, respectively (see Fig. 1). We observe fourth order convergence only in the signal emitted throughout the merger. In the first and second after-merger pulse we still obtain second order convergence whereas we start to loose it from the third reflection on. Thus, one important implication of these results is that one *can* extract reliable information from these simulations, as they do converge at least during the first reflections.

We now consider information obtained from the first three cycles. In is apparent from the waveform itself (left Panel in Fig. 1) that there is a broadening of the merger pulse upon each interaction with the hole. A possible explanation relies on superradiant amplification of the low-frequency part of the waveform, though more work is necessary to pinpoint the reason for this. That the pulse contains frequencies both in the superradiant regime and outside of it is clearly

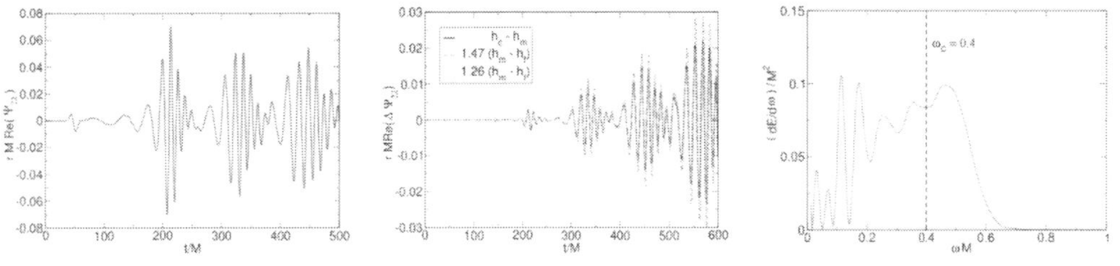

Figure 1. Left: the dominant $l = m = 2$ mode of Ψ_4. Center: convergence analysis. Right: energy spectrum of first emitted pulse.

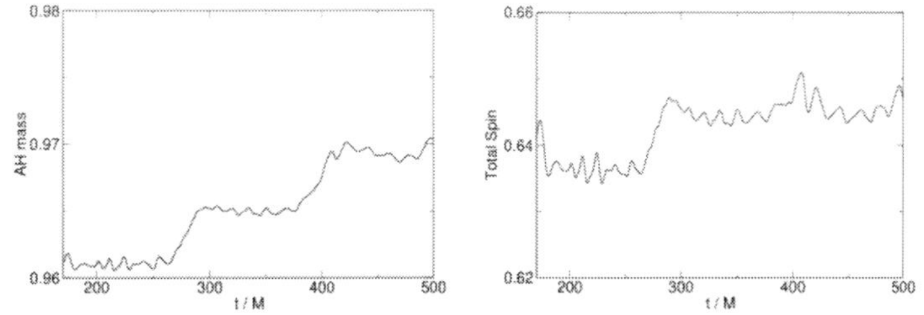

Figure 2. Time evolution of the mass (left) and total spin (right) of the final BH after the merger of the inspiralling BH binary.

seen in the right panel of Fig. 1, where we show the spectrum of the signal emitted throughout the merger. We estimate the critical frequency for superradiance to be around $M\omega_c \sim 0.4$.

3.2. AH data

Although we can get rough estimates of radiated quantities by considering the Newman Penrose scalar Ψ_4 these results should be taken with a grain of salt. Due to our setup we extract the gravitational wave signal only at $r_{ex} = 40M$ which is not yet in the "wave-zone" where the requirement of asymptotically flatness and extraction at spatial infinity is approximately satisfied. Therefore we now turn to look at the AH properties of the final BH, which is a locally defined quantity. The time evolution of AH mass (assuming a Kerr black hole as final state) is depicted in Fig. 2.

The mass of the AH increases upon each interaction with the gravitational radiation pulse. We estimate that in the course of the first and second interaction about $\approx 15\%$ of the incident pulse is absorbed by the BH. In Fig. 2 we also show the time evolution of the total spin J of the rotating BH, obtained assuming a Kerr BH. Not shown here is the dimensionless spin parameter, which we estimate to be around $J/M^2 \sim 0.69$ at the time of formation of the final black hole (therefore in good agreement with simulations of the analogous setup using outgoing boundary conditions [21]). In the course of the first interaction between the spinning BH and the ingoing gravitational wave angular momentum is absorbed. However, during the second interaction the spin of the central BH remains approximately constant. This process can be explained if we consider two competing phenomena, namely absorption of energy and angular momentum of high frequency modes of the radiation by the BH on the one hand and amplification of low frequency modes due to the superradiance effect on the other hand. This behaviour is in qualitative agreement with linearized studies [22], but clearly deserves a more thorough investigation.

4. Conclusions and Outlook

The dynamics of BHs in generic spacetimes is a fascinating, yet terribly complex problem. By considering a BH binary in a "boxed" spacetime we have performed the first successful steps towards a full numerical simulation of BHs in AdS spacetimes. We have studied an initial configuration corresponding to the inspiral of an equal-mass, non-spinning binary yielding a spinning final BH, plus gravitational radiation.

Our results are consistent with expectations, namely a pulse of radiation travelling back and forth between the wall and the black hole, part being absorbed and part being amplified by the BH. Our results show some evidence for the absorption of energy and angular momentum during each interaction with the BH, but they are also consistent with parts (presumably the low-frequency part) being amplified. In future work we plan to investigate this stability studies further by considering a highly spinning, final BH produced by the inspiral of spinning BHs.

Perhaps the most important conclusion of the present work is that these simulations *can* be done and represent the first step to a full numerical evolution of BHs in AdS spacetimes.

Acknowledgments

The authors wish to thank David Hilditch for useful suggestions and discussions. We also thank the participants of the V Iberian Cosmology Meeting, the XII Marcel Grossman Meetings and the Spanish Relativity Meeting 2009 for useful feedback. V.C acknowledges financial support from Fundação Calouste Gulbenkian through a short-term scholarship. V.C and C.H. are supported by a "Ciência 2007" research contract. H.W.'s work was supported by FCT-Portugal through grant SFRH/BD/46061/2008. M.Z. is funded by FCT through grant SFRH/BD/43558/2008. A.N. is funded by FCT through grant SFRH/BPD/47955/2008. This work was partially supported by FCT - Portugal through projects PTDC/FIS/64175/2006, PTDC/FIS/098025/2008, PTDC/FIS/098032/2008, CERN/FP/109306/2009, CERN/FP/109290/2009, NSF grants PHY-090003, PHY-0900735 and the Fairchild foundation to Caltech. Computations were performed at Magerit in Madrid, part of the Barcelona Supercomputing Center.

References

[1] Mateos D 2007 *Class. Quantum Grav.* **24** S713
[2] Arkani-Hamed N, Dimopoulos S and Dvali G R 1998 *Phys. Lett.* B **429** 263
[3] Randall L and Sundrum R 1999 *Phys. Rev. Lett.* **83** 4690
[4] Wald R M 1980 *J. Math. Phys.* **21** 2802
[5] Ishibashi A and Wald R M 2003 *Class. Quantum Grav.* **20** 3815
[6] Ishibashi A and Wald R M 2004 *Class. Quantum Grav.* **21** 2981
[7] Yoshino H and Shibata M 2009 *Phys. Rev.* D **80** 084025
[8] Zilhão M, Witek H, Sperhake U, Cardoso V, Gualtieri L, Herdeiro C and Nerozzi A *Preprint* arXiv:1001.2302
[9] Press W H and Teukolsky S A 1972 *Nature* **238** 211
[10] Cardoso V and Dias O J C 2004 *Phys. Rev.* D **70** 084011
[11] Cardoso V, Dias O J C, Lemos J P S and Yoshida S 2004 *Phys. Rev.* D **70** 044039
[12] Sperhake U 2007 *Phys. Rev.* D **76** 104015
[13] Brandt S and Bruegmann B 1997 *Phys. Rev. Lett.* **78** 3606
[14] Ansorg M, Bruegmann B and Tichy W 2004 *Phys. Rev.* D **70** 064011
[15] Baumgarte Th W and Shapiro S L 1999 *Phys. Rev.* D **59** 024007
[16] Shibata M and Nakamura T 1995 *Phys. Rev.* D **52** 5428
[17] Campanelli M, Lousto C O, Marronetti P and Zlochower Y 2006 *Phys. Rev. Lett.* **96** 111101
[18] Baker J G, Centrella J, Choi D I, Koppitz M and van Meter J 2006 *Phys. Rev. Lett.* **96** 111102
[19] Thornburg J 2004 *Class. Quantum Grav.* **21** 743
[20] Thornburg J 1996 *Phys. Rev.* D **54** 4899
[21] Berti E, Cardoso V, Gonzalez J A, Sperhake U, Hannam M, Husa S and Bruegmann B 2007 *Phys. Rev.* D **76** 064034
[22] Teukolsky S A and Press W H 1974 *Astrophys. J.* **193** 443

Maximally inhomogeneous Gödel-Farnsworth-Kerr generalizations

Lode Wylleman
Faculty of Applied Sciences IR16, UGent (Ghent University), Galglaan 2, 9000 Gent, Belgium
E-mail: lode.wylleman@ugent.be

Abstract. It is pointed out that physically meaningful aligned Petrov type D perfect fluid space-times with constant zero-order Riemann invariants are either the homogeneous solutions found by Gödel (isotropic case) and Farnsworth and Kerr (anisotropic case), or new inhomogeneous generalizations of these with non-constant rotation. We present the construction of the line element and the local geometric properties for the latter.

1. Introduction

The discovery by Milson and Pelavas [1, 2] of a Petrov type N pure radiation space-time family for which the theoretically maximal Karlhede bound $q = 7$ is reached, reopened the question whether the upper bounds for given Weyl-Petrov and/or Ricci-Segre type are sharp as well. In any case, a *necessary* condition is that the space-time is curvature homogeneous of order zero (further denoted by CH_0), i.e., its zero-order Cartan-Riemann invariants are all constant. It is well known that ample families of pure radiation metrics satisfy this property [1]. Here we revise the situation for electrovac fields and their Einstein space limits, and present a theorem classifying all CH_0 genuine (i.e. non-Einstein space) perfect fluids, in the case where the Weyl tensor is of aligned Petrov type D. We will focus on the physically relevant models, which turn out to be exhausted by the celebrated homogeneous Gödel universe, its anisotropic generalizations found by Farnsworth and Kerr, and a new family constituted by generalizations of these known solutions, the members of which have *non-constant* rotation. We mention the construction of good coordinates, summarize the local properties and end with concrete and more general conclusions of this investigation.

2. Inhomogeneous CH_0perfect fluid solutions of Petrov type D

In general, CH_0 Petrov type D space-times are characterized by the existence of a Weyl principal complex null frame $(k^a, l^a, m^a, \overline{m}^a)$ relative to which

$$\Psi_0 = \Psi_1 = \Psi_3 = \Psi_4 = 0, \quad \Psi_2 = const \neq 0. \tag{1}$$

k^a and l^a spanning the Weyl principal null directions (PND's).

All double aligned Petrov type D, non-null Einstein-Maxwell 'electrovac' fields have been classified and integrated by Debever, Kamran and McLenaghan [3, 4] and independently by

[1] I thank Michael Bradley for reminding me of this after the talk

Garcia [5]. From their results, or from the invariant approaches in [6, 7] (see also [8] for a more recent account in the vacuum case, making use of the GHP formalism) one readily infers that the *only* CH_0 solutions in this class are given by

$$\begin{aligned} ds^2 &= \frac{dx^2}{P(x)} + P(x)d\phi^2 + \frac{dy^2}{Q(y)} - Q(y)d\psi^2, \\ P(x) &= 1-(\Lambda+\Phi_0)x^2, \quad Q(y) = 1-(\Lambda-\Phi_0)y^2. \end{aligned}$$

They are gravito-electric ($\Psi_2 = -\Lambda/3$), homogeneous, have a complete group G_6 of isometries and are attributed to Levi-Civita [9], Robinson [10] and Bertotti [11], the last author giving the more general case with $\Lambda \neq 0$. Putting the electromagnetic field parameter Φ_0 equal to zero one gets all CH_0 Einstein spaces (notice that $\Lambda = 0$ gives the Petrov type O Minkowski space-time), having the same isometry group.

A Petrov type D space-time represents an *aligned perfect fluid* if its Einstein tensor has the structure

$$G_{ab} = (w+p)u_a u_b + p g_{ab}, \quad w+p \neq 0 \quad u^a = \frac{qk^a + l^a}{\sqrt{2q}} \tag{2}$$

where $q > 0$, $w = w' + \Lambda$ and $p = p' + \Lambda$ are the (geometric or effective) energy density, resp. pressure of the fluid (in which the cosmological constant Λ has been absorbed) and u^a is the fluid 4-velocity, lying in the PND-plane. Space-times of this nature are CH_0 if and only if the conditions (1) hold and w and p are constant. Then, by energy-momentum conservation and $w+p \neq 0$, u^a is non-accelerating and non-expanding. The *only explicitly known* CH_0 examples are given by

$$\begin{aligned} A^2 ds^2 = \quad & du^2 - 2(dt + e^x dy)^2 + \frac{e^v}{\cosh v}(\cos t\, dx + \sin t\, e^x dy)^2 \\ & + \frac{e^{-v}}{\cosh v}(-\sin t\, dx + \cos t\, e^x dy)^2 \end{aligned} \tag{3}$$

where A and v are constant. For $v = 0$ one obtains the G_5 shearfree homogeneous Gödel universe [12], and for $v \neq 0$ the anisotropic and shearing, yet still homogeneous generalizations discovered by Farnsworth and Kerr [13]. Let us list the local geometric properties of these space-times:

(A) the vector field $\mathbf{v} \equiv \frac{q\mathbf{k}-\mathbf{l}}{\sqrt{2q}} = \frac{\partial}{\partial u}$ is covariantly constant;

(B) the vorticity and shear of u^a are given by

$$\omega^a = \omega\, v^a, \quad \omega = a \cosh v \, (= \mathrm{i}\sqrt{2q}\,\rho = \mathrm{i}\sqrt{2/q}\,\mu), \tag{4}$$

$$\sigma_{ab} = V\overline{m}_{(a}\overline{m}_{b)} + \overline{V}m_{(a}m_{b)}, \quad V = a \sinh v\, e^{\mathrm{i}\varphi_V} \,(= -\sqrt{2q}\,\sigma = \sqrt{2/q}\,\overline{\lambda}), \tag{5}$$

where the relation with the NP spin coefficients ρ, μ, σ and λ has been added, and where $e^{\mathrm{i}\varphi_V}$ is a spin gauge field;

(C) the space-time represents gravito-electric 'stiff dust', satisfying

$$A^2 \equiv p = w = -3\Psi_2 = \omega_a\omega^a - \sigma_{ab}\sigma^{ab} = \omega^2 - V\overline{V}. \tag{6}$$

Notice that perfect fluids with $p = w$ satisfy the dominant energy condition only when $\Lambda \leq 0$ and, when w is moreover constant, may be interpreted as e.g. dust space-times ($\Lambda = -w$) or stiff fluids ($\Lambda = 0$).

The question arises whether the Gödel-Farnsworth-Kerr space-times are the only CH_0 aligned Petrov type D perfect fluids. In contrast to the Einstein space case mentioned above, however, homogeneity of the curvature invariants does not imply homogeneity of the space-time here. One can deduce the following classification result [14]:

Theorem. Any CH_0 aligned Petrov type D perfect fluid satisfies property (C). It is either an unphysical member $w = p = 3\Psi_2 < 0$ of Ellis' LRS II non-rotating dust family [15] or the Stephani [16]-Barnes [17] rotating dust family, or it satisfies properties (A) and (B) as well.

Let us focus on the physically relevant class of space-times satisfying (A)-(C). When ω is constant one recovers the homogeneous Gödel-Farnsworth-Kerr solutions; when ω is non-constant, however, new inhomogeneous solutions arise as follows (see [18] for more details). At each point the variables ω, V and $\overline{V}$ are constrained by the hyperbolic equation $\omega^2 - V\overline{V} = A^2$. Thus they can be parametrized as in (4)-(5), but this does not give suitable coordinates. A better choice is the parametrization

$$\omega = A\frac{\cosh x}{\cos y}, \qquad V = A\frac{\sin y + \mathrm{i}\sinh x}{\cos y}.$$

Also, the Jacobi identities allow to set the NP spin coefficients α, β, γ and ϵ to zero. On rectifying the vector field $\mathbf{u} = a\,\partial_t$, the integration of the Cartan equations is quite straightforward. On using x, y, t and u as coordinates, where $\mathbf{v} = \frac{\partial}{\partial u}$, one obtains

$$\begin{aligned} A^2\mathrm{d}s^2 = \quad & \mathrm{d}u^2 - \left(\mathrm{d}t + \frac{1}{2}\frac{\partial F(x,y)}{\partial x}\mathrm{d}y - \frac{1}{2}\frac{\partial F(x,y)}{\partial y}\mathrm{d}x\right)^2 \\ & + e^{F(x,y)}e^{x}\left(\cos\left(t - \frac{y}{2}\right)\mathrm{d}x + \sin\left(t - \frac{y}{2}\right)\mathrm{d}y\right)^2 \\ & + e^{F(x,y)}e^{-x}\left(-\sin\left(t + \frac{y}{2}\right)\mathrm{d}x + \cos\left(t + \frac{y}{2}\right)\mathrm{d}y\right)^2, \end{aligned} \tag{7}$$

$F(x, y)$ being a solution of

$$\frac{\partial^2 F(x,y)}{\partial x^2} + \frac{\partial^2 F(x,y)}{\partial y^2} = 4\cosh(x)e^{F(x,y)}. \tag{8}$$

We emphasize that this is the general line element for inhomogeneous aligned Petrov type D perfect fluids with positive effective energy density w.

Finally, denote t_k for the number of functionally independent components of the Riemann tensor and its first k covariant derivatives, relative to the canonically fixed frame at step k in the Karlhede space-time classification algorithm [19], and let q be the Karlhede bound. As there is at least one Killing vector field $\frac{\partial}{\partial u}$ one has $t_q \leq 3$. The CH_0 assumption means precisely $t_0 = 0$, and by direct calculation one finds that $t_1 = 1$ and $2 \leq t_2$, whence $2 \leq t_2 \leq t_3 \leq 3$. Further investigation then shows that $t_2 = t_3$, such that $q = 3$ for any member of the new family. The generic situation is $t_2 = 3$, in which case $\frac{\partial}{\partial u}$ is the only Killing vector field; only a very specific subfamily satisfies $t_2 = 2$, in which case the complete isometry group is Abelian G_2.

3. Conclusions

For physically relevant CH_0 aligned Petrov type D perfect fluids the constant energy density w equals the pressure p on the one hand, and the difference between the vorticity and shear amplitudes on the other. In contrast to the homogeneous Gödel-Farnsworth-Kerr models, these amplitudes can be non-constant, resulting in the non-constancy of higher-order curvature invariants and, correspondingly, a dramatic drop of the isometry group dimension of the relevant 3D part of the metric. A special choice of (non-invariantly defined) coordinates leads to the metric (7)-(8).

Put in a much broader context, the results exemplify a *generation technique*, valid for any metric-based gravitation theory in any space-time dimension: investigate whether a set of (possibly higher-order) invariant relations, valid for a well-known family of space-times, singles out this family. If not, new interesting families may arise.

References

[1] Milson R and Pelavas N 2008 *Class. Quantum Grav.* **25** 012001
[2] Milson R and Pelavas 2009 *Int. J. Geom. Meth. Mod. Phys.* **6** 99
[3] Debever R Kamran N and McLenaghan R G 1982 *Bull. Acad. Roy. Belg. Cl. Sci.* **68** 592
[4] Debever R Kamran N and McLenaghan R G 1984 *J. Math. Phys.* **25** 1955
[5] García D A 1984 *J. Math. Phys.* **25** 1951
[6] Debever R and McLenaghan R G 1981 *J. Math. Phys.* **22** 1711
[7] Czapor S R and McLenaghan R G 1982 *J. Math. Phys.* **23** 2159
[8] Edgar S B, Garcia-Parrado Gómez-Lobo A and Martin-Garcia J M 2009 *Class. Quantum Grav.* **26** 105022
[9] Levi-Civita T 1917 *Rend. R. Acad. Lincei, Cl. Sci. Fis. Mat. Nat.* **26** 519
[10] Robinson I 1959 *Bull. Acad. Pol. Sci. Ser. Sci. Math. Astron. Phys.* **7** 351
[11] Bertotti B 1959 *Phys. Rev.* **116** 1331
[12] Gödel K 1949 *Rev. Mod. Phys.* **21** 447
[13] Farnsworth D L and Kerr R P 1966 *J. Math. Phys.* **7** 1625
[14] Wylleman L Karlhede classification of aligned Petrov type *D* curvature homogeneous perfect fluids (in preparartion)
[15] Ellis G F R 1967 *J. Math. Phys.* **8** 1171
[16] Stephani H 1987 *Class. Quantum Grav.* **4** 125
[17] Barnes B 1999 *Class. Quantum Grav.* **16** 919
[18] Wylleman L Inhomogeneous perfect fluid generalizations of the Gödel universe (in preparation)
[19] Karlhede A 1980 *Gen. Rel. Grav.* **12** 693

Spanish Relativity Meeting (ERE 2009) IOP Publishing
Journal of Physics: Conference Series **229** (2010) 012074 doi:10.1088/1742-6596/229/1/012074

Numerical relativity in higher dimensions

Miguel Zilhão[1⋆], Helvi Witek[2], Ulrich Sperhake[3], Vitor Cardoso[2,4], Leonardo Gualtieri[5], Carlos Herdeiro[1] and Andrea Nerozzi[2]

[1] Departamento de Física e Centro de Física do Porto, Faculdade de Ciências da Universidade do Porto, Rua do Campo Alegre 687, 4169-007 Porto, Portugal
[2] Centro Multidisciplinar de Astrofísica - CENTRA, Departamento de Física, Instituto Superior Técnico - IST, Av. Rovisco Pais 1, 1049-001 Lisboa, Portugal
[3] California Institute of Technology, Pasadena, CA 91109, USA
[4] Department of Physics and Astronomy, The University of Mississippi, University, MS 38677-1848, USA
[5] Dipartimento di Fisica, Università di Roma "Sapienza" & Sezione, INFN Roma1, P.A. Moro 5, 00185, Roma, Italy

E-mail: mzilhao@fc.up.pt, helvi.witek@ist.utl.pt, vitor.cardoso@ist.utl.pt, Leonardo.Gualtieri@roma1.infn.it, crherdei@fc.up.pt, andrea.nerozzi@ist.utl.pt, sperhake@tapir.caltech.edu

Abstract. We give a status report on our project targeted at performing numerical simulations of a head-on collision of non-spinning black holes in higher dimensional non-compact space-times. These simulations should help us understand black objects in higher dimensions and their stability properties. They are also relevant for the problem of black hole formation and evaporation in particle accelerators and cosmic rays. We use the symmetries of the system to reduce the problem to an effective 3+1 problem, allowing the use of existing numerical codes. As a simple application of the formalism, we present the results for the evolution of a five dimensional single black hole space-time.

1. Introduction

Modern theories of supergravity and strings naturally encompass extra dimensions, and as such, the study and understanding of higher dimensional space-times plays here an important role. To study highly dynamical and non-linear phenomena in gravity, numerical relativity is the only tool available. For four dimensional general relativity, the past decade has seen remarkable achievements since the 2005 breakthroughs [1–3]. The higher dimensional case, however, is mostly unexplored, and only more recently some pioneering works started to appear [4–7]. There are a lot of interesting problems to explore in this field, and we give a few examples in the following.

If the scale of quantum gravity is near the TeV, ultra-relativistic scattering of point particles should be rather insensitive to the structure of the particles, and as such these type of phenomena may be described by black hole scattering [8]. Furthermore, there is now the exciting possibility of seeing black hole production at the LHC [9]. These black holes would quickly evaporate by Hawking radiation, emitting high-energy particles which may be detected. Thus, studies of gravitational radiation are extremely important, and numerical relativity plays here a major role.

Journal of Physics: Conference Series **229** (2010) 012074 doi:10.1088/1742-6596/229/1/012074

Another related and interesting application of black hole physics to high-energy particle systems is suggested by the AdS/CFT conjecture [10], with the possibility of using this duality to understand properties of the quark-gluon plasma at the RHIC by studying black hole collisions in AdS [11]. Numerical relativity in AdS space-times is difficult, and has so far only been studied in very special cases [12, 13] (see also Helvi Witek's contribution).

We here summarise a framework introduced in [14] that allows us to study D dimensional axially symmetric space-times. Symmetry has been a crucial ingredient in solving the Einstein field equations numerically ever since the early attempts of Hahn and Lindquist [15], Eppley [16] and Smarr [17]. Our formalism explores the symmetry of higher dimensional head-on collisions of black holes in order to reduce the space-time dimensionality, in the spirit of the Kaluza-Klein reduction, thereby reducing the problem to an effective four dimensional one with "matter" terms.

2. An effective $3+1$ dimensional system

To make use of the existing numerical codes, we are interested in making a $4+(D-4)$ split of the D dimensional space-time. We start by noting that the most general D dimensional metric compatible with $SO(D-3)$ isometry, in $D \geq 6$, is

$$d\bar{s}^2 = g_{\mu\nu}dx^\mu dx^\nu + \lambda(x^\mu)d\Omega_{D-4}\,, \tag{2.1}$$

where $\mu,\nu = 0,\ldots,3$ and $d\Omega_{D-4}$ is the line element on a unit S^{D-4} sphere. For such a metric element, the D dimensional Einstein vacuum equations yield the following system

$$R_{\mu\nu} = \frac{D-4}{2\lambda}\left(\nabla_\mu\partial_\nu\lambda - \frac{1}{2\lambda}\partial_\mu\lambda\partial_\nu\lambda\right), \tag{2.2}$$

$$\nabla^\mu\partial_\mu\lambda = 2(D-5) - \frac{D-6}{2\lambda}\partial_\mu\lambda\partial^\mu\lambda\,. \tag{2.3}$$

In these equations, all operators are covariant with respect to the four dimensional metric $g_{\mu\nu}$. The energy-momentum tensor is[1]

$$T_{\mu\nu} = \frac{D-4}{16\pi\lambda}\left[\nabla_\mu\partial_\nu\lambda - \frac{1}{2\lambda}\partial_\mu\lambda\partial_\nu\lambda - (D-5)g_{\mu\nu} + \frac{D-5}{4\lambda}g_{\mu\nu}\partial_\alpha\lambda\partial^\alpha\lambda\right]. \tag{2.4}$$

We can thus view our system as an effective four dimensional problem with a source given by the above energy-momentum tensor, a sort of fake "matter"[2]. With this four dimensional perspective, we introduce the metric γ_{ij} and extrinsic curvature K_{ij} of the three dimensional hypersurface Σ, and perform the usual $3+1$ splitting of the space-time (see, e.g [18, 19]).

The resulting system is

$$(\partial_t - \mathcal{L}_\beta)\,\gamma_{ij} = -2\alpha K_{ij}\,, \tag{2.5a}$$

$$\begin{aligned}(\partial_t - \mathcal{L}_\beta)\,K_{ij} = {} & -D_i\partial_j\alpha + \alpha\left({}^{(3)}R_{ij} + KK_{ij} - 2K_{ik}K^k{}_j\right)\\ & -\alpha\frac{D-4}{2\lambda}\left(D_i\partial_j\lambda - 2K_{ij}K_\lambda - \frac{1}{2\lambda}\partial_i\lambda\partial_j\lambda\right),\end{aligned} \tag{2.5b}$$

$$(\partial_t - \mathcal{L}_\beta)\,\lambda = -2\alpha K_\lambda\,, \tag{2.5c}$$

$$\frac{1}{\alpha}(\partial_t - \mathcal{L}_\beta)\,K_\lambda = -\frac{1}{2\alpha}D^k\lambda D_k\alpha + (D-5) + KK_\lambda + \frac{D-6}{\lambda}K_\lambda^2 - \frac{D-6}{4\lambda}D^k\lambda D_k\lambda - \frac{1}{2}D^k\partial_k\lambda\,, \tag{2.5d}$$

where D_i is the covariant derivative with respect to the three metric γ_{ij}.

[1] We use the standard form of the Einstein equations $G_{\mu\nu} = 8\pi T_{\mu\nu}$ and choose geometrised units throughout.

[2] We dub the source terms of the lower dimensional Einstein equations as matter although its energy-momentum tensor is not that of canonical matter.

3. Initial data

For the initial data, we follow the approach outlined by Yoshino *et al* [20, 21]. For Brill-Linquist initial data of two black holes with mass parameters μ_1 and μ_2 in D dimensions, the $(D-1)$ dimensional spatial metric element has the form

$$\gamma_{ab}dx^a dx^b = \psi^{\frac{4}{D-3}}\left(dz^2 + d\rho^2 + \rho^2 d\theta^2\right) + \psi^{\frac{4}{D-3}}\rho^2 \sin^2\theta^2 d\Omega^2_{D-4}\,, \tag{3.1}$$

$$\psi = 1 + \frac{\mu_1^2}{4\left[(z-z_1)^2+\rho^2\right]^{\frac{D-3}{2}}} + \frac{\mu_2^2}{4\left[(z-z_2)^2+\rho^2\right]^{\frac{D-3}{2}}}\,, \tag{3.2}$$

where $a, b = 1, \ldots, D-1$, $\theta \in [0, \pi]$, $-\infty < z < +\infty$ and $\rho \in [0, \infty[$. Introducing "incomplete" cartesian coordinates as

$$x = \rho\cos\theta, \qquad y = \rho\sin\theta\,, \tag{3.3}$$

where $-\infty < x < +\infty$ and $0 \leq y < +\infty$, we can then write the initial data as

$$\gamma_{ij}dx^i dx^j = \psi^{\frac{4}{D-3}}\left[dx^2 + dy^2 + dz^2\right]\,, \tag{3.4}$$

and

$$\lambda = y^2\psi^{\frac{4}{D-3}}\,, \tag{3.5}$$

$$K_{ij} = 0\,, \tag{3.6}$$

$$K_\lambda = 0\,, \tag{3.7}$$

which close the system (2.5).

4. Numerical results in $D = 5$

For the numerical implementation, we write the evolution equations in the Baumgarte, Shapiro, Shibata and Nakamura (BSSN) formulation [22, 23].

We have performed numerical simulations in $D = 5$ for the evolution of a single black hole. The evolution is stable and the constraints are preserved in this time evolution. In Fig. 1 we exhibit the Hamiltonian constraint and the y component of the momentum constraint for evolution time $t = 28\mu$. One can see that there is some noise, but the overall convergence is still very good.

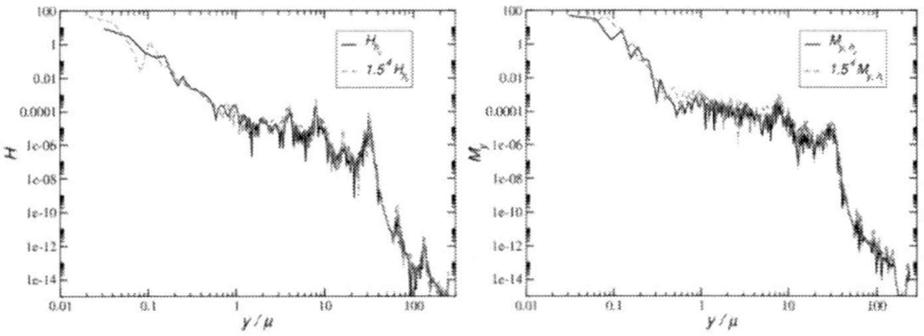

Figure 1. Hamiltonian constraint (left panel) and y-component of the momentum constraint (right panel) at time $t = 28\mu$, for the evolution of a single Tangherlini black hole in five dimensions.

5. Final remarks

The formalism we have introduced allows us to reduce the head-on collision of non-spinning black holes in any dimension to an effective $3+1$ system with source terms. This enables us to study higher dimensional systems (with enough symmetry) using the existing numerical codes. A key advantage of our approach is that the very same code should work for *any* dimension.

We are currently implementing and testing the formalism using the LEAN code [24], developed by one of us (U.S.), and have already found good numerical results for the evolution of a single black hole in $D = 5$. For more details see [14].

Acknowledgments

M.Z. and H.W. are funded by FCT through grants SFRH/BD/43558/2008 and SFRH/BD/46061/2008. V.C. acknowledges financial support from Fundação Calouste Gulbenkian through a short-term scholarship. V.C. and C.H. are supported by a "Ciência 2007" research contract. A.N. is funded by FCT through grant SFRH/BPD/47955/2008. This work was partially supported by FCT - Portugal through projects PTDC/FIS/64175/2006, PTDC/FIS/098025/2008, PTDC/FIS/098032/2008, CERN/FP/109306/2009, CERN/FP/109290/2009, NSF grants PHY-090003, PHY-0900735 and the Fairchild foundation to Caltech. Computations were performed at TeraGrid in Texas and Magerit in Madrid. The authors thankfully acknowledge the computer resources, technical expertise and assistance provided by the Barcelona Supercomputing Center - Centro Nacional de Supercomputación.

References

[1] Pretorius F 2005 *Phys. Rev. Lett.* **95** 121101
[2] Campanelli M, Lousto C O, Marronetti P and Zlochower Y 2006 *Phys. Rev. Lett.* **96** 111101
[3] Baker J G, Centrella J, Choi D I, Koppitz M and van Meter J 2006 *Phys. Rev. Lett.* **96** 111102
[4] Choptuik M, Lehner L, Olabarrieta I, Petryk R, Pretorius F and Villegas H 2003 *Phys. Rev.* D **68** 044001
[5] Sorkin E and Choptuik M W 2009 *Preprint* 0908.2500
[6] Sorkin E 2009 *Preprint* 0911.2011
[7] Yoshino H and Shibata M 2009 *Phys. Rev.* D **80** 084025
[8] Banks T and Fischler W 1999 *Preprint* hep-th/9906038
[9] Dimopoulos S and Landsberg G L 2001 *Phys. Rev. Lett.* **87** 161602
[10] Maldacena J M 1998 *Adv. Theor. Math. Phys.* **2** 231-252
[11] Nastase H 2005 *Preprint* hep-th/0501068
[12] Pretorius F and Choptuik M W 2000 *Phys. Rev.* D **62** 124012
[13] Witek H, Cardoso V, Gualtieri L, Herdeiro C, Nerozzi A, Sperhake U and Zilhão M 2010 *to appear*
[14] Zilhão M, Witek H, Sperhake U, Cardoso V, Gualtieri L, Herdeiro C and Nerozzi A 2010 *Preprint* 1001.2302
[15] Hahn S G and Lindquist R W 1964 *Annals of Physics* **29** 304-331
[16] Eppley K R 1975 *The numerical evolution of the collision of two black holes* Ph.D. thesis AA(Princeton Univ., NJ.)
[17] Smarr L 1977 *Eighth Texas Symposium on Relativistic Astrophysics* (*New York Academy Sciences Annals* vol 302) ed M D Papagiannis pp 569–+
[18] Gourgoulhon E 2007 *Preprint* gr-qc/0703035
[19] Alcubierre M 2008 *Introduction to 3+1 numerical relativity* International series of monographs on physics (Oxford: Oxford Univ. Press)
[20] Yoshino H, Shiromizu T and Shibata M 2005 *Phys. Rev.* D **72** 084020
[21] Yoshino H, Shiromizu T and Shibata M 2006 *Phys. Rev.* D **74** 124022
[22] Baumgarte T W and Shapiro S L 1999 *Phys. Rev.* D **59** 024007
[23] Shibata M and Nakamura T 1995 *Phys. Rev.* D **52** 5428-5444
[24] Sperhake U 2007 *Phys. Rev.* D **76** 104015

The relativistic Brownian motion: interdisciplinary applications

A Aragonés-Muñoz*, **A Sandoval-Villalbazo**

Departamento de Física y Matemáticas, Universidad Iberoamericana, Prolongación Paseo de la Reforma 880, México D. F. 01219, México

Abstract. Relativistic Brownian motion theory will be applied to the study of analogies between physical and economic systems, emphasizing limiting cases in which Gaussian distributions are no longer valid. The characteristic temperatures of the particles will be associated with the concept of variance, and this will allow us to choose whether the pertinent distribution is classical or relativistic, while working specific situations. The properties of particles can be interpreted as economic variables, in order to study the behavior of markets in terms of Levy financial processes, since markets behave as stochastic systems. As far as we know, the application of the Jüttner distribution to the study of economic systems is a new idea.

1. Introduction

Brownian motion corresponds to a random displacement of matter in a thermal bath, produced by collisions between particles. In non-relativistic physics, a Gaussian distribution is used to describe the velocities of particles. In other contexts, the use of classical functions to describe fluctuations is based on the Central Limit Theorem (CLT). In relativistic systems, significant deviations are observed with respect to Gaussian distributions. A Lèvy process, used a lot in economics and finance, corresponds to a continuous-time stochastic process that starts at $t = 0$ and admits a CLT. The most common Lèvy processes are the Wiener process and the Poisson process. As we can see, in physics and in economics and finance the use of stochastic processes is very common, and it is important to mention that in a stochastic process exists time dependence and randomness[1].

Econophysics is a study area which uses the mathematical methods of physics to develop important knowledge in economics and finance. Specially, models that use classical distribution functions have been developed in this area. For example, the Boltzmann-Gibbs distribution of money[2]. In the core of the curious and very interesting comparison between physics and econophysics models relies the validity of conservation laws. E.g. in a collision there is conservation of momentum and energy; as well as in a money exchange, there is money conservation. What we are trying to show in this paper, is that relativistic Brownian motion, can also be used to develop models in econophysics, so it deserves a deeper look. Using a Jüttner distribution, important results are obtained when compared to its Maxwell-Boltzmann distribution counterparts[3].

2. Comparison of the Mean Square Displacement for the Non-relativistic and the Relativistic Cases

For a massive particle, of mass m, immersed in a relativistic thermal bath, a relativistic distribution function, like the Jüttner distribution, is appropriate to describe the Brownian motion. This is because the particles can't move at velocities higher than the speed of light c[3].

The mean square displacement for the non-relativistic case at temperature T and time t is given by

$$\langle [r(t)]^2 \rangle = \frac{6KT}{\mu} t. \tag{1}$$

In Eq. 1 μ is the friction coefficient and K is the Boltzmann constant. In the relativistic case the mean square displacement is now given by

$$\langle [r(t)]^2 \rangle = \frac{2}{\mu} \left[3KT + mc^2 \frac{K_1\left(\frac{1}{z}\right)}{K_2\left(\frac{1}{z}\right)} \right] t, \tag{2}$$

where $z = KT/mc^2$ and K_i represents the modified Bessel function of type i.

Fig. 1 shows a stochastic process compatible with a Brownian motion. The model consists of random successive steps, moving right or left at each step. The value of the parameter z was 0.1 and 2000 steps were used. The most important observation of this graphic is that for the Jüttner's mean square displacement the fluctuations are smaller than their Maxwell-Boltzmann's counterpart. Exporting this observation to financial systems a conclusion is that Jüttnerian stochastic processes are less uncertain than the Gaussian ones. Markets don't fluctuate infinitely, as is implied by Maxwell-Boltzmann's statistics. As shown in Fig. 2, for large values of z the characteristic speed of the particles of the thermal bath in the non-relativistic case tends to infinity, while the relativistic value tends to be constant.

3. Final Remarks

This study was made for a non-degenerate relativistic single component gas, where the Brownian particles are immersed. The models in physics help to understand the financial markets because they also behave in a random character. A very interesting interaction between relativistic

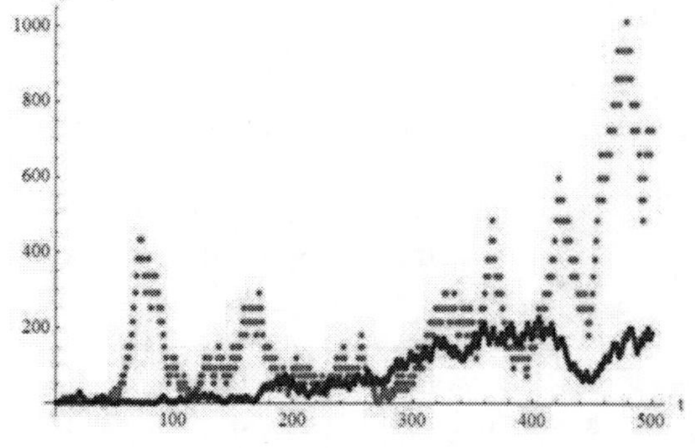

Figure 1. This graphic shows the behavior of the mean square displacement in a one dimensional random walk process in a relativistic (blue) and in the non-relativistic (red) cases. The Gaussian random walk shows larger fluctuations compared to the Jüttnerian case, for the same thermal bath temperatures. In the case of price fluctuations this behavior possesses a potential interest.

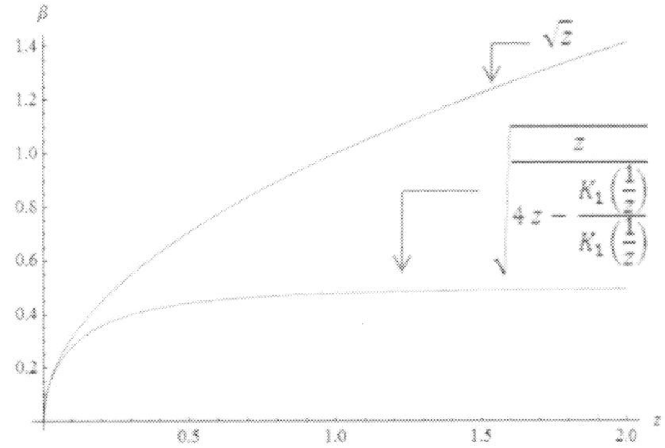

Figure 2. This graphic shows the behavior of the characteristic velocity of the particles of the thermal bath. One can see that for a larger value of z the non-relativistic speed grows to infinity, while the relativistic speed tends to a constant value.

physics and economics and finance is a fact which will help us to understand in a better way the modelling of money and wealth.

References

[1] Shreve S E 2008 *Stochastic Calculus for Finance II. Continuous-Time Models* (Springer) pp 83-116 125-189
[2] Yakovenko V M 2007 Econophysics, Statistical Mechanics Approach to *Preprint* arXiv:0709.3662
[3] Sandoval-Villalbazo A, Aragonés-Muñoz A A and García-Perciante A L 2008 The Simple Non-degenerate Relativistic Gas: Statistical Properties and Brownian Motion *Preprint* arXiv:0802.1023
[4] Stanley H E *et al* 2007 *An Introduction to Econophysics* (Cambridge University Press) pp 14-33
[5] Chacón-Acosta G *et al* 2007 Fokker-Planck-type equations for a simple gas and for a semirelativistic Brownian motion from a relativistic kinetic theory *Phys. Rev.* E **76** 021201

Spanish Relativity Meeting (ERE 2009) IOP Publishing
Journal of Physics: Conference Series **229** (2010) 012076 doi:10.1088/1742-6596/229/1/012076

Geodesic deviation equation in Bianchi cosmologies

D L Cáceres[1]*, L Castañeda[2] and J M Tejeiro

Observatorio Astronómico Nacional, Universidad Nacional de Colombia

E-mail: [1]`dlcaceresu@bt.unal.edu.co`, [2]`lcastanedac@unal.edu.co`

Abstract. We present the Geodesic Deviation Equation (GDE) for the Friedmann Robertson Walker(FRW) universe and we compare it with the equation for Bianchi type I model. We justify consider this cosmological model due to the recent importance the Bianchi Models have as alternative models in cosmology. The main property of these models, solutions of Einstein Field Equations (EFE) is that they are homogeneous as the FRW model but they are not isotropic. We can see this because they have a non-null Weyl tensor, which is zero for FRW model. We study some consequences of this Weyl tensor in the GDE.

1. Introduction

Bianchi cosmological models are homogeneous solutions of EFE. These models generalize FRW model, although they are homogeneous, as FRW one, generally they are anisotropic. Standard Cosmological Model has survived almost all observational tests [1], however Bianchi cosmologies haven't been discarded. Today the Universe is highly isotropic, but in early times it couldn't been so, and it is an open question how anisotropies died. Standard Model fails to explain quadrupole anomaly [2], obtained from Cosmic Microwave Background data. This anomaly have a plausible explanation with Bianchi models, for that reason these models are considered today [2]. Thus we considered the idea of studying Geodesic Deviation Equation(GDE) in some of these models.

2. 1+3 Covariant description

We assume a fluid description for the cosmological model. The 4-velocity u^α of preferred world-lines is normalized, $u_\alpha u^\alpha = -1$. Given this 4-velocity, we define the projection tensor $h_{\alpha\beta}$, the expansion scalar Θ, the shear $\sigma_{\alpha\beta}$ and the vorticity $\omega_{\alpha\beta}$ [3], [4]. In FRW case $\sigma_{\alpha\beta} = \omega_{\alpha\beta} = 0$. These quantities are related with Weyl tensor. In FRW case Weyl tensor is zero, so it is conformally flat, but it is not the case in Bianchi models. In the $1+3$ covariant description we can split the Weyl tensor into an electric $E_{\alpha\beta}$ and a magnetic part $H_{\alpha\beta}$ [4], [3].

3. Deviation Equation

In a space-time torsion free, if V^α is the normalized tangent vector field of a fiducial geodesic, parameterized by an affine parameter ν, with $V_\alpha V^\alpha = \varepsilon$, with $\varepsilon = +1, 0, -1$ if the geodesics are spacelike, null or timelike respectively; if η^α is the deviation vector for the congruence, then the evolution of the separation or the deviation vector η^α evolves according to GDE [5]:

$$\frac{\delta^2 \eta^\alpha}{\delta \nu^2} = -R^\alpha_{\beta\gamma\delta} V^\beta \eta^\gamma V^\delta \tag{1}$$

where $\frac{\delta T^{a...}_{b...}}{\delta \nu} = V^\gamma \nabla_\gamma T^{a...}_{b...}$ is the covariant derivative along the geodesic.

Now, let's see how is this equation in Bianchi type I model, given by the metric[6], [7]:

$$ds^2 = dt^2 - a^2(t)dx^2 - b^2(t)dy^2 - c^2(t)dz^2 \tag{2}$$

If we define $\tau(t) := a(t)b(t)c(t)$, from the energy-momentum tensor conservation we get an analogous expression to a known Friedmann equation [6], [7]:

$$\frac{\ddot{\tau}}{\tau} = \frac{3\kappa}{2}(\mu - p) \tag{3}$$

where μ is the Energy density and p is the pressure. As we see in the GDE, there is a difference if space-time has a nonzero Weyl tensor. Ellis and van Elst in [5] consider this equation for FRW space-time.

For Bianchi type I model vorticity and Magnetic part of Weyl tensor are equal to zero, $\omega_{\alpha\beta} = 0$, $H_{\alpha\beta} = 0$. The Electric and shear tensors are non-null and diagonal. For that reason, if we define $F = E_{\beta\gamma} V^\beta \eta^\gamma$ and $E = -V_\alpha u^\alpha$, if we suppose a perfect fluid, $q^\alpha = 0$, $\pi_{\alpha\beta} = 0$ we have that GDE for this space is:

$$\begin{aligned} R^\alpha_{\beta\gamma\delta} V^\beta \eta^\gamma V^\delta = & [\varepsilon \frac{1}{3}(\mu + \Lambda) + \frac{1}{2}(\mu + p)E^2]\eta^\alpha \\ & + F(2u^\alpha + \eta^\alpha - V^\alpha) + E^\alpha_\gamma \eta^\gamma (\varepsilon + 2E^2) \end{aligned} \tag{4}$$

where Λ is the cosmological constant. The first part of this expression is the force term representing the perfect part of the fluid [4] and only gives isotropic deviation. In FRW case we only have this term and reflects spatial isotropy of space-time, deviation vector only changes in magnitude but not in direction. The other terms change the direction of η^α.

If we suppose an equation of state $p = w\epsilon$, ($w = 0$ for dust, $w = \frac{1}{3}$ for radiation), then, for $\Lambda = 0$ we get[6]:

$$\tau = At^{\frac{2}{1+w}} \tag{5}$$

Given τ as a function of t we can get $a_(t)$, $b(t)$ and $c(t)$. The electric Weyl tensor tends to zero. It is diagonal and its components decay. For this model the vorticity and the Magnetic part of Weyl tensor are null. The EFE imply that $q^\alpha = 0$. This means we have a fluid without viscosity. However, we can have an imperfect fluid because the model allows to have an anisotropic pressure $\pi_{\alpha\beta}$. But this anisotropic pressure tensor must be diagonal, following EFE's. Given τ as a function of t we have expressions for $a(t)$, $b(t)$ and $c(t)$. The evolution of the three scale factors for $w = 0$ is shown in Figure 1 and for $w = \frac{1}{3}$ in figure 2. We can see that in radiation case anisotropy is greater than in matter dominated case, but in both cases anisotropy dies.

The σ scalar defined by $\sigma^2 = \frac{1}{2}\sigma^{\alpha\beta}\sigma_{\alpha\beta}$ also measure the degree of anisotropy in the model. In Figure 4 we illustrate the evolution of σ^2.

Journal of Physics: Conference Series **229** (2010) 012076 doi:10.1088/1742-6596/229/1/012076

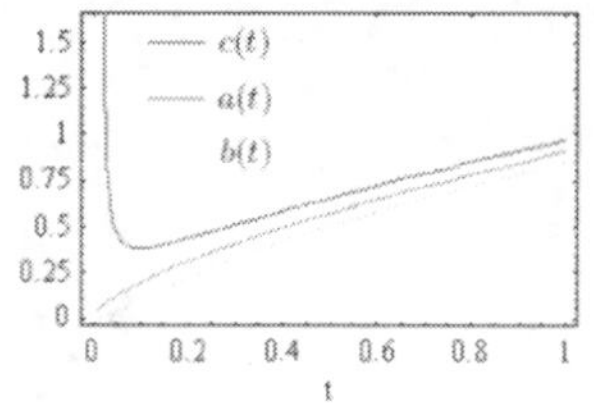

Figure 1. Scale factors for $w = 0$, $A = \frac{3}{4}$.

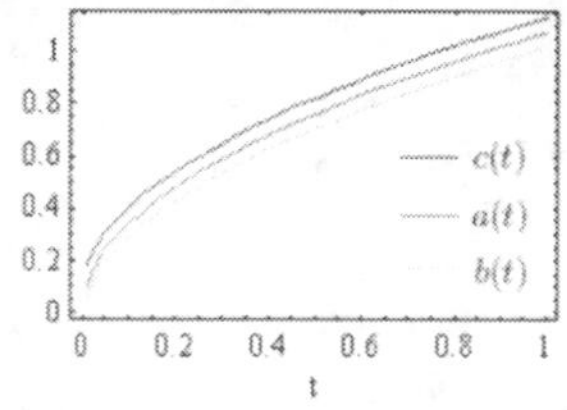

Figure 2. Scale factors for $w = \frac{1}{3}$, $A = \frac{3}{4}$.

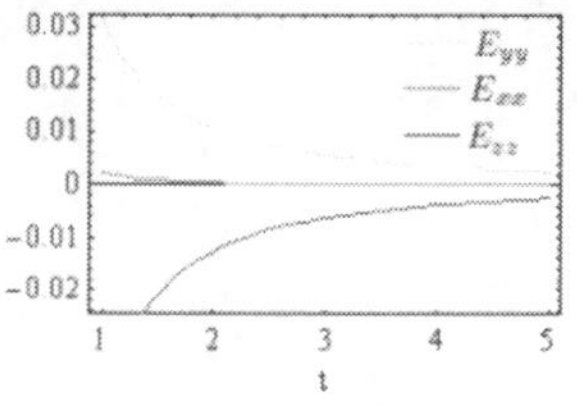

Figure 3. Electric Weyl tensor for $A = \frac{3}{4}$, $w = \frac{1}{3}$.

4. Conclusions

The evolution of the Electric Weyl tensor show us that this model tends to isotropy as t grows, the model isotropizes. Also the evolution of shear scalar. In FRW only the magnitude η will change along a geodesic, while its spatial orientation remain fixed [5]. So, in early times the deviation is anisotropic, the orientation is important, for example, along the z axis deviation will grow, while along x and y axis the effect of Weyl Tensor is to contract the deviation, because there are the other terms that are also present in FRW. Here we considered a perfect fluid, $\pi_{\alpha\beta} = 0$. Along x axis, the effect of anisotropy is minimal.

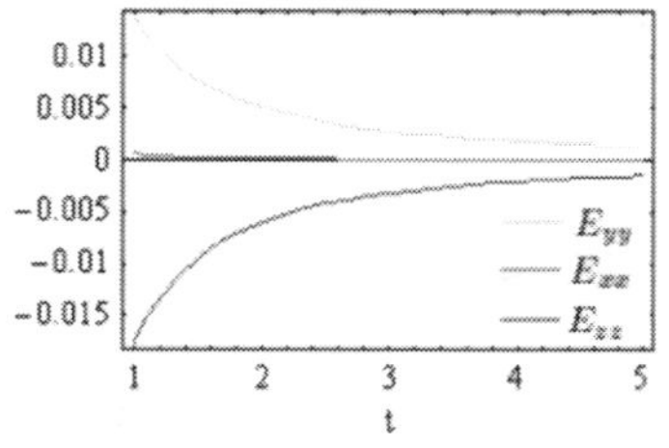

Figure 4. Electric Weyl tensor for $A = \frac{3}{4}$, $w = \frac{1}{3}$.

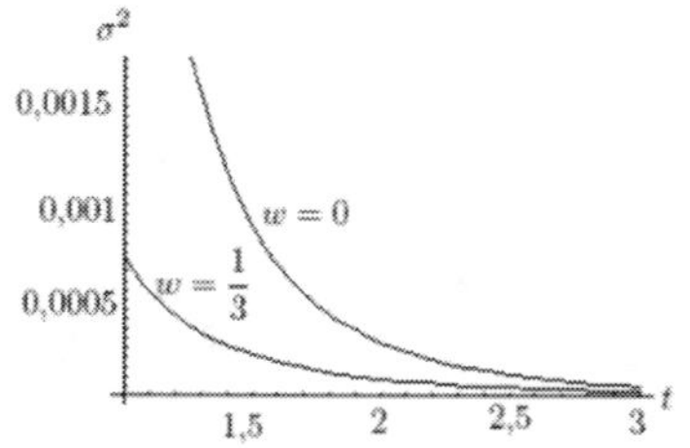

Figure 5. Shear scalar evolution for $A = \frac{3}{4}$.

References

[1] Spergel D N *et al.* 2007 *Astrophys. J. S.* **170** 377 (*Preprint* gr-qc/0603449)
[2] Rodrigues D C 2008 *Phys. Rev.* D **77** 023534
[3] Wainwright J *et al.* 1997 *Dynamical Systems in Cosmology* (Cambridge: Cambridge University Press)
[4] Ellis G F R and van Elst H 1998 Cosmological models, cargese lectures (*Preprint* gr-qc/9812046)
[5] Ellis G F R and van Elst H 1997 Deviation of geodesics in flrw spacetime geometries (*Preprint* gr-qc/9709060)
[6] Saha B 2004 Anisotropic cosmological models with a perfect fluid and a λ term (*Preprint* gr-qc/0411080)
[7] Kumar S and Singh C P *Astrophys. Space Sci.* **312** 57
[8] Pirani F A E 1956 *Acta. Phys. Polon.* **15** 389
[9] Tegmark M *et al.* 2003 *Phys. Rev.* D **68** 123523 (*Preprint* gr-qc/0302496)

Journal of Physics: Conference Series **229** (2010) 012077 doi:10.1088/1742-6596/229/1/012077

Remarks on the extreme limit of the double–Kerr solution in the equilibrium case

I Cabrera–Munguia[1], **V S Manko**[1*] **and E Ruiz**[2]

[1] Departamento de Física, Centro de Investigación y de Estudios Avanzados del IPN, A.P. 14-740, 07000 México D.F., Mexico

[2] Instituto Universitario de Física Fundamental y Matemáticas, Universidad de Salamanca, 37008 Salamanca, Spain

E-mail: rcabrera@fis.cinvestav.mx

Abstract. The Ernst complex potential and entire metric describing two extreme Kerr constituents in equilibrium are worked out in a closed analytical form.

1. Introduction

In the paper [1] it is shown how to obtain the analytical solution of the extended double–Kerr (EDK) equilibrium problem. This problem can be solved by using quantities X_i which fulfil the relation $X_i\tilde{X}_i = 1$. The general formulae obtained in [1] can be used, by taking appropriately the limit, for finding an exact solution for two balancing extreme Kerr objects, which will be the objective of the present communication. The case of two extreme objects emerges when the equation $e(z) + \tilde{e}(z) = 0$ [2] has two different real roots of the multiplicity two. However, by simply setting $\alpha_1 = \alpha_2$ and $\alpha_3 = \alpha_4$ in the general formulae one can only get the indeterminate expressions, and this suggests the application of the l'Hôpital rule for performing the limit. For our purpose of obtaining the extreme solution we shall use the expansions in terms of the small quantity ε that will finally lead us to the desired result. In the next section we will present the general Manko–Ruiz solution of the EDK equilibrium problem, and in section 3 we will work out the Ernst potential and the corresponding entire metric for two balancing extreme Kerr constituents.

2. Manko–Ruiz solution of the EDK equilibrium problem

The Ernst complex potential $\mathcal{E}$ [3] of the EDK solution and the corresponding metric functions f, γ, ω which enter the stationary axially symmetric line element

$$ds^2 = f^{-1}[e^{2\gamma}(d\rho^2 + dz^2) + \rho^2 d\varphi^2] - f(dt - \omega d\varphi)^2, \tag{1}$$

in terms of the quantities α_i and X_i have the form [1, 2]

$$\mathcal{E} = \frac{E_+}{E_-},\ f = \frac{E_+\bar{E}_- + \bar{E}_+E_-}{2E_-\bar{E}_-},\ e^{2\gamma} = \frac{E_+\bar{E}_- + \bar{E}_+E_-}{2\lambda\bar{\lambda}r_1r_2r_3r_4},\ \omega = \frac{2i(G\bar{E}_- - \bar{G}E_-)}{E_+\bar{E}_- + \bar{E}_+E_-},$$

$$E_\pm = \Lambda \pm \Gamma,\ G = -\sigma\Lambda + \bar{\sigma}\Gamma + z\Gamma + \theta - \nu,\ \Lambda = \sum_{1\le i<j\le 4} \lambda_{ij}r_ir_j,\ \Gamma = \sum_{i=1}^{4} \nu_i r_i,$$

$$\theta = \sum_{1\le i<j\le 4} \lambda_{ij}(\alpha_i+\alpha_j)r_ir_j, \quad \nu = \sum_{i=1}^{4} \nu_i(\alpha_{i'}+\alpha_{j'}+\alpha_{k'})r_i, \quad \lambda = \sum_{1\le i<j\le 4} \lambda_{ij},$$

$$\lambda_{ij} := (-1)^{i+j}(\alpha_i-\alpha_j)(\alpha_{i'}-\alpha_{j'})X_iX_j, \qquad (i',j'\neq i,j;\ i'<j') \tag{2}$$

$$\nu_i := (-1)^i(\alpha_{i'}-\alpha_{j'})(\alpha_{i'}-\alpha_{k'})(\alpha_{j'}-\alpha_{k'})X_i, \qquad (i',j',k'\neq i;\ i'<j'<k')$$

$$X_i := \frac{(\alpha_i-\bar\beta_1)(\alpha_i-\bar\beta_2)}{(\alpha_i-\beta_1)(\alpha_i-\beta_2)}, \quad r_i := \sqrt{\rho^2+(z-\alpha_i)^2}, \quad \sigma := \beta_1+\beta_2.$$

The general analytical solution of the EDK equilibrium problem defines a family of two Kerr sources which are in equilibrium due to balance of the gravitational attraction and spin–spin repulsion forces, and it is given by the following values of X_i [4]:

$$X_1 = \frac{\upsilon_1-\phi}{\phi^{-1}-\upsilon_1}, \quad X_2 = \frac{1-\upsilon_1\phi}{\upsilon_1\phi^{-1}-1}, \quad X_3 = \frac{1+i\upsilon_4\phi}{1-i\upsilon_4\phi^{-1}}, \quad X_4 = \frac{-\phi+i\upsilon_4}{\phi^{-1}+i\upsilon_4} \tag{3}$$

$$\upsilon_1 := \epsilon_1\left[\frac{(\alpha_1-\alpha_3)(\alpha_1-\alpha_4)}{(\alpha_2-\alpha_3)(\alpha_2-\alpha_4)}\right]^{1/2}, \quad \upsilon_4 := \epsilon_4\left[\frac{(\alpha_1-\alpha_4)(\alpha_2-\alpha_4)}{(\alpha_1-\alpha_3)(\alpha_2-\alpha_3)}\right]^{1/2}, \quad \epsilon_1=\pm1,\ \epsilon_4=\pm1,$$

where the arbitrary complex constant ϕ is subjected to the constraint $\phi\bar\phi = 1$.

3. The metric of two balancing extreme Kerr particles

In the case of two extreme objects the pair of roots α_1, α_2 must tend to the common value α, and the other pair α_3, α_4 to $-\alpha$. Therefore, we can set $\alpha_1 = \alpha+\varepsilon$, $\alpha_2 = \alpha-\varepsilon$ and, besides, $\alpha_3 = -\alpha+\varepsilon$, $\alpha_4 = -\alpha-\varepsilon$. Then, after expanding the quantities X_i and r_i in the first order in ε and setting $\phi = p + \mathrm{i}q$, $p^2+q^2 = 1$, we substitute the resulting expressions into (2) and take the limit $\varepsilon \to 0$; introducing now the prolate ellipsoidal coordinates (x, y) via the formulae

$$x = \frac{1}{2\alpha}(r_+ + r_-), \quad y = \frac{1}{2\alpha}(r_+ - r_-), \quad r_\pm = \sqrt{\rho^2+(z\pm\alpha)^2}, \tag{4}$$

we finally arrive at the following expressions for $\mathcal{E}$, f, γ and ω:

$$\begin{aligned}
\mathcal{E} &= \frac{\Lambda_1+\Gamma_1+2\mathrm{i}\epsilon(\Lambda_2+\Gamma_2)}{\Lambda_1-\Gamma_1+2\mathrm{i}\epsilon(\Lambda_2-\Gamma_2)},\\
f &= \frac{N}{D}, \quad e^{2\gamma} = \frac{N}{(a+3)^2(x^2-y^2)^4}, \quad \omega = \frac{4\epsilon\alpha(x-1)(y^2-1)W}{(a+3)N},\\
\Lambda_1 &= (a+3)(x^2-y^2)^2 - 2(a-1)(x^2y^2-1),\\
\Lambda_2 &= xy[(a+1)(x^2-1)+(3-a)(y^2-1)] + 2b(x^2+y^2-2x^2y^2),\\
\Gamma_1 &= 2x[2x^2-2+(a^2-a-2)(y^2-1)] - 2by[(a+1)(x^2-1)-2y^2+2],\\
\Gamma_2 &= (a-1)[(a+2)y(x^2-1)+(bx-2y)(y^2-1)],\\
D &= (\Lambda_1-\Gamma_1)^2 + 4(\Lambda_2-\Gamma_2)^2,\\
N &= [(a+3)(x^2-y^2)^2 - 2(a-1)(x^2-1)(y^2-1)]^2\\
&- 4(x^2-1)(1-y^2)[(a+1)(x^2-y^2)+4y(y-bx)]^2,\\
W &= (a-1)[(a+3)(x^2-y^2)^2 - 2(a-1)(x^2-1)(y^2-1)]\\
&\times \{2(a+1)-(a+2)(a-1)x-(a+3)[(a+2)x^2-2y^2+by(x+1)]\}\\
&- 2(x+1)[(a+1)(x^2-y^2)+4y(y-bx)]\\
&\times \Big(4[(ax-by)^2+(bx-y)^2] - (a+3)\{2(x-by)(x^2-y^2)\\
&+ (a-1)[ax(y^2+1)-by(x^2+1)]\}\Big),\\
\epsilon &= \pm1, \quad a \equiv p+q, \quad b \equiv p-q.
\end{aligned} \tag{5}$$

It is worth pointing out that the above metric also represents the extreme limit of the well-known Kramer–Neugebauer solution [5] for two Kerr black holes.

4. The Ernst potential on the symmetry axis

The Ernst potential on the upper part of the symmetry axis is obtainable by setting $y = 1$ in the expression for $\mathcal{E}$ in (5) and taking into account that in this case $x = \alpha z$:

$$e(z) = e_+/e_-, \tag{6}$$
$$e_\pm = (a+3)z^2 \pm 2\alpha[2 \pm i\epsilon(a+1)]z - \alpha^2\{1 \pm 2b + a(3 \pm 2b) \pm 2i\epsilon[(a+2)(a-1) \pm 2b]\}.$$

This axis formula allows one to calculate the corresponding Geroch–Hansen relativistic multipole moments which are related to some physical quantities of the system such as the total mass and total angular momentum of the extreme objects. It is also helpful for understanding the equatorial symmetry/antisymmetry properties of the solution.

Acknowledgements

This work was partially supported by Project 45946–F from CONACyT, Mexico, by Project FIS2006-05319 from Ministerio de Ciencia y Tecnología, Spain, and by the Junta de Castilla y León under the “Programa de Financiación de la Actividad Investigadora del Grupo de Excelencia GR-234”, Spain.

References

[1] Manko V S, Ruiz E and Sanabria–Gómez J D 2000 *Class. Quantum Grav.* **17** 3881
[2] Manko V S and Ruiz E 1998 *Class. Quantum Grav.* **15** 2007
[3] Ernst F J 1968 *Phys. Rev.* **167** 1175
[4] Manko V S and Ruiz E 2001 *Class. Quantum Grav.* **18** L11
[5] Kramer D and Neugebauer G 1980 *Phys. Lett.* A **75** 259

Journal of Physics: Conference Series **229** (2010) 012078 doi:10.1088/1742-6596/229/1/012078

An anti-Schwarzshild solution: wormholes and scalar-tensor solutions

José P Mimoso* **and Francisco S N Lobo**

Centro de Astronomia e Astofísica da Universidade de Lisboa, Avenida Professor Gama Pinto 2, P-1649-003 Lisbon, Portugal

E-mail: `jpmimoso@cii.fc.ul.pt`, `flobo@cii.fc.ul.pt`

Abstract. We investigate a static solution with an hyperbolic nature, characterised by a pseudo-spherical foliation of space. This space-time metric can be perceived as an anti-Schwarzschild solution, and exhibits repulsive features. It belongs to the class of static vacuum solutions termed "a degenerate static solution of class A" (see [1]). In the present work we review its fundamental features, discuss the existence of generalised wormholes, and derive its extension to scalar-tensor gravity theories in general.

1. Introduction

We consider a largely ignored metric which belongs to a class of vacuum solutions referred to as degenerate solutions of class A [1] given by

$$ds^2 = -e^{\mu(r)}\,\mathrm{d}t^2 + e^{\lambda(r)}\,\mathrm{d}r^2 + r^2\,(\mathrm{d}u^2 + \sinh^2 u\,\mathrm{d}v^2)\,, \tag{1}$$

which are axisymmetric solutions [2], but where the usual 2-dimensional spheres are replaced by pseudo-spheres, $\mathrm{d}\sigma^2 = \mathrm{d}u^2 + \sinh^2 u\,\mathrm{d}v^2$, i.e., by surfaces of negative, constant curvature. These are still surfaces of revolution around an axis, and v represents the corresponding rotation angle. For the vacuum case we get

$$e^{\mu(r)} = e^{-\lambda(r)} = \left(\frac{2\mu}{r} - 1\right)\,, \tag{2}$$

where μ is a constant [1, 2]. We immediately see that the static solution holds for $r < 2\mu$ and that there is a coordinate singularity at $r = 2\mu$ (note that $|g|$ neither vanishes nor becomes ∞ at $r = 2\mu$)[3]. This is the complementary domain of the exterior Schwarzschild solution. In the region $r > 2\mu$, as in the latter solution, the g_{tt} and g_{rr} metric coefficients swap signs. Defining $\tau = r$ and $\rho = t$, we obtain $\mathrm{d}s^2 = -\mathrm{d}\tilde{\tau}^2 + A^2(\tilde{\tau})\,\mathrm{d}\rho^2 + B^2(\tilde{\tau})\,(\mathrm{d}u^2 + \sinh^2 u\,\mathrm{d}v^2)$, with the following parametric definitions $\tilde{\tau} = -\tau + 2\mu\ln|\tau - 2\mu|$, $A^2 = 2\mu/\tau - 1$ and $B^2(\tau) = \tau^2$, which is a particular case of a Bianchi III axisymetric universe.

Using pseudo-spherical coordinates $\{x = r\sinh u\cos v,\ y = r\sinh u\sin v,\ z = r\cosh u,\ w = b(r)\}$, the spatial part of the metric (1) can be related to the hyperboloid $w^2 + x^2 + y^2 - z^2 = (b^2/r^2 - 1)\,r^2$ embedded in a 4-dimensional flat space. We then have

$$\mathrm{d}w^2 + \mathrm{d}x^2 + \mathrm{d}y^2 - \mathrm{d}z^2 = \left[(b'(r))^2 - 1\right]\mathrm{d}r^2 + r^2\left(\mathrm{d}u^2 + \sinh^2 u\,\mathrm{d}v^2\right)\,. \tag{3}$$

where the prime stands for differentiation with respect to r, and $b(r) = \mp 2\sqrt{2\mu}\sqrt{2\mu - r}$. We can recast metric (1) into the following

$$ds^2 = -\tan^2\left[\ln(\bar{r})^{\mp 1}\right] d\tau^2 + \left(\frac{2\mu}{\bar{r}}\right)^2 \cos^4\left[\ln(\bar{r})^{\mp 1}\right]\left[d\bar{r}^2 + \bar{r}^2\left(du^2 + \sinh^2 u\, dv^2\right)\right], \qquad (4)$$

which is the analogue of the isotropic form of the Schwarzschild solution. The spatial surfaces are conformally flat, but the flat metric is not euclidean. Indeed, the 3–dim spatial metric $d\sigma^2 = d\bar{r}^2 + r^2(du^2 + \sinh^2 u\, dv^2)$ is foliated by 2-dimensional surfaces of negative curvature, since $R^2{}_{323} = -\sinh^2 u < 0$, and it corresponds to $d\sigma^2 = dx^2 + dy^2 - dz^2$. We thus cannot recover the usual Newtonian weak-field limit.

Analysing the "radial" motion of test particles, we have the following equation $\dot{r}^2 + \left(\frac{2\mu}{r} - 1\right)\left(1 + \frac{h^2}{r^2\sinh^2 u_*}\right) = \epsilon$ where ϵ and h are constants of motion defined by $\epsilon = (2\mu/r - 1)\,\dot{t} = \text{const}_t$ and $h^2 = r^2 \sinh^2 u_*\, \dot{v} = \text{const}_v$, for fixed $u = u_*$. The former and latter constants represent the energy and angular momentum per unit mass, respectively. We thus define the potential $2V(r) = \left(\frac{2\mu}{r} - 1\right)\left(1 + \frac{h^2}{r^2\sinh^2 u_*}\right)$. This potential is manifestly repulsive, crosses the r-axis at $r = 2\mu$, and for sufficiently high values of h it has a minimum at $r_\pm = (h^2 \mp \sqrt{h^4 - 12\mu^2 h^2})/(2\mu)$. However this minimum falls outside the $r = 2\mu$ divide. So a test particle is subject to a repulsive potential forcing it to inevitably cross the event horizon at $r = 2\mu$ attracted either by some mass at the minimum or by masses at infinity. In [2] it is hinted that the non-existence of a clear Newtonian analogue is related to the existence of mass sources at ∞, but no definite conclusions were drawn.

2. Alter-ego of Morris-Thorne wormholes

A natural extension of the solution (2) would be to add exotic matter to obtain static and pseudo-spherically symmetric traversable wormhole solutions [4]. Consider the metric (1) given by $\mu(r) = 2\Phi(r)$ and $\lambda(r) = -\ln[1 - b(r)/r]$. The coordinate r decreases from a constant value μ to a minimum value r_0, representing the location of the throat of the wormhole, where $b(r_0) = r_0$, and then it increases from r_0 back to the value μ. The condition $(b/r - 1) \geq 0$ imposes that $b(r) \geq r$, contrary to the Morris-Thorne counterpart [5].

The solution provides the following stress-energy scenario

$$\rho(r) = -\frac{1}{8\pi}\frac{b'}{r^2}, \qquad p_r(r) = \frac{1}{8\pi}\left[\frac{b}{r^3} + 2\left(\frac{b}{r} - 1\right)\frac{\Phi'}{r}\right], \qquad (5)$$

$$p_t(r) = \frac{1}{8\pi}\left(\frac{b}{r} - 1\right)\left[\Phi'' + (\Phi')^2 + \frac{b'r + b - r}{2r(b - r)}\Phi' + \frac{b'r - b}{2r^2(b - r)}\right], \qquad (6)$$

in which $\rho(r)$ is the energy density, $p_r(r)$ is the radial pressure, $p_t(r)$ is the pressure measured in the tangential directions. Note that the radial pressure is always positive at the throat, i.e, $p_r = 1/(8\pi r_0^2)$, contrary to the Morris-Thorne wormhole, where a radial tension at the throat is needed to sustain the wormhole. In addition to this, the mathematics of embedding imposes that $b'(r_0) > 1$ at the throat, which implies a negative energy density at the throat (see [4] for more details). This condition is another significant difference to the Morris-Thorne wormhole, where the existence of negative energy densities at the throat is not a necessary condition. Several interesting equations of state were considered in [4], and we refer the reader to this work for more details.

3. Pseudo-spherical scalar-tensor solution

A theorem by Buchdahl [6] establishes the reciprocity between any static solution of Einstein's vacuum field equations and a one-parameter family of solutions of Einstein's equations with a (massless) scalar field. In the conformally transformed Einstein frame, note that scalar-tensor gravity theories are described by $\tilde{S} = \int \sqrt{-\tilde{g}} \left[\left[\tilde{R} - \frac{1}{2}(\nabla\varphi)^2\right] + (16\pi G_N/\Phi^{-2}(\varphi))\ L_{Matter}\right]$. In this representation we have GR plus a massles scalar field which is now coupled to the matter fields. Different scalar-tensor theories correspond to different couplings. In the absence of matter we can use Buchdahl's theorem. So, given the metric (1), we derive the corresponding scalar-tensor solution

$$ds^2 = -\left(\frac{2\mu}{r}-1\right)^B dt^2 + \left(\frac{2\mu}{r}-1\right)^{-B} dr^2 + \left(\frac{2\mu}{r}-1\right)^{1-B} r^2\,(du^2+\sinh^2 u\,dv^2)\,, \quad (7)$$

$$\varphi(r) = \sqrt{\frac{C^2(2\omega+3)}{16\pi}\varphi_0}\,\ln\left(\frac{2\mu}{r}-1\right)\,, \quad (8)$$

where $C^2 = (1-B^2)/(2\omega+3)$ and $-1 \le B \le 1$. This clearly reduces to our anti-Schwarzschild metric (1) in the GR limit when $B = 1$, and hence $C = 0$ implying that $G = \Phi^{-1}$ is constant. Reverting $\varphi = \int \sqrt{\Phi_0(2\omega+3)/(16\pi)}\,d\ln(\Phi/\Phi_0)$, and the conformal transformation, $g_{ab} = (2\mu/r-1)^{-C}\,\tilde{g}_{ab}$, we can recast this solution in the original frame in which the scalar-field is coupled to the geometry and the content is vacuum, the so-called Jordan frame.

The $r = 2\mu$ limit is no longer just a coordinate singularity, but rather a true singularity as it can be verified from the analysis of the curvature invariants. In the Einstein frame this occurs because the energy density of the scalar field diverges likewise in the Schwarzschild case[7]. Of paramount importance is that, once again, the ST-solution has no Newtonian limit (as its GR limit does not have one). This implies that the usual Parametrized Post-Newtonian formalism that assesses the departures of modified gravity theories from GR does not hold for this class of metrics (see [8]).

4. Discussion

We have outlined the exotic features of the vacuum static solution with a pseudo-spherical foliation of space. We have revealed the existence of generalised wormholes, and derived its extension to scalar-tensor gravity theories. A fundamental feature of these solutions is the absence of a Newtonian weak field limit, which reminds us of a quotation from John Barrow [9]

> *The miracle of general relativity is that a purely mathematical assembly of second-rank tensors should have anything to do with Newtonian gravity in any limit.*

Acknowledgments

The authors are grateful to Raül Vera and Guillermo A. González for helpful discussions. JPM also acknowledges the LOC members, Ruth, Raül, Jesús and José for a very enjoyable conference.

References

[1] Stephani H, Kramer D, MacCallum M A H, Hoenselaers C and Herlt E 2003 *Exact solutions of Einstein's field equations Second edition* (Cambridge, UK: Cambridge University Press)
[2] Bonnor W B and Martins M A P 1991 *Class. Quantum Grav.* **8** 727
Martins M A P 1996 *Gen. Rel. Grav.* **28** 1309
[3] Anchordoqui L A, Edelstein J D, Nunez C and Birman G S *Preprint* gr-qc/9509018
[4] Lobo F S N and Mimoso J P *Preprint* arXiv:0907.3811 [gr-qc]
[5] Morris M S and Thorne K S 1988 *Am. J. Phys.* **56** 395
[6] Buchdahl H A 1959 *Phys. Rev.* **115** 1325
[7] Wands D 1993 Phd thesis, pp. 20, (University of Sussex)
[8] Mimoso J P and Lobo F S N "Vacuum solutions with pseudo-spherical symmetry," in preparation.
[9] Barrow J D 1991 Gravitation and Hot Big-Bang Cosmology, in *Lecture notes in physics, 383* (Berlin, Germany: Springer)

Journal of Physics: Conference Series **229** (2010) 012079 doi:10.1088/1742-6596/229/1/012079

Polar oscillations in magnetars

Hajime Sotani* **and Kostas D Kokkotas**

Theoretical Astrophysics, University of Tübingen, 72076 Tübingen, Germany

E-mail: `sotani@astro.auth.gr`

Abstract. We study polar Alfvén oscillations of relativistic stars endowed with a strong global poloidal dipole magnetic field. Here we focus only on the axisymmetric oscillations which are studied by evolving numerically the 2D perturbation equations. We show that the spectrum of the polar Alfvén oscillations is discrete in contrast to that of axial Alfvén ones. We also show that the typical fluid modes are not significantly affected by the presence of the magnetic field.

1. Introduction

Up to now at least three giant flares have been detected in three different SGRs. The timing analysis of the decaying tail of the latest two giant flares revealed the existence of QPOs [1]. The most promising mechanism to produce these QPOs are the magnetar models. The attempts to explain the above QPO frequencies by using the crust torsional oscillations have been based both in Newtonian and in general relativistic dynamics (e.g., [2, 3]). These attempts were partially successful, but it has been also realized the difficulty to explain all the observed frequencies by using only the crust torsional oscillations. While, in an attempt to explain the QPOs without the crust torsional oscillations, Levin [4] using a toy model suggested that the torsional Alfvén oscillations could not form a discrete spectrum but instead they should form a continuum, while Sotani *et al.* [5] showed that the spectrum of the torsional Alfvén oscillations of relativistic magnetars forms a continuum and that there exist two distinct families as the so called "lower" and "upper" QPOs. They also figure out that the observed QPO frequencies in SGRs can be explained by using the frequencies of torsional Alfvén and the crust torsional oscillations.

One could actually argue that the features of torsional Alfvén oscillations are quite well understood. However, there are only a few studies of polar oscillations of the magnetized stars and it is still unknown whether spectrum of the polar Alfvén oscillations is discrete or continuous. Moreover, some of the higher QPO frequencies observed may be related to polar and not to axial Alfvén oscillations. Finally, polar oscillations are related to the emission of gravitational waves. So, in order to shed some light to the various features of polar Alfvén oscillations, we performed 2D numerical evolutions of the perturbation equation of relativistic magnetar models endowed with a global poloidal dipole magnetic field. The more detailed study can be seen in [6].

2. Equilibrium configuration

We assume that the background models of the strongly magnetized stars are spherically symmetric and non-rotating. This choice is justified since all known magnetars rotate very slowly and even for the case of magnetars the magnetic energy is much smaller than the gravitational

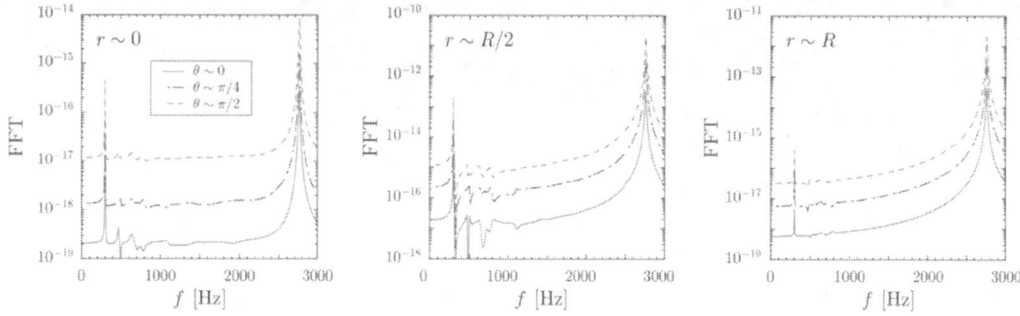

Figure 1. The FFT of the perturbation function v for a stellar model with $B = 10^{16}$ G.

binding energy. On this stellar model we superimpose a dipole magnetic field. We assume that the star consists of a perfect fluid and in addition we adopt "the ideal MHD approximation".

3. Numerical Results

For the time evolution of the perturbation equations, we use the iterated Crank-Nicholson scheme. In order to eliminate the spurious higher frequencies, we also add a 4th-order Kreiss-Oliger dissipation into the evolution equations. As numerical code tests, we compare the stellar oscillation frequencies without magnetic field with the previous results in [7].

The various features of polar oscillations are examined in two ways. The first involves checking of the FFT amplitude at various points inside the star. If the spectrum is continuous the peaks in FFT will depend on the position, while if the peaks are independent of the observer's position then they correspond to a discrete spectrum. The second check involves the study of the phase of oscillation for each peak frequency in FFT, i.e., for the discrete spectrum the phase should be constant throughout the star. Figure 1 shows the FFT of the perturbation function v at various points inside the star with $B = 10^{16}$ G, where v is corresponding to the displacement vector in the θ direction. In this figure the three different lines correspond to different angular positions while the different panels correspond to different radial positions. It should be noted that the FFT of the other perturbation function, w which denotes the displacement vector in the radial direction, shows that the peaks correspond exactly to the same frequencies as in figure 1 and obviously the peaks are independent of the observer's position. The peak at 2754 Hz corresponds to the ${}_2f$ mode, while the lower peak at 300 Hz corresponds to the polar Alfvén oscillation mode ${}_2a_0$. Next we study the phase of the specific Alfvén oscillation mode ${}_2a_0$. In figure 2 we show the phase of the oscillation mode ${}_2a_0$ for a magnetar with $B = 10^{16}$ G. The left and right panels correspond to the phases for the perturbation functions w and v, respectively. We observe that the phases for both functions w and v are almost constant except for some regions around $\theta \sim \pi/3$ for the function w and a part of $r/R \sim 0.8$ for the function v. We identify that these two regions with somehow strange behavior of the phase are related to the due to numerical loss of accuracy in the calculation of the effective amplitudes, where the eigenfunctions are almost zero because these are nodal points (see the figures 3 and 4 in [6]). Thus we can argue that the phase of the oscillation modes is independent from the position. With the above two tests we conclude that the Alfvén oscillations of polar parity are described by a discrete spectrum, in contrast to the spectrum of the axial Alfvén oscillations. This feature seems to be very similar to the case of inertial modes of rotating stars. That is, the spectrum of the axial inertial modes seems to be continuous, at least in the slow rotation approximation, while the polar modes admit a discrete spectrum.

Additionally, we examine how the fluid modes depend on the magnetic field strength. In

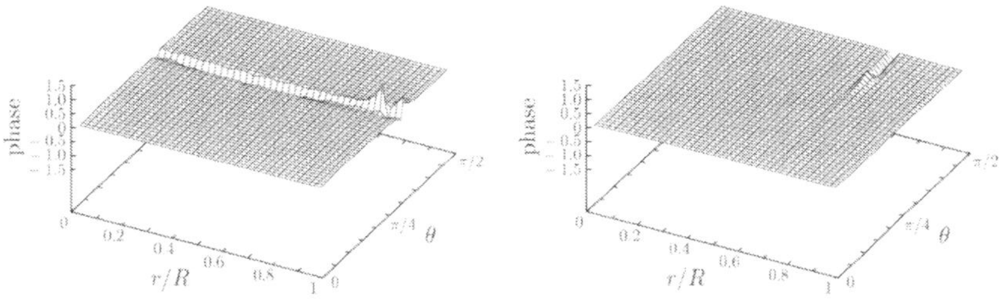

Figure 2. The phase of the fundamental polar Alfvén mode with $\ell = 2$ for a magnetar with $B = 10^{16}$ G. The left and right panels correspond to the perturbation functions w and v.

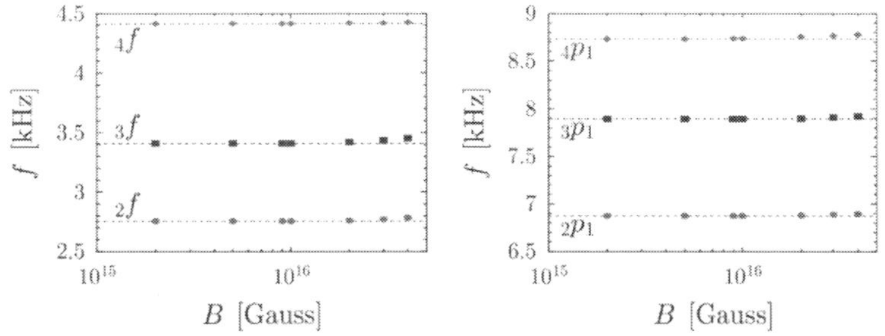

Figure 3. The frequencies of the f and p_1 modes with harmonic indexes $\ell = 2$, 3 and 4.

figure 3, we plot the frequencies of the f and p_1 modes with harmonic indexes $\ell = 2$, 3 and 4 as functions of magnetic field strength. The marks correspond to frequencies of the modes of the magnetized stars while the horizontal dashed lines are the frequencies for the non-magnetized ones. From this figure, we can see that the effect of magnetic field on the frequencies of fluid modes can be traced only when the magnetic field becomes stronger than a few times 10^{16} G.

4. Conclusion

In this work, we study the polar type oscillations of strongly magnetized neutron stars. By using the 2D time evolutions of the perturbation equations, we estimate the Alfvén oscillation modes as well as the fluid modes. Actually, by examining the FFT amplitude at various points inside the star and phase of the corresponding frequencies, we conclude that the Alfvén oscillations with polar parity are described by a spectrum consisting only by discrete modes. This is the first study of its kind for axisymmetric polar Alfvén oscillations, it remains open to understand the form of the spectrum for non-axisymmetric perturbations while more complicated forms of the magnetic field geometry should be assumed. For example, the inclusion of a toroidal component as in [8] will affect the form of the oscillation spectrum.

References

[1] Watts A L and Strohmayer T E 2006 *Adv. Space Res.* **40** 1446
[2] Lee U 2007 *Mon. Not. R. Astron. Soc.* **374** 1015
[3] Sotani H, Kokkotas K D and Stergioulas N 2007 *Mon. Not. R. Astron. Soc.* **375** 261
[4] Levin Y 2006 *Mon. Not. R. Astron. Soc.* **368** L35
[5] Sotani H, Kokkotas K D and Stergioulas N 2008 *Mon. Not. R. Astron. Soc.* **385** L5
[6] Sotani H and Kokkotas K D 2009 *Mon. Not. R. Astron. Soc.* **395** 1163
[7] Sotani H and Kokkotas K D 2004 *Phys. Rev.* D **70** 084026
[8] Sotani H, Colaiuda A and Kokkotas K D 2008 *Mon. Not. R. Astron. Soc.* **385** 2161

Spanish Relativity Meeting (ERE 2009) IOP Publishing
Journal of Physics: Conference Series **229** (2010) 012080 doi:10.1088/1742-6596/229/1/012080

Black hole entropy for the general area spectrum

Tomo Tanaka[1*] **and Takashi Tamaki**[2]
[1]Department of Physics, Waseda University, Okubo 3-4-1, Tokyo 169-8555, Japan
[2]Department of Physics, Nihon University, Fukushima 963-8642, Japan

E-mail: [1]tomo@gravity.phys.waseda.ac.jp, [2]tamaki@ge.ce.nihon-u.ac.jp

Abstract. We consider the possibility that the horizon area is expressed by the general area spectrum in loop quantum gravity when we leave off the semiclassical consideration. To check this idea, we calculate the number of degrees of freedom in spin-network states related to its area. We obtain that logarithm of this number is proportional to its area as in previous works where the simplified area formula has been used. Our result shows that we should be careful in justifying (or falsifying) the area spectrum if we respect to leave off the semiclassical consideration.

One of the most mysterious things about black holes is their entropy S which is *not* proportional to their volume *but* to their horizon area A. This was first pointed out related to the first law of black hole thermodynamics. The famous relation $S = A/4$ has been established by the discovery of the Hawking radiation. Recently, its statistical origin has been discussed in the candidate theories of quantum gravity, such as string theory [1], or loop quantum gravity (LQG) [2], etc. It has been discussed that LQG can describe its statistical origin independent of black hole species because of its background independent formulation [3]. For this reason, we concentrate on LQG here.

Quantum states in LQG are described by spin-network [4], and basic ingredients of the spin-network are edges, which are lines labelled by spin $j(j = 0, 1/2, 1, 3/2, \cdots)$ reflecting the SU(2) nature of the gauge group, and vertices which are intersections between edges. For three edges having spin j_1, j_2, and j_3 that merge at an arbitrary vertex, we have following conditions.

$$j_1 + j_2 + j_3 \in \mathbb{N}, \tag{1}$$

$$j_i \leqq j_j + j_k, \ (i, j, k \text{ different from each other.}). \tag{2}$$

These conditions guarantee the gauge invariance of the spin-network.

Using this formalism, general expressions for the spectrum of the area and the volume operators can be derived [5, 6]. For example, the area spectrum A_j is

$$A_j = 4\pi\gamma \textstyle\sum_i \sqrt{2j_i^u(j_i^u + 1) + 2j_i^d(j_i^d + 1) - j_i^t(j_i^t + 1)} \tag{3}$$

where γ is the Barbero-Immirzi parameter related to an ambiguity in the choice of canonically conjugate variables [7]. The sum is added up all intersections between a surface and edges. Here, the indices u, d, and t mean edges upper side, down side, and tangential to the surface, respectively (We can determine which side is upper or down side arbitrarily).

In [2], it was proposed that black hole entropy is obtained by counting the number of degrees of freedom about j when we fix the horizon area where a simplified area formula is used. This simplified area formula is obtained by assuming that there are no tangential edges on black hole horizon, that is $j_i^t = 0$. We obtain $j_i^u = j_i^d := j_i$ by using the condition (2). Then, we consider the degrees of freedom about j satisfying

$$A_j = 8\pi\gamma \sum_i \sqrt{j_i(j_i+1)} = A. \tag{4}$$

The standard procedure is to impose the Bekenstein-Hawking entropy-area law $S = A/4$ for macroscopic black holes in order to fix the value of γ. Ashtekar et al. in [3] extended this idea using the isolated horizon framework (ABCK framework) [8]. Error in counting in this original work has been corrected in [9, 10].

These rely on semiclassical consideration since it is difficult to express black holes in full quantum gravity. Interesting idea to circumvent this difficulty is to define the horizon as a boundary which is accessible to the external observer at spatial infinity [11]. If we respect this view point, it is valuable to examine (3) as a horizon area. We can also discuss this possibility as follows. Imagine to describe it by spin-network states. If we suppose that its entropy is proportional to its area and is related to number of degrees of freedom in j (or $m = -j, -j+1, \cdots, j$ considered in [3]), it is evident that (3) can gain larger number of states than (4) for the fixed area. Is there any reason that nature does not choose this possibility? [12] also discussed using (3) as expressing the horizon area from the consideration of dynamical situations.

One of the criteria to check (3) as a horizon area is examining the number of degrees of freedom related to (3). We notice that both counting the horizon freedom in [3] and j freedom in [2] obtain the entropy proportional to the area (4). Based on this observation, we assume that counting j freedom can be a good indicator in judging the area spectrum (3) as a horizon area by investigating whether or not the entropy is proportional to the area (3).

1. Consideration of the general area spectrum

In the case for (4), it has been shown that isolated horizon conditions do not affect the number of states in the limit $A \to \infty$. Based on this observation, we consider only degrees of freedom about its area (3) as a first step. Then in this case, we also denote number of states as $N(A)$ which is defined as

$$N(A) := \Big\{ (j_1^u, j_1^d, j_1^t, \cdots, j_n^u, j_n^d, j_n^t) | j_i^u, j_i^d, j_i^t \in 0, \mathbb{N}/2, \ \mathbf{j}_i := (j_i^u, j_i^d, j_i^t) \neq (0,0,0)$$

$$j_i^u, j_i^d, j_i^t \text{ should satisfy (1) and (2).} \ \sum_i \sqrt{2j_i^u(j_i^u+1) + 2j_i^d(j_i^d+1) - j_i^t(j_i^t+1)} = \frac{A}{4\pi\gamma} \Big\}.$$

Then, we perform counting as follows. First, we count the case $j^t \in 0, \mathbb{N}$ to simplify the problem and denote the corresponding number of states as $N(A)_{\text{even}}$. Then, we have $j^u + j^d := n \in \mathbb{N}$ by (1). If we fix n, we can classify the possible j^u, j^d, j^t as follows, which is one of the most important parts in this paper. First, we have $(j^u, j^d) = (\frac{n}{2} \pm \frac{s}{2}, \frac{n}{2} \mp \frac{s}{2})$ (double-sign corresponds) for $0 \leqq s \leqq n$, $s \in 0, \mathbb{N}$ to satisfy (2). Then, for each s, possible value of j^t is $j^t = s, s+1, \cdots, n$ to satisfy (2).

Then, in the same way, we can count the case $j^t \in n + \frac{1}{2}, (n = 0, 1, 2, \cdots)$ and denote the corresponding number of states as $N(A)_{\text{odd}}$. In summary, number of states $N(A)$ can be written as

$$N(A) = \sum_{n=1}^{\infty} \left[\sum_{s=1}^{n} \sum_{t=s}^{n} 2\theta(A - 4\pi\gamma) N(A - x(n,s,t)) + \sum_{t=0}^{n} \theta(A - 4\pi\gamma) N(A - x(n, s=0, t)) \right]$$
$$+ \sum_{n=0}^{\infty} \sum_{s=0}^{n} \sum_{t=s}^{n} 2\theta(A - 2\pi\gamma\sqrt{3}) N(A - y(n,s,t)) \ . \tag{5}$$

For $A \to \infty$, by assuming the relation $N(A) = Ce^{\frac{A\gamma_M}{4\gamma}}$, where C is a constant and substituting to the recursion relation (5), we obtain the beautiful formula as a generalization of the case (4) as,

$$1 = \sum_{n=1}^{\infty} \left[\sum_{s=1}^{n} \sum_{t=s}^{n} 2\theta(A - 4\pi\gamma) \exp(-\gamma_M x(n,s,t)/4\gamma) + \sum_{t=0}^{n} \theta(A - 4\pi\gamma) \right.$$
$$\left. \times \exp(-\gamma_M x(n, s=0, t)/4\gamma) \right] + \sum_{n=0}^{\infty} \sum_{s=0}^{n} \sum_{t=s}^{n} 2\theta(A - 2\pi\gamma\sqrt{3}) \exp\left(-\gamma_M y(n,s,t)/4\gamma\right) . (6)$$

If we require $S = A/4$, we have $\gamma = \gamma_M = 0.7274\cdots$. This means that even if we use (3) as the horizon spectrum, we can reproduce the entropy formula $S = A/4$ by adjusting the Barbero-Immirzi parameter. This is nontrivial and is our main result in this paper.

2. Conclusion and Discussion

In this paper, we checked the possibility of the general area formula as a horizon area by counting the j freedom. It is important that its entropy is proportional to its area. Then, it is natural to ask what the area formula should be in describing the horizon area. There are many possibilities examining the area spectrum. For example, we have not yet established the black hole thermodynamics in LQG which is one of the most important topics to be investigated. There is an idea that black hole evaporation process should also be described by using the general area formula [12]. Therefore, whether we can establish the generalized second law of black hole thermodynamics might be one of the criteria in judging which area formula is appropriate. For this purpose, black hole states in full quantum gravity is required since the exact counting is required. Though we do not take care of the topology of the horizon, discussing the difference caused by the topology is important as considered for the simplified area formula [13]. The covariant entropy bound [14] is also important which has been discussed in the LQG context recently [15]. Confirming LQG in many independent methods would be the holy grail of the theory.

References

[1] Strominger A and Vafa C 1996 *Phys. Lett.* B **379** 99
Maldacena J M and Strominger A 1996 *Phys. Rev. Lett.* **77** 428
[2] Rovelli C 1996 *Phys. Rev. Lett.* **77** 3288
[3] Ashtekar A, Baez J, Corichi A and Krasnov K 1998 *Phys. Rev. Lett.* **80** 904
Ashtekar A, Baez J and Krasnov K 2000 *Adv. Theor. Math. Phys.* **4** 1
[4] Rovelli C and Smolin L 1995 *Phys. Rev.* D **52** 5743
[5] Rovelli C and Smolin L 1995 *Nucl. Phys.* B **442** 593; Erratum, *ibid.* **456** 753
[6] Ashtekar A and Lewandowski J 1997 *Class. Quantum Grav.* **14** A55
[7] Barbero J F 1995 *Phys. Rev.* D **51** 5507
Immirzi G 1997 *Nucl. Phys. Proc. Suppl.* B **57** 65
[8] Ashtekar A, Corichi A and Krasnov K 1999 *Adv. Theor. Math. Phys.* **3** 419
[9] Domagala M and Lewandowski J 2004 *Class. Quantum Grav.* **21** 5233
[10] Meissner K A 2004 *Class. Quantum Grav.* **21** 5245
[11] Krasnov K and Rovelli C *Prerpint* gr-qc/0905.4916
[12] Ansari M H 2007 *Nucl. Phys.* B **783** 179; *ibid.*, **795** 635
[13] Kloster S, Brannlund J and DeBenedictis A 2008 *Class. Quantum Grav.* **25** 065008
[14] Bousso R 1999 *J. High Energy Phys.* JHEP07(1999)004
[15] Ashtekar A and Wilson-Ewing E 2008 *Phys. Rev.* D **78** 064047

AUTHOR INDEX

Aceña, Andrés E. ... 78
Agulló, I. ... 253
Alexeyev, S. O. ... 265
Alho, Artur ... 82
Åman, Jan E. ... 19, 261
Aragonés-Muñoz, A. ... 321
Atrio-Barandela, F. ... 13
Bäckdahl, Thomas ... 86
Bandos, Igor A. ... 90
Barceló, C. ... 94
Barrau, A. ... 265
Bengtsson, Ingemar ... 19
Bergshoeff, Eric ... 25
Biamonte, Jacob D. ... 102
Blanco, Oihane F. ... 106
Bona, Carles ... 110
Bona-Casas, Carles ... 110
Borunda, Mónica ... 114
Bouhmadi-López, Mariam ... 118
Bradley, M. ... 122, 217
Bretón, Nora ... 31
Brito, I. ... 126
Brizuela, David ... 281
Cabrera-Munguia, I. ... 327
Cáceres, D. L. ... 324
Cardoso, Vitor ... 309, 317
Carot, J. ... 126, 201
Carrasco, Alberto ... 130
Castañeda, L. ... 324
Charters, Tiago ... 225
Chimento, L. P. ... 134, 293
Chrusciel, P. T. ... 142
Cordero-Carrión, Isabel ... 138
Cortier, J. ... 142
Costa, João Lopes ... 205
Costa, L. Filipe O. ... 146
Costa, Miguel S. ... 269
Cruz, Norman ... 281
Cuchí, J. E. ... 150
De La Cruz-Dombriz, A. ... 154
Del Campo, Sergio ... 67
Diez-Tejedor, Alberto ... 158
Dobado, A. ... 154
Ebeling, H. ... 13
Edgar, S. B. ... 162, 217
Emparan, Roberto ... 1
Ernst, F. J. ... 221
Fernández, Daniel ... 166
Fernández-Jambrina, Leonardo ... 170
Finazzi, S. ... 94
Fodor, G. ... 122
Font, José A. ... 37
Ghosh, Suman ... 178
Gómez-Lobo, Alfonso García-Parrado ... 162, 174
González, Guillermo A. ... 182
González-Morales, Alma X. ... 186
Gualtieri, Leonardo ... 309, 317
Hartmann, Betti ... 43
Herberthson, Magnus ... 49
Herdeiro, Carlos ... 146, 269, 309, 317
Herrera, J. ... 190
Herrera, Ramón ... 67
Herrero, Alicia ... 193
Hohm, Olaf ... 25
Ibáñez, José María ... 138
Iguchi, Hideo ... 55
Jiménez, Jose Beltrán ... 98
Kashlinsky, A. ... 13
Kocevski, D. ... 13
Kokkotas, Kostas D. ... 333
Kotake, Kei ... 61
Kothawala, Dawood ... 197
Lapiedra, Ramon ... 233
Lazkoz, R. ... 170, 293
Liberati, S. ... 94
Llosa, Josep ... 201
Lobo, Francisco S. N. ... 330
López, L. A. ... 209
Lübbe, Christian ... 213
Manko, V. S. ... 221, 289, 327
Maroto, A. L. ... 98, 154
Mars, Marc ... 130
Martín, J. ... 150
Martín-García, José M. ... 162, 174
Masip, Manuel ... 114
Mimoso, José P. ... 225, 330
Minguzzi, Ettore ... 229
Mishima, Takashi ... 55
Molina, A. ... 150
Montero, Pedro J. ... 37
Morales-Lladosa, Juan Antonio ... 138, 193, 233
Morita, Masaaki ... 237
Navarro-Salas, J. ... 253
Nerozzi, A. ... 241, 309, 317
Nozawa, Masato ... 245
Núñez, Darío ... 186
Nungesser, Ernesto ... 249
Olmo, G. J. ... 253
Parker, L. ... 253
Pascual-Sánchez, J.-F. ... 257
Pavón, Diego ... 67
Pidokrajt, Narit ... 261
Ramos, M. P. Machado ... 217
Rannu, K. A. ... 265
Rebelo, Carmen ... 269
Richarte, Martín G. ... 134
Rodríguez-Meza, M. A. ... 273
Rostworowski, Andrzej ... 277
Rozas-Fernández, Alberto ... 281
Ruiz, E. ... 150, 221, 289, 327

AUTHOR INDEX

Sáez-Gómez, Diego 285
San Miguel, A. 257
Sánchez, Miguel 106
Sánchez-Mondragón, J. 289
Sandoval-Villalbazo, A. 321
Sendra, I. 293
Senovilla, José M. M. 19, 106
Shibata, Masaru 37
Silk, Joseph 7
Sotani, Hajime 297, 333
Sperhake, Ulrich 309, 317
Svítek, Otakar 301
Takahashi, Hirotaka 237
Tamaki, Takashi 336
Tanaka, Tomo 336
Tejeiro, J. M. 324
Tod, Paul 73
Townsend, Paul 25
Urrestilla, Jon 43
Van Den Bergh, Norbert 305
Vaz, E. G. L. R. 126
Vera, Raül 182
Vicente, F. 257
Witek, Helvi 309, 317
Wylleman, Lode 313
Zilhão, Miguel 309, 317

Institute of Physics
Dirac House, Temple Back
Bristol BS1 6BE UK

CURRAN ASSOCIATES INC
proceedings
.com